Endotoxins: Structure, Function and Recognition

SUBCELLULAR BIOCHEMISTRY

Recent Volumes in this Series

Volume 33 **Bacterial Invasion into Eukaryotic Cells**
Tobias A. Oelschlaeger and Jorg Hacker
Volume 34 **Fusion of Biological Membranes and Related Problems**
Edited by Herwig Hilderson and Stefan Fuller
Volume 35 **Enzyme-Catalyzed Electron and Radical Transfer**
Andreas Holzenburg and Nigel S. Scrutton
Volume 36 **Phospholipid Metabolism in Apoptosis**
Edited by Peter J. Quinn and Valerian E. Kagan
Volume 37 **Membrane Dynamics and Domains**
Edited by P.J. Quinn
Volume 38 **Alzheimer's Disease: Cellular and Molecular Aspects of Amyloid beta**
Edited by R. Harris and F. Fahrenholz
Volume 39 **Biology of Inositols and Phosphoinositides**
Edited by A. Lahiri Majumder and B.B. Biswas
Volume 40 **Reviews and Protocols in DT40 Research**
Edited by Jean-Marie Buerstedde and Shunichi Takeda
Volume 41 **Chromatin and Disease**
Edited by Tapas K. Kundu and Dipak Dasgupta
Volume 42 **Inflammation in the Pathogenesis of Chronic Diseases**
Edited by Randall E. Harris
Volume 43 **Subcellular Proteomics**
Edited by Eric Bertrand and Michel Faupel
Volume 44 **Peroxiredoxin Systems**
Edited by Leopold Flohd J. Robin Harris
Volume 45 **Calcium Signalling and Disease**
Edited by Ernesto Carafoli and Marisa Brini
Volume 46 **Creatine and Creatine Kinase in Health and Disease**
Edited by Gajja S. Salomons and Markus Wyss
Volume 47 **Molecular Mechanisms of Parasite Invasion**
Edited by Barbara A. Burleigh and Dominique Soldati-Favre
Volume 48 **The Coronin Family of Proteins**
Edited by Christoph S. Clemen, Ludwig Eichinger and Vasily Rybakin
Volume 49 **Lipids in Health and Disease**
Edited by Peter J. Quinn and Xiaoyuan Wang
Volume 50 **Genome Stability and Human Diseases**
Edited by Heinz Peter Nasheuer
Volume 51 **Cholesterol Binding and Cholesterol Transport Proteins**
Edited by J. Robin Harris
Volume 52 **A Handbook of Transcription Factors**
Edited by Tim Hughes

Xiaoyuan Wang · Peter J. Quinn
Editors

Endotoxins: Structure, Function and Recognition

Springer

Editors
Xiaoyuan Wang
Jiangnan University
State Key Laboratory of Food Science and Technology
214122 Wuxi, Jiangsu
China, People's Republic
xwang@jiangnan.edu.cn

Peter J. Quinn
Department of Biochemistry
King's College London
150 Stamford Street
London SE1 9NH
United Kingdom
p.quinn@kcl.ac.uk

ISSN 0306-0225
ISBN 978-90-481-9077-5 e-ISBN 978-90-481-9078-2
DOI 10.1007/978-90-481-9078-2
Springer Dordrecht Heidelberg London New York

Library of Congress Control Number: 2010928909

Printed on acid-free paper

Springer is part of Springer Science+Business Media (www.springer.com)

Frontispiece

Preface

Endotoxins are potentially toxic compounds produced by Gram-negative bacteria including some pathogens. Unlike exotoxins, which are secreted in soluble form by live bacteria, endotoxins are comprised of structural components of bacteria. Endotoxins can cause sepsis characterized by a whole-body inflammatory state. Sepsis can lead to low blood pressure, multiple organ dysfunction syndrome and death. The way endotoxins interact with the host cells is fundamental to understanding the mechanism of sepsis. This Volume presents a description of endotoxins according to their genetic constitution, structure, function and mode of interaction with host cells. Recent research on these aspects of endotoxins has served to illuminate hitherto undescribed functions of the innate immune system.

Endotoxins are components of the surface of nearly all gram-negative bacteria. Their structure has been characterized by a variety of biophysical and analytical techniques. They fall into the category of lipopolysaccharides because they consist of a polysaccharide chain linked to a lipid moiety referred to as lipid A. Lipid A is primarily responsible for toxicity. Endotoxin is synthesized in the cytosol of bacteria and subsequently translocated across the inner membranes, the periplasmic space and the outer membranes ultimately to be exposed on the surface of the bacterium. There are a multiplicity of genes required for the biosynthesis of endotoxin in bacteria, and a number of other genes specifically required for the transport of endotoxin from cytosol to the surface of the bacterium. Because endotoxins are unique to bacteria inhibitors of enzymes and transport proteins involved in the biosynthesis and translocation of endotoxins in bacteria are ideal targets for antibiotic attack against bacteria. Studies of the biosynthesis and translocation of endotoxins have provided considerable insight on how glycolipids are synthesized and the sequential function of cytoplasmic and membrane proteins in transporting hydrophobic molecules across cell membranes. Although most of the genes required for the biosynthesis of endotoxins are conserved, additional genes in bacteria, especially pathogens, have been identified that are involved in modifying the structure of lipid A. It seems that bacteria exploit this process to change the structure of their endotoxins to combat the host immune system. This has been revealed by recent studies of the relationship between the infectivity of the bacteria and the detailed structure of lipid A.

Host cells targeted by endotoxin on the surface of bacteria have developed defense mechanisms for protection against gram-negative bacteria. The first line of defense is the binding of endotoxin to lipopolysaccharide binding protein in the serum and its transfer to CD14 on the cell membrane. CD14 in turn transfers endotoxin to MD2, which associates with Toll-like receptor-4, a trans-membrane protein present in several cell types of the immune system including macrophages and dendritic cells. Upon binding endotoxin, Toll-like receptor-4 undergoes a conformational change, which triggers a signaling cascade inside the cell to secrete pro-inflammatory cytokines and nitric oxide that lead to sepsis. The signal transduction events activated by Toll-like receptor-4 have been the subject of much interest in recent years. Many biochemical details of these events have been characterised, including novel adaptor proteins, protein kinases and transcription factors. To gain new insights into the role of the innate immune system in infection by gram-negative bacteria, this volume brings together contributions from researchers in the forefront of this subject. It is divided into two sections, the first dealing with how endotoxins are synthesized and end up on the bacterial surface and secondly, how endotoxins activate Toll-like receptor-4 and, in turn, how Toll-like receptor-4 generates the molecular signals leading to infectious and inflammatory diseases.

Wuxi, China Xiaoyuan Wang
London, UK Peter J. Quinn
January, 2010

Contents

Contributors

K. Brandenburg Division of Biophysics and Division of Immunbiophysics, Forschungszentrum Borstel, Parkallee 10, D-23845 Borstel, Germany, kbranden@fz-borstel.de

Russell W. Carlson Complex Carbohydrate Research Center, University of Georgia, 315 Riverbend Rd., Athens, GA 30602, USA, rcarlson@ccrc.uga.edu

Alan S. Cross Center for Vaccine Development, University of Maryland School of Medicine, 685 W. Baltimore Street, HSF 1, Suite 480, Baltimore, MD 21201, USA, across@medicine.umaryland.edu

Sunil A. David Department of Medicinal Chemistry, University of Kansas, Multidisciplinary Research Building, Room 320D, 2030 Becker Drive, Lawrence, KS 66047, USA, sdavid@ku.edu

R. Phillip Dellinger Critical Care Medicine, Cooper University Hospital, Camden, NJ 08103, USA, dellinger-phil@cooperhealth.edu

Jeak Ling Ding Department of Biological Sciences, Faculty of Science, National University of Singapore, 14 Science Drive 4, Singapore 117543, dbsdjl@nus.edu.sg

J. M. Dow Department of Microbiology, BIOMERIT Research Centre, BioSciences Institute, University College Cork, Cork, Ireland

G. Erbs Department of Plant Biology and Biotechnology, Faculty of Life Sciences, University of Copenhagen, Thorvaldsensvej 40, 1871 Frederiksberg, Denmark

Robert K. Ernst Department of Microbial Pathogenesis, School of Dentistry, University of Maryland – Baltimore, 650 W. Baltimore Street - 8 South, Baltimore, MD, 21201, USA, rkernst@umaryland.edu

L. Scott Forsberg Complex Carbohydrate Research Center, University of Georgia, 315 Riverbend Rd., Athens, GA 30602, USA

Christopher B. Fox Infectious Disease Research Institute, 1124 Columbia St, Ste 400, Seattle, WA 98104, USA

Martin Friede World Health Organization, Initiative for Vaccine Research, 20 Avenue Appia, CH-12211, Geneva 27, Switzerland

John S. Gunn Department of Molecular Virology, Immunology and Medical Genetics, Center for Microbial Interface Biology, The Ohio State University, 460 W. 12th Avenue, Columbus, OH 43210-1214, USA, gunn.43@osu.edu

T. Gutsmann Division of Biophysics and Division of Immunbiophysics, Forschungszentrum Borstel, Parkallee 10, D-23845 Borstel, Germany

Bow Ho Department of Microbiology, Yong Loo Lin School of Medicine, National University of Singapore, 5 Science Drive 2, Singapore 117597

Xiaoqing Hu State Key Laboratory of Food Science and Technology, Jiangnan University, 1800 Lihu Avenue, Wuxi 214122, China

Gregory C. Ireton Infectious Disease Research Institute, 1124 Columbia St, Ste 400, Seattle, WA 98104, USA, gireton@idri.org

Elmar L. Kannenberg Complex Carbohydrate Research Center, University of Georgia, 315 Riverbend Rd., Athens, GA 30602, USA

Anne F. McGettrick School of Biochemistry and Immunology, Trinity College Dublin, Dublin 2, Ireland, mcgettra@tcd.ie

Antonio Molinaro Dipartimento di Chimica Organica e Biochimica, Università di Napoli Federico II, Complesso Universitario Monte Santangelo, Via Cintia 4, Napoli I-80126, Italy, molinaro@unina.it

Anthony P. Moran Laboratory of Molecular Biochemistry, Microbiology, School of Natural Sciences, National University of Ireland, Galway, Ireland, anthony.moran@nuigalway.ie

Samuel M. Moskowitz Simches Research Center, Massachusetts General Hospital, 185 Cambridge Street, CPZN-3-830, Boston, MA 02114, USA

M.-A. Newman Department of Plant Biology and Biotechnology, Faculty of Life Sciences, University of Copenhagen, Thorvaldsensvej 40, 1871 Frederiksberg, Denmark, mari@life.ku.dk

Luke A.J. O'Neill School of Biochemistry and Immunology, Trinity College Dublin, Dublin 2, Ireland

Peter J. Quinn Department of Biochemistry, King's College London, 150 Stamford Street, London SE1 9NH, UK, p.quinn@kcl.ac.uk

Jean-Sebastien Rachoin Critical Care Medicine, Cooper University Hospital, Camden, NJ 08103, USA

Steven G. Reed Infectious Disease Research Institute, 1124 Columbia St, Ste 400, Seattle, WA 98104, USA

Peter R. Reeves School of Molecular and Microbial Biosciences, University of Sydney, Sydney, NSW, Australia

Susan M. Richards Department of Molecular Virology, Immunology and Medical Genetics, Center for Microbial Interface Biology, The Ohio State University, 460 W. 12th Avenue, Columbus, OH 43210-1214, USA

Christa A. Schorr Critical Care Medicine, Cooper University Hospital, Camden, NJ 08103, USA

A.B. Schromm Division of Biophysics and Division of Immunbiophysics, Forschungszentrum Borstel, Parkallee 10, D-23845 Borstel, Germany

Feng Shi State Key Laboratory of Food Science and Technology, Jiangnan University, 1800 Lihu Avenue, Wuxi 214122, China

Diptesh Sil Department of Medicinal Chemistry, University of Kansas, Multidisciplinary Research Building, Room 320D, 2030 Becker Drive, Lawrence, KS 66047, USA

Alba Silipo Dipartimento di Chimica Organica e Biochimica, Università di Napoli Federico II, Complesso Universitario Monte Santangelo, Via Cynthia 4, I-80126 Napoli, Italy

Kristi L. Strandberg Department of Molecular Virology, Immunology and Medical Genetics, Center for Microbial Interface Biology, The Ohio State University, 460 W. 12th Avenue, Columbus, OH 43210-1214, USA

Kathy Triantafilou Infection and Immunity Group, School of Life Sciences, University of Sussex, Falmer, Brighton, BN1 9QG, UK, K.Triantafilou@sussex.ac.uk

Martha Triantafilou Infection and Immunity Group, School of Life Sciences, University of Sussex, Falmer, Brighton, BN1 9QG, UK

Lei Wang TEDA School of Biological Sciences and Biotechnology, Nankai University, 23 Hongda Street, TEDA, Tianjin 300457, China, wanglei@nankai.edu.cn

Quan Wang TEDA School of Biological Sciences and Biotechnology, Nankai University, 23 Hongda Street, TEDA, Tianjin 300457, China

Xiaoyuan Wang State Key Laboratory of Food Science and Technology, Jiangnan University, 1800 Lihu Avenue, Wuxi 214122, China, xwang65@gmail.com

Chan Zhang State Key Laboratory of Food Science and Technology, Jiangnan University, 1800 Lihu Avenue, Wuxi 214122, China

About the Editors

Xiaoyuan Wang is the Cheung Kong Professor in the State Key Laboratory of Food Science and Technology at Jiangnan University. He received his PhD from University of London in 2000, and completed postdoctoral training at University of Texas Houston Medical School in 2002. Then he worked as a Research Associate in Dr. Christian Raetz's group at Duke University Medical School until 2007. During these years Dr. Wang had been using a combined molecular genetic and biochemical approach to study the structure, function, and assembly of phospholipids and lipopolysaccharides in membranes of Gram-negative bacteria. In 2007 Dr. Wang joined Jiangnan University, and was appointed Cheung Kong Chair Professor by the Ministry of Education in the same year. His research interests now include metabolic engineering of industrial microorganism, the detection and control of food-borne pathogens.

Peter J Quinn was Professor of Biochemistry at King's College London during 1989-2009 and is currently Emeritus Professor of Biochemistry. He has held visiting Professorships at Pittsburgh, Nagoya and Tsinghua Universities. His primary research interest is biological membranes and their constituents. The approach in his research has been to apply a range of biophysical methods including real-time synchrotron X-ray diffraction, neutron scattering, differential scanning calorimetry, freeze-fracture electron microscopy, nuclear magnetic resonance spectroscopy, laser flash photolysis and Fourier transform infrared spectroscopy to address questions concerned with relationships between biomembrane structure and function. Professor Quinn received his PhD from the University of Sydney and was awarded a DSc from the University of London in 1980.

α-L-Ara4N	4-amino-4-deoxy-α-L-arabinose
Und-P-α-L-Ara4N	undecaprenyl phosphate-L-Ara4N
galacturonic acid	GalA

1.1 Introduction

The discovery of endotoxin in the late nineteenth century was based on the demonstration that heat-killed cholera bacteria were themselves toxic rather than causing toxicity by secretion of a product from the living organism. Secreted toxins became broadly known as exotoxins, and the toxic materials of bacteria as endotoxins. The historical aspects of the role of endotoxins in bacterial pathogenesis (Beutler and Rietschel, 2003) and their chemical characterization as lipopolysaccharide (LPS) (Raetz et al., 2007; Raetz and Whitfield, 2002) have been the subject of some comprehensive reviews. LPS can be divided into three parts: lipid A, core sugars and O-antigen repeats. Lipid A represents the hydrophobic component of LPS which locates in the outer leaflet of the outer membrane, while core sugars and O-antigen repeats are displayed on the surface of bacteria. Lipid A is known to be responsible for the toxic effects of infections with Gram-negative bacteria (Galanos et al., 1985). The detailed structure of LPS varies from one bacterium to another, and this variation could affect the virulence of bacteria (Wilkinson, 1996). The biosynthetic pathway and export mechanism of LPS has been well characterized in *Escherichia coli*. They are shared to most Gram-negative bacteria, but the exact structures of LPS differ in different bacteria.

LPS can stimulate the transmembrane protein Toll-like receptor 4 (TLR4), a LPS receptor found on the surface of many immune cells such as monocytes, macrophages, neutrophils and dendritic cells (Akira et al., 2006; Poltorak et al., 1998). TLR4 functions as a dimer, and depends on a small protein MD-2 for the recognition of LPS (Triantafilou and Triantafilou, 2002). Other proteins such as CD14 and LBP facilitate the presentation of LPS to MD-2 (Carpenter and O'Neill, 2007; Zhang et al., 1999). After activated by LPS, TLR4 recruits intracellular adapter molecules such as MyD88, Mal, Trif, and Tram (Yamamoto et al., 2002, 2003) which in turn activate other molecules, including protein kinases IRAK1, IRAK4, TBK1, and IKKi, to amplify the signal, and result in the induction or suppression of genes that orchestrate the inflammatory response.

The response from the host immune system depends on both the severity of infection and the particular structure of LPS of the invading bacteria. High concentrations of LPS can induce fever, increase heart rate, and lead to septic shock and death (Parillo, 1993). However, in relatively low concentrations some LPSs could be active immuno-modulators, which can induce non-specific resistance to the invading microbe. Some Gram-negative pathogens synthesize LPS molecules that are poorly recognized by human TLR4, these include *Helicobacter pylori* (Suda et al., 2001), *Francisella tularensis* (Ancuta et al., 1996), and *Chlamydia trachomatis* (Heine et al., 2003). The phosphate groups and the length and number of fatty acyl chains

About the Editors

Xiaoyuan Wang is the Cheung Kong Professor in the State Key Laboratory of Food Science and Technology at Jiangnan University. He received his PhD from University of London in 2000, and completed postdoctoral training at University of Texas Houston Medical School in 2002. Then he worked as a Research Associate in Dr. Christian Raetz's group at Duke University Medical School until 2007. During these years Dr. Wang had been using a combined molecular genetic and biochemical approach to study the structure, function, and assembly of phospholipids and lipopolysaccharides in membranes of Gram-negative bacteria. In 2007 Dr. Wang joined Jiangnan University, and was appointed Cheung Kong Chair Professor by the Ministry of Education in the same year. His research interests now include metabolic engineering of industrial microorganism, the detection and control of food-borne pathogens.

Peter J Quinn was Professor of Biochemistry at King's College London during 1989-2009 and is currently Emeritus Professor of Biochemistry. He has held visiting Professorships at Pittsburgh, Nagoya and Tsinghua Universities. His primary research interest is biological membranes and their constituents. The approach in his research has been to apply a range of biophysical methods including real-time synchrotron X-ray diffraction, neutron scattering, differential scanning calorimetry, freeze-fracture electron microscopy, nuclear magnetic resonance spectroscopy, laser flash photolysis and Fourier transform infrared spectroscopy to address questions concerned with relationships between biomembrane structure and function. Professor Quinn received his PhD from the University of Sydney and was awarded a DSc from the University of London in 1980.

Part I
Structure and Properties of Endotoxins

Chapter 1
Endotoxins: Lipopolysaccharides of Gram-Negative Bacteria

Xiaoyuan Wang and Peter J. Quinn

Abstract Endotoxin refers lipopolysaccharide that constitutes the outer leaflet of the outer membrane of most Gram-negative bacteria. Lipopolysaccharide is comprised of a hydrophilic polysaccharide and a hydrophobic component known as lipid A which is responsible for the major bioactivity of endotoxin. Lipopolysaccharide can be recognized by immune cells as a pathogen-associated molecule through Toll-like receptor 4. Most enzymes and genes related to the biosynthesis and export of lipopolysaccharide have been identified in *Escherichia coli*, and they are shared by most Gram-negative bacteria based on available genetic information. However, the detailed structure of lipopolysaccharide differs from one bacterium to another, suggesting that additional enzymes that can modify the basic structure of lipopolysaccharide exist in bacteria, especially some pathogens. These structural modifications of lipopolysaccharide are sometimes tightly regulated. They are not required for survival but closely related to the virulence of bacteria. In this chapter we will focus on the mechanism of biosynthesis and export of lipopolysaccharide in bacteria.

Keywords Outer membrane · Lipopolysaccharide · Endotoxin · Lipid A · LPS biosynthesis

Abbreviations

LPS	lipopolysaccharide
TLR4	Toll-like receptor 4
Kdo	3-deoxy-D-manno-octulosonic acid
Hep	L-glycero-D-manno-heptose
CAMPs	cationic antimicrobial peptides

X. Wang (✉)
State Key Laboratory of Food Science and Technology, Jiangnan University, 1800 Lihu Avenue, Wuxi 214122, China
e-mail: xwang65@gmail.com

X. Wang, P.J. Quinn (eds.), *Endotoxins: Structure, Function and Recognition*, Subcellular Biochemistry 53, DOI 10.1007/978-90-481-9078-2_1,

α-L-Ara4N	4-amino-4-deoxy-α-L-arabinose
Und-P-α-L-Ara4N	undecaprenyl phosphate-L-Ara4N
galacturonic acid	GalA

1.1 Introduction

The discovery of endotoxin in the late nineteenth century was based on the demonstration that heat-killed cholera bacteria were themselves toxic rather than causing toxicity by secretion of a product from the living organism. Secreted toxins became broadly known as exotoxins, and the toxic materials of bacteria as endotoxins. The historical aspects of the role of endotoxins in bacterial pathogenesis (Beutler and Rietschel, 2003) and their chemical characterization as lipopolysaccharide (LPS) (Raetz et al., 2007; Raetz and Whitfield, 2002) have been the subject of some comprehensive reviews. LPS can be divided into three parts: lipid A, core sugars and O-antigen repeats. Lipid A represents the hydrophobic component of LPS which locates in the outer leaflet of the outer membrane, while core sugars and O-antigen repeats are displayed on the surface of bacteria. Lipid A is known to be responsible for the toxic effects of infections with Gram-negative bacteria (Galanos et al., 1985). The detailed structure of LPS varies from one bacterium to another, and this variation could affect the virulence of bacteria (Wilkinson, 1996). The biosynthetic pathway and export mechanism of LPS has been well characterized in *Escherichia coli*. They are shared to most Gram-negative bacteria, but the exact structures of LPS differ in different bacteria.

LPS can stimulate the transmembrane protein Toll-like receptor 4 (TLR4), a LPS receptor found on the surface of many immune cells such as monocytes, macrophages, neutrophils and dendritic cells (Akira et al., 2006; Poltorak et al., 1998). TLR4 functions as a dimer, and depends on a small protein MD-2 for the recognition of LPS (Triantafilou and Triantafilou, 2002). Other proteins such as CD14 and LBP facilitate the presentation of LPS to MD-2 (Carpenter and O′Neill, 2007; Zhang et al., 1999). After activated by LPS, TLR4 recruits intracellular adapter molecules such as MyD88, Mal, Trif, and Tram (Yamamoto et al., 2002, 2003) which in turn activate other molecules, including protein kinases IRAK1, IRAK4, TBK1, and IKKi, to amplify the signal, and result in the induction or suppression of genes that orchestrate the inflammatory response.

The response from the host immune system depends on both the severity of infection and the particular structure of LPS of the invading bacteria. High concentrations of LPS can induce fever, increase heart rate, and lead to septic shock and death (Parillo, 1993). However, in relatively low concentrations some LPSs could be active immuno-modulators, which can induce non-specific resistance to the invading microbe. Some Gram-negative pathogens synthesize LPS molecules that are poorly recognized by human TLR4, these include *Helicobacter pylori* (Suda et al., 2001), *Francisella tularensis* (Ancuta et al., 1996), and *Chlamydia trachomatis* (Heine et al., 2003). The phosphate groups and the length and number of fatty acyl chains of lipid A play important roles on TLR4 activation (Persing et al., 2002; Rietschel et al., 1994). The *E. coli* lipid A, containing two phosphate groups and six acyl chains composed of 12 or14 carbons, is a powerful activator of the innate immune system (Golenbock et al., 1991).

Since the structure of LPS is closely related to the survival and virulence of Gram-negative bacteria, this chapter will focus on the mechanism of LPS biosynthesis and export to the surface of bacteria.

1.2 Biosynthesize of LPS on the Surfaces of Inner Membrane

The biosynthesis of LPS has been intensively studied in order to develop methods to control Gram-negative pathogens and to cure septic shock. LPS molecules are major constituents of the outer leaflet of the outer membranes in most Gram-negative bacteria. They are essential for the survival of bacteria, including some pathogens that cause human diseases such as septic shock. Although LPS distributes on the surface of bacterial cells, its synthesis is actually initiated in the cytoplasm. How LPS is synthesized in the cytoplasm and exported to the surface of bacteria has been characterized in *E. coli*. The biosynthesis of LPS is initiated from a small molecule, UDP-*N*-acetylglucosamine (UDP-GlcNAc). Multiplicities of enzymes sequentially function to convert UDP-GlcNAc into core-lipid A, and culminating in LPS. Among the three parts of LPS, the structure of lipid A is more widely conserved in different bacteria than that of core sugars or O-antigen, so are the enzymes involved in the biosynthesis of lipid A, core sugars and O-antigen.

1.2.1 Beginning in the Cytoplasm to Form Kdo$_2$-Lipid A

For studies of LPS biosynthesis, *E. coli* has been the most favoured Gram-negative bacterium. The first stage of LPS biosynthesis is to synthesize Kdo$_2$-lipid A (Doerrler, 2006; Raetz and Whitfield, 2002). This involves nine enzymes (Table 1.1), takes place in the cytoplasm at first and end up on the cytoplasmic surface of the inner membrane. The initial building block of lipid A is UDP-GlcNAc. The first three reactions are catalyzed by soluble enzymes LpxA, LpxC and LpxD, resulting in the addition of two 3-OH fatty acid chains to the 2- and 3- positions of the UDP-GlcNAc to form UDP-diacyl-GlcN (Fig. 1.1).

The first reaction catalyzed by LpxA is reversible; therefore, the second reaction catalyzed by LpxC is a committed step. Recently, LpxA, LpxC and LpxD have been isolated and their structure characterized by X-ray diffraction and NMR methods (Barb et al., 2007; Buetow et al., 2007; Williams and Raetz, 2007; Bartling and Raetz, 2009). Both LpxA and LpxD are acyltransferase and their active forms are homotrimers. LpxC, however, is a Zn^{2+}-dependent deacetylase which has no sequence homology with other deacetylases. This makes LpxC a promising target for development of novel antibiotics (Barb et al., 2009). The active site of *E. coli*

Table 1.1 Information on nine enzymes required for the biosynthesis of Kdo_2-lipid A in *E. coli*

Enzyme	Gene	Function	Substate	Donor	Reference
LpxA	*lpxA*	Acyltransferase	UDP-GlcNAc	R-3-hydroxymyristoyl ACP	Williams and Raetz (2007)
LpxC	*lpxC*	Deacetylase	UDP-3-O-(acyl)-GlcNAc	None	Barb et al. (2007)
LpxD	*lpxD*	Acyltransferase	UDP-3-O-(acyl)-GlcN	R-3-hydroxymyristoyl ACP	Buetow et al. (2007)
LpxH	*lpxH*	Pyrophosphatase	UDP-2,3-diacyl-GlcN	None	Babinski et al. (2002a, b)
LpxB	*lpxB*	Disaccharide synthase	UDP-2,3-diacyl-GlcN; Lipid X	None	Crowell et al. (1986; 1987)
LpxK	*lpxK*	4′-Kinase	Disaccharide 1-phosphate	None	Garrett et al. (1997; 1998)
KdtA	*kdtA*	Kdo transferase	Lipid IV_A	CMP-Kdo	Brozek et al. (1989)
LpxL	*lpxL*	Acyltransferase	Kdo_2-lipid IV_A	Lauroyl ACP	Brozek and Raetz (1990)
LpxM	*lpxM*	Acyltransferase	Kdo_2-penta-lipid A	Myristoyl ACP	Brozek and Raetz (1990)

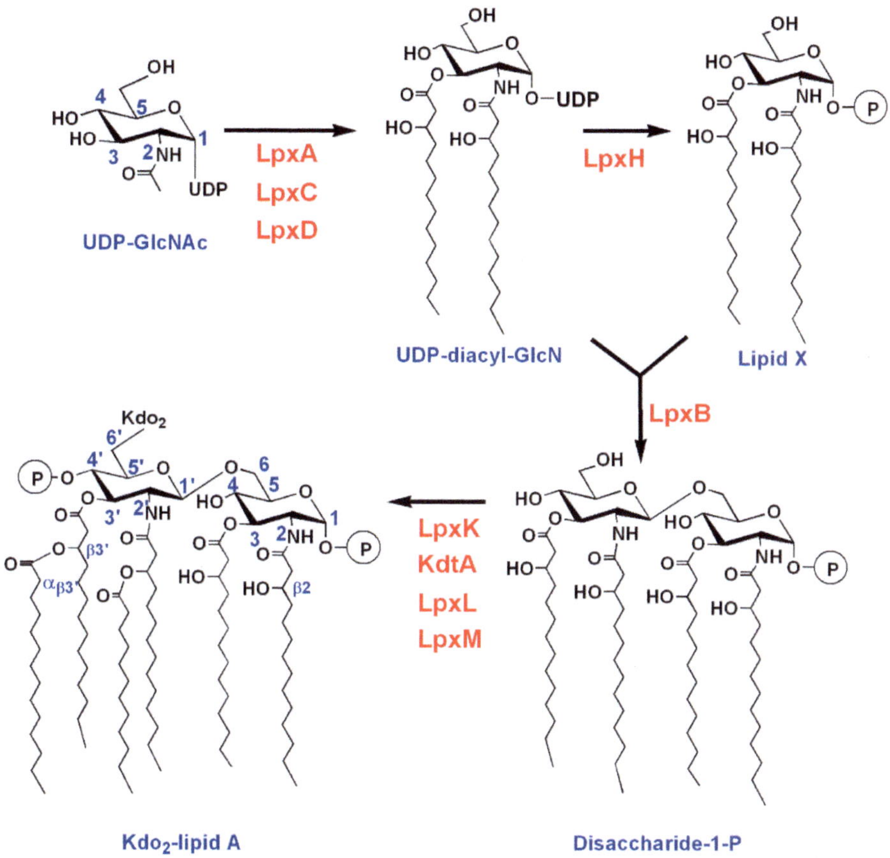

Fig. 1.1 Structure and biosynthetic pathway of Kdo_2-lipid A in *E. coli*. Each reaction is catalyzed by a single enzyme. The names of the enzyme and substrate are highlit. The carbon position and the carbon number of fatty acid chains in lipid A are labeled. The genes encoding the enzymes of Kdo_2-lipid A biosynthesis are present in single copy and highly conserved among bacteria (Raetz et al., 2007; Raetz and Whitfield, 2002)

LpxA functions as a precise hydrocarbon ruler and is manifested by incorporation of C14 hydroxyacyl chains at a rate two orders of magnitude faster than C12 or C16 chains. This explains why most of fatty acids of lipid A contain 14 carbons (Fig. 1.1).

The UDP-diacyl-GlcN is next hydrolyzed by LpxH to form lipid X (Babinski et al., 2002a, b), which is further condensed with its precursor UDP-diacyl-GlcN by LpxB to form disaccharide-1-P (Crowell et al., 1986, 1987; Metzger and Raetz, 2009). Both LpxH and LpxB enzymes are peripheral membrane proteins, while enzymes that catalyze next reactions in the pathway, LpxK, KdtA, LpxL and LpxM, are all integral proteins in the inner membrane. LpxK is a kinase that phosphorylates the 4′-position of the disaccharide-1-P to form lipid IVA (Garrett et al., 1997, 1998). KdtA is a bifunctional enzyme that incorporates two 3-deoxy-D-manno-octulosonic acid (Kdo) residues at the 6′-position of the lipid IVA, using a sugar nucleotide CMP-Kdo as the donor (Brozek et al., 1989). The resulting Kdo_2-lipid IVA undergoes further reactions catalyzed by LpxL and LpxM to form Kdo_2-lipid A (Fig. 1.1). LpxL adds a secondary lauroyl residue and LpxM a myristoyl residue to the distal glucosamine unit, respectively (Brozek and Raetz, 1990). These subsequent acylations do not depend on Kdo in vivo (Klein et al., 2009). The nine enzymes involved in the biosynthesis of Kdo_2-lipid A all have relatively high specificity for their respective substrates (Table 1.1). For example, LpxA, LpxD, LpxL and LpxM are all acyltransferases, but they selectively catalyze different substrates and employ different acyl donors.

LpxD, FabZ, LpxA and LpxB are encoded by a gene cluster *lpxD-fabZ-lpxA-lpxB* in *E. coli* and several other bacteria (Mohan et al., 1994; Steeghs et al., 1997). Proteins LpxA, LpxB and LpxD catalyze early steps in the lipid A pathway using (3R)-hydroxyacyl-ACP as a donor; while FabZ catalyzes the dehydration of (3R)-hydroxyacyl-ACP to trans-2-acyl-ACP (Heath and Rock, 1996), which is further utilized as a fatty acid donor in the biosynthesis of phospholipids. Therefore, this gene cluster could be important for regulating the proportions of LPS and phospholipids in the bacterial membranes. Another gene cluster *msbA-lpxK* also exists in many Gram-negative bacteria, and these two genes are even found to be fused together in some marine bacteria (Venter et al., 2004). MsbA is known as a specific transporter for LPS, while LpxK is a kinase that adds a phosphate group to the 4′-position of lipid A (Garrett et al., 1997, 1998). Why these two genes are always in the same cluster is not clear.

1.2.2 Connecting the Core Oligosaccharides to Lipid A

The core oligosaccharides are sequentially assembled on lipid A at the cytoplasmic surface of the inner membrane in a process that involves a number of membrane-associated glycosyltransferases, using nucleotide sugars as donors. The biosynthesis of core oligosaccharides is rapid and efficient, suggesting that the glycosyltransferases function as a coordinated complex. Core oligosaccharides can be divided into two structurally distinct regions: the inner core which connects to lipid A; the outer core which connects to the O-antigen. Although the structure of lipid A is highly conserved, the structure of the core oligosaccharides shows more variations. The inner core oligosaccharides typically contain residues of Kdo and L-glycero-D-manno-heptose (Hep). The Kdo residue is the most conserved component found in the core region of LPS. The outer core oligosaccharides show more structural diversity than those of the inner core. Structures of core oligosaccharides in *E. coli* strains R1, R2, R3, R4, and K-12 are different (Muller-Loennies et al., 2002, 2003), but the basic backbones are all a linear oligosaccharide of six units. The common sugars found in the core oligosaccharides are Kdo, Hep, D-glucose and D-Galactose.

Genes required for the biosynthesis of core oligosaccharides exist in three operons: *gmhD* and *waaQ* operons in *E. coli* and *Salmonella* (Roncero and Casadaban, 1992). The *gmhD* operon contains four genes *gmhD-waaF-waaC-waaL* that are required for the biosynthesis of inner core oligosaccharides (Schnaitman and Klena, 1993). The *gmhD*, *waaF* and *waaC* genes encode proteins involved in the biosynthesis and transfer of Hep, whereas the *waaL* gene encodes a ligase enzyme required for the attachment of O-antigen to the core-lipid A (Whitfield et al., 1997). The *waaQ* operon contains 7–9 genes that code for enzymes responsible for the biosynthesis of outer core oligosaccharides and their modification.

1.2.3 Synthesizing the O-antigen at the Cytoplasmic Surface of the Inner Membrane

Similarly to the core oligosaccharides, O-antigen is synthesized at the cytoplasmic surface of the inner membrane. Using the sugar nucleotides as donors, the units of O-antigen are assembled by glycosyltransferase enzymes on the membrane-bound carrier, undecaprenyl phosphate which is also used for synthesis of peptidoglycan and capsular polysaccharides. The *rfb* gene cluster in both *E. coli* and *S. enterica* encodes the enzymes required for the synthesis of the sugar-nucleotide precursors that are unique to O-antigens, the glycosyltransferases and polymerases needed for the assembly of the O-antigen and the components required for the transfer of O-antigen polymers across the inner membrane (Raetz and Whitfield, 2002). The O-antigens of LPS exhibit considerable diversity. The unit structures of O-antigen differ in the monomer type as well as the position and stereochemistry of the O-glycosidic linkages. The connection of units in O-antigen may be linear or branched. The O-antigen repeats can be homopolymers or heteropolymers.

1.2.4 Crossing the Inner Membrane

After synthesized at the cytoplasmic surface of the inner membrane, the core-lipid A and the O-antigen are flipped separately to the periplasmic surface of the inner membrane (Doerrler, 2006).

The flip of core-lipid A is carried out by a membrane protein MsbA (Doerrler et al., 2004; Doerrler and Raetz, 2002). MsbA is highly conserved in Gram-negative bacteria and shares homology with the multidrug resistance proteins of eukaryotes. It works as a homodimer and each monomer contains six transmembrane helices and a cytosolic ATP-binding domain (Ward et al., 2007). The flip efficiency of MsbA depends on the exact structure of LPS. In wild type *E. coli*, the best substrate for MsbA is core-lipid A, precursors of core-lipid A usually are not flipped. When core-lipid A is not available, MsbA could flip precursors of core-lipid A, such as Kdo_2-lipid A produced in a heptose-deficient mutant of *E. coli*

(Raetz et al., 2006). Recently, Klein et al. (2009) constructed a *kdtA* mutant of *E. coli* that can grow at low temperature and synthesize lipid IV_A, the minimal LPS structure that can be synthesized so far in vivo. The slow growth appears to be due to defects in the export of lipid IV_A by MsbA. The fast synthesis of lipid IV_A and slow flip of lipid IV_A by MsbA at high temperatures could cause the accumulation of lipid IV_A in the cytosol, leading to cell death. Lowering the growth temperature or overexpressing MsbA help the survival of *kdtA* mutant. The *kdtA* mutants grown at 21°C can synthesize lipid IV_A as well as its pentaacylated and hexaacylated derivatives, indicating in vivo late acylation of lipid A can occur without Kdo (Klein et al., 2009). Further study by Reynolds and Raetz (2009) showed that the optical substrates for MsbA are penta- and hexaacylated lipid A. Once flipped by MsbA, any structures of LPS can always end up on the surface of bacteria.

The O antigen is translocated across the membrane mostly by the Wzy-dependent pathways (Raetz and Whitfield, 2002). The Wzy-dependent pathway requires the membrane proteins Wzx, Wzy and Wzz. Wzx protein is a flippase for the O-antigen across the inner membranes (Alaimo et al., 2006; Liu et al., 1996). Wzx proteins from different bacteria have similar hydropathy profiles (Macpherson et al., 1995) and can complement each other in the translocation of different O-antigen sugar precursors, but no sequence homology or conserved residues are found amongst Wzx proteins (Feldman et al., 1999). Wzx proteins might function by recognizing the first sugar phosphate bound to the undecaprenyl-phosphate (Marolda et al., 2004). ATP-binding domains do not exist in the primary sequences of Wzx proteins (Marolda et al., 1999, 2004).

1.2.5 Assembling LPS at the Periplasmic Surface of the Inner Membrane

After flipped to the periplasmic face of the inner membranes, O-antigen is polymerized by Wzy and Wzz and ligated to terminal sugar residues of the core-lipid A in a reaction mediated by WaaL, resulting in a nascent LPS (Abeyrathne et al., 2005) (Fig. 1.2). Wzy is an O-antigen polymerase; Wzz is a membrane protein with two transmembrane helices that flank a large periplasmic domain. Wzz functions as a molecular ruler to determine the O-antigen chain length (Larue et al., 2009). WaaL proteins from different bacteria all have 12 transmembrane helices, but show significant divergence in their primary amino acid sequence (Raetz et al., 2007). They have conserved amino acids in two adjacent periplasmic loops which might interact with undecaprenyl-diphosphate, the common component in all WaaL substrates (Perez et al., 2008). Although WaaL catalyses the formation of a glycosidic bond, it bears no relationship with classical glycosyltransferases, which modify sugar nucleotide substrates. WaaL could be a good target for searching novel LPS biosynthesis inhibitors because O-antigens of pathogenic bacteria are usually required for resistance to complement-mediated killing (Joiner, 1988).

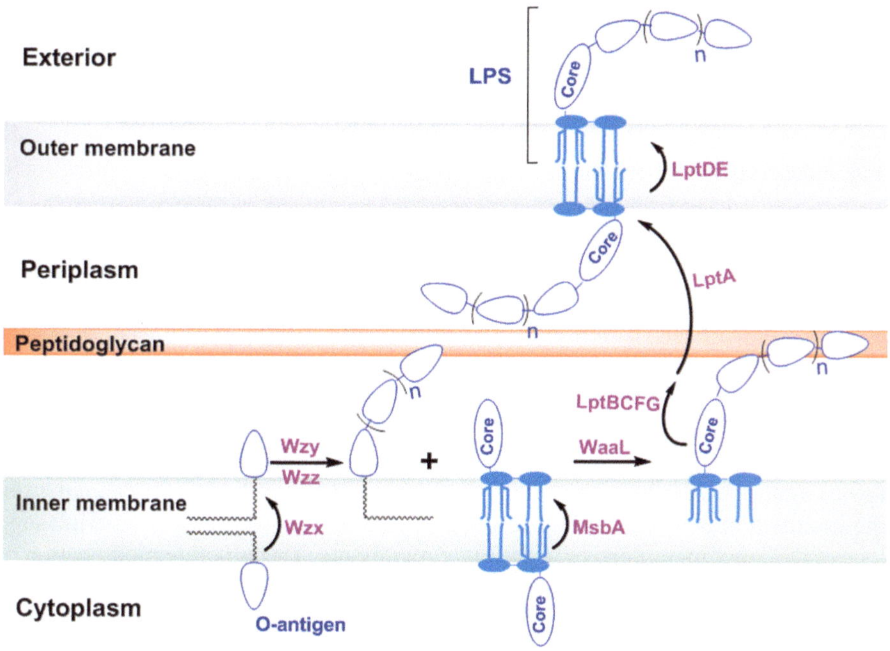

Fig. 1.2 Export of LPS and its precursors in *E. coli*. O-antigen oligosaccharides is assembled separately on undecaprenyl diphosphate, flipped from the cytoplasmic face to the periplasmic face of the inner membrane by the transporter Wzx, and polymerized on the periplasmic face of the inner membrane by Wzy and Wzz. The ABC transporter MsbA flips the core-lipid A from the inner surface to the outer surface of the inner membrane. Then the polymerized O-antigen is transferred to the core-lipid A in the periplasm by WaaL. The protein LptA, LptB, LptC, LptF and LptG might shuttle the nascent LPS from the periplasmic face of the inner membrane to the inner layer of the outer membrane. The outer membrane proteins LptD and LptE are required for the assembly of LPS into the outer surface of the outer membrane (Ma et al., 2008; Wu et al., 2006)

1.3 Export of LPS to the Surface of Bacteria

Nascent LPS molecules are synthesized in periplasm, and shuttled to the inner surface of the outer membrane by proteins LptA, LptB, LptC, LptF and LptG, where protein complex LptD and LptE assemble LPS into the outer surface of the outer membrane (Fig. 1.2). LptA is a periplasmic protein; LptB is a cytosolic protein with ATP binding cassette; LptC, LptF and LptG are inner membrane proteins; and LptD and LptE are outer membrane proteins (Ruiz et al., 2008; Sperandeo et al., 2007, 2008). Depletion of any of these proteins leads to abnormal membrane structures. For example, when LptA, LptB, or both were depleted, LPS was found to accumulate in the periplasm (Sperandeo et al., 2007). Some of these proteins may function as complexes (Bos et al., 2004a). The ABC transporter LptBFG, functioning with LptC and LptA, translocates LPS to the inner leaflet of the outer membrane

(Sperandeo et al., 2007, 2008). In the outer membrane, nascent LPS is exported to the outer leaflet by complex LptD and LptE (Bos et al., 2004b; Ma et al., 2008; Wu et al., 2006).

1.4 Structural Modification of LPS

After synthesized, the structure of LPS can be modified. The LPS modification study is focused on its most conserved part lipid A. Various structures of lipid A have been found in different bacteria (Fig. 1.3). The most conserved part of lipid A is its backbone, disaccharide of glucosamine. The groups connecting to the backbone of lipid A could be modified. Table 1.2 lists the known enzymes that modify the structure of LPS and their genes. The modification of LPS can occue in the hydrophilic polysaccharide region as well as the hydrophobic acyl chain domain. Orthologs of the genes required for the biosynthesis of lipid A in *E. coli* exist in most Gram-negative bacteria, suggesting that lipid A synthesis is separated from the modifications in vivo. Modifications of LPS usually occur at the periplasmic face of the inner membrane or in the outer membrane. The structure modification of LPS might help the bacteria to resist the cationic antimicrobial peptides (CAMPs) released by the host immune system, or to evade recognition by the innate immune receptor TLR4.

1.4.1 Regulation of LPS Modification

PhoP-PhoQ is a two-component system that governs virulence, mediates the adaptation to Mg^{2+}-limiting environment and regulates numerous cellular activities in Gram-negative bacteria (Gibbons et al., 2005; Guo et al., 1998; Soncini et al., 1996). It consists of an inner membrane sensor PhoQ and a cytoplasmic regulator PhoP. PhoQ contains an acidic patch on the surface of its periplasmic domain. Mg^{2+} bridges the acidic patch with anionic phospholipid polar head groups to maintain a repressed regulatory state (Bader et al., 2005; Cho et al., 2006). The PhoP–PhoQ system can also be activated when the bacterium is exposed to CAMPs (Bader et al., 2003, 2005; Martin-Orozco et al., 2006). The activation of the PhoP–PhoQ system can lead to the activation or repression of over 40 genes (Alpuche Aranda et al., 1992; Gooderham and Hancock, 2009). PmrA-PmrB two-component system is also required for *S. enterica* virulence in mice (Gunn et al., 2000). It is usually induced by high Fe^{3+}, the specific signal recognized by the sensor PmrB (Wosten et al., 2000). It can also be induced by low Mg^{2+}, which is detected by the sensor PhoQ of the PhoP-PhoQ system (Garcia Vescovi et al., 1996). The activation by low Mg^{2+}

Fig. 1.3 (continued) the 2′-positon but no phosphate. Additional sugars were also found connecting to the backbone of lipid A in *F. tularensis* (**B**), *S. typhimuriu* (**C**) and *R. etli* (**D**). The carbon position and the carbon number of fatty acid chains in lipid A are labeled

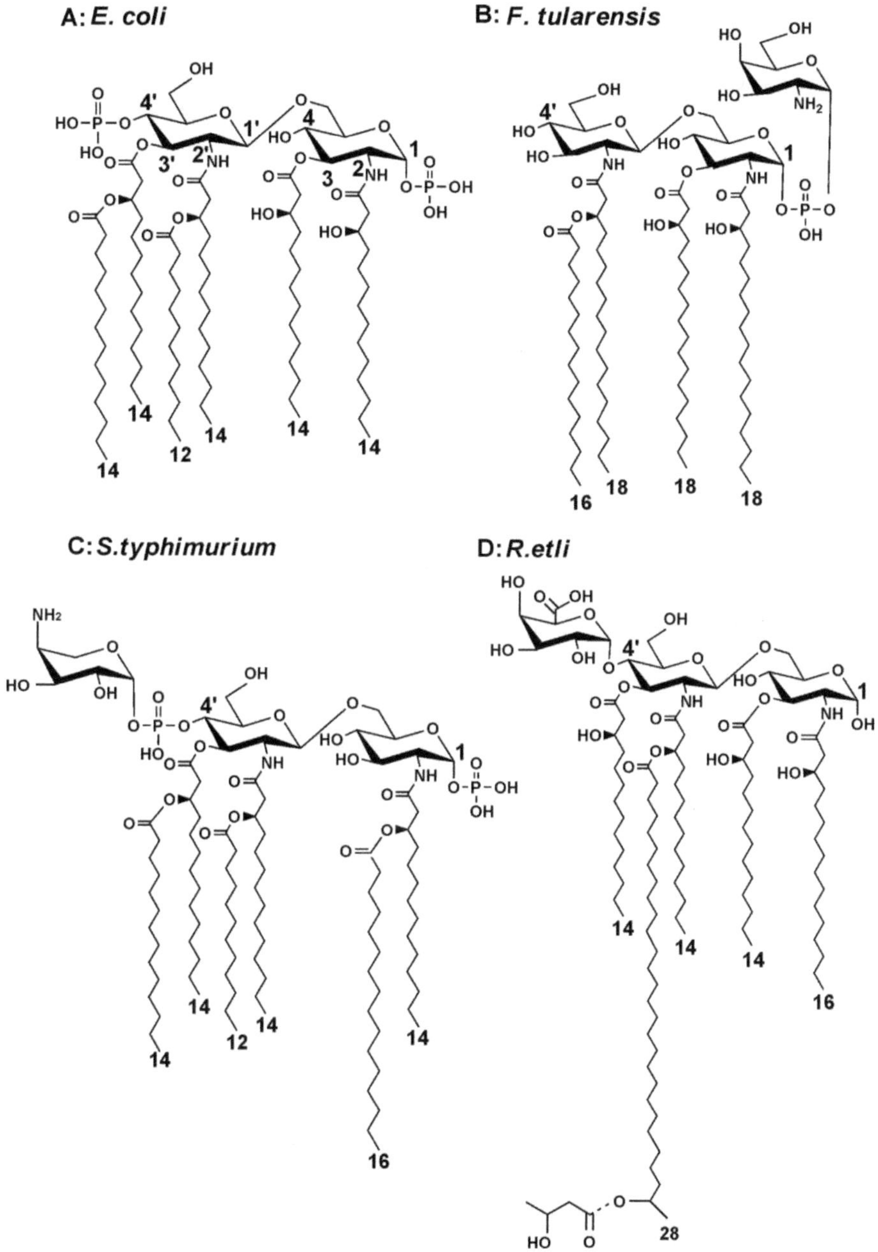

Fig. 1.3 Various structures of lipid A in different bacteria. The most conserved part of lipid A is its backbone, disaccharide of glucosamine. The groups connecting to the backbone of lipid A could be different from one bacterium to another. (**A**) In *E. coli* lipid A there are two phosphates and six fatty acid chains connecting to the backbone. (**B**) In *F. tularensis* lipid A there are only one phosphate and four fatty acid chains. (**C**) In the lipid A of *S. typhimurium* there is an additional second fatty acid chain at 2-position. (**D**) In *R. etli* lipid A there is a very long fatty acid chain at

Table 1.2 Enzymes involved in the structural modification of LPS in Gram-negative bacteria

Enzyme	Gene	Function
LpxE	*lpxE*	Remove the phosphate group from the 1-position of lipid A (Wang et al., 2004).
LpxF	*lpxF*	Remove the phosphate group from the 4′-position of lipid A (Wang et al., 2006a).
LpxO	*lpxO*	Add an OH group to the $\alpha_{\beta 3'}$-position of lipid A (Gibbons et al., 2000, 2008).
ArnT	*arnT*	Transfer the L-Ara4N unit to lipid A (Trent et al., 2001b).
LpxR	*lpxR*	Remove the 3′-acyloxyacyl moiety of lipid A (Reynolds et al., 2006).
PagL	*pagL*	Remove the 3-O-linked acyl chain of lipid A (Rutten et al., 2006).
PagP	*pagP*	Transfer a palmitate to the β2-position of lipid A (Bishop, 2008; Hwang et al., 2004).
LpxXL	*lpxXL*	Add a very long fatty acid chain to the $\beta_{2'}$-position of lipid A (Haag et al., 2009).
LpxT	*lpxT*	Transfer a phosphate group to the 1-phosphate of lipid A (Touze et al., 2008).
LpxQ	*lpxQ*	Oxidize the proximal glucosamine of lipid A to form an aminogluconate unit (Que-Gewirth et al., 2003).
LmtA	*lmtA*	Catalyze the methylation of 1-phosphate of lipid A (Boon Hinckley et al., 2005).
RgtA	*rgtA*	Add a GalA moiety to the distal unit of Kdo (Kanjilal-Kolar et al., 2006).
RgtB	*rgtB*	Add a GalA moiety to the distal unit of Kdo (Kanjilal-Kolar et al., 2006).
RgtC	*rgtC*	Add a GalA moiety to the mannose residue of core oligosaccharide of LPS (Kanjilal-Kolar et al., 2006).
EptA	*eptA*	Add a phosphoethanolamine to 1-position of lipid A (Lee et al., 2004).
EptB	*eptB*	Add a pEtN moiety to the distal unit of Kdo (Reynolds et al., 2005).

requires PhoP, PhoQ, PmrA and PmrB proteins (Soncini et al., 1996) as well as the PhoP-activated PmrD protein (Kox et al., 2000). In *E. coli*, the PmrA-PmrB pathway cannot be triggered by the PhoP-PhoQ system because the PmrD is not functional (Winfield and Groisman, 2004).

Some modifications of LPS are under control of the PhoP-PhoQ system and/or PmrA-PmrB system (Guo et al., 1997). The best example of the regulation of PmrA is the *arn* operon (Breazeale et al., 2002, 2003; Gunn et al., 1998; Trent et al., 2001a) and *ugd* gene (Groisman et al., 1997). Protein products encoded by these genes can synthesize and incorporate a 4-amino-4-deoxy-α-L-arabinose (α-L-Ara4N) into the lipid A part of LPS (Breazeale et al., 2003; Gunn et al., 1998; Zhou et al., 2001). This modification can assist the bacteria resist the antibiotic polymyxin B (Roland et al., 1993). The *arn* operon contains *arnB-arnC-arnA-arnD-arnT-arnE-arnF* genes that encode seven enzymes, ArnB, ArnC, ArnA, ArnD, ArnT, ArnE and ArnF, respectively (Breazeale et al., 2005). Ugd initiates the pathway by converting UDP-glucose to UDP-glucuronic acid. The C-terminal domain of ArnA catalyzes the oxidative decarboxylation of UDP-glucuronic acid to generate UDP-4-keto-pyranose. ArnB then catalyzes a transamination using glutamic acid as the amine donor to form UDP-L-Ara4N. Subsequently, the *N*-terminal domain of ArnA uses *N*-10-formyltetrahydrofolate to synthesize *N*-formylate UDP-α-L-Ara4N, which is, in turn, transferred by ArnC to undecaprenyl phosphate. Then ArnD catalyzes deformylation of this substrate to undecaprenyl phosphate-α-L-Ara4N (Und-P-α-L-Ara4N). ArnE and ArnF flip the Und-P-α-L-Ara4N from the cytoplasmic face to the periplasmic face of the inner membrane (Yan et al., 2007), where ArnT transfers the L-Ara4N unit to the core-lipid A (Fig. 1.3c).

1.4.2 Modifications in the Hydrophobic Region of LPS

Membrane proteins PgaP, PagL, LpxR and LpxO have been reported to modify the fatty acyl chain region of LPS. PagP is a palmitoyl transferase which locates in the outer membrane; it transfers a palmitate from glycerophospholipids to the β2-position of lipid A (Fig. 1.3c), resulting in a hepta-acylated structure (Ahn et al., 2004). PagP is regulated by PhoP-PhoQ system. It was originally identified in *Salmonella* as a protein that is important for resistance to certain CAMPs. The hepta-acylated structure of lipid A might prevent the insertion of CAMPs. PagP has been well characterized in both *E. coli* and *Salmonella*, and its structure has been determined by both NMR spectroscopy and X-ray crystallography (Bishop, 2008; Hwang et al., 2004). PagL is a lipase that removes the 3-O-linked acyl chain of lipid A (Fig. 1.3c) but plays no role in antimicrobial peptide resistance (Kawasaki et al., 2004). Like PagP, PagL is also located in the outer membrane. The *pagL* mutant of *S. typhimurium* displays no obvious phenotypes in a murine model. Although PagL is under the control of the PhoP-PhoQ system it is not active in the outer membrane of *Salmonella* when grown under Mg^{2+}-limiting conditions. PagL might be post-translationally inhibited within the outer membrane because it could be activated in mutants of *Salmonella* that were unable to modify their lipid A with L-Ara4N. PagL from *Pseudomonas aeruginosa* consists of an eight-stranded beta-barrel with the axis tilted by approximately 30 degrees with respect to the lipid bilayer. It contains an active site with a Ser-His-Glu catalytic triad and an oxyanion hole that comprises the conserved Asn (Rutten et al., 2006). Molecules of lipid A, PagL and PagP all locate in the outer membrane of bacteria, which facilitates the rapid modification of the lipid A structure. Modification of the acylation pattern of *Salmonella* lipid A by either PagP or PagL also results in attenuation of lipid A signaling through the TLR4 pathway and, therefore, may promote evasion of the innate immune system during infection (Kawasaki et al., 2004). LpxR is another outer membrane protein that removes the 3′-acyloxyacyl moiety of *Salmonella* lipid A (Reynolds et al., 2006). Orthologs of *Salmonella* LpxR can be found in various Gram-negative bacteria such as *H. pylori*, *Y. enterocolitica*, *E. coli* O157:H7, and *V. cholerae*. LpxR usually remains inactive in *Salmonella* outer membrane, but appears to be activated in *H. pylori* since the major lipid A species of *H. pylori* is completely 3′-*O*-deacylated. LpxR is not regulated by either PhoP-PhoQ or PmrA-PmrB, but requires the divalent cation Ca^{2+} for enzymatic activity. The crystal structure of *S. typhimurium* LpxR revealed that it is a 12-stranded beta-barrel and its active site is located between the barrel wall and an alpha-helix formed by an extracellular loop (Rutten et al., 2009). LpxO is an inner membrane protein that can generate a 2-OH at the $\alpha_{\beta 3'}$-position of *Salmonella* lipid A (Gibbons et al., 2000). This hydroxylation is independent of MsbA transport, indicating a cytoplasmic active site for LpxO. LpxO is not regulated by either the PhoP-PhoQ or the PmrA-PmrB systems.

Different Gram-negative bacteria could make LPS containing different length of the fatty acid chains. For example, the fatty acyl chains of *E. coli* lipid A are 12 or 14 carbons long, while that of *F. novicida* lipid A are 16–18 carbons long (Wang et al., 2006b; Shaffer et al., 2007). The lipid A molecules of *Rhizobium etli*, a legume symbiont and *Brucella abortus*, a phylogenetically related mammalian pathogen, are unusually modified with a very-long-chain fatty acid which has 28 carbons and is attached to lipid A (Fig. 1.3d) by acyltransferase LpxXL (Basu et al., 2002). LpxXL plays an important role in bacterial development (Haag et al., 2009). This unusual lipid A modification could be crucial for the chronic infection of both *S. meliloti* and *B. abortus*. The fatty acid chains in LPS are also related to the infectivity of bacteria. *Yersinia pestis* causes infection through flea bites. In fleas which have a body temperature around 21–27°C, *Y. pestis* synthesizes lipid A containing six fatty acid chains, but in the human host (37°C) *Y. pestis* synthesizes lipid A containing four fatty acid chains (Montminy et al., 2006). The lipid A with six fatty acid chains can activate the immune system through TLR4, but the lipid A with four fatty acid chains cannot (Golenbock et al., 1991). Therefore, *Y. pestis* can escape attack by the immune system because of its unique molecular structure of lipid A.

1.4.3 Modifications in the Hydrophilic Region of LPS

Except for the fatty acid region of LPS, the hydrophilic region of LPS can also be modified. For example, lipid A usually contains two phosphate groups which impart net negative charges to the molecule. The negative charges of lipid A allow the binding of positively charged CAMPs produced by the immune system. To evade the attack by the immune system some bacterial pathogens have evolved less negatively-charged variations of lipid A by removing or decorating the phosphate groups at the 1- and 4′-positions. The decoration includes the addition of amine-containing residues such as α-L-Ara4N and phosphoethanolamine. These modifications result in resistance to CAMPs and are controlled by the PmrA-PmrB two-component system.

Two genes *lpxE* and *lpxF* encoding the lipid A phosphatases have been identified in *F. novicida* (Wang et al., 2004, 2006a). LpxE selectively removes the phosphate group at the 1-position of lipid A, while LpxF selectively removes the phosphate group at the 4′-position (Fig. 1.3b). Orthlogs of LpxE also exist in *R. etli* and in *H. pylori* (Karbarz et al., 2003; Tran et al., 2004). The mutant of *F. novicida* lacking *lpxF* synthesizes a lipid A molecule with an additional phosphate group at 4′-position and an additional fatty acid group at 3′-position when compared with the wild type lipid A. The *lpxF* mutant of *F. novicida* is avirulent in a mouse infection model and is hypersensitive to CAMPs. Following short-term intraperitoneal injection, the *lpxF* mutant bacteria triggers the production of a subset of cytokines, whereas wild-type cells do not (Wang et al., 2007). The lipid A of *lpxF* mutant does not activate TLR4, and *lpxF* mutant cells do not trigger the production of TNF. The *lpxF* mutant of *F. novicida* no longer infects host mice (Wang et al., 2007), suggesting that the phosphate group on lipid A is closely related to the infectivity of bacteria. The removal of phosphate groups to reduce the overall negative charge of lipid A occurs in several bacterial pathogens or endosymbionts. For example, *R. etli* lipid A does not contain phosphate (Que et al., 2000b), while *F. tularensis* lipid A contains only one phosphate group (Wang et al., 2006b). The absence of a phosphate group would greatly decrease the surface negative charge of these bacteria.

Another strategy that bacteria employ to decrease the surface negative charge is the addition of amino groups at 1- or 4′-phosphates of lipid A. EptA encodes a protein necessary for addition of phosphoethanolamine to the 1-phosphate of lipid A (Lee et al., 2004). ArnT is an amino-arabinose transferase found in *S. typhimurium* and transfers L-Arn4N to the 4′-phosphate of lipid A (Trent et al., 2001b). Under some conditions, the positions of phosphoethanolamine and L-Ara4N substituents can be reversed, and lipid A species with two phosphoethanolamine units or two L-Ara4N moieties may be present. The expression of ArnT and EptA is controlled by PmrA. Another example is the galactosamine attached to the 1-phosphate of *F. novicida* lipid A, it is added by an enzyme encoded by an ortholog gene of *arnT* (Wang et al., 2006b). The pathway for the synthesis and incorporation of the galactosamine to lipid A has been characterized in *F. novicida* (Wang et al., 2009; Song et al., 2009). The 1-position of lipid A can also be modified by enzymes LpxT, LmtA and LpxQ. LpxT adds a second phosphate group at 1-phosphate of lipid A, using undecaprenyl pyrophosphate as the substrate donor, therefore one-third of the lipid A in *E. coli* contains a diphosphate unit at 1-position (Touze et al., 2008). LmtA is a membrane enzyme in *Leptospira interrogans* that transfers a methyl group from *S*-adenosylmethionine to the 1-phosphate of lipid A (Boon Hinckley et al., 2005). LpxQ can oxidize the proximal glucosamine of *Rhizobium* lipid A in the presence of O2 to form an aminogluconate unit (Que-Gewirth et al., 2003).

Several proteins have been found to modify the core region of LPS. For example, the inner core of *R. leguminosarum* LPS is modified with three galacturonic acid (GalA) moieties, two on the distal Kdo unit and one on the mannose residue. The three GalA transferases RgtA, RgtB, and RgtC have been characterized (Kanjilal-Kolar et al., 2006). Reconstitution experiments with the individual genes demonstrated that the activity of RgtA precedes and is necessary for the subsequent activity of RgtB, which is followed by the activity of RgtC. Another protein that could modify the Kdo unit is a pEtN transferase EptB (Reynolds et al., 2005). EptB could only be detected in membranes of *E. coli* grown in the presence of 5–50 mM Ca^{2+} (Kanipes et al., 2001), due to the PhoP-PhoQ dependent activation of sRNA MgrR production (Overgaard et al., 2009). MgrR could promote degradation of the *eptB* mRNA, resulting in decreased EptB-dependent LPS modification.

In the presence of high Ca^{2+} concentrations, however, phosphatase activity of PhoQ is activated, leading to dephosphorylation of PhoP and, in turn, to silencing of *mgrR* transcription.

1.5 Conclusion

More enzymes have been identified that modify the inner core and lipid A regions of LPS. Diverse biochemical structures of lipid A have been found on the outer surface of different bacteria (Wilkinson, 1996). Some modifications to the lipid A structure are regulated by two-component regulatory systems in response to specific environmental stimuli (Guo et al., 1997) while other bacteria appear to modify their lipid A constitutively (Wang et al., 2006b). As the major component of the outer membrane, LPS is essential for the survival of most Gram-negative bacteria. Therefore, the enzymes involved in the biosynthesis and transport of lipid A and LPS have become targets for the development of new antibiotics. At present, the first three enzymes LpxA, LpxC and LpxD of the LPS biosynthetic pathway have been purified, and their structures have been characterised by X-ray diffraction and NMR methods (Buetow et al., 2007; Coggins et al., 2003; Williams and Raetz, 2007). Based on the structural information from these proteins, research into developing new antibiotics has been initiated (Barb et al., 2007, 2009).

LPS can cause diseases such as septic shock, multiple organ dysfunction and failure. Understanding the biochemistry of LPS modifications and their impact on pathogenesis could lead to novel treatment options for these diseases. By modifying the LPS structures, we could develop new LPS immune adjuvant or antagonists (Hawkins et al., 2004; Persing et al., 2002; Stover et al., 2004), or improve the traditional Gram-negative bacterial live vaccines.

Acknowledgements Funding was provided by grants from the National Natural Science Foundation of China (NSFC 30770114, NSFC30870074), the Program of State Key Laboratory of Food Science and Technology (SKLF-MB-200801), the 111 project (111-2-06), the Basic Research Programs of Jiangsu Province (BK2009003), and the Human Science Frontier Programme (RGP0016/2005C).

References

Abeyrathne, P.D., Daniels, C., Poon, K.K., Matewish, M.J., Lam, J.S. Functional characterization of WaaL, a ligase associated with linking O-antigen polysaccharide to the core of *Pseudomonas aeruginosa* lipopolysaccharide. J Bacteriol 187 (2005) 3002–3012.

Ahn, V.E., Lo, E.I., Engel, C.K., Chen, L., Hwang, P.M., Kay, L.E., Bishop, R.E., Prive, G.G. A hydrocarbon ruler measures palmitate in the enzymatic acylation of endotoxin. EMBO J 23 (2004) 2931–2941.

Akira, S., Uematsu, S., Takeuchi, O. Pathogen recognition and innate immunity. Cell 124 (2006) 783–801.

Alaimo, C., Catrein, I., Morf, L., Marolda, C.L., Callewaert, N., Valvano, M.A., Feldman, M.F., Aebi, M. Two distinct but interchangeable mechanisms for flipping of lipid-linked oligosaccharides. EMBO J 25 (2006) 967–976.

Alpuche Aranda, C.M., Swanson, J.A., Loomis, W.P., Miller, S.I. *Salmonella typhimurium* activates virulence gene transcription within acidified macrophage phagosomes. Proc Natl Acad Sci USA 89 (1992) 10079–10083.

Ancuta, P., Pedron, T., Girard, R., Sandstrom, G., Chaby, R. Inability of the *Francisella tularensis* lipopolysaccharide to mimic or to antagonize the induction of cell activation by endotoxins. Infect Immun 64 (1996) 2041–2046.

Babinski, K.J., Kanjilal, S.J., Raetz, C.R.H. Accumulation of the lipid A precursor UDP-2,3-diacylglucosamine in an *Escherichia coli* mutant lacking the *lpxH* gene. J Biol Chem 277 (2002a) 25947–25956.

Babinski, K.J., Ribeiro, A.A., Raetz, C.R.H. The *Escherichia coli* gene encoding the UDP-2,3-diacylglucosamine pyrophosphatase of lipid A biosynthesis. J Biol Chem 277 (2002b) 25937–25946.

Bader, M.W., Navarre, W.W., Shiau, W., Nikaido, H., Frye, J.G., McClelland, M., Fang, F.C., Miller, S.I. Regulation of *Salmonella typhimurium* virulence gene expression by cationic antimicrobial peptides. Mol Microbiol 50 (2003) 219–230.

Bader, M.W., Sanowar, S., Daley, M.E., Schneider, A.R., Cho, U., Xu, W., Klevit, R.E., Le Moual, H., Miller, S.I. Recognition of antimicrobial peptides by a bacterial sensor kinase. Cell 122 (2005) 461–472.

Barb, A.W., Jiang, L., Raetz, C.R., Zhou, P. Structure of the deacetylase LpxC bound to the antibiotic CHIR-090: time-dependent inhibition and specificity in ligand binding. Proc Natl Acad Sci USA 104 (2007) 18433–18438.

Barb, A.W., Leavy, T.M., Robins, L.I., Guan, Z., Six, D.A., Zhou, P., Hangauer, M.J., Bertozzi, C.R., Raetz, C.R. Uridine-based inhibitors as new leads for antibiotics targeting *Escherichia coli* LpxC. Biochemistry 48 (2009) 3068–3077.

Bartling, C.M., Raetz, C.R. Crystal structure and acyl chain selectivity of *Escherichia coli* LpxD, the *N*-acyltransferase of lipid a biosynthesis. Biochemistry 48 (2009) 8672–8683.

Basu, S.S., Karbarz, M.J., Raetz, C.R.H. Expression cloning and characterization of the C28 acyltransferase of lipid A biosynthesis in *Rhizobium leguminosarum*. J Biol Chem 277 (2002) 28959–28971.

Beutler, B., Rietschel, E.T. Innate immune sensing and its roots: the story of endotoxin. Nat Rev Immunol 3 (2003) 169–176.

Bishop, R.E. Structural biology of membrane-intrinsic beta-barrel enzymes: sentinels of the bacterial outer membrane. Biochim Biophys Acta 1778 (2008) 1881–1896.

Boon Hinckley, M., Reynolds, C.M., Ribeiro, A.A., McGrath, S.C., Cotter, R.J., Lauw, F.N., Golenbock, D.T., Raetz, C.R. A *Leptospira interrogans* enzyme with similarity to yeast Ste14p that methylates the 1-phosphate group of lipid A. J Biol Chem 280 (2005) 30214–30224.

Bos, M.P., Tefsen, B., Geurtsen, J., Tommassen, J. Identification of an outer membrane protein required for the transport of lipopolysaccharide to the bacterial cell surface. Proc Natl Acad Sci USA 101 (2004a) 9417–9422.

Bos, M.P., Tefsen, B., Geurtsen, J., Tommassen, J. Identification of an outer membrane protein required for the transport of lipopolysaccharide to the bacterial cell surface. Proc Natl Acad Sci USA 101 (2004b) 9417–9422.

Breazeale, S.D., Ribeiro, A.A., McClerren, A.L., Raetz, C.R.H. A formyltransferase required for polymyxin resistance in *Escherichia coli* and the modification of lipid A with 4-amino-4-deoxy-L-arabinose. Identification and function of UDP-4-deoxy-4-formamido-L-arabinose. J Biol Chem 280 (2005) 14154–14167.

Breazeale, S.D., Ribeiro, A.A., Raetz, C.R.H. Oxidative decarboxylation of UDP-glucuronic acid in extracts of polymyxin-resistant *Escherichia coli.* J Biol Chem 277 (2002) 2886–2896.

Breazeale, S.D., Ribeiro, A.A., Raetz, C.R.H. Origin of lipid A species modified with 4-amino-4-deoxy-L-arabinose in polymyxin-resistant mutants of *Escherichia coli.* J Biol Chem 279 (2003) 24731–24739.

Brozek, K.A., Hosaka, K., Robertson, A.D., Raetz, C.R.H. Biosynthesis of lipopolysaccharide in *Escherichia coli*: cytoplasmic enzymes that attach 3-deoxy-D-manno-octulosonic acid to lipid A. J Biol Chem 264 (1989) 6956–6966.

Brozek, K.A., Raetz, C.R.H. Biosynthesis of lipid A in *Escherichia coli*: acyl carrier protein-dependent incorporation of laurate and myristate. J Biol Chem 265 (1990) 15410–15417.

Buetow, L., Smith, T.K., Dawson, A., Fyffe, S., Hunter, W.N. Structure and reactivity of LpxD, the *N*-acyltransferase of lipid A biosynthesis. Proc Natl Acad Sci USA 104 (2007) 4321–4326.

Carpenter, S., O′Neill, L.A. How important are Toll-like receptors for antimicrobial responses? Cell Microbiol 9 (2007) 1891–1901.

Cho, U.S., Bader, M.W., Amaya, M.F., Daley, M.E., Klevit, R.E., Miller, S.I., Xu, W. Metal bridges between the PhoQ sensor domain and the membrane regulate transmembrane signaling. J Mol Biol 356 (2006) 1193–1206.

Coggins, B.E., Li, X., McClerren, A.L., Hindsgaul, O., Raetz, C.R.H., Zhou, P. Structure of the LpxC deacetylase with a bound substrate-analog inhibitor. Nat Struct Biol 10 (2003) 645–651.

Crowell, D.N., Anderson, M.S., Raetz, C.R.H. Molecular cloning of the genes for lipid A disaccharide synthase and UDP-*N*-acetylglucosamine acyltransferase in *Escherichia coli*. J Bacteriol 168 (1986) 152–159.

Crowell, D.N., Reznikoff, W.S., Raetz, C.R.H. Nucleotide sequence of the *Escherichia coli* gene for lipid A disaccharide synthase. J Bacteriol 169 (1987) 5727–5734.

Doerrler, W.T. Lipid trafficking to the outer membrane of Gram-negative bacteria. Mol Microbiol 60 (2006) 542–552.

Doerrler, W.T., Gibbons, H.S., Raetz, C.R.H. MsbA-dependent translocation of lipids across the inner membrane of *Escherichia coli*. J Biol Chem 279 (2004) 45102–45109.

Doerrler, W.T., Raetz, C.R.H. ATPase activity of the MsbA lipid flippase of *Escherichia coli*. J Biol Chem 277 (2002) 36697–36705.

Feldman, M.F., Marolda, C.L., Monteiro, M.A., Perry, M.B., Parodi, A.J., Valvano, M.A. The activity of a putative polyisoprenol-linked sugar translocase (Wzx) involved in *Escherichia coli* O antigen assembly is independent of the chemical structure of the O repeat. J Biol Chem 274 (1999) 35129–35138.

Galanos, C., Lüderitz, O., Rietschel, E.T., Westphal, O., Brade, H., Brade, L., Freudenberg, M., Schade, U., Imoto, M., Yoshimura, H., Kusumoto, S., Shiba, T. Synthetic and natural *Escherichia coli* free lipid A express identical endotoxic activities. Eur J Biochem 148 (1985) 1–5.

Garcia Vescovi, E., Soncini, F.C., Groisman, E.A. Mg^{2+} as an extracellular signal: environmental regulation of *Salmonella* virulence. Cell 84 (1996) 165–174.

Garrett, T.A., Kadrmas, J.L., Raetz, C.R.H. Identification of the gene encoding the *Escherichia coli* lipid A 4′ kinase. Facile synthesis of endotoxin analogs with recombinant LpxK. J Biol Chem 272 (1997) 21855–21864.

Garrett, T.A., Que, N.L., Raetz, C.R.H. Accumulation of a lipid A precursor lacking the 4′-phosphate following inactivation of the *Escherichia coli lpxK* gene. J Biol Chem 273 (1998) 12457–12465.

Gibbons, H.S., Kalb, S.R., Cotter, R.J., Raetz, C.R.H. Role of Mg^{++} and pH in the modification of *Salmonella* lipid A following endocytosis by macrophage tumor cells. Mol Microbiol 55 (2005) 425–440.

Gibbons, H.S., Lin, S., Cotter, R.J., Raetz, C.R.H. Oxygen requirement for the biosynthesis of the S-2-hydroxymyristate moiety in *Salmonella typhimurium* lipid A. Function of LpxO, a new $Fe^{(II)}$/alpha-ketoglutarate-dependent dioxygenase homologue. J Biol Chem 275 (2000) 32940–32949.

Gibbons, H.S., Reynolds, C.M., Guan, Z., Raetz, C.R. An inner membrane dioxygenase that generates the 2-hydroxymyristate moiety of *Salmonella* lipid A. Biochemistry 47 (2008) 2814–2825.

Golenbock, D.T., Hampton, R.Y., Qureshi, N., Takayama, K., Raetz, C.R.H. Lipid A-like molecules that antagonize the effects of endotoxins on human monocytes. J Biol Chem 266 (1991) 19490–19498.

Gooderham, W.J., Hancock, R.E. Regulation of virulence and antibiotic resistance by two-component regulatory systems in *Pseudomonas aeruginosa*. FEMS Microbiol Rev 33 (2009) 279–294.

Groisman, E.A., Kayser, J., Soncini, F.C. Regulation of polymyxin resistance and adaptation to low-Mg^{2+} environments. J Bacteriol 179 (1997) 7040–7045.

Gunn, J.S., Lim, K.B., Krueger, J., Kim, K., Guo, L., Hackett, M., Miller, S.I. PmrA-PmrB-regulated genes necessary for 4-aminoarabinose lipid A modification and polymyxin resistance. Mol Microbiol 27 (1998) 1171–1182.

Gunn, J.S., Ryan, S.S., Van Velkinburgh, J.C., Ernst, R.K., Miller, S.I. Genetic and functional analysis of a PmrA-PmrB-regulated locus necessary for lipopolysaccharide modification, antimicrobial peptide resistance, and oral virulence of *Salmonella enterica* serovar *typhimurium*. Infect Immun 68 (2000) 6139–6146.

Guo, L., Lim, K.B., Gunn, J.S., Bainbridge, B., Darveau, R.P., Hackett, M., Miller, S.I. Regulation of lipid A modifications by *Salmonella typhimurium* virulence genes *phoP-phoQ*. Science 276 (1997) 250–253.

Guo, L., Lim, K.B., Poduje, C.M., Daniel, M., Gunn, J.S., Hackett, M., Miller, S.I. Lipid A acylation and bacterial resistance against vertebrate antimicrobial peptides. Cell 95 (1998) 189–198.

Haag, A.F., Wehmeier, S., Beck, S., Marlow, V.L., Fletcher, V., James, E.K., Ferguson, G.P. The *Sinorhizobium meliloti* LpxXL and AcpXL proteins play important roles in bacteroid development within alfalfa. J Bacteriol 191 (2009) 4681–4686.

Hawkins, L.D., Christ, W.J., Rossignol, D.P. Inhibition of endotoxin response by synthetic TLR4 antagonists. Curr Top Med Chem 4 (2004) 1147–1171.

Heath, R.J., Rock, C.O. Roles of the FabA and FabZ beta-hydroxyacyl-acyl carrier protein dehydratases in *Escherichia coli* fatty acid biosynthesis. J Biol Chem 271 (1996) 27795–27801.

Heine, H., Muller-Loennies, S., Brade, L., Lindner, B., Brade, H. Endotoxic activity and chemical structure of lipopolysaccharides from *Chlamydia trachomatis* serotypes E and L2 and *Chlamydophila psittaci* 6BC. Eur J Biochem 270 (2003) 440–450.

Hwang, P.M., Bishop, R.E., Kay, L.E. The integral membrane enzyme PagP alternates between two dynamically distinct states. Proc Natl Acad Sci USA 101 (2004) 9618–9623.

Joiner, K.A. Complement evasion by bacteria and parasites. Annu Rev Microbiol 42 (1988) 201–230.

Kanjilal-Kolar, S., Basu, S.S., Kanipes, M.I., Guan, Z., Garrett, T.A., Raetz, C.R. Expression cloning of three *Rhizobium leguminosarum* lipopolysaccharide core galacturonosyltransferases. J Biol Chem 281 (2006) 12865–12878.

Kanipes, M.I., Lin, S., Cotter, R.J., Raetz, C.R. Ca^{2+}-induced phosphoethanolamine transfer to the outer 3-deoxy-D-manno-octulosonic acid moiety of *Escherichia coli* lipopolysaccharide. A novel membrane enzyme dependent upon phosphatidylethanolamine. J Biol Chem 276 (2001) 1156–1163.

Karbarz, M.J., Kalb, S.R., Cotter, R.J., Raetz, C.R.H. Expression cloning and biochemical characterization of a *Rhizobium leguminosarum lipid A 1-phosphatase*. J Biol Chem 278 (2003) 39269–39279.

Kawasaki, K., Ernst, R.K., Miller, S.I. 3-*O*-deacylation of lipid A by PagL, a PhoP/PhoQ-regulated deacylase of *Salmonella typhimurium*, modulates signaling through Toll-like receptor 4. J Biol Chem 279 (2004) 20044–20048.

Klein, G., Lindner, B., Brabetz, W., Brade, H., Raina, S. *Escherichia coli* K-12 suppressor-free mutants lacking early glycosyltransferases and late acyltransferases: minimal lipopolysaccharide structure and induction of envelope stress response. J Biol Chem 284 (2009) 15369–15389.

Kox, L.F., Wosten, M.M., Groisman, E.A. A small protein that mediates the activation of a two-component system by another two-component system. EMBO J 19 (2000) 1861–1872.

Larue, K., Kimber, M.S., Ford, R., Whitfield, C. Biochemical and structural analysis of bacterial O-antigen chain length regulator proteins reveals a conserved quaternary structure. J Biol Chem 284 (2009) 7395–7403.

Lee, H., Hsu, F.F., Turk, J., Groisman, E.A. The PmrA-regulated pmrC gene mediates phosphoethanolamine modification of lipid A and polymyxin resistance in *Salmonella enterica.* J Bacteriol 186 (2004) 4124–4133.

Liu, D., Cole, R.A., Reeves, P.R. An O-antigen processing function for Wzx (RfbX): a promising candidate for O-unit flippase. J Bacteriol 178 (1996) 2102–2107.

Ma, B., Reynolds, C.M., Raetz, C.R. Periplasmic orientation of nascent lipid A in the inner membrane of an *Escherichia coli* LptA mutant. Proc Natl Acad Sci USA 105 (2008) 13823–13828.

Macpherson, D.F., Manning, P.A., Morona, R. Genetic analysis of the rfbX gene of *Shigella flexneri.* Gene 155 (1995) 9–17.

Marolda, C.L., Feldman, M.F., Valvano, M.A. Genetic organization of the O7-specific lipopolysaccharide biosynthesis cluster of *Escherichia coli* VW187 (O7:K1). Microbiology 145 (1999) 2485–2495.

Marolda, C.L., Vicarioli, J., Valvano, M.A. Wzx proteins involved in biosynthesis of O antigen function in association with the first sugar of the O-specific lipopolysaccharide subunit. Microbiology 150 (2004) 4095–4105.

Martin-Orozco, N., Touret, N., Zaharik, M.L., Park, E., Kopelman, R., Miller, S., Finlay, B.B., Gros, P., Grinstein, S. Visualization of vacuolar acidification-induced transcription of genes of pathogens inside macrophages. Mol Biol Cell 17 (2006) 498–510.

Metzger, L.E., Raetz, C.R. Purification and characterization of the lipid A disaccharide synthase (LpxB) from *Escherichia coli*, a peripheral membrane protein. Biochemistry 48 (2009) 11559–11571.

Mohan, S., Kelly, T.M., Eveland, S.S., Raetz, C.R.H., Anderson, M.S. An *Escherichia coli* gene (*fabZ*) encoding *R*-3-hydroxymyristoyl acyl carrier protein dehydrase. Relation to *fabA* and suppression of mutations in lipid A biosynthesis. J Biol Chem 269 (1994) 32896–32903.

Montminy S.W., Khan N., McGrath S., Walkowicz M.J., Sharp F., Conlon J.E., et al. Virulence factors of *Yersinia pestis* are overcome by a strong lipopolysaccharide response. Nat Immunol 7 (2006) 1066–1073.

Muller-Loennies, S., Lindner, B., Brade, H. Structural analysis of deacylated lipopolysaccharide of *Escherichia coli* strains 2513 (R4 core-type) and F653 (R3 core-type). Eur J Biochem 269 (2002) 5982–5991.

Muller-Loennies, S., Lindner, B., Brade, H. Structural analysis of oligosaccharides from lipopolysaccharide (LPS) of *Escherichia coli* K12 strain W3100 reveals a link between inner and outer core LPS biosynthesis. J Biol Chem 278 (2003) 34090–34101.

Overgaard, M., Johansen, J., Moller-Jensen, J., Valentin-Hansen, P. Switching off small RNA regulation with trap-mRNA. Mol Microbiol 73 (2009) 790–800.

Parillo, J.E. Pathogenic mechanisms of septic shock. N Engl J Med 328 (1993) 1471–1477.

Perez, J.M., McGarry, M.A., Marolda, C.L., Valvano, M.A. Functional analysis of the large periplasmic loop of the *Escherichia coli* K-12 WaaL O-antigen ligase. Mol Microbiol 70 (2008) 1424–1440.

Persing, D.H., Coler, R.N., Lacy, M.J., Johnson, D.A., Baldridge, J.R., Hershberg, R.M., Reed, S.G. Taking toll: lipid A mimetics as adjuvants and immunomodulators. Trends Microbiol 10 (2002) S32–37.

Poltorak, A., He, X., Smirnova, I., Liu, M.Y., Huffel, C.V., Du, X., Birdwell, D., Alejos, E., Silva, M., Galanos, C., Freudenberg, M., Ricciardi-Castagnoli, P., Layton, B., Beutler, B. Defective LPS signaling in C3H/HeJ and C57BL/10ScCr mice: mutations in Tlr4 gene. Science 282 (1998) 2085–2088.

Que, N.L.S., Ribeiro, A.A., Raetz, C.R.H. Two-dimensional NMR spectroscopy and structures of six lipid A species from *Rhizobium etli* CE3; detection of an acyloxyacyl residue in each component and origin of the aminogluconate moiety. J Biol Chem 275 (2000b) 28017–28027.

Que-Gewirth, N.L.S., Karbarz, M.J., Kalb, S.R., Cotter, R.J., Raetz, C.R.H. Origin of the 2-amino-2-deoxy-gluconate unit in *Rhizobium leguminosarum* lipid A. Expression cloning of the outer membrane oxidase LpxQ. J Biol Chem 278 (2003) 12120–12129.

Raetz, C.R., Garrett, T.A., Reynolds, C.M., Shaw, W.A., Moore, J.D., Smith, D.C., Jr., Ribeiro, A.A., Murphy, R.C., Ulevitch, R.J., Fearns, C., Reichart, D., Glass, C.K., Benner, C., Subramaniam, S., Harkewicz, R., Bowers-Gentry, R.C., Buczynski, M.W., Cooper, J.A., Deems, R.A., Dennis, E.A. Kdo_2-lipid A of *Escherichia coli*, a defined endotoxin that activates macrophages via TLR-4. J Lipid Res 47 (2006) 1097–1111.

Raetz, C.R., Reynolds, C.M., Trent, M.S., Bishop, R.E. Lipid A modification systems in Gram-negative bacteria. Annu Rev Biochem 76 (2007) 295–329.

Raetz, C.R.H., Whitfield, C. Lipopolysaccharide endotoxins. Annu Rev Biochem 71 (2002) 635–700.

Reynolds, C.M., Kalb, S.R., Cotter, R.J., Raetz, C.R. A phosphoethanolamine transferase specific for the outer 3-deoxy-D-manno-octulosonic acid residue of *Escherichia coli* lipopolysaccharide. Identification of the eptB gene and Ca^{2+} hypersensitivity of an eptB deletion mutant. J Biol Chem 280 (2005) 21202–21211.

Reynolds, C.M., Raetz, C.R. Replacement of lipopolysaccharide with free lipid A molecules in *Escherichia coli* mutants lacking all core sugars. Biochemistry 48 (2009) 9627–9640.

Reynolds, C.M., Ribeiro, A.A., McGrath, S.C., Cotter, R.J., Raetz, C.R., Trent, M.S. An outer membrane enzyme encoded by *Salmonella typhimurium* lpxR that removes the 3′-acyloxyacyl moiety of lipid A. J Biol Chem 281 (2006) 21974–21987.

Rietschel, E.T., Kirikae, T., Schade, F.U., Mamat, U., Schmidt, G., Loppnow, H., Ulmer, A.J., Zähringer, U., Seydel, U., Di Padova, F., Schreier, M., Brade, H. Bacterial endotoxin: molecular relationships of structure to activity and function. FASEB J 8 (1994) 217–225.

Roland, K.L., Martin, L.E., Esther, C.R., Spitznagel, J.K. Spontaneous *pmrA* mutants of *Salmonella typhimurium* LT2 define a new two-component regulatory system with a possible role in virulence. J Bacteriol 175 (1993) 4154–4164.

Roncero, C., Casadaban, M.J. Genetic analysis of the genes involved in synthesis of the lipopolysaccharide core in *Escherichia coli* K-12: three operons in the rfa locus. J Bacteriol 174 (1992) 3250–3260.

Ruiz, N., Gronenberg, L.S., Kahne, D., Silhavy, T.J. Identification of two inner-membrane proteins required for the transport of lipopolysaccharide to the outer membrane of *Escherichia coli*. Proc Natl Acad Sci USA 105 (2008) 5537–5542.

Rutten, L., Geurtsen, J., Lambert, W., Smolenaers, J.J., Bonvin, A.M., de Haan, A., van der Ley, P., Egmond, M.R., Gros, P., Tommassen, J. Crystal structure and catalytic mechanism of the LPS 3-*O*-deacylase PagL from *Pseudomonas aeruginosa*. Proc Natl Acad Sci USA 103 (2006) 7071–7076.

Rutten, L., Mannie, J.P., Stead, C.M., Raetz, C.R., Reynolds, C.M., Bonvin, A.M., Tommassen, J.P., Egmond, M.R., Trent, M.S., Gros, P. Active-site architecture and catalytic mechanism of the lipid A deacylase LpxR of *Salmonella typhimurium*. Proc Natl Acad Sci USA 106 (2009) 1960–1964.

Schnaitman, C.A., Klena, J.D. Genetics of lipopolysaccharide biosynthesis in enteric bacteria. Microbiol Rev 57 (1993) 655–682.

Shaffer, S.A., Harvey, M.D., Goodlett, D.R., Ernst R.K. Structural heterogeneity and environmentally regulated remodeling of *Francisella tularensis* subspecies *novicida* lipid A characterized by tandem mass spectrometry. J Am Soc Mass Spectrom 18 (2007) 1080–1092.

Soncini, F.C., Garcia Vescovi, E., Solomon, F., Groisman, E.A. Molecular basis of the magnesium deprivation response in *Salmonella typhimurium*: identification of PhoP-regulated genes. J Bacteriol 178 (1996) 5092–5099.

Song F., Guan Z., Raetz C.R.H. Biosynthesis of undecaprenyl phosphate–galactosamine and undecaprenyl phosphate–glucose in *Francisella novicida*. Biochemistry 48 (2009) 1173–1182.

Sperandeo, P., Cescutti, R., Villa, R., Di Benedetto, C., Candia, D., Deho, G., Polissi, A. Characterization of lptA and lptB, two essential genes implicated in lipopolysaccharide transport to the outer membrane of *Escherichia coli*. J Bacteriol 189 (2007) 244–253.

Sperandeo, P., Lau, F.K., Carpentieri, A., De Castro, C., Molinaro, A., Deho, G., Silhavy, T.J., Polissi, A. Functional analysis of the protein machinery required for transport of lipopolysaccharide to the outer membrane of *Escherichia coli*. J Bacteriol 190 (2008) 4460–4469.

Steeghs, L., Jennings, M.P., Poolman, J.T., van der Ley, P. Isolation and characterization of the *Neisseria meningitidis lpxD-fabZ-lpxA* gene cluster involved in lipid A biosynthesis. Gene 190 (1997) 263–270.

Stover, A.G., Da Silva Correia, J., Evans, J.T., Cluff, C.W., Elliott, M.W., Jeffery, E.W., Johnson, D.A., Lacy, M.J., Baldridge, J.R., Probst, P., Ulevitch, R.J., Persing, D.H., Hershberg, R.M. Structure–activity relationship of synthetic toll-like receptor 4 agonists. J Biol Chem 279 (2004) 4440–4449.

Suda, Y., Kim, Y.M., Ogawa, T., Yasui, N., Hasegawa, Y., Kashihara, W., Shimoyama, T., Aoyama, K., Nagata, K., Tamura, T., Kusumoto, S. Chemical structure and biological activity of a lipid A component from *Helicobacter pylori* strain 206. J Endotoxin Res 7 (2001) 95–104.

Touze, T., Tran, A.X., Hankins, J.V., Mengin-Lecreulx, D., Trent, M.S. Periplasmic phosphorylation of lipid A is linked to the synthesis of undecaprenyl phosphate. Mol Microbiol 67 (2008) 264–277.

Tran, A.X., Karbarz, M.J., Wang, X., Raetz, C.R.H., McGrath, S.C., Cotter, R.J., Trent, M.S. Periplasmic cleavage and modification of the 1-phosphate group of *Helicobacter pylori* lipid A. J Biol Chem 279 (2004) 55780–55791.

Trent, M.S., Ribeiro, A.A., Doerrler, W.T., Lin, S., Cotter, R.J., Raetz, C.R.H. Accumulation of a polyisoprene-linked amino sugar in polymyxin resistant mutants in *Salmonella typhimurium* and *Escherichia coli*: structural characterization and possible transfer to lipid A in the periplasm. J Biol Chem 276 (2001a) 43132–43144.

Trent, M.S., Ribeiro, A.A., Lin, S., Cotter, R.J., Raetz, C.R.H. An inner membrane enzyme in *Salmonella typhimurium* and *Escherichia coli* that transfers 4-amino-4-deoxy-L-arabinose to lipid A: Induction in polymyxin resistant mutants and role of a novel lipid-linked donor. J Biol Chem 276 (2001b) 43122–43131.

Triantafilou, M., Triantafilou, K. Lipopolysaccharide recognition: CD14, TLRs and the LPS-activation cluster. Trends Immunol 23 (2002) 301–304.

Venter, J.C., Remington, K., Heidelberg, J.F., Halpern, A.L., Rusch, D., Eisen, J.A., Wu, D., Paulsen, I., Nelson, K.E., Nelson, W., Fouts, D.E., Levy, S., Knap, A.H., Lomas, M.W., Nealson, K., White, O., Peterson, J., Hoffman, J., Parsons, R., Baden-Tillson, H., Pfannkoch, C., Rogers, Y.H., Smith, H.O. Environmental genome shotgun sequencing of the Sargasso Sea. Science 304 (2004) 66–74.

Wang, X., Karbarz, M.J., McGrath, S.C., Cotter, R.J., Raetz, C.R.H. MsbA transporter-dependent lipid A 1-dephosphorylation on the periplasmic surface of the inner membrane: topography of *Francisella novicida* LpxE expressed in *Escherichia coli*. J Biol Chem 279 (2004) 49470–49478.

Wang, X., McGrath, S.C., Cotter, R.J., Raetz, C.R. Expression cloning and periplasmic orientation of the *Francisella novicida* lipid A 4′-phosphatase LpxF. J Biol Chem 281 (2006a) 9321–9330.

Wang, X., Ribeiro, A.A., Guan, Z., Abraham, S.N., Raetz, C.R. Attenuated virulence of a *Francisella* mutant lacking the lipid A 4′-phosphatase. Proc Natl Acad Sci USA 104 (2007) 4136–4141.

Wang, X., Ribeiro, A.A., Guan, Z., McGrath, S.C., Cotter, R.J., Raetz, C.R. Structure and biosynthesis of free lipid A molecules that replace lipopolysaccharide in *Francisella tularensis* subsp. *novicida*. Biochemistry 45 (2006b) 14427–14440.

Wang, X., Ribeiro, A.A., Guan, Z., Raetz, C.R. Identification of undecaprenyl phosphate-beta-D-galactosamine in *Francisella novicida* and its function in lipid A modification. Biochemistry 48 (2009) 1162–1172.

Ward, A., Reyes, C.L., Yu, J., Roth, C.B., Chang, G. Flexibility in the ABC transporter MsbA: alternating access with a twist. Proc Natl Acad Sci USA 104 (2007) 19005–19010.

Whitfield, C., Amor, P.A., Koplin, R. Modulation of the surface architecture of Gram-negative bacteria by the action of surface polymer:lipid A-core ligase and by determinants of polymer chain length. Mol Microbiol 23 (1997) 629–638.

Wilkinson, S.G. Bacterial lipopolysaccharides—themes and variations. Prog Lipid Res 35 (1996) 283–343.

Williams, A.H., Raetz, C.R. Structural basis for the acyl chain selectivity and mechanism of UDP-*N*-acetylglucosamine acyltransferase. Proc Natl Acad Sci USA 104 (2007) 13543–13550.

Winfield, M.D., Groisman, E.A. Phenotypic differences between *Salmonella* and *Escherichia coli* resulting from the disparate regulation of homologous genes. Proc Natl Acad Sci USA 101 (2004) 17162–17167.

Wosten, M.M., Kox, L.F., Chamnongpol, S., Soncini, F.C., Groisman, E.A. A signal transduction system that responds to extracellular iron. Cell 103 (2000) 113–125.

Wu, T., McCandlish, A.C., Gronenberg, L.S., Chng, S.S., Silhavy, T.J., Kahne, D. Identification of a protein complex that assembles lipopolysaccharide in the outer membrane of *Escherichia coli*. Proc Natl Acad Sci USA 103 (2006) 11754–11759.

Yamamoto, M., Sato, S., Hemmi, H., Sanjo, H., Uematsu, S., Kaisho, T., Hoshino, K., Takeuchi, O., Kobayashi, M., Fujita, T., Takeda, K., Akira, S. Essential role for TIRAP in activation of the signalling cascade shared by TLR2 and TLR4. Nature 420 (2002) 324–329.

Yamamoto, M., Sato, S., Hemmi, H., Uematsu, S., Hoshino, K., Kaisho, T., Takeuchi, O., Takeda, K., Akira, S. TRAM is specifically involved in the Toll-like receptor 4-mediated MyD88-independent signaling pathway. Nat Immunol 4 (2003) 1144–1150.

Yan, A., Guan, Z., Raetz, C.R. An undecaprenyl phosphate-aminoarabinose flippase required for polymyxin resistance in *Escherichia coli*. J Biol Chem 282 (2007) 36077–36089.

Zhang, F.X., Kirschning, C.J., Mancinelli, R., Xu, X.P., Jin, Y., Faure, E., Mantovani, A., Rothe, M., Muzio, M., Arditi, M. Bacterial lipopolysaccharide activates nuclear factor-kappaB through interleukin-1 signaling mediators in cultured human dermal endothelial cells and mononuclear phagocytes. J Biol Chem 274 (1999) 7611–7614.

Zhou, Z., Ribeiro, A.A., Lin, S., Cotter, R.J., Miller, S.I., Raetz, C.R. Lipid A modifications in polymyxin-resistant *Salmonella typhimurium*: PMRA-dependent 4-amino-4-deoxy-L-arabinose, and phosphoethanolamine incorporation. J Biol Chem 276 (2001) 43111–43121.

Chapter 2
Purification and Characterization of Lipopolysaccharides

Xiaoyuan Wang, Chan Zhang, Feng Shi, and Xiaoqing Hu

Abstract Lipopolysaccharides are the major components on the surface of most Gram-negative bacteria, and recognized by immune cells as a pathogen-associated molecule. They can cause severe diseases like sepsis and therefore known as endotoxins. Lipopolysaccharide consists of lipid A, core oligosaccharide and O-antigen repeats. Lipid A is responsible for the major bioactivity of endotoxin. Because of their specific structure and amphipathic property, purification and analysis of lipopolysaccharides are difficult. In this chapter, we summarize the available approaches for extraction, purification and analysis of lipopolysaccharides.

Keywords Lipopolysaccharide · LPS · Lipid A · Extraction · Purification · Analysis

Abbreviations

LPS	lipopolysaccharide
R-LPS	rough type LPS
S-LPS	smooth type LPS
PCP	phenol, chloroform and petroleum ether
RNase	ribonuclease
DNase I	deoxyribonuclease I
SDS	sodium dodecyl sulfate
SDS-PAGE	sodium dodecyl sulfate-polyacylamide gel electrophoresis
DOC	deoxycholate
CE	capillary electrophoresis
FT-ICR MS	Fourier-transform ion cyclotron resonance mass spectrometer
ESI MS	electrospray ionization mass spectrometry
EtBr	ethidium bromide
TAE	triethylamine

X. Wang (✉)
State Key Laboratory of Food Science and Technology, Jiangnan University, 1800 Lihu Avenue, Wuxi 214122, China
e-mail: xwang65@gmail.com

X. Wang, P.J. Quinn (eds.), *Endotoxins: Structure, Function and Recognition*, Subcellular Biochemistry 53, DOI 10.1007/978-90-481-9078-2_2,

TLR2	Toll-like receptor 2
TLR4	Toll-like receptor 4
TLC	thin layer chromatography
CFU	colony forming unit

2.1 Introduction

Lipopolysaccharide (LPS) locates on the cell surface of most Gram-negative bacteria. It consists of lipid A, core oligosaccharide, and O-antigen repeats. The detailed structure of LPS differs from one bacterium to another and the structure of lipid A part is more conserved than that of the core oligosaccharide and O-antigen repeats. According to their sizes and structures, LPS can be roughly divided into three groups: smooth type LPS (S-LPS) which contains various sizes of O-antigen repeats; rough type LPS (R-LPS) which contains various sizes of core oligosaccharide but no O-antigens; and free lipid A. Because of their structure variation, different techniques have been developed to extract, purify and analyze LPS.

2.2 Extraction of Lipopolysaccharides

Because LPS molecules in different bacteria have different structures and amphipathic properties, there is no universal method for extraction of LPS. Several methods have been developed; each favors some specific groups of LPS. For example, the phenol–water extraction method favors S-LPS extraction, but not R-LPS (Hickman and Ashwell, 1966; Kasai and Nowotny, 1967); the ether extraction method favors the extraction of R-LPS (Galanos et al., 1969). Both methods for large scale or micro-scale extractions have been developed. Agents used by different extraction methods are listed in Table 2.1.

2.2.1 Large Scale Extraction of Lipopolysaccharides

Since the discovery of LPS (Shear, 1941), various methods for the extraction of LPS have been developed. These include extraction with trichloroacetic acid (Ribi et al., 1961), with ether (Galanos et al., 1969), with water (Robert et al., 1967), with pyridine (Goebel et al., 1945), with phenol (Westphal and Jann, 1965), with butanol (Morrison and Leive, 1975), and with sodium dodecyl sulfate (SDS) (Darveau and Hancock, 1983). A few of these methods can chemically alter the structure of LPS (Nowotny et al., 1966; Tsang et al., 1974; Wober and Alaupovic, 1971). Among these methods, the most used are the phenol–water extraction (Westphal and Jann, 1965) and the ether extraction (Galanos et al., 1969). The former is most efficient for the extraction of S-LPS, while the latter for R-LPS. A method which could efficiently extract both S-LPS and R-LPS has also been developed (Darveau and Hancock, 1983).

Table 2.1 Agents used in extraction and purification of LPS

Agent	Function	Reference
Phenol and water (45:50, v/v)	Separate LPS from contaminated proteins, polysaccharides, and nucleic acids	Westphal and Jann (1965)
PCP extract solution (2:5:8, v/v/v)/ (5:5:8, v/v/v)	Separate R-LPS away from S-LPS as well as proteins, nucleic acids and polysaccharides	Galanos et al. (1969) Helander et al. (1992)
Butanol	Extract LPS	Morrison and Leive (1975)
Phenol–chloroform mixture	Extract LPS	Kido et al. (1990)
Bligh–Dyer solvent system	Separate phospholipids away from LPS; purify lipid A	Bligh and Dyer (1959)
RNase A DNase I	Remove contaminated nucleic acids in LPS	Johnson and Perry (1976)
Pronase Proteinase K	Remove contaminated proteins in LPS	Johnson and Perry (1976)
EDTA	Dissociate LPS from bacterial membranes; Chelate metal ions such as Mg^{2+} and Zn^{2+}	Darveau and Hancock (1983) Hardy et al. (1998)
SDS	Disaggregate LPS and denature proteins	Darveau and Hancock (1983) Caroff et al. (1988)
Deoxycholate Triton X-100 TEA	Disaggregate LPS	Morrison and Leive (1975) Peterson and McGroarty (1985) Hirschfeld et al. (2000)
NaAc (pH 4.5) HCl (0.1 M)	Cleave the ketosidic bond between the lipid A and Kdo in the LPS	Osborn (1963) Morrison and Leive (1975)
4M NaOH	Deacylate lipid A	Holst (2000)

2.2.1.1 Phenol–Water Extraction

The mixture of phenol and water (45:50, v/v) can be used to extract LPS (Westphal and Jann, 1965). This mixture is a single phase above 65°C but separates into two phases below 65°C. LPS and proteins can be extracted from bacteria by this mixture above 65°C. When cooled down, phase separation occurs. The phenol phase mainly contains proteins, while the water phase contains LPS, polysaccharides, and nucleic acids. The following protocol is the most used for phenol–water extraction of LPS (Johnson and Perry, 1976).

(1) Suspend 5 g dry bacteria in 25 mL of 50 mM sodium phosphate (pH 7.0) containing 5 mM EDTA. Add 100 mg of egg lysozyme to the suspension and stir overnight at 4°C.
(2) Stir the suspension at 37°C for 20 min. Add 50 mM sodium phosphate (pH 7.0) containing 20 mM $MgCl_2$ until the total volume reaches 100 mL. Add RNase A

and DNase I to final concentrations of 1 μg/mL. Incubate the suspension at 37°C for 1 h, then heat at 60°C for another 1 h.
(3) Heat the suspension to 70°C in a water bath. Add an equal volume of 90% (w/v) phenol preheated to 70°C, and mix thoroughly. The resulting mixture is rapidly cooled by stirring in an ice water bath.
(4) Centrifuge the mixture at 18,000×*g* for 15 min at 4°C. Phase separation occurs. Collect the aqueous phase which contains LPS.
(5) Dialyze the aqueous phase against distilled water until no detectable phenol odor remains. Spin the dialyzate at 1,100×*g* for 5 min. Discard the pellets and centrifuge the supernatant fractions at 105,000×*g* for 2 h. The gel-like pellets are LPS. Re-suspend the LPS pellets in distilled water and lyophilized.

As the extracted LPS is always contaminated with proteins, specially lipoproteins, additional hydrolysis by proteinase and other purification processes have been integrated into the phenol–water extraction protocol (Hitchcock and Brown, 1983). The following protocol applies proteinase degradation prior to the phenol extraction (Apicella, 2008).

(1) Suspend 500 mg dry bacterial cells in 15 mL of 10 mM Tris–HCl buffer (pH 8.0), containing 2% SDS, 4% 2-mercaptoethanol, and 2 mM $MgCl_2$ in a 65°C water bath. Add 1 mL of proteinase K solution (100 μg/mL) to the cell mixture and keep the sample at 65°C for 1 h. Subsequently place the sample in a 37°C water bath overnight.
(2) Add 2 mL of 3 M sodium acetate to the cell suspension and mix thoroughly. Add 40 mL of cold absolute ethanol to the cell suspension and allow precipitate to form at –20°C.
(3) Centrifuge the mixture at 4,000×*g* for 15 min and discard the supernatant.
(4) Raise the precipitate in 9 mL of distilled water. Add 1 mL of 3 M sodium acetate and vortex. Add 20 mL of cold absolute ethanol and vortex again. The suspension is allowed to precipitate at –20°C.
(5) Suspend the precipitated LPS in 9 mL of 10 mM Tris–HCl (pH 7.4), and add 0.5 mL of DNase I (100 μg/mL) and 0.5 mL of RNase (25 μg/mL). Incubate at 37°C for 4 h.
(6) Place the LPS mixture in a 65°C water bath for 30 min, add an equal volume of 90% phenol preheated to 65°C, and incubate at 65°C for 15 min. Place the mixture in an ice bath and cool down to 4°C.
(7) Centrifuge the cooled mixture at 6,000×*g* for 15 min. Remove the aqueous layer and re-extract the phenol layer with an equal volume of distilled water. Combine the aqueous extractions and dialyze against multiple changes of distilled water over 2 days. After dialysis, the LPS can be lyophilized.

2.2.1.2 Ether Extraction

One major limitation of the phenol–water extraction method is that R-LPS frequently partitions into the phenol phase (Hickman and Ashwell, 1966; Kasai and

Nowotny, 1967). Therefore, the ether extraction method was developed for the extraction of R-LPS (Galanos et al., 1969). This method uses a monophasic solution consisting of liquid phenol, chloroform and petroleum ether (PCP). In such a solution, R-LPS is completely soluble, but S-LPS, proteins, nucleic acids and polysaccharides are insoluble and can be excluded from the extracts. The ratio of phenol, chloroform and petroleum ether in PCP is normally 2:5:8 (v/v/v), but 5:5:8 (v/v/v) also works (Helander et al., 1992). The following is a brief protocol for the ether extraction of LPS:

(1) Cells should be harvested and washed with distilled water. Using saline or any other salt solution may result in a very poor yield of LPS. The washed cells are dried in vacuo to a constant weight.
(2) Place 50 g of dry bacteria in 200 mL of PCP, consisting of aqueous phenol (90 g of dry phenol and 11 mL of water), chloroform and petroleum ether (b.p. 40°C ~ 60°C) at a volume ratio of 2:5:8. Homogenize the suspension for 2 min at 5°C ~ 20°C.
(3) Centrifuge at 5,000×*g* for 15 min, the supernatant containing LPS is filtered into a round flask. The pellet is re-extracted with the same amount of PCP, stirred and centrifuged as above. The supernatant is pooled into the first extract.
(4) Remove the petroleum ether and chloroform in the extracts by a rotary evaporator. Transfer the remaining extracts into a centrifuge tube and add water until LPS precipitates.
(5) Centrifuge at 3,000×*g* for 10 min to get the precipitated LPS. Wash the precipitates three times with 80% phenol, three times with ether, and dry in vacuo.
(6) Dissolve the LPS precipitates in 50 mL of distilled water, warm to 45°C, and shake to get a viscous solution. Centrifuge the solution at 100,000×*g* for 4 h. Re-dissolve the LPS pellets in water and freeze-dried.

2.2.1.3 EDTA Promoted Extraction

Since size heterogeneity of LPS molecules can exist within a single organism, a extraction method applied to both S-LPS and R-LPS extractions has been developed (Darveau and Hancock, 1983). In this method, SDS and EDTA are applied to precipitate contaminants, such as peptidoglycan and proteins. The following is the brief protocol of this method:

(1) Suspend 500 mg of dry bacterial cells in 15 mL of the solution containing 10 mM Tris–HCl (pH 8.0), 2 mM $MgCl_2$, 100 μg/mL pancreatic DNase I and 25 μg/mL pancreatic RNase A. Break the cells with a French Press. Add additional DNase I and RNase A to final concentrations of 200 and 50 μg/mL, respectively. Then incubate the suspension at 37°C for 2 h.
(2) After digestion, EDTA, SDS and 10 mM Tris–HCl (pH 8.0) are added to the total volume of 25 mL containing 0.1 M EDTA and 2% SDS at pH 9.5.

Vortex the solution and centrifuge at 50,000×*g* for 30 min at 20°C to remove peptidoglycan.

(3) Add pronase to the supernatant to a final concentration of 200 μg/mL. Incubate overnight at 37°C with constant shaking. Remove precipitates by centrifugation.

(4) Add two volumes of 0.375 M $MgCl_2$ dissolved in 95% ethanol and mix. Cool the mixture to 0°C and centrifuge at 12,000×*g* for 15 min at 4°C. Resuspend the pellets in 25 mL of 10 mM Tris–HCl (pH 8.0) containing 2% SDS and 0.1 M EDTA, and sonicate. Lower the pH of the solution to pH 7.0 by adding 4 M HCl. Incubate the solution at 85°C for 30 min to denature SDS-resistant outer membrane proteins. Cool down and raise the pH to 9.5 by adding 4 M NaOH. Add pronase to a concentration of 25 μg/mL, and incubate overnight at 37°C with constant agitation.

(5) Precipitate LPS with 2 volumes of 0.375 M $MgCl_2$ dissolved in 95% ethanol at 0°C, followed by centrifugation at 12,000×*g* for 15 min at 4°C. Resuspend the pellet in 15 mL of 10 mM Tris–HCl (pH 8.0), sonicate and centrifuge at 1,000×*g* for 5 min to remove any Mg^{2+}-EDTA crystals. Centrifuge the supernatant at 200,000×*g* at 15°C for 2 h in the presence of 25 mM $MgCl_2$, and collect the LPS pellets.

2.2.2 *Micro-extraction of Lipopolysaccharides*

Several methods for small-scale LPS preparation have been developed (Hitchcock, 1984; Inzana and Pichichero, 1984; Kido et al., 1990) because sometimes there are only small amount of bacteria available or only small amounts of LPS are needed. The purity of LPS extracted by these methods may not be high, but it is good enough for analysis. The following is the simplest protocol: Cells harvested from 1.5 ml culture are resuspended in 50 μL of lysing buffer containing 2% SDS, 4% 2-mercaptoethanol, 10% glycerol, 1 M Tris–HCl (pH 6.8), and heated at 100°C for 10 min. Then 25 μg of proteinase K in 10 μL of lysing buffer is added and incubated at 60°C for 1 h. The LPS prepared this way can be detected by acrylamide gel electrophoresis or western blots (Hitchcock, 1984).

2.2.2.1 Phenol–Water Micro-extraction

The conventional phenol–water extraction for LPS is time consuming and not practical for comparison of LPS from many clinical isolates. Therefore, a rapid micro-extraction method has been developed as the following. Briefly, 2×10^9 colony forming unit (CFU) of bacteria are washed once in phosphate-buffered saline (pH 7.2) containing 0.15 mM $CaCl_2$ and 0.5 mM $MgCl_2$ and resuspended in 300 μL of distilled water. An equal volume of hot phenol (90%, 68°C) is added, and the mixture is stirred vigorously at 68°C. The mixture is chilled, and the phenol–water phases are separated by centrifugation, and the aqueous phase is removed. 300 μL of distilled water is added to the phenol phase, and the extraction is repeated. The aqueous phases are pooled and sodium acetate is added to the final concentration of 0.5 M. 10 volumes of 95% ethanol is added. After incubation overnight at –20°C, the insoluble crude LPS can be collected by centrifugation. The crude LPS is re-dissolved in 100 μL of distilled water, and the precipitation is repeated. The final LPS precipitates can be re-dissolved in 50 μL of distilled water and stored at –20°C. This method has been used to extract LPS from many clinical isolates simultaneously, to examine the occurrence and transmission of strains during an outbreak of invasive *H. influenzae* type b disease (Inzana and Pichichero, 1984).

2.2.2.2 Phenol–Chloroform Micro-extraction

LPS extracted by the following protocol can be visualized by ethidium bromide (EtBr) in less than 30 min (Kido et al., 1990). A 1.5-mL overnight culture of bacterial cells is centrifuged. The bacterial pellet is suspended in 100 μL of triethylamine (TAE) buffer and mixed with 200 μL of alkaline solution containing 3 g of SDS, 0.6 g of Trizma base, and 6.4 mL of 2 M NaOH in 100 mL of H_2O. The mixture is heated at 60°C for 70 min and then mixed with phenol–chloroform (1:1, v/v). Centrifuge at 16,000×*g* for 10 min, the supernatant is mixed with 200 μL of H_20 and 50 μL of 3 M sodium acetate (pH 5.2). LPS is precipitated by adding 2 volumes of ethanol. The precipitate is dissolved in 200 μL of 50 mM Tris hydrochloride (pH 8.0)-100 mM sodium acetate and precipitated with 2 volumes of ethanol. The final precipitation of LPS is dissolved in 50 μL of H_2O, and the concentration of LPS is about 2 mg/mL.

2.2.3 *Extraction of Lipid A*

2.2.3.1 Extraction of Lipid A from LPS

Extraction of lipid A from LPS takes advantage of the acido-labile ketosidic bond between the lipid A and the ketodeoxyoctanoate (Kdo) in the LPS. Acid and heat are sufficient to disrupt the linkage. The lipid A is insoluble in water and therefore can be readily collected by centrifugation. Harsh hydrolysis such as 0.1 M hydrochloric acid at 100°C, and milder hydrolysis treatment with 1% acetic acid have been used to liberate lipid A moiety from LPS molecules (Fensom and Meadow, 1970; Morrison and Leive, 1975; Oertelt et al., 2001; Osborn, 1963). The harsh hydrolytic conditions could result in partially dephosphorylation and O-deacylation of lipid A (Karibian et al., 1995). This could affect the biological activities of the lipid A, but is useful for the extraction of monophosphoryl lipid A (Qureshi et al., 1982). Milder hydrolysis conditions, such as sodium acetate at pH 4.5, have been proved to be efficient to cleave the lipid A-polysaccharide bond (Rosner et al., 1979). When the hydrolysis is ineffective, 1% SDS can be added to the system (Caroff et al., 1988). The lipid A can be extracted from the hydrolytic reaction mixture using the solvent of chloroform and methanol (2:1, v/v).

2.2.3.2 Extraction of Lipid A from Bacteria

Extracting lipid A from purified LPS is considerably laborious and time-consuming. Therefore, a method for extraction of lipid A from bacteria was developed (Babinski et al., 2002; Gibbons et al., 2008; Wang et al., 2006a). Two extraction solutions are needed in this extraction method. One is a single-phase solution containing $CHCl_3/MeOH/H_2O$ (1:2:0.8, v/v/v) (Bligh and Dyer, 1959), in which glycerophospholipids and free lipid A are soluble. The other is two-phase solution consisting of $CHCl_3/MeOH/H_2O$ (2:2:1.8, v/v/v) in which lipid A is soluble in the hydrophobic phase, and proteins and nucleic acids are soluble in the hydrophilic phase. Addition of 1% SDS (Caroff et al., 1988) and sonic irradiation (Zhou et al., 1999) can increase the yield of lipid A. The following is a typical protocol for extraction of lipid A from bacteria.

(1) Inoculate 400 mL of cultures with overnight cultured cells, starting at an initial A_{600} of 0.02. The cells are grown to A_{600} of 1.2, harvested and washed twice with 100 mL of 50 mM HEPES buffer (pH 7.5). The washed cell pellets are re-suspended in 24 mL of phosphate-buffered saline (pH 7.4).
(2) Add 30 mL of chloroform and 60 mL of methanol to the cell suspension, converting to a single-phase Bligh-Dyer mixture which contains chloroform, methanol, and water (1:2:0.8, v/v/v).
(3) Incubate at room temperature for 60 min with constant stirring. Centrifuge at $1{,}000 \times g$ for 20 min. The recovered debris is washed once with 50 mL of single-phase Bligh–Dyer mixture. The insoluble pellets are collected by centrifugation at $1{,}000 \times g$ for 20 min.
(4) The washed pellets containing LPS are re-suspended in 27 mL of 12.5 mM sodium acetate (pH 4.5), heated at 100°C for 30 min, containing 1% SDS if needed. After a brief sonic irradiation, the pH is readjusted to 4.5 by adding glacial acetic acid.
(5) Cool down, and convert the suspension into a two-phase Bligh–Dyer system by addition of 30 mL of chloroform and 30 mL of methanol.
(6) Mix thoroughly by stirring for 30 min. Centrifuge at $1{,}000 \times g$ for 20 min. Collect the lower phase containing the released lipid A. A second extraction of the remaining upper phases is done by addition of 30 mL pre-equilibrated lower phase. The lower phase is collected by centrifugation and combined to the first lower phase. The combined lower phases are passed through a funnel plugged with glass wool to remove insoluble cell debris.
(7) Dry the pooled lower phases by rotary evaporation to get the crude lipid A.

2.2.3.3 Micro-extraction of Lipid A

As the typical molecule in Gram-negative bacteria, lipid A can be used to detect the bacteria. Detection methods need to be rapid and sensitive; therefore, a few rapid methods for lipid A micro-extraction from whole bacteria have been developed.

The following two detection methods can be done in a working day with a detection limit of 10–100 μg.

(1) SDS-promoted hydrolysis method (Zhou et al., 1999): 5 mg cells are suspended in 200 μL of 1% SDS in 10 mM sodium acetate (pH 4.5) and incubated at 100°C for 1 hr. The reaction mixture is dried. SDS is removed by washing the mixture with 50 μL of distilled water and 250 μL of acidified ethanol (prepared by adding 100 μL 4 M HCl to 20 mL 95% ethanol) followed by centrifugation (10,000×*g* for 5 min). The samples are then washed twice with 400 μL of 95% ethanol. The lipid A is extracted from the pellets with 200 μL of solvent of chloroform and methanol (1:1, v/v), followed by centrifugation at 8,000×*g* for 5 min. The supernatant containing lipid A is dried under a stream of nitrogen.
(2) Isobutyric acid-ammonium hydroxide promoted micro-extraction (El Hamidi et al., 2005; Tirsoaga et al., 2007a): Suspend 100 mg of the cells sample in 400 μL of solution containing isobutyric acid and 1 M ammonium hydroxide (5:3, v/v), and incubate at 100°C for 2 h. Cool the mixture in ice water and centrifuge at 2,000×*g* for 15 min at 4°C. The supernatant is then diluted with water (1:1, v/v) and lyophilized. Wash the sampled with 400 μL of methanol and centrifuged at 10,000×*g* for 5 min. Dissolve and extract the insoluble lipid A once with 200 μL of solvent of chloroform and methanol (1:1, v/v), followed by centrifugation at 8,000×*g* for 5 min. Dry the supernatant containing lipid A under a stream of nitrogen.

Due to phase variation, there are fluctuations in expression levels of certain enzymes in bacteria, therefore, not all colonies or cells make the same structure of lipid A species. A micro-extraction method for extraction of lipid A from a single colony has been developed (Zhou et al., 2009). This method uses microwave-assisted enzymatic digestion and sodium acetate hydrolysis, suitable to analyze lipid A from both cell samples and an individual colony. Because the clean up of SDS is very time-consuming, and the contaminated SDS would seriously interfere with the analysis by mass spectrometry, the proteinase K, instead of SDS, is used to disrupt the cells. Using this method, the entire process for lipid A preparation only takes about 2 h with a detection limit to 1 μg.

(1) Lipid A isolation from cells. The lyophilized cells (5 mg) are placed in a 1.5-mL eppendorf tube and suspended in 200 μL of 50 mM sodium acetate buffer (pH 4.5) containing proteinase K (60 μg/mL). Under microwave irritation at 50 W, the enzymatic digestion is carried out for 5 min at 58°C. The suspension is then kept for 1 h at 100°C. The reaction mixture was centrifuged at 10,000×*g* for 5 min. The pellets are then washed twice with 400 μL methanol and centrifuged at 10,000×*g* for 5 min. Finally the lipid A is dissolved and extracted once with 200 μL of solvent of chloroform and methanol (1:1, v/v), followed by centrifugation at 8,000×*g* for 5 min. The supernatant containing lipid A is dried under a stream of nitrogen.

(2) Lipid A isolation from a single colony. Bacteria of each single colony are carefully scraped off the culture plate and collected in a 1.5-mL eppendorf tube containing 1% phenol in PBS. Vortex and spin down. Wash the cell pellets with deionized water, and suspend the pellets in 100 μL of 50 mM sodium acetate buffer (pH 4.5) containing proteinase K (60 μg/mL). After microwave irritation, the suspension is heated at 100°C for 1 h. The mixture is centrifuged at 10,000×*g* for 5 min and the pellets are washed twice with methanol. Lipid A is extracted from the pellets with 100 μL solvent of chloroform and methanol (1:1, v/v). Centrifuge at 8,000×*g* for 5 min, the supernatant containing lipid A is dried under a stream of nitrogen.

2.2.3.4 Extraction of Free Lipid A

Some wild strains or mutants of Gram-negative bacteria can make free lipid A (Reynolds and Raetz, 2009; Wang et al., 2006b). Free lipid A can be extracted the single Bligh–Dyer system described as above, together with glycerophospholipids. Briefly, 400 mL of bacterial cells are harvested and washed twice with phosphate-buffered saline. The cell pellets are suspended in 190 mL of a single-phase Bligh–Dyer mixture and incubate at room temperature for 1 h with occasionally mixing. Centrifuge at 4,000×*g* for 20 min, the supernatant containing glycerophospholipids and free lipid A is converted to a two-phase Bligh–Dyer system by the addition of 50 mL of chloroform and 50 mL of water. Concentrated HCl can be added to adjust the pH of the upper phase to 1.5, which may increase the recovery yield of lipids (Nishijima and Raetz, 1979; Raetz et al., 1985). The mixture is vortexed and centrifuged at 4,000×*g* for 20 min. The lower phase is collected, and the upper water–methanol phase is re-extracted by the addition of pre-equilibrated lower phase. Finally, the lower phases containing lipid A are combined and dried by rotary evaporation. The lipid A in this sample needs to be purified away from glycerophospholipids by using DEAE chromatography.

2.3 Purification of Lipopolysaccharides

LPS extracted directly from bacteria are often contaminated with other macromolecules in the bacterial cells, such as phospholipids, nucleic acids, lipoproteins, capsular polysaccharides, and peptidoglycan. Cellular responses to the contaminated LPS may be confounded. For example, LPS and peptidoglycan can cause the similar host response to bacterial infection through Toll-like receptor 4 (TLR4) and Toll-like receptor 2 (TLR2), respectively (Girardin et al., 2003). Cells overexpressing TLR2 are extremely sensitive to minor contaminants of peptidoglycan in LPS (Hirschfeld et al., 2000). Therefore, the contaminants must be removed from LPS preparations. Several methods have been developed to dissociate LPS-contaminant aggregates and purify LPS. These methods are usually efficient for elimination of the major contaminants. The efficiency of a given method depends on the physico-chemical interactions occurring between LPS and the contaminants.

2.3.1 Removal of Contaminants

Phospholipids are the major components in membranes; therefore they are one of the major contaminants in the initial extraction of LPS. The degree of contaminated phospholipids can be detected by thin layer chromatography (TLC) or mass spectrometry. To remove the contaminated phospholipids, the LPS samples could be washed with chloroform–methanol mixture (1:2, v/v). Phospholipids are soluble in this solvent, but LPS not.

Lipoproteins are also important contaminants in the LPS samples. The degree of contamination by lipoproteins can be estimated by UV absorbance at the appropriate wavelengths. Lipoproteins in the LPS samples can be degraded by proteinase K.

Peptidoglycans are contaminants in the LPS that can activate the similar host response as LPS, therefore have to be removed completely from the LPS sample. Different dissociation agents, such as SDS and mineral acid have been used to disassociate and remove peptidoglycan (Hirschfeld et al., 2000). Their efficiency was estimated using HEK293 cells overexpressing TLR2 and transfected with an NF-κB-dependent luciferase reporter gene (Tirsoaga et al., 2007b). All treatments can decrease the TLR2 activation compared to that of the starting LPS sample, and the acid treatment gave the lowest TLR2 activation and the best-purified LPS. The following is the procedure for acid-promoted LPS purification (Tirsoaga et al., 2007b):

(1) LPS preparations (10 mg) are suspended in 0.5 mL 1 M hydrochloric acid at a concentration of 20 mg/mL. The suspension is sonicated in an ultrasonic bath for 2 min, and then 6 mL of chloroform and 4 mL of methanol are added in order to obtain a final ratio of the volumes of chloroform, methanol, and 1 M hydrochloric acid of 3:2:0.25 (v/v/v).
(2) The suspension is sonicated in an ultrasonic bath for 2 min and then centrifuged at 2,000×*g* for 10 min. The supernatant and the pellet are separated.
(3) The supernatant is evaporated under vacuum. The pellet is dried under a stream of nitrogen, and the extraction is repeated twice, replacing acid with water. The LPS residue is suspended in water (5–10 mg/mL) and ultracentrifuged at 300,000×*g* for 45 min at 4°C. The purified LPS can be recovered by lyophilizing the pellets.

Methods that remove individual contaminant such as phospholipids, nucleic acids, lipoproteins, and peptidoglycan can be combined into a single protocol as shown below.

(1) Resuspend the LPS preperation in 10 mM Tris–HCl (pH 8.0). Add RNase to a final concentration of 25 μg/mL, and DNase I to a final concentration of 100 μg/mL. Incubate the suspension at 37°C for 2 h.

(2) Add Proteinase K to a final concentration of 100 μg/mL, and incubate the suspension at 37°C for 2 h.
(3) Add 5 mL of water-saturated phenol. Vortex the mixtures thoroughly, then centrifuge at 3,000×*g* for 30 min at room temperature.
(4) Dialyze the aqueous fraction for 12 h against distilled water at 4°C. After dialysis, centrifuge the aqueous solution at 17,000×*g* for 20 min at room temperature to remove any insoluble material.
(5) Lyophilize the supernatant to get purified LPS.

2.3.2 Gel Filtration Chromatography of LPS

The gel filtration chromatography is employed as an important technique to isolate heterogeneous LPS extracted from bacterial cells, especially for isolation of truncated LPS without lipid A moiety. The truncated LPS can be obtained by hydrolysis with 1% acetic acid at 100°C for 1 h, or 0.1 M NaOH in 99% ethanol at 37°C for 45 min (Chester and Meadow, 1975; Muller-Seitz et al., 1968; Prehm et al., 1975). The former releases the polysaccharide from the lipid A moiety with exclusive spitting of the extremely acid-labile Kdo linkage (Luderitz et al., 1966; Muller-Seitz et al., 1968); the latter removes ester-linked fatty acids from the lipid A, reducing non-polar interactions between LPS components and facilitating their separation (Chester and Meadow, 1975).

A series of Sephadex columns have been widely used in separation of intact or truncated LPS. The selection of the column is based on the repeating constitution of O-antigen in LPS. The intact LPS can be separated by Sephadex G-200 and eluted with detergent containing materials (Morrison and Leive, 1975; Peterson and McGroarty, 1985; Rivera et al., 1988). Sephadex G-200 was also used to separate the alkaline hydrolyzed LPS which can be eluted with pyridine/0.05 M acetate buffer (Chester and Meadow, 1975). Sephadex G-25, 50, and 75 have been applied to fractionate acetic-acid hydrolyzed LPS with distilled water or pyridine acetate buffer elution (Byrd and Kadis, 1989; Carlson, 1984; Koval and Meadow, 1977; Kropinski et al., 1982; Lacroix et al., 1993; Prehm et al., 1975; Temple et al., 1986). Biogel P6 has been used to fractionate the partially degraded LPS with 1% acetic acid, eluted with pyridine/0.05 M acetate buffer (Koval and Meadow, 1977). Sepharose 4B has been used to fractionate LPS from *E. coli* O111:B4, eluted with 0.12 M Tris buffer (Morrison and Leive, 1975).

To disaggregate the LPS molecule, the elution buffer always contains detergents, such as deoxycholate (DOC), SDS, and Triton X-100. Two kinds of elution buffer have been reported to use for Sephadex G chromatography (Morrison and Leive, 1975; Peterson and McGroarty, 1985). One is 10 mM Tris buffer (pH 8.0) containing 0.2 M NaCl, 0.25% DOC, 1 mM EDTA, and 0.02% NaN_3. The other is 0.12 M Tris–HCl buffer (pH 8.1), containing SDS or Triton X-100, and 1 mM β-mercaptoethanol. The latter elution buffer is also applied for Sepharose. DOC can be eliminated by dialyzing the pooled fractions extensively with the elution buffer without DOC. EDTA and high concentration of monovalent cations in column and dialysis buffers are formulated to remove multivalent cations from the sample.

2.3.3 Ion-Exchange Chromatography of LPS

Due to the negative charges from the phosphate moiety in the lipid A and core oligosaccharide regions, LPS can be fractionated on the column of anion-exchange chromatography, and analyzed with mass spectrometry and NMR (Lukasiewicz et al., 2006; Muller-Loennies et al., 2003; Oertelt et al., 2001; Vilches et al., 2007; Yildirim et al., 2003; Zahringer et al., 2004; Zdorovenko et al., 2004). This method is efficient for purification of LPS less hydrophobic. To decrease its hydrophobicity, LPS can be hydrolyzed to remove some fatty acyl chains of lipid A. LPS can be O-deaylated by mild hydrolysis at 37°C for 30 min, or N-deacylated in 4 M KOH at 120°C for 16 h.

2.3.4 Capillary Electrophoresis of LPS

Capillary electrophoresis (CE) is a high-resolution separation technique widely used for the analysis of complex mixtures of low-mass molecules like peptides, saccharides and nucleotides. It can be used for LPS purification because LPS contains highly polar groups such as phosphates, phosphoethanolamins, carboxylates, and amino groups. Compared with gel electrophoresis, CE has better heat dissipation, which lead to sharper zone definition and higher efficiency (Jorgenson and Lukacs, 1981). The coupling of CE to electrospray mass spectrometry (CE-ES-MS) has provided unparalleled resolution and identification of glycoform populations and substituted groups present in LPS (Kelly et al., 1996; Li et al., 2004; Li et al., 2005a, b; Li et al., 2007). The Fourier-transform ion cyclotron resonance mass spectrometer (FT-ICR MS) coupling with CE has also been introduced for detection species in heterogeneous native R-LPS, giving high resolution and mass accuracy of LPS (Hubner and Lindner, 2009).

2.3.5 Purification of Lipid A

DEAE-cellulose ion-exchange chromatography, which is dependent on the phosphate, pyrophosphate or phosphoethanolamine groups of the lipid A moiety, is widely used for lipid A isolation (El Hamidi et al., 2005; Raetz and Kennedy, 1973). TLC, as a lipids detection method, also can be applied for lipid A isolation (Zhou et al., 1999). In addition, the chromatography techniques based on the molecular polarity discrepancy, have been developed. For instance, Bio-Sil column had been reported to separate the lipid A components from *Rhizobium etli*, eluted stepwise with different volume of solvents of increasing polarity as $CHCl_3$/MeOH (95:5 v/v), $CHCl_3$/MeOH (90:10 v/v), $CHCl_3$/MeOH (85:15 v/v), and $CHCl_3$/MeOH (2:1 v/v) (Que et al., 2000). A C_{18} reverse column had been reported to separate crude lipid A of *Rhizobium trifolii* ANU843, eluted with water, water/ methanol (2:1, v/v), methanol/water (2:1, v/v), methanol/water (4:1, v/v), pure methanol, and pure chloroform (Hollingsworth and Lill-Elghanian, 1989). Moreover, a series of separation protocols have been widely and successfully used for monophosphoryl lipid A isolation (Qureshi et al., 1982; Qureshi et al., 1985; Qureshi et al., 1988; Qureshi et al., 1997). Briefly, the monophosphoryl lipid A is converted into the free acid form by passing through Chelex 100 (Na^+) and Dowex 50 (H^+) columns, and then methylated with diazomethane. HPLC technique can be applied for monophosphoryl lipid A isolation (Qureshi et al., 1997). The following protocol can be used for purification of lipid A from 400 mL of bacterial cultures.

(1) 5 mL of DEAE-cellulose column (2.5×13 cm) in the acetate form is equilibrated with the solvent $CHCl_3/MeOH/H_2O$ (2:3:1, v/v/v).
(2) Dissolve the extracted crude lipid A sample in 8 mL of $CHCl_3/MeOH/H_2O$ (2:3:1, v/v/v) and load onto the column. Collect the run-through as a single fraction.
(3) Wash the column with 25 mL of $CHCl_3/MeOH/H_2O$ (2:3:1, v/v/v), also collect as a single fraction. The various lipid A components are then eluted by increasing the NH_4Ac concentration (30 mM, 60 mM, 120 mM, 240 mM and 480 mM) of the aqueous portion with the 4 column volumes of $CHCl_3/MeOH/NH_4Ac$ (2:3:1, v/v/v). Collect each 2 mL of elute in a tube.
(4) 20 μL portions of each fraction are spotted directly onto Silica Gel 60 TLC plates to monitor the lipid A elution profile. The factions containing selective lipid A components are converted to two-phase Bligh–Dyer system by addition of the appropriate amounts of chloroform and water. After mixing, the phases are separated by centrifugation at 4,000×*g* for 20 min at room temperature. The lower phases containing lipid A are pooled and dried by rotary evaporation.

The obtained lipid A can be further purified by using preparative TLC to remove the minor phospholipids contamination (Zhou et al., 1999). The following is the brief procedure.

(1) Dissolve the lipid A samples in appropriate volume of the chloroform and methanol solution (4:1, v/v), and apply the samples to a Silica Gel 60 analytical TLC plate, developed in a solvent system.
(2) When the plates are drying at room temperature, the lipid A can be seen as transiently visible white bands. Mark the bands with a pencil and scrap them off with a clean razor blade.
(3) Dry the collected debris and dissolve in 3.8 mL of single-phase Bligh–Dyer mixture.

(4) Mix the suspension thoroughly, and convert it into a two-phase Bligh–Dyer system by adding 2 mL of chloroform, 1 mL of methanol, and 1.9 mL of water.
(5) Vortex and centrifuge at 4°C for 10 min. The lower phase containing lipid A is collected, and dried under a stream of nitrogen. To remove residual silica chips and other possible contaminants, the purified sample can be passed through a DEAE-cellulose column.

2.3.6 Micro-purification Method

The micro-purification method is based on elution of LPS from polyacylamide gels with zinc-imidazole staining (Hardy et al., 1998; Pupo et al., 2000). It is arising from the micro-purification of nucleic acids and proteins with negatively zinc-imidazole staining in SDS-polyscylamide gels (Castellanos-Serra et al., 1996; Castellanos-Serra et al., 1997). As the structural and biological characterization have not been influenced by zinc-imidazation detection, the LPS samples can be purified and recovered simply by passive elution from SDS-polyacylamide gels (Hardy et al., 1997; Hardy et al., 1998; Pupo and Hardy, 2007). After 3 h of elution for 32 μM average size gel micro-particles in water, the recovery of rough and semi-smooth type LPS is about 70–80% (Hardy et al., 1998). Because of the high aggregation in the gel, the low solubility in the water and diffusion rate from the gel to the water, the recovery of smooth LPS with high-molecular-mass is usually from 5 to 10%. Replace distilled water with 1% SDS or DOC eluted for 3 h, or 5% TEA eluted for 2 min, the yield and reproducibility of smooth LPS are increased. However, compared with 1% SDS and DOC, the using of 5% TEA as the elution buffer is more convenient because of its easy removal by simple evaporation (Pupo et al., 2000).

2.4 Analysis of Lipopolysaccharides

Understanding the role of LPS in bacterial pathogenesis has come about largely through analysis of strains that contain mutations in genes responsible for LPS biosynthesis. Therefore, a rapid and sensitive screening method is required that can differentiate minor changes in LPS composition or structure. The analytical method most commonly used is sodium dodecyl sulfate – polyacrylamide gel electrophoresis (SDS-PAGE), followed by silver staining (Dubray and Bezard, 1982; Fomsgard et al., 1990; Munford et al., 1980; Tsai and Frasch, 1982). SDS-PAGE of ^{3}H-,^{14}C-, or ^{33}P-labeled LPS has been developed (Logan and Trust, 1984; Munford et al., 1980). A fluorescent dye, EB, generally used to stain nucleic acids in gels, has been applied for staining of LPS gels (Kido et al., 1990). Due to the drawbacks of silver staining, such as the irreversible fixation and chemical modification of the LPS, sensitive reverse staining with zinc-imidazole salts have been developed (Hardy et al., 1997; Hardy et al., 1998).

2.4.1 Electrophoresis of LPS

Alteration or heterogeneity of LPS structure is most often assessed by alteration of electrophoretic band profiles using SDS-PAGE (Amano et al., 1998; Jann et al., 1975; Peterson and McGroarty, 1985). Several different formulations of mini-gels have been developed with different acylamide concentrations (Kim et al., 1994; Apicella et al., 1994). To obtain a better resolution of LPS, tricine has been applied for both gel and electrophoresis buffer instead of glycine (Lesse et al., 1990). To eliminate the aggregation of LPS, SDS can be replaced by sodium DOC in electrophoresis (Komuro and Galanos, 1988). In order to discern minor differences of the low molecular weight bands of LPS in polyacrylamide gels, a bilayer stacking gel was developed (Inzana and Apicella, 1999). The key elements for this SDS-PAGE are shown below.

(1) Gel preparation: Pour the preparation gel with 14% acylamide to 4 cm from the top of the plates and overlay with distilled water containing 0.1% SDS. After polymerization for 2 h, discard the water layer and rinse the top of the gel. Then add 3% acylamide solution of stacking gel A onto the separating gel to a level just above the length of the well comb, and overlay with 0.1% SDS. After 2 h, rinse the top of the stacking gel A with distilled water, followed by overlay with the stacking gel with 3% acylamide solution of stacking gel B. The compositions in the solution of stacking gel B are identical to that of the stacking gel A, except that the 0.125 M Tris–HCl (pH 6.8) is replaced with 0.025 M Tris-0.192 M glycine buffer. Insert the well comb to the point that the bottom of the comb is contact with the top of the stacking gel A. Place overnight at the room temperature.
(2) Dye buffer: Prepare 2×solubilization buffer, consisted of 0.59 mL of store buffer (0.06 M Tris–HCl (pH 6.8), 1 mM EDTA, and 2% SDS), 0.4 mL of glycerol, 0.08 mL of 2-mercaptoethanol, and 0.04 mL of saturated bromophenol blue.
(3) Reservoir buffer: Fill the upper reservoir with 0.25 M Tris-1.92 M glycine (pH 8.0), and the remaining buffer is added to the lower reservoir.
(4) LPS samples preparation: Dissolve 10 μg of LPS samples in 10 μL distilled water, and mix with 10 μL of freshly prepared 2×solubilization buffer. Incubate the sample in boiling bath for 5 min. Load the samples to the wells of the prepared gel.
(5) Electrophoresis: Samples are migrated at 9 mA in both the stacking gels, and 12 mM in the separating gel. When the dye is reached to the bottom of the separating gel, stop the electrophoresis, and strip the gel from the plate and stain.

Because of long time and large sample consuming in large PAGE, liner gradient gel electrophoresis was carried out to analysis smaller molecular LPS. This electrophoresis system was also applied to separate high molecular LPS with tight and even bands (Noda et al., 2000). The key elements for the linear gradient gel electrophoresis of LPS are shown below.

(1) Gel preparation: Apply 6% stacking gel with gel buffer of 1.25 M Tris–HCl (pH 6.8), and a resolving gel buffer of 1.88 M Trsi–HCl (pH 8.88) is for the linear gradient gel. The linear gradient gel is made with gradient former (10 mL of chamber), a peristaltic pump, and a mini gel casting unit. Degas and chill the monomer solutions prior to prepare the linear gradient gel. To form a linear gradient in the casting unit, the 30% stock acylamide solution diluted with 50% glycerol for the higher concentration of the acylamide solutions is stored in the gradient former. The water made to be the lower concentrations of the solutions is in other former. The polyethylene tube connected with gradient former and the pump is filled with butanol saturated with water. After the polymerization is initiated, pump the solutions in the chambers into the unit at a flow rate of 0.4 mL/min.
(2) Reservoir buffer: 0.025 M Tris, 0.192 M glycine, and 2% SDS.
(3) Samples preparation: Dissolve the LPS samples in 60 mM Tris–HCl (pH 6.8) buffer, containing 1 mM EDTA and 2% SDS, to a final concentration of 0.1% (w/v). After diluted with 34 mM Tris–HCl (pH 6.8) buffer containing 39% glycerol and 3.9% saturated BPB in water, apply the mixtures on the prepared gel.
(4) Electrophoresis procedure: The electrophoresis is run at 200 V with a constant voltage at the room temperature. When the BPB is reached a point of 3 mm from the bottom of the separating gel, the electrophoresis is terminated. Strip the gel from the plate and stain.

2.4.2 Staining Methods

2.4.2.1 Silver Stain

The polysaccharide part of the LPS molecule is the reactive residue in the silver stain, because oxidization of the hexoses present makes aldehyde groups available for subsequent reaction with the silver nitrate. However, the number of fatty acids in lipid A part of LPS may be responsible for retention of LPS fractions in the SDS-PAGE during the initial fixing or/and oxidization steps. The certain LPS fractions with partly deacylated S-LPS are not efficiently detectable by traditional silver stain technique. Therefore, a modified silver stain method is improved by omitting the fixing step and extending the oxidation time of the staining procedure (Fomsgard et al., 1990). Less than 1 μg of LPS is sufficient for visualization in polyacrylamide gels by silver staining. The following is a modified silver staining procedure (Fomsgard et al., 1990).

(1) The LPS in the SDS-PAGE gel is oxidized with the solution containing 0.7% periodic acid, 40% ethanol, and 5% acetic acid at 22°C for 20 min without prior fixation.
(2) Wash the gel three times with distilled water for 5 min to clear the oxidation buffer.

(3) Stain the gel for 10 min with freshly prepared staining solution prepared as shown below: Mix 4 mL of concentrated ammonium hydroxide with 56 mL of 0.1 M sodium hydroxide. Add 200 mL water, and 10 ml of 20% (w/v) silver nitrate in drops with constant stirring. The final volume is adjusted to 300 mL with water.
(4) Wash the gel three times with distilled water for 5 min.
(5) The color is developed by reduction in 200 mL solution containing 10 mg of citric acid and 0.1 mL of 37% formaldehyde. Stop the color reaction by exposure to 10% acetic acid for 1 min followed by repeated washings in distilled water.

2.4.2.2 Ethidium Bromide Stain

The EB-staining method was particularly suitable for staining LPS possessing acidic O-specific polysaccharides, which were poorly visualized by silver staining. The high-molecular-weight LPS containing long O-specific polysaccharide chains is more strongly stained by EB, compared to low-molecular weight LPS. The stain protocol (Kido et al., 1990) is as the following: After electrophoresis, remove the gel from glass plate, and immerse the gel in a 30 μg/mL of EtBr solution for 10 s. To destain, place the gel in distilled water with constant gently shaking for 30 min. LPS bands are observed by using a transilluminator at wave length of 302 nm.

2.4.2.3 Zinc-Imidazole Stain

Silver staining has several drawbacks: the sensitivity is decreased resulting from high background; the method is time-consuming, toxic and expensive; and the most important is the irreversible fixation and chemical modification of the LPS. Therefore, the sensitive reverse stain technique using zinc and imidazole salts has been developed to recover LPS from gel slices for subsequent structural and biological studies (Hardy et al., 1997; Hardy et al., 1998). In order to develop a staining background and leave the zinc-LPS complex as transparent and colorless bands, the imidazole is applied. The staining protocol (Hardy et al., 1997) is as the following.

(1) After electrophoresis, incubate the gel in boiling distilled water for 15 min three times to significantly remove SDS and other electrophoresis-associated chemicals.
(2) Incubate the gel in 10 mM zinc sulfate for 15 min.
(3) Soak the gel under agitation in 0.2 M imidazole for 3 min. Then a homogeneous white background precipitates along the face of the gel, except the zones containing LPS which remain transparent and colorless. The appearance

of the negative staining patterns should be observed by placing the gel a few centimeters over a dark background.
(4) To stop the reaction, rinse the gel three times with distilled water for 1 min after an adequate image contrast attained.

2.4.3 *Immunoblotting Method*

The immunoblotting technique has been carried out to characterize the immunochemistry of LPS, because of the ability of the O-antigen in LPS binding specific antibodies. The antiserum is often chosen as the first antibody, followed with peroxidase-conjugated anti-rabbit IgG as the second antibody (Fomsgaard et al., 1988; Perez et al., 1985; Prendergast et al., 1998). The specific O-antigen has the ability to bind monoclonal antibodies, which as the primary antibody, and visualized with alkaline phosphatase-conjugated secondary antibody (D'Haeze et al., 2007; Yokota et al., 2000). After electrophoresis, the LPS-containing gel is soaked in transfer buffer, and transferred to a nitrocellulose membrane. Visualization of nitrocellulose blots is performed sequentially with first antibody and secondary antibody. Finally, the bound antibody is detected with color-generating agents.

2.4.4 *Analysis of Lipid A*

TLC is a simple and rapid method to separate lipid A species. After chromatography, such lipid A molecule can be colored by spraying with chromogenic agent.

The lipids are dissolved in $CHCl_3$/MeOH (4:1, v/v) and spotted onto a Silica Gel 60 TLC plate, and developed in the solvent $CHCl_3$/MeOH/H_2O/NH_4OH (40:25:4:2, v/v/v/v) or chloroform/pyridine/88% formic acid/MeOH/H_2O (60:35:10:5:2, v/v/v/v/v). After drying, the lipids can be detected by charring at 145°C with spray 10% sulfuric acid in ethanol (Que et al., 2000). Autoradiography techniques had also been developed for analysis of lipid A, which can greatly increase the detection sensitivity (Brozek et al., 1989; Clementz et al., 1997). The purified lipid A can also be analyzed by ESI MS (Wang et al., 2006b).

2.5 Conclusions

As an important membrane anchored molecule in Gram-negative bacteria that can activate the immune response, LPS and lipid A are largely needed for research to understand infection mechanism of bacterial pathogens. Methods for extraction and purification of LPS and lipid A still need to be modified to increase the yield and purity, so do the methods for analysis of LPS.

Acknowledgements Funding was provided by grants from the National Natural Science Foundation of China (NSFC 30770114, NSFC30870074), the Program of State Key Laboratory of Food Science and Technology (SKLF-MB-200801), the 111 project (111-2-06), the Basic Research Programs of Jiangsu Province (BK2009003).

References

Amano, K.I., Williams, J.C., Dasch, G.A. Structural properties of lipopolysaccharides from *Rickettsia typhi* and *Rickettsia prowazekii* and their chemical similarity to the lipopolysaccharide from *Proteus vulgaris* OX19 used in the Weil-Felix test. Infect Immun 66 (1998) 923–926.

Apicella, M.A. Isolation and characterization of lipopolysaccharides. Methods Mol Biol 431 (2008) 3–13.

Apicella, M.A., Griffiss, J.M., Schneider, H. Isolation and characterization of lipopolysaccharides, lipooligosaccharides, and lipid A. Methods Enzymol 235 (1994) 242–252.

Babinski, K.J., Kanjilal, S.J., Raetz, C.R. Accumulation of the lipid A precursor UDP-2,3-diacylglucosamine in an *Escherichia coli* mutant lacking the *lpxH* gene. J Biol Chem 277 (2002) 25947–25956.

Bligh, E.G., Dyer, W.J. A rapid method of total lipid extraction and purification. Can J Biochem Physiol 37 (1959) 911–917.

Brozek, K.A., Hosaka, K., Robertson, A.D., Raetz, C.R. Biosynthesis of lipopolysaccharide in *Escherichia coli*. Cytoplasmic enzymes that attach 3-deoxy-D-manno-octulosonic acid to lipid A. J Biol Chem 264 (1989) 6956–6966.

Byrd, W., Kadis, S. Structures and sugar compositions of lipopolysaccharides isolated from seven *Actinobacillus pleuropneumoniae* serotypes. Infect Immun 57 (1989) 3901–3906.

Carlson, R.W. Heterogeneity of *Rhizobium* lipopolysaccharides. J Bacteriol 158 (1984) 1012–1017.

Caroff, M., Tacken, A., Szabo, L. Detergent-accelerated hydrolysis of bacterial endotoxins and determination of the anomeric configuration of the glycosyl phosphate present in the "isolated lipid A" fragment of the *Bordetella pertussis* endotoxin. Carbohydr Res 175 (1988) 273–282.

Castellanos-Serra, L.R., Fernandez-Patron, C., Hardy, E., Huerta, V.A. procedure for protein elution from reverse-stained polyarcylamide gels applicable at the low picomole level: An alternative route to the preparation of low abundance proteins for microanalysis. Electrophoresis 17 (1996) 1564–1572.

Castellanos-Serra, L.R., Fernandez-Patron, C., Hardy, E., Santana, H., Huerta, V. High yield elution of proteins from sodium dodecyl sulfate-polyacrylamide gels at the low-picomole level. Application to N-terminal sequencing of a scarce protein and to in-solution biological activity analysis of on-gel renatured proteins. J Protein Chem 16 (1997) 415–419.

Chester, I.R., Meadow, P.M. Heterogeneity of the lipopolysaccharide from *Pseudomonas aeruginosa*. Eur J Biochem 58 (1975) 273–282.

Clementz, T., Zhou, Z., Raetz, C.R. Function of the *Escherichia coli msbB* gene, a multicopy suppressor of htrB knockouts, in the acylation of lipid A. Acylation by MsbB follows laurate incorporation by HtrB. J Biol Chem 272 (1997) 10353–10360.

D'Haeze, W., Leoff, C., Freshour, G., Noel, K.D., Carlson, R.W. *Rhizobium etli* CE3 bacteroid lipopolysaccharides are structurally similar but not identical to those produced by cultured CE3 bacteria. J Biol Chem 282 (2007) 17101–17113.

Darveau, R.P., Hancock, R.E.W. Procedure for isolation of bacterial lipopolysaccharides from both smooth and rough *Pseudomonas aeruginosa* and *Salmonella typhimurium* strains. J Bacteriol 155 (1983) 831–838.

Dubray, G., Bezard, G. A highly sensitive periodic acid-silver stain for 1,2-diol groups of glycoproteins and polysaccharides in polyacrylamide gels. Anal Biochem 119 (1982) 325–329.

El Hamidi, A., Tirsoaga, A., Novikov, A., Hussein, A., Caroff, M. Microextraction of bacterial lipid A: easy and rapid method for mass spectrometric characterization. J Lipid Res 46 (2005) 1773–1778.

Fensom, A.H., Meadow, P.M. Evidence for two regions in the polysaccharide moiety of the lipopolysaccharide of *Pseudomonas aeruginosa* 8602. FEBS Lett 9 (1970) 81–84.

Fomsgaard, A., Conrad, R.S., Galanos, C., Shand, G.H., Hoiby, N. Comparative immunochemistry of lipopolysaccharides from typable and polyagglutinable *Pseudomonas aeruginosa* strains isolated from patients with cystic fibrosis. J Clin Microbiol 26 (1988) 821–826.

Fomsgard, A., Freudenberg, M.A., Galanos, C. Modification of the silver staining technique to detect lipopolysaccharide in polyacrylamide gels. J Clin Microbiol 28 (1990) 2627–2631.

Galanos, C., Lüderitz, O., Westphal, O. A new method for the extraction of R lipopolysaccharides. Eur J Biochem 9 (1969) 245–249.

Gibbons, H.S., Reynolds, C.M., Guan, Z., d Raetz, C.R. An inner membrane dioxygenase that generates the 2-hydroxymyristate moiety of *Salmonella* lipid A. Biochemistry 47 (2008) 2814–2825.

Girardin, S.E., Boneca, I.G., Carneiro, L.A., Antignac, A., Jéhanno, M., Viala, J., Tedin, K., Taha, M.K., Labigne, A., Zähringer, U., Coyle, A.J., DiStefano, P.S., Bertin, J., Sansonetti, P.J., Philpott, D.J. Nod1 detects a unique muropeptide from Gram-negative bacterial peptidoglycan. Science 300 (2003) 1584–1587.

Goebel, W.F., Binkley, F., Perlman, E. Studies on the Flexner group of dysentery bacilli. I. The specific antigens of *Shigella paradysenteriae*. J Exp Med 81 (1945) 315–330.

Hardy, E., Pupo, E., Castellanos-Serra, L., Reyes, J., Fernández-Patrón, C. Sensitive reverse staining of bacterial lipopolysaccharides on polyacrylamide gels by using zinc and imidazole salts. Anal Biochem 244 (1997) 28–32.

Hardy, E., Pupo, E., Santana, H., Guerra, M., Castellanos-Serra, L.R. Elution of lipopolysaccharides from polyacrylamide gels. Anal Biochem 259 (1998) 162–165.

Helander, I.M., Hurme, R., Haikara, A., Moran, A.P. Separation and characterization of two chemically distinct lipopolysaccharides in two *Pectinatus* species. J Bacteriol 174 (1992) 3348–3354.

Hickman, J., Ashwell, G. Isolation of a bacterial lipopolysaccharide from *Xanthomonas campestris* containing 3-acetamido-3,6-dideoxy-D-galactose and D-rhamnose. J Biol Chem 241 (1966) 1424–1428.

Hirschfeld, M., Ma, Y., Weis, J.H., Vogel, S.N., Weis, J.J. Cutting edge: repurification of lipopolysaccharide eliminates signaling through both human and murine toll-like receptor 2. J Immunol 165 (2000) 618–622.

Hitchcock, P.J. Analyses of gonococcal lipopolysaccharide in whole-cell lysates by sodium dodecyl sulfate-polyacrylamide gel electrophoresis: stable association of lipopolysaccharide with the major outer membrane protein (protein I) of *Neisseria gonorrhoeae*. Infect Immun 46 (1984) 202–212.

Hitchcock, P.J., Brown, T.M. Morphological heterogeneity among *Salmonella* lipopolysaccharide chemotypes in silver-stained polyacrylamide gels. J Bacteriol 154 (1983) 269–277.

Hollingsworth, R.I., Lill-Elghanian, D.A. Isolation and characterization of the unusual lipopolysaccharide component, 2-amino-2-deoxy-2-*N*-(27-hydroxyoctacosanoyl)-3-*O*-(3-hydroxy-tetradecanoyl)-gluco-hexuronic acid, and its de-*O*-acylation product from the free lipid A of *Rhizobium trifolii* ANU843. J Biol Chem 264 (1989) 14039–14042.

Hubner, G., Lindner, B. Separation of R-form lipopolysaccharide and lipid A by CE-Fourier-transform ion cyclotron resonance MS. Electrophoresis 30 (2009) 1808–1816.

Inzana, T.J., Apicella, M.A. Use of a bilayer stacking gel to improve resolution of lipopolysaccharides and lipooligosaccharides in polyacrylamide gels. Electrophoresis 20 (1999) 462–465.

Inzana, T.J., Pichichero, M.E. Lipopolysaccharide subtypes of *Haemophilus influenzae* type b from an outbreak of invasive disease. J Clin Microbiol 20 (1984) 145–150.

Jann, B., Reske, K., Jann, K. Heterogeneity of lipopolysaccharides. Analysis of polysaccharide chain lengths by sodium dodecylsulfate-polyacrylamide gel electrophoresis. Eur J Biochem 60 (1975) 239–246.

Johnson, K.G., Perry, M.B. Improved techniques for the preparation of bacterial lipopolysaccharides. Can J Microbiol 22 (1976) 29–34.

Jorgenson, J.W., Lukacs, K.D. Free-zone electrophoresis in glass capillaries. Clin Chem 27 (1981) 1551–1553.

Karibian, D., Deprun, C., Caroff, M. Use of plasma desorption mass spectrometry in structural analysis of endotoxins: effects on lipid A of different acid treatments. Prog Clin Biol Res 392 (1995) 103–111.

Kasai, N., Nowotny, A. Endotoxic glycolipid from a heptoseless mutant of *Salmonella minnesota*. J Bacteriol 94 (1967) 1824–1836.

Kelly, J., Masoud, H., Perry, M.B., Richards, J.C., Thibault, P. Separation and characterization of O-deacylated lipooligosaccharides and glycans derived from *Moraxella catarrhalis* using capillary electrophoresis-electrospray mass spectrometry and tandem mass spectrometry. Anal Biochem 233 (1996) 15–30.

Kido, N., Ohta, M., Kato, N. Detection of lipopolysaccharides by ethidium bromide staining after sodium dodecyl sulfate-polyacrylamide gel electrophoresis. J Bacteriol 172 (1990) 1145–1147.

Kim, J.J., Phillips, N.J., Gibson, B.W., Griffiss, J.M., Yamasaki, R. Meningococcal group A lipooligosaccharides (LOS): preliminary structural studies and characterization of serotype-associated and conserved LOS epitopes. Infect Immun 62 (1994) 1566–1575.

Komuro, T., Galanos, C. Analysis of *Salmonella* lipopolysaccharides by sodium deoxycholate-polyacrylamide gel electrophoresis. J Chromatogr 450 (1988) 381–387.

Koval, S.F., Meadow, P.M. The isolation and characterization of lipopolysaccharide-defective mutants of *Pseudomonas aeruginosa* PAC1. J Gen Microbiol 98 (1977) 387–398.

Kropinski, A.M., Kuzio, J., Angus, B.L., Hancock, R.E. Chemical and chromatographic analysis of lipopolysaccharide from an antibiotic-supersusceptible mutant of *Pseudomonas aeruginosa*. Antimicrob Agents Chemother 21 (1982) 310–319.

Lacroix, R.P., Duncan, J.R., Jenkins, R.P., Leitch, R.A., Perry, J.A., Richards, J.C. Structural and serological specificities of *Pasteurella haemolytica* lipopolysaccharides. Infect Immun 61 (1993) 170–181.

Lesse, A.J., Campagnari, A.A., Bittner, W.E., Apicella, M.A. Increased resolution of lipopolysaccharides and lipooligosaccharides utilizing tricine-sodium dodecyl sulfate-polyacrylamide gel electrophoresis. J Immunol Methods 126 (1990) 109–117.

Li, J., Cox, A.D., Hood, D., Moxon, E.R., Richards, J.C. Application of capillary electrophoresis-electrospray-mass spectrometry to the separation and characterization of isomeric lipopolysaccharides of *Neisseria meningitidis*. Electrophoresis 25 (2004) 2017–2025.

Li, J., Cox, A.D., Hood, D.W., Schweda, E.K., Moxon, E.R., Richards, J.C. Electrophoretic and mass spectrometric strategies for profiling bacterial lipopolysaccharides. Mol Biosyst 1 (2005a) 46–52.

Li, J., Dzieciatkowska, M., Hood, D.W., Cox, A.D., Schweda, E.K., Moxon, E.R., Richards, J.C. Structural characterization of sialylated glycoforms of *H. influenzae* by electrospray mass spectrometry: fragmentation of protonated and sodiated O-deacylated lipopolysaccharides. Rapid Commun Mass Spectrom 21 (2007) 952–960.

Li, J., Martin, A., Cox, A.D., Moxon, E.R., Richards, J.C., Thibault, P. Mapping bacterial glycolipid complexity using capillary electrophoresis and electrospray mass spectrometry. Methods Enzymol 405 (2005b) 369–397.

Logan, S.M., Trust, T.J. Structural and antigenic heterogeneity of lipopolysaccharides of *Campylobacter jejuni* and *Campylobacter coli*. Infect Immun 45 (1984) 210–216.

Luderitz, O., Staub, A.M., Westphal, O. Immunochemistry of O and R antigens of *Salmonella* and related *Enterobacteriaceae*. Bacteriol Rev 30 (1966) 192–255.

Lukasiewicz, J., Niedziela, T., Jachymek, W., Kenne, L., Lugowski, C. Structure of the lipid A-inner core region and biological activity of *Plesiomonas shigelloides* O54 (strain CNCTC 113/92) lipopolysaccharide. Glycobiology 16 (2006) 538–550.

Morrison, D.C., Leive, L. Fractions of lipopolysaccharide from *Escherichia coli* O111:B4 prepared by two extraction procedures. J Biol Chem 250 (1975) 2911–2919.

Muller-Loennies, S., Lindner, B., Brade, H. Structural analysis of oligosaccharides from lipopolysaccharide (LPS) of *Escherichia coli* K12 strain W3100 reveals a link between inner and outer core LPS biosynthesis. J Biol Chem 278 (2003) 34090–34101.

Muller-Seitz, E., Jann, B., Jann, K. Degradation studies on the lipopolysaccharide from *E. coli* 071:K?:H12. Separation and investigation of O-specific and core polysaccharides. FEBS Lett 1 (1968) 311–314.

Munford, R.S., Hall, C.L., Rick, P.D. Size heterogeneity of *Salmonella typhimurium* lipopolysaccharides in outer membranes and culture supernatant membrane fragments. J Bacteriol 144 (1980) 630–640.

Nishijima, M., Raetz, C.R. Membrane lipid biogenesis in *Escherichia coli*: identification of genetic loci for phosphatidylglycerophosphate synthetase and construction of mutants lacking phosphatidylglycerol. J Biol Chem 254 (1979) 7837–7844.

Noda, K., Kubota, K., Yamasaki, R. Separation of lipooligosaccharides by linear gradient gel electrophoresis. Anal Biochem 279 (2000) 18–22.

Nowotny, A., Cundy, K.R., Neale, N.L., Nowotny, A.M., Radvany, R., Thomas, S.P., Tripodi, D.J. Relation of structure to function in bacterial O-antigens. IV. Fractionation of the components. Ann NY Acad Sci 133 (1966) 586–603.

Oertelt, C., Lindner, B., Skurnik, M., Holst, O. Isolation and structural characterization of an R-form lipopolysaccharide from *Yersinia enterocolitica* serotype O:8. Eur J Biochem 268 (2001) 554–564.

Osborn, M.J. Studies on the Gram-Negative Cell Wall. I. Evidence for the Role of 2-Keto- 3-Deoxyoctonate in the Lipopolysaccharide of *Salmonella Typhimurium*. Proc Natl Acad Sci USA 50 (1963) 499–506.

Perez, G.I., Hopkins, J.A., Blaser, M.J. Antigenic heterogeneity of lipopolysaccharides from *Campylobacter jejuni* and *Campylobacter fetus*. Infect Immun 48 (1985) 528–533.

Peterson, A.A., McGroarty, E.J. High-molecular-weight components in lipopolysaccharides of *Salmonella typhimurium*, *Salmonella minnesota*, and *Escherichia coli*. J Bacteriol 162 (1985) 738–745.

Prehm, P., Stirm, S., Jann, B., Jann, K. Cell-wall lipopolysaccharide from *Escherichia coli* B. Eur J Biochem 56 (1975) 41–55.

Prendergast, M.M., Lastovica, A.J., Moran, A.P. Lipopolysaccharides from *Campylobacter jejuni* O:41 strains associated with Guillain-Barre syndrome exhibit mimicry of GM1 ganglioside. Infect Immun 66 (1998) 3649–3655.

Pupo, E., López, C.M., Alonso, M., Hardy, E. High-efficiency passive elution of bacterial lipopolysaccharides from polyacrylamide gels. Electrophoresis 21 (2000) 526–530.

Pupo, E., Hardy, E. Isolation of smooth-type lipopolysaccharides to electrophoretic homogeneity. Electrophoresis 28 (2007) 2351–2357.

Que, N.L., Lin, S., Cotter, R.J., Raetz, C.R. Purification and mass spectrometry of six lipid A species from the bacterial endosymbiont *Rhizobium etli*. Demonstration of a conserved distal unit and a variable proximal portion. J Biol Chem 275 (2000) 28006–28016.

Qureshi, N., Honovich, J.P., Hara, H., Cotter, R.J., Takayama, K. Location of fatty acids in lipid A obtained from lipopolysaccharide of *Rhodopseudomonas sphaeroides* ATCC 17023. J Biol Chem 263 (1988) 5502–5504.

Qureshi, N., Kaltashov, I., Walker, K., Doroshenko, V., Cotter, R.J., Takayama, K., Sievert, T.R., Rice, P.A., Lin, J.S., Golenbock, D.T. Structure of the monophosphoryl lipid A moiety obtained from the lipopolysaccharide of *Chlamydia trachomatis*. J Biol Chem 272 (1997) 10594–10600.

Qureshi, N., Mascagni, P., Ribi, E., Takayama, K. Monophosphoryl lipid A obtained from lipopolysaccharides of *Salmonella minnesota* R595. Purification of the dimethyl derivative by high performance liquid chromatography and complete structural determination. J Biol Chem 260 (1985) 5271–5278.

Qureshi, N., Takayama, K., Ribi, E. Purification and structural determination of nontoxic lipid A obtained from the lipopolysaccharide of *Salmonella typhimurium*. J Biol Chem 257 (1982) 11808–11815.

Raetz, C.R., Kennedy, E.P. Function of cytidine diphosphate-diglyceride and deoxycytidine diphosphate-diglyceride in the biogenesis of membrane lipids in *Escherichia coli*. J Biol Chem 248 (1973) 1098–1105.

Raetz, C.R., Purcell, S., Meyer, M.V., Qureshi, N., Takayama, K. Isolation and characterization of eight lipid A precursors from a 3-deoxy-D-manno-octylosonic acid-deficient mutant of *Salmonella typhimurium*. J Biol Chem 260 (1985) 16080–16088.

Reynolds, C.M., Raetz, C.R. Replacement of lipopolysaccharide with free lipid A molecules in *Escherichia coli* mutants lacking all core sugars. Biochemistry 48 (2009) 9627–9640.

Ribi, E., Haskins, W.T., Laudy, M., Milner, K.C. Preparation and host-reactive properties of endotoxin with low content of nitrogen and lipid. J Exp Med 114 (1961) 647–663.

Rivera, M., Bryan, L.E., Hancock, R.E., McGroarty, E.J. Heterogeneity of lipopolysaccharides from *Pseudomonas aeruginosa*: analysis of lipopolysaccharide chain length. J Bacteriol 170 (1988) 512–521.

Robert, N.A., Gray, G.W., Wilkinson, S.G. Release of lipopolysaccharide during the preparation of cell walls of *Pseudomonas aeruginosa*. Biochim Biophys Acta 135 (1967) 1068–1071.

Rosner, M.R., Tang, J., Barzilay, I., Khorana, H.G. Structure of the lipopolysaccharide from an *Escherichia coli* heptose-less mutant. I. Chemical degradations and identification of products. J Biol Chem 254 (1979) 5906–5917.

Shear, M.J. Effect of concentrate from *B. prodigiosus* filtrate on subcutaneous primary induced mouse tumors. Cancer Res 1 (1941)732–741.

Temple, G.S., Ayling, P.D., Wilkinson, S.G. Isolation and characterization of a lipopolysaccharide-specific bacteriophage of *Pseudomonas aeruginosa*. Microbios 45 (1986) 81–91.

Tirsoaga, A., El Hamidi, A., Perry, M.B., Caroff, M., Novikov, A. A rapid, small-scale procedure for the structural characterization of lipid A applied to *Citrobacter* and *Bordetella* strains: discovery of a new structural element. J Lipid Res 48 (2007a) 2419–2427.

Tirsoaga, A., Novikov, A., Adib-Conquy, M., Werts, C., Fitting, C., Cavaillon, J.M., Caroff, M. Simple method for repurification of endotoxins for biological use. Appl Environ Microbiol 73 (2007b) 1803–1808.

Tsai, C.M., Frasch, C.E. A sensitive silver stain for detecting lipopolysaccharides in polyacrylamide gels. Anal Biochem 119 (1982) 115–119.

Tsang, J.C., Wang, C.S., Alaupovic, P. Degradative effect of phenol on endotoxin and lipopolysaccharide preparations from *Serratia marcescens*. J Bacteriol 117 (1974) 786–795.

Vilches, S., Canals, R., Wilhelms, M., Salo, M.T., Knirel, Y.A., Vinogradov, E., Merino, S., Tomas, J.M. *Mesophilic Aeromonas* UDP-glucose pyrophosphorylase (GalU) mutants show two types of lipopolysaccharide structures and reduced virulence. Microbiology 153 (2007) 2393–2404.

Wang, X., McGrath, S.C., Cotter, R.J., Raetz, C.R. Expression cloning and periplasmic orientation of the *Francisella novicida* lipid A 4′-phosphatase LpxF. J Biol Chem 281 (2006a) 9321–9330.

Wang, X., Ribeiro, A.A., Guan, Z., McGrath, S.C., Cotter, R.J., Raetz, C.R. Structure and biosynthesis of free lipid A molecules that replace lipopolysaccharide in *Francisella tularensis* subsp. novicida. Biochemistry 45 (2006b) 14427–14440.

Westphal, O., Jann, K. Bacterial lipopolysaccharides. Extraction with phenol water and further applications of the procedure. In Whistler, R.L. (ed), Methods in Carbohydrate Chemistry. Academic, New York, NY (1965), pp. 83–91.

Wober, W., Alaupovic, P. Studies on the protein moiety of endotoxin from Gram-negative bacteria. Characterization of the protein moiety isolated by acetic acid hydrolysis of endotoxin from *Serratia marcescens* 08. Eur J Biochem 19 (1971) 357–367.

Yildirim, H.H., Hood, D.W., Moxon, E.R., Schweda, E.K. Structural analysis of lipopolysaccharides from *Haemophilus influenzae* serotype f. Structural diversity observed in three strains. Eur J Biochem 270 (2003) 3153–3167.

Yokota, S.I., Amano, K.I., Shibata, Y., Nakajima, M., Suzuki, M., Hayashi, S., Fujii, N., Yokochi, T. Two distinct antigenic types of the polysaccharide chains of *Helicobacter pylori* lipopolysaccharides characterized by reactivity with sera from humans with natural infection. Infect Immun 68 (2000) 151–159.

Zahringer, U., Lindner, B., Knirel, Y.A., van den Akker, W.M., Hiestand, R., Heine, H., Dehio, C. Structure and biological activity of the short-chain lipopolysaccharide from *Bartonella henselae* ATCC 49882T. J Biol Chem 279 (2004) 21046–21054.

Zdorovenko, E.L., Vinogradov, E., Zdorovenko, G.M., Lindner, B., Bystrova, O.V., Shashkov, A.S., Rudolph, K., Zahringer, U., Knirel, Y.A. Structure of the core oligosaccharide of a rough-type lipopolysaccharide of *Pseudomonas syringae* pv. phaseolicola. Eur J Biochem 271 (2004) 4968–4977.

Zhou, P., Chandan, V., Liu, X., Chan, K., Altman, E., Li, J. Microwave-assisted sample preparation for rapid and sensitive analysis of *Helicobacter pylori* lipid A applicable to a single colony. J Lipid Res 50 (2009) 1936–1944.

Zhou, Z., Lin, S., Cotter, R.J., Raetz, C.R. Lipid A modifications characteristic of *Salmonella typhimurium* are induced by NH_4VO_3 in *Escherichia coli* K12. Detection of 4-amino-4-deoxy-L-arabinose, phosphoethanolamine and palmitate. J Biol Chem 274 (1999) 18503–18514.

Chapter 3
Endotoxins: Relationship Between Structure, Function, and Activity

K. Brandenburg, A.B. Schromm, and T. Gutsmann

Abstract Endotoxins as amphiphilic components of the outer layer of the outer membrane of Gram-negative bacteria exert their immunostimulatory activity after release from bacterial cells. Thus, the characterization of the physicochemical properties of this glycolipid in physiological fluids is of utmost importance for an understanding of cell activation processes. Here, the essential physicochemical parameters describing endotoxins such as critical micellar concentration, acyl chain fluidity, intramolecular conformation, supramolecular structures, and size as well as morphology of the aggregates are discussed and assessed with respect to their importance for an understanding of the interaction mechanisms with immunorelevant cells. The reviewed data clearly indicate that knowledge of these parameters is essential for understanding the bioactivity of not only endotoxins, but also endotoxin-like amphiphiles.

Keywords Lipopolysaccharide · Biophysical analysis · FTIR spectroscopy · Small-angle X-ray scattering · Molecular conformation

Abbreviations

LPS	lipopolysaccharide
CMC	critical micellar concentration
EM	electron microscopy
SAXS	small-angle X-ray scattering
LBP	lipopolysaccharide-binding protein

K. Brandenburg (✉)
Division of Biophysics and Division of Immunbiophysics, Forschungszentrum Borstel, Parkallee 10, D-23845 Borstel, Germany
e-mail: kbranden@fz-borstel.de

X. Wang, P.J. Quinn (eds.), *Endotoxins: Structure, Function and Recognition*, Subcellular Biochemistry 53, DOI 10.1007/978-90-481-9078-2_3,

3.1 Introduction

It was found early that changes of the chemical primary structure of the poly- or oligosaccharide part of lipopolysaccharides (LPS, endotoxins) has only minor influence on the biological activities and it was concluded that the glycolipid anchor of LPS, lipid A, represents the endotoxic principle (Rietschel et al., 1971). These and other findings led to a variety of investigations which all were directed to attribute the different chemical groups of this macromolecule to its biological activity and to determine the biologically active unit (Rietschel et al., 1990; Loppnow et al., 1993; Zähringer et al., 1996). It was found that relatively strong variations in the core oligosaccharide and, even more, in the O-antigen structure between different bacterial strains and genera (Holst et al., 1996; Holst, 1999) were indicative of the role of the sugar part as defense factor of the bacteria, which contributes to the permeability barrier of the cell. Interestingly, although lipid A is the endotoxic principle, because it exerts qualitatively all the characteristic biological activities of LPS in mammals, it was found that the presence of additional sugars leads to an increase of bioactivity by at least hundred-fold (Rietschel et al., 1991; Schromm et al., 2000). As the added sugars, whether only the two monosaccharides 2-keto-3-deoxyoctonate (Kdo) as in the case of deep rough mutant LPS Re or the complete core oligosaccharide as in the case of LPS Ra, are by themselves completely inactive, one can assume that the physico-chemical behaviour of whole aggregates of these amphiphiles is influenced by the absence or presence of the different core oligosaccharides, which will be discussed below.

3.2 Physicochemical Characterisation of Endotoxins

LPS is a glycolipid with an amphiphilic character similar to other membrane constituents such as phospholipids. Above a critical concentration – critical micellar concentration (CMC), LPS forms multimers, i.e., aggregates in aqueous environments, depending on the hydrophobicity of the contributing molecules. So far, the CMC values for lipid A and LPS have not been exactly determined.. Measurements with partial structures such as the precursor lipid IVa (a tetraacylated compound) indicate very low CMC values for lipid A $< 10^{-8}$ M, (Maurer et al., 1991; Seydel et al., 1999), and the data by Takayama et al. (1990) indicate for deep rough mutant LPS Re (lipid A plus two Kdo monosaccharides) CMC values between 10^{-7} and 10^{-8}. M. Aurell and Wistrom (1998) have published values in the μM range for chemically completely different endotoxins, free lipid A and LPS from rough mutants to smooth LPS from wild-type strains. These values, however, are likely to reflect the sensitivity of the fluorescence and light scattering technique used rather than real CMC values confirmed by our own unpublished data involving also phospholipids.

Values in the nM range were found by Sasaki and White (2008), who reported a decrease of the "hydrodynamic radius" of highly purified LPS Re around 10 nM at 37°C. The authors, however, unfortunately did not test the detection limit of the method by using, for example, Latex particles, so that the given value can be assumed as upper limit of the actual CMC. Aurell et al. (1999) published data on LPS Re as well as smooth type LPS with tapping mode atomic force microscopy, and found down to very low concentrations of 10 pg/ml (in the pM range) aggregates, without any visible signs of breaking up into single LPS monomers. Summarizing these data, it can be assumed that a reasonable estimate for the CMC is the pM range, in which most biological reactivities disappear, which gives also evidence that aggregates are of biological relevance.

The aggregate structure of amphiphiles like LPS is not only determined by the primary chemical structure, but also influenced by ambient conditions such as temperature, pH, water content, and concentration of mono- and divalent cations. According to the theory introduced by Israelachvili, the type of aggregate structure of amphiphilic molecules can be estimated by a simple geometric model, which relates the resulting structure to the ratio of the effective cross-sectional areas, a_o, of the hydrophilic polar and, a_h, the hydrophobic apolar regions, respectively (Israelachvili, 1991). For this, a dimensionless shape parameter S was introduced, which is defined as $S = v/(a_o \cdot l_c) = a_h/a_o$ (v volume per molecule of the hydrophobic moiety, l_c length of the fully extended hydrophobic portion). This allows a prediction of the supramolecular aggregate structure. Furthermore, the molecular shape of a given amphiphilic molecule within a supramolecular aggregate depends on the fluidity (inversely correlated to the state of order) of its acyl chains, which can adopt two main phase states, the highly ordered gel phase (β) and the unordered liquid-crystalline (α) phase. Between these two states a lipid-specific reversible phase transition at a temperature T_m takes place, which depends on the length and the degree of saturation of the acyl chains as well as on the conformation and the charge density of and its distribution within the headgroup region. This phase transition can take place within one and the same aggregate structure, but can also be accompanied by a change of the aggregate structure. It is known that not all aggregate structures can occur within both phases. Thus, the H_{II} structure has been observed only in the liquid crystalline phase, because in the gel phase the acyl chains are not flexible enough to adapt to the high curvature of the acyl chains in the H_{II} structure. It was assumed earlier that also cubic structures are restricted to the fluid phase (Luzzati et al., 1986), this, however, could not be confirmed in the case of glycolipids from bacterial origin (Brandenburg et al., 1990, 1992).

From the statements above, it is therefore not surprising, that due to the variability in the acylation patterns and in the chemical structure and length of the sugar moiety, a complex phase behavior and structural polymorphism is observed for LPS. A whole body of experimental work has been published with respect to the gel to liquid crystalline phase transition of LPS and free lipid A (Brandenburg and Seydel, 1984, 1990; Naumann et al., 1989; Brandenburg and Seydel, 1998). It was found that enterobacterial LPS adopts a phase transition at 30–35°C for the LPS from rough mutant strains (lowest value for LPS Re and highest value for LPS Ra), and around 45°C for free lipid A. This means that for the parent LPS the hydrocarbon chain moiety of the lipid A part is in a highly fluid state at 37°C, whereas it is relatively rigid for free lipid A. These findings could be interpreted as being the reason for the considerably lower biological activity of free lipid A as compared to LPS (Schromm et al., 2000; Brandenburg et al., 2003; Garidel et al., 2005). This interpretation is strongly supported by the observation that LPS and free lipid A from *Coxiella burnetii*, which are both in a highly fluid state at 37°C, exhibit almost identical biological activity (Toman et al., 2004). It should be noted that the acyl chain fluidity is not a determinant of bioactivity in contrast to the central importance of the aggregate structure (see below). However, it can modulate the response in biological systems as illustrated in Fig. 3.1 showing the interleukin-6 inducing activity and the order parameter (inverse proportional to the fluidity) for LPS Re in various salt forms. Clearly, the samples with high order parameter (low fluidity or rigidity of the acyl chains) have lowest IL-6 inducing activity and vice versa.

A critical region of the shape parameter (see above) is the range around $S=1$, where various phases may coexist and in which small changes of parameters such as hydration and ion concentration may lead to a phase transition. Since frequently the packing parameter cannot be estimated with a sufficient precision, physical techniques such as small-angle scattering with X-rays (SAXS) or neutrons (SANS) must be applied for a reliable determination of the structures.

A systematic study on the structural polymorphism of lipid A and LPS Re with synchrotron radiation small-angle X-ray scattering showed, that at lower water content (<70 w%) and higher concentration of divalent cations such as Mg^{2+} ([lipid A]:[Mg^{2+}] < 3:1 molar ratio), only multilamellar structures are formed (Brandenburg et al., 1990, 1992). In contrast, at high water content and

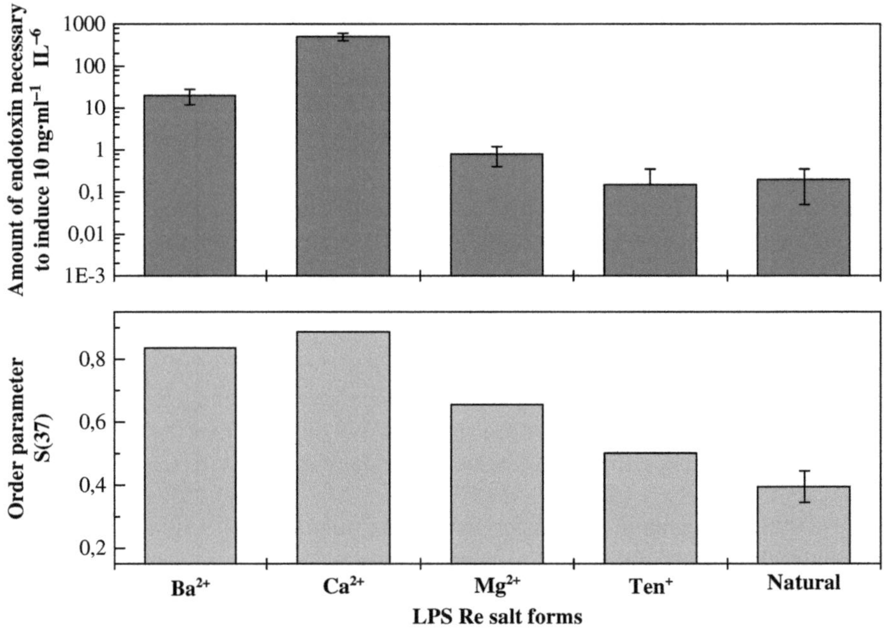

Fig. 3.1 Amount of LPS necessary to induce 10 ng/ml interleukin-6 in human mononuclear cells (*top*) and order parameter of the acyl chains (*bottom*) for LPS Re from *S. minnesota* R595 in different salt forms (adapted from Garidel et al., 2005)

lower cation concentrations, nonlamellar, cubic inverted structures are formed, in particular under near physiological conditions. These cubic structures have been shown to be of essential importance for the expression of bioactivity of endotoxins (Brandenburg, 1993; Brandenburg et al., 1993, 2003) Interestingly, the observed cubic structures were essentially of the bicontinous type (Brandenburg et al., 1990) of symmetry Q^{224} (Pn3m), Q^{229} (Im3m) and Q^{230} (Ia3d), which was confirmed independently 18 years later (Reichelt et al., 2008). The analysis of synthetic hexaacyl lipid A, in which the phosphate groups were substituted by carboxymethyl groups, showed similar results to those of "normal" lipid A (Seydel et al., 2005). The investigation of synthetic triacylated monosaccharide lipid A part structures corresponding to the non-reducing moiety of enterobacterial lipid A (oxyacyl group linked to position 3 and an unbranched fatty acid to position 2 of the glucosamine), exhibited a non-lamellar, probably cubic phase, whereas for the reversed linkage a multilamellar phase was observed (Brandenburg et al., 2002). This could be correlated to different inclination angles of the backbone with respect to the direction of the acyl chains and, most interestingly, also to different biological activities (see later paragraphs). Lipid A from non-enterobacterial source such as from some phototropic strains showed a preference for multilamellar structures (Brandenburg et al., 1993 ; Schromm et al., 1998, 2000) and this was similarly the case for enterobacterial lipid A with one or two cleaved acyl chains (Schromm et al., 2000). This has direct impact on the expression of biological activity (Section 2.4).

In further investigations on other rough mutant LPS Rd through Ra (Seydel et al., 1993) from enterobacterial strains, non-lamellar aggregate structures were also found under near physiological conditions. No multilamellar structures were found, except at low water content and high divalent cation concentrations.

3.3 Morphology and Size Distribution of Endotoxin Aggregates

The first who studied the morphology of LPS aggregates was the group of Shands (1971), who found long filaments and ribbon-like structures for LPS from wild-type strains, as seen by electron microscopy. These investigations as well as those from Risco et al. (1993) were hampered by the fact that wild-type LPS is a heterogenous mixture of various compounds, and it is difficult to decide which fraction within this mixture corresponds to the biologically active fraction, which seem to be a LPS Ra or Rb according to Jiao et al. (1989). Risco et al. (1993) have applied freeze-fracture in comparison to negative staining EM and have found that the former technique gave mainly spherical or elliptical particles of rather homogeneous size, whereas the latter technique showed a more heterogenous population. From these data the authors conclude that negative staining gives rise to artefactual forms due to the preparation technique. These findings are in accordance with the SAXS data cited above, in which at low water content only multilamellar structures were observed. Negative staining leads to dried samples measured in the electron microscope, thus, the sample preparation irreversibly changes their morphologies.

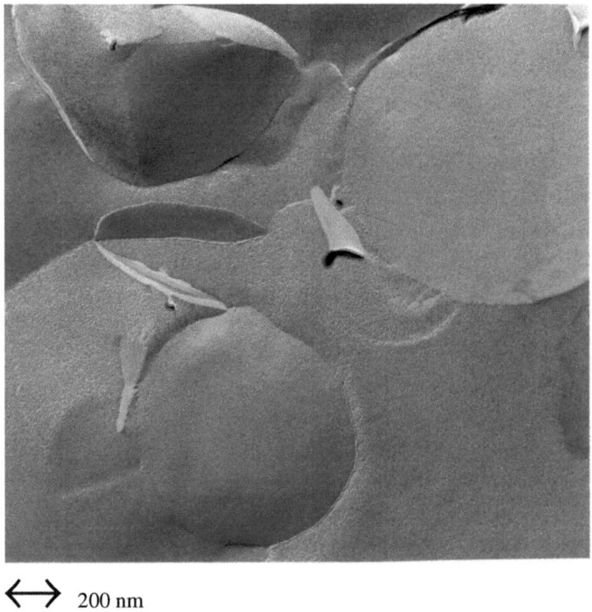

Fig. 3.2 Freeze fracture electron micrograph of lipid A extracted from LPS from *S. minnesota* strain R595 (kindly performed by W. Richter, Jena)

Freeze-fracture EM was applied in recent investigations to analyse the morphologies of lipid A and LPS from rough mutants. As examples of these investigations, micrographs were published for deep rough mutant Re from *Salmonella enterica* sv. Minnesota R595 and a mutant LPS Ra with a complete core oligosaccharide from *S. enterica* sv. Minnesota R60 (Andrä et al., 2007; Chen et al., 2007). It was shown that the former sample exhibited spherical-like particles, but no closed liposomes, rather "open egg-shells" were formed. Interestingly, with increasing the sugar (oligosaccharide) chain length the morphology was completely converted, giving rise to ribbon-like fibrillary structures (Chen et al., 2007). The authors found diameters of the fibrilles in the range of 10–12 nm, which would correspond to the bilayer repeat of LPS Ra (Seydel et al., 1993). Interestingly, lipid A exhibits very large aggregates with diameters of more than 1 μm (Fig. 3.2 and unpublished data). Aggregate sizes of various LPS were also analysed by Laser light scattering techniques. Typical sizes for aggregates of LPS rough mutants of 100–600 nm, as found by Howe et al. (2007). Interpretation of these data is hampered by the fact that the complex geometry of the aggregates deviates considerably from spherical forms, which cannot be accounted for by the evaluation software.

3.4 Intramolecular Conformation

It seems to be important to study the intramolecular conformation of the LPS and lipid A molecules within the aggregate. In this way one might understand the way the molecule is arranged, which may be necessary for an understanding of the interaction with target molecules such as binding and receptor proteins.

It was found that enterobacterial lipid A has a strong inclination of the diglucosamine backbone with respect to the direction of the hydrocarbon chains of 40° to 55°, while less acylated lipid A (penta- and tetraacyl) as well as some non-enterobacterial lipid A exhibited a much lower angle (<20°). These informations were deduced from measurements by applying Fourier-transform infrared (FTIR) spectroscopy with polarized IR light using an attenuated total reflectance unit (Seydel et al., 2000). As example, this is illustrated for two biologicallyactive hexaacyl lipid A from *E. coli* and the biologically inactive lipid A from *Rhodobacter capsulatus*. Figure 3.3 depicting low dichroic values R for the former and high R for the latter lipid A, from which an inclination of the backbone of active lipid A with respect to the membrane plane can be calculated. These findings apparently express the packing constraints of the hydrocarbon chains with respect to the available cross-sectional areas, and determine also the ability of lipid A to act as agonists or antagonists (see later). Therefore, if also the SAXS data presented above are included, it can be generalized that the backbones of lipid A with a conical shape

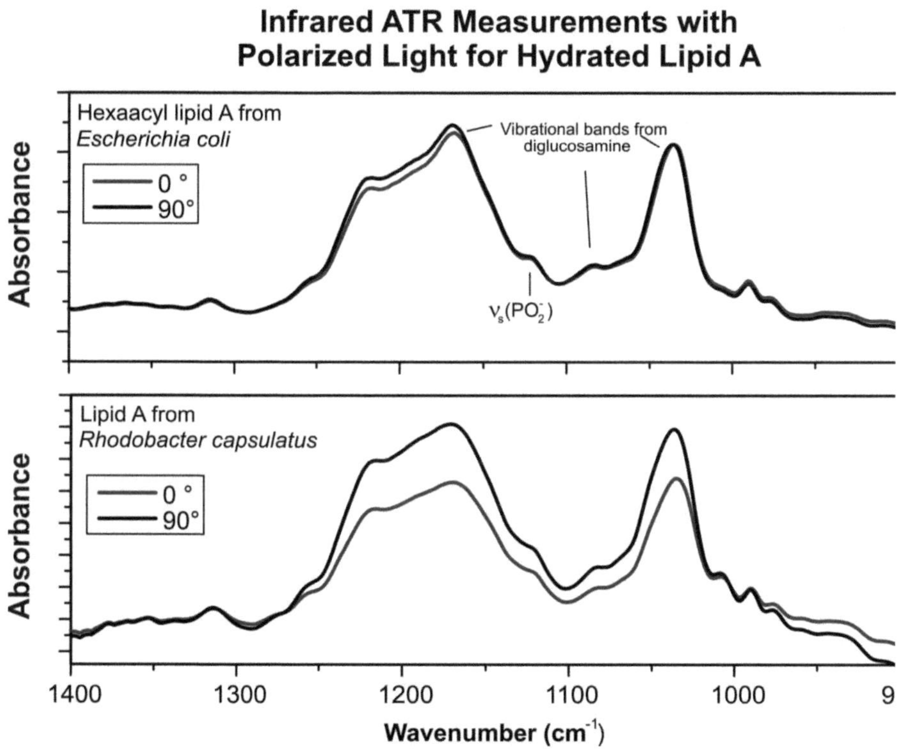

Fig. 3.3 Infrared spectra with attenuated total reflectance using polarized light for lipid A from *E.coli* and from *Rhodobacter capsulatus*. The two lipid A have identical backbone structures, but differences in the acylation pattern. As can be seen, the vibrational bands corresponding to diglucosamine ring vibrations have different dichroic ratios, which is indicative for a different inclination of the backbones with respect to the membrane normal

(cubic inverted structure) are strongly inclined, and those with a cylindrical shape (multilamellar structure) have – if at all – only a slight inclination. It has to be emphasized, that this molecular conformation only holds for the single molecules within an aggregate, a monomer outside the aggregate would not be exposed to the intermolecular forces, which comes from the cation bridging of adjacent molecules.

Molecular modelling techniques, based on the primary chemical structure, allow to calculate the potential conformations a given molecule is thermodynamically likely to adopt. These calculations on the conformation of hexaacyl lipid A components of *S. enterica* sv. Minnesota suggests an angle of 45° between the plane of the bisphosphoryl disaccharide backbone and the membrane normal (perpendicular to the direction of the acyl chains). The fatty acids occupy positions lying almost exactly on a hexagonal lattice (Kastowsky et al., 1991, 1993). Furthermore, a length of 2.6 nm and a cross-section of 0.6–0.8 nm for the smaller and 1.2–1.6 nm for the longer side of the rectangular cross-section of the acyl chains was caculated. From this model, a dense packing of the acyl chains and a tilt of the diglucosamine axis with respect to the membrane surface of 53.7° with respect to the reducing side of the diglucosamine was derived, with the 1-phosphate emerging in the hydrophobic moiety of the neighbouring molecule, and the 4′-phosphate sticking out into the aqueous phase. It was found later, however, in infrared spectroscopic experiments that the 1-phosphate is surrounded by water and the 4′-phosphate is buried in the backbone-close hydrophobic region, probably facing the 3-hydroxyl groups of the neighbouring molecules (Brandenburg et al., 1997; Seydel et al., 2000). Despite the fact that in this configuration the packing of the acyl chains is not optimal, this model could be verified by the observations, that several lipid A analogues and partial structures do not have a hexagonal dense packing and that lipid A and LPS have strongly reduced enthalpy changes (ΔH_c) of the gel-to-liquid crystalline phase transition as compared to saturated phospholipids with the same acyl chain length (Brandenburg and Blume, 1987).

3.5 Biophysical Mechanisms of Agonism and Antagonism

The bioactivity of endotoxins should be discussed in the context of the interaction of LPS molecules with the membrane of immunologically relevant cells, since the interaction of endotoxin molecules with membrane-associated proteins is an important step in the initiation of biological effects. In our model of endotoxin activation, LPS intercalates in a first step, transported by the LPS-binding protein (LBP), into target immune cell membranes (Schromm et al., 1996; Gutsmann et al., 2000, 2001a, b). Within the membrane, LPS forms domains due to the differences in chemical structures with respect to the membrane lipids. These domains, which may laterally diffuse within the target cell membrane, represent a strong disturbance of the membrane architecture because of the conical shape of the lipid A moiety of the contributing LPS molecules. At the site of a signalling protein, this may lead to cell activation (Blunck et al., 2001).

Nomura et al. (2008) performed an interaction study of highly purified LPS Re with phospholipid membranes by applying ^{31}P-NMR spectroscopy, and found that the LPS became part of the phospholipid membranes (phosphatidylcholine: phosphatidylglycerol 9:1 mixtures) in the form of aggregates, i.e., they formed domains within the target cell membrane. These findings are in excellent agreement with the here described model of cell activation.

Another aspect important for the biological activity of LPS is the mechanism of antagonism, i.e. the inhibitory effect of endotoxically non-active compounds that are able to inhibit cell activation by enterobacterial LPS. There are lipid A variants, which exhibit – despite an identical backbone to that of enterobacterial lipid A – a significantly reduced or completely absent biological activity due to the changed acyl chain moiety such as pentaacyl lipid A or synthetic tetraacyl lipid A 406 (Mayer and Weckesser, 1984; Loppnow et al., 1990). For a better understanding of the mechanisms of antagonism, the chemical composition of several natural lipid A preparations was determined, derived from *E. coli*. Interestingly, in these isolates beside the hexaacylated lipid A also significant amounts of penta- and tetraacylated molecules were found (Mueller et al., 2004). Despite the antagonizing effect of the latter compounds, their occurrence in these preparations did obviously not reduce the biological activity. To understand this phenomenon, separate aggregates of either synthetic compound 506 (hexaacyl lipid A) or 406 (tetraacyl lipid A precursor IVa) were prepared. Addition of these separate aggregates to human *immune* cells at different molar ratios showed the well described antagonistic effect of compound 406. In contrast, when 506 and 406 were prepared in molecularly mixed aggregates containing both compounds, these mixed aggregates showed higher endotoxic activity than that of the pure compound 506 up to an admixture of 20% of compound 406 (Mueller et al., 2004). These observations can be understood by assuming that the active unit of endotoxins is the aggregate as outlined above. The addition of highly fluid molecularly incorporated 406 leads to a decrease of the binding energy of 506 molecules within the aggregates, which allows a better accessibility to binding proteins of the host. This interpretation is in accordance with the finding that the acyl chain order of 506 is relatively high, corresponding to a high gel to liquid crystalline phase transition temperature T_m around 45°C (see above, (Brandenburg and Seydel, 1984)). Compound 406 has a much lower T_m <20°C (Brandenburg et al., 1997), which leads to a significant decrease in acyl chain order of the mixture, and thus to a higher ability to interact with target structures. These findings could be extended by applying other negatively charged non-LPS phospholipid compounds such as cardiolipin (CL), phosphatidylglycerol (PG) and phosphatidylinositol (PI), and a typical experiment is given in Fig. 3.4 (Mueller et al., 2005a, b) As can be seen, all three phospholipids added from 10-fold up to 1000-fold excess to LPS considerably reduced the ability of human mononuclear cells to secret tumor-necrosis-factorα (TNF-α). In contrast, cell activation by TNF-α itself was not influenced by the addition of the phospholipids.

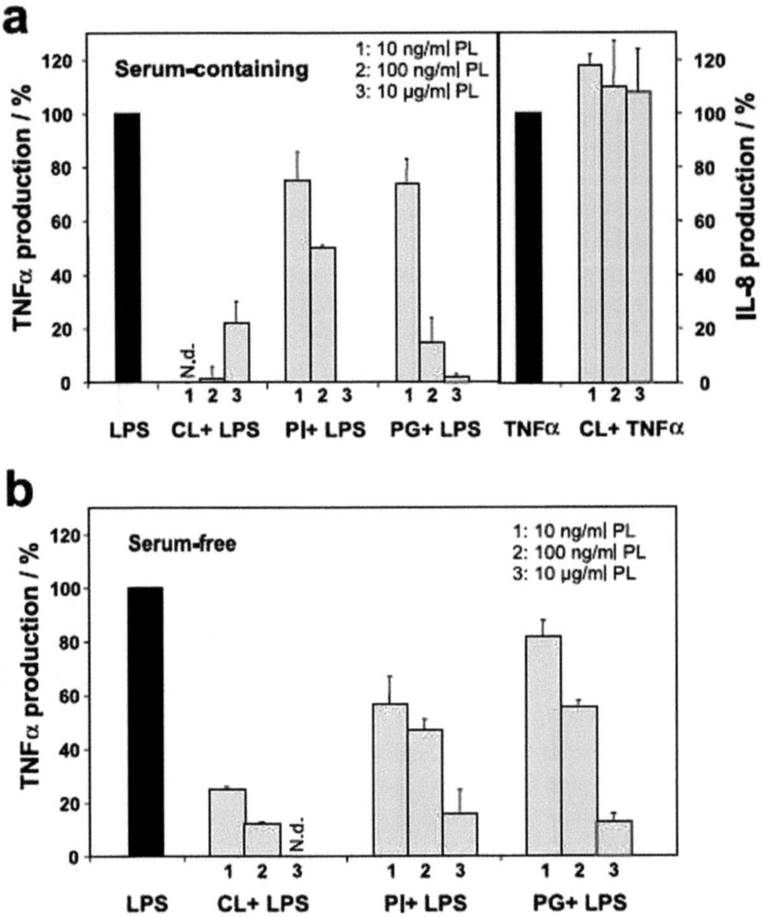

Fig. 3.4 Cytokine production by human macrophages after stimulation with 1 ng/ml LPS or 100 ng/ml TNF-α in the absence or preasence of cardiolipin (CL), phosphatidylinositol (PI), and phosphatidylglycerol (PG) at the indicated concentrations under serum-containing (**a**) and serum-free (**b**) conditions. Adapted from Mueller et al. (2005a) with permission

3.6 Endotoxically Active Unit

There has been a long lasting controversy regarding the biologically-active units of endotoxins, highly aggregated, small aggregates, or monomers. Takayama and co-workers (1990) have found that aggregates with low apparent molecular weight or even monomers are more active than large aggregates, whereas Shnyra et al. (1993) found the opposite. In recent publications the group of Weiss demonstrated that monomers produced by a complex interplay of LBP, soluble CD14, serum albumin and MD-2, play a decisive role for cell activation (Teghanemt et al., 2007; Gioannini et al., 2004). In these and other papers, however, problems may arise from the use of metabolically labelled radiaoactive ^{3}H-LPS, which is used for an understanding of the different steps of cell activation. The radioactivity used (ca. 1.5 $10^6/\mu g$, (Kitchens and Munford, 1998) or 1000 cpm/ng, (Tobias et al., 1997)) may lead to severe radiation damage. Thus, for example, it is known that ^{3}H-labelled compounds decompose at a rate of 2–3% per month per ^{3}H group.

In our picture of cell activation, LPS aggregates are the biologically-active units initiating the complex transport and signal transduction cascade. In the transmembrane signal transduction a number of proteins are involved, including LBP, CD14, and TLR4/MD2. Our concept of the aggregate as the bioactive unit is however compatible with an importance of LPS monomers at a later step within the signalling cascade, reflecting the complexity of molecular processes underlying the innate immune recognition of bacterial LPS.

Various findings, also applying endotoxin-neutralizing peptides, indicate that smaller aggregates are more favourable for the recognition by immune cells than

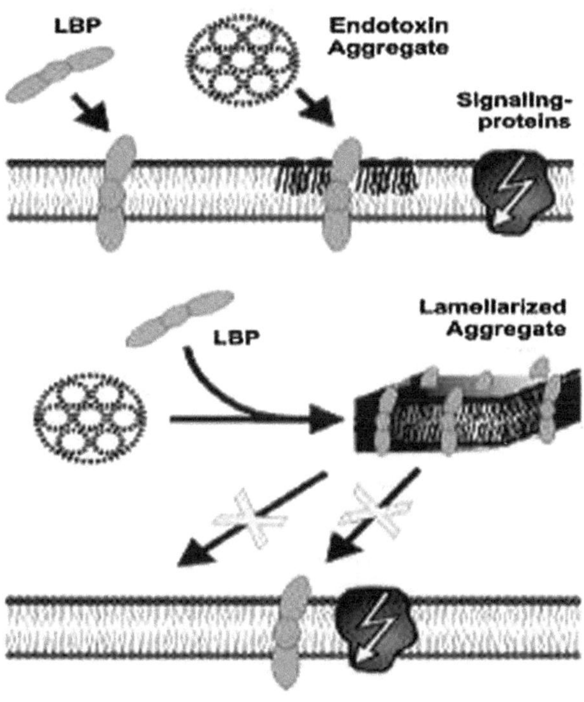

Fig. 3.5 Model of cell activation initiated by LBP. Dual role of the LPS-binding protein LBP: On the one hand, LBP mediates and enhances the activation of immune cells induced by LPS. This process seems to be influenced by LBP which is intercalated in or associated with the immune cell membrane (*upper* cartoon). An interaction of membrane-bound LBP (mLBP) with LPS can led to an intercalation of LPS into the phospholipid matrix. On the other hand, the interaction between soluble LBP (sLBP) and LPS aggregates induces a multilamellarization of the aggregates and finally to a neutralization of the endotoxin because of the inhibition of the binding to signaling proteins of the immune cell (*lower* cartoon)

larger ones, because a more direct interaction with binding proteins is possible. It should be mentioned in this context that also non-LPS structures with six hydrocarbon chains and two phosphate groups, but a serine-like backbone rather than the diglucosamine part of lipid A adopt cubic inverted structures and express full endotoxic activity (Brandenburg et al., 2004; Seydel et al., 2003). In a similar way, also bacterial lipopeptides (Schromm et al., 2007), lipopeptides from mycoplasma (Brandenburg et al., 2003) as well as mycobacteria (*M. tuberculosis*) (Schromm et al., 2010) may form non-lamellar cubic aggregate structures connected with considerable endotoxic activity.

Summarizing these findings, a general formulation of the "endotoxic conformation", or better: "endotoxic supramolecular conformation" may be presented here, which does not necessarily need LPS chemical structures. A necessary and sufficient condition for endotoxic activity is (i) the existence of a conical shape of the single monomers within the aggregates with a higher cross-section of the hydrophobic than the hydrophilic moiety, (ii) the existence of charges, which may be either negative (the normal case) or positive, and (iii) the transport of these aggregates into target cell membranes by themselves, driven by hydrophobic forces or by the action of binding proteins such as LBP. A schematic scheme of cell activation according to this model is shown in Fig. 3.5 (deduced from (Mueller et al., 2005b)).

Acknowledgement The authors are indebted to the German ministry BMBF (project: 01GU0824) and the DFG (project SCHR621/2-3), for financial support, and to Dr. Walter Richter (Universitätsklinikum Jena) for performing the freeze-fracture electron microscopy measurements.The authors would like to thank Prof. Ulrich Seydel, former head of the Division of Biophysics, for continous support and development of new concepts in the biophysics of endotoxins.

References

Andrä, J., Howe, J., Garidel, P., Rössle, M., Richter, W., Leiva-Leon, J., Moriyon, I., Bartels, R., Gutsmann, T., Brandenburg, K. Mechanism of interaction of optimized Limulus-derived cyclic peptides with endotoxins: thermodynamic, biophysical and microbiological analysis. Biochem J 406 (2007) 297–307.

Aurell, C.A., Hawley, M.E., Wistrom, A.O. Direct visualization of Gram-negative bacterial lipopolysaccharide self-assembly. Mol Cell Biol Res Commun 2 (1999) 42–46.

Aurell, C.A., Wistrom, A.O., Critical aggregation concentrations of Gram-negative bacterial lipopolysaccharides (LPS). Biochem Biophys Res Commun 253 (1998) 119–123.

Blunck, R., Scheel, O., Müller, M., Brandenburg, K., Seitzer, U., Seydel, U., New insights into endotoxin-induced activation of macrophages: Involvement of a K^+ channel in transmembrane signaling. J Immunol 166 (2001) 1009–1015.

Brandenburg, K. Fourier transform infrared spectroscopy characterization of the lamellar and nonlamellar structures of free lipid A and Re lipopolysaccharides from *Salmonella minnesota* and *Escherichia coli*. Biophys J 64 (1993) 1215–1231.

Brandenburg, K., Andrä, J., Müller, M., Koch, M.H.J., Garidel, P. Physicochemical properties of bacterial glycopolymers in relation to bioactivity. Carbohydr Res 338 (2003) 2477–2489.

Brandenburg, K., Blume, A. Investigations into the thermotropic phase behaviour of natural membranes extracted from Gram-negative bacteria and artificial membrane systems made from lipopolysaccharides and free lipid A. Thermochim Acta 119 (1987) 127–142.

Brandenburg, K., Hawkins, L., Garidel, P., Andrä, J., Müller, M., Heine, H., Koch, M.H.J., Seydel, U. Structural polymorphism and endotoxic activity of synthetic phospholipid-like amphiphiles. Biochemistry 43 (2004) 4039–4046.

Brandenburg, K., Koch, M.H.J., Seydel, U. Phase diagram of lipid A from *Salmonella minnesota* and *Escherichia coli* rough mutant lipopolysaccharide. J Struct Biol 105 (1990) 11–21.

Brandenburg, K., Seydel, U. Investigation into the fluidity of lipopolysaccharide and free lipid A membrane systems by Fourier-transform infrared spectroscopy and differential scanning calorimetry. Eur J. Biochem 191 (1990) 229–236.

Brandenburg, K., Koch, M.H.J., Seydel, U. Phase diagramm of deep rough mutant lipopolysaccharide from *Salmonella minnesota* R595. J Struct Biol 108 (1992) 93–106.

Brandenburg, K., Kusumoto, S., Seydel U. Conformational studies of synthetic lipid A analogues and partial structures by infrared spectroscopy. Biochim Biophys Acta 1329 (1997) 193–201.

Brandenburg, K., Matsuura, M., Heine, H., Müller, M., Kiso, M., Ishida, H., Koch, M.H.J., Seydel, U. Biophysical characterization of triacyl monosaccharide lipid A partial structures in relation to bioactivity. Biophys J 83 (2002) 322–333.

Brandenburg, K., Mayer, H., Koch, M.H.J., Weckesser, J., Rietschel, E.Th., Seydel, U. Influence of the supramolecular structure of free lipid A on its biological activity. Eur J Biochem 218 (1993) 555–563.

Brandenburg, K., Seydel, U. Physical aspects of structure and function of membranes made from lipopolysaccharides and free lipid A. Biochim Biophys Acta 775 (1984) 225–238.

Brandenburg, K., Seydel, U. Infrared spectroscopy of glycolipids. Chem Phys Lipids 96 (1998) 23–40.

Brandenburg, K., Wagner, F., Müller, M., Heine, H., Andrä, J., Koch, M.H.J., Zähringer, U., Seydel, U. Physicochemical characterization and biological activity of a glycoglycerolipid from Mycoplasma fermentans. Eur J Biochem 270 (2003) 3271–3279.

Chen, X., Howe, J., Andra, J., Rossle, M., Richter, W., da Silva, A.P., Krensky, A.M., Clayberger, C., Brandenburg, K. Biophysical analysis of the interaction of granulysin-derived peptides with enterobacterial endotoxins. Biochim Biophys Acta 1768 (2007) 2421–2431.

Garidel, P., Rappolt, M., Schromm, A.B., Howe, J., Lohner, K., Andrä, J., Koch, M.H., Brandenburg, K. Divalent cations affect chain mobility and aggregate structure of lipopolysaccharide from *Salmonella minnesota* reflected in a decrease of its biological activity, Biochim Biophys Acta 1715 (2005) 122–131.

Gioannini, T.L., Teghanemt, A., Zhang, D., Coussens, N.P., Dockstader, W., Ramaswamy, S., Weiss, J.P. Isolation of an endotoxin-MD-2 complex that produces Toll-like receptor 4-dependent cell activation at picomolar concentrations. Proc Natl Acad Sci USA 101 (2004) 4186–4191.

Gutsmann, T., Haberer, N., Carroll, S.F., Seydel, U., Wiese, A. Interaction between lipopolysaccharide (LPS), LPS-binding protein (LBP), and planar membranes, Biol Chem 382 (2001a) 425–434.

Gutsmann, T., Mueller, M., Carroll, S.F., MacKenzie, R.C., Wiese, A., Seydel, U. Dual role of lipopolysaccharide (LPS)-binding protein in neutralization of LPS and enhancement of LPS-induced activation of mononuclear cells. Infect Immun 69 (2001b) 6942–6950.

Gutsmann, T., Schromm, A.B., Koch, M.H.J., Kusumoto, S., Fukase, K., Oikawa, M., Seydel, U., Brandenburg, K. Lipopolysaccharide-binding protein-mediated interaction of lipid A from different origin with phospholipid membranes. Phys Chem Chem Phys 2 (2000) 4521–4528.

Holst, O. Chemical structure of the core region of lipopolysaccharides. In Brade, H., Opal, S.M., Vogel, S.N., Morrison, D.C. (eds), Endotoxin in Health and Disease, Marcel Dekker, Inc., New York, NY (1999), pp. 115–154.

Holst, O., Ulmer, A.J., Brade, H., Flad, H.D., Rietschel, E.T. Biochemistry and cell biology of bacterial endotoxins. FEMS Immunol Med Microbiol 16 (1996) 83–104.

Howe, J., Hammer, M., Alexander, C., Rossle, M., Fournier, K., Mach, J.P., Waelli, T., Gorczynski, R.M., Ulmer, A.J., Zahringer, U., Rietschel, E.T., Brandenburg, K. Biophysical characterization of the interaction of endotoxins with hemoglobins. Med Chem 3 (2007) 13–20.

Israelachvili, J.N., Intermolecular and surface forces. In Israelachvili, J.N. (ed), Academic, London (1991), pp. 366–394.

Jiao, B., Freudenberg, M., Galanos, C. Characterization of the lipid A component of genuine smooth-form lipopolysaccharide. Eur J Biochem 180 (1989) 515–518.

Kastowsky, M., Gutberlet, T., Bradaczek, H. Comparison of X-ray powder-diffraction data of various bacterial lipopolysaccharide structures with theoretical model conformations. Eur J Biochem 217 (1993) 771–779.

Kastowsky, M., Sabisch, A., Gutberlet, T., Bradaczek, H. Molecular modelling of bacterial deep rough mutant lipopolysaccharide of *Escherichia coli*, Eur J Biochem 197 (1991) 707–716.

Kitchens, R.L., Munford, R.S. CD14-dependent internalization of bacterial lipopolysaccharide (LPS) is strongly influenced by LPS aggregation but not by cellular responses to LPS. J Immunol 160 (1998) 1920–1928.

Loppnow, H., Flad, H.-D., Rietschel, E.Th., Brade, H. The active principle of bacterial lipopolysaccharides (endotoxins) for cytokine induction. In Schlag, G., Redl, H.(eds), Pathophysiology of Shock, Sepsis, and Organ. Springer-Verlag, Heidelberg (1993), pp. 405–416.

Loppnow, H., Libby, P., Freudenberg, M.A., Kraus, J.H., Weckesser, J., Mayer, H. Cytokine induction by lipopolysaccharide (LPS) corresponds to the lethal toxicity and is inhibited by nontoxic Rhodobacter capsulatus LPS. Infect Immun 58 (1990) 3743–3750.

Luzzati, V., Gulik, A., Gulik-Krzywicki, T., Tardieu, A. Lipid polymorphism revisited: Structural aspects and biological implications. Lipids and Membranes: Past, Present, and Future. Elsevier, Amsterdam (1986), pp. 137–151.

Maurer, N., Glatter, O., Hofer, M. Determination of size and structure of lipid IV a vesicles by quasi-elastic light scattering and small-angle X-ray scattering. J Appl Cryst 24 (1991) 832–835.

Mayer, H., Weckesser, J. 'Unusual' lipid A's: structures, taxonomical relevance and potential value for endotoxin research. In Rietschel, E.T. (ed), Handbook of Endotoxin, Vol.1: Chemistry of Endotoxin. Elsevier, Amsterdam (1984), p. 221.

Mueller, M., Brandenburg, K., Dedrick, R., Schromm, A.B., Seydel, U. Phospholipids inhibit lipopolysaccharide (LPS)-induced cell activation: a role for LPS-binding protein. J Immunol 174 (2005a) 1091–1096.

Mueller, M., Lindner, B., Dedrick, R., Schromm, A.B., Seydel, U. Endotoxin: physical requirements for cell activation. J Endotoxin Res 11 (2005b) 299–303.

Mueller, M., Lindner, B., Kusumoto, S., Fukase, K., Schromm, A.B., Seydel, U. Aggregates are the biologically active units of endotoxin. J Biol Chem 279 (2004) 26307–26313.

Naumann, D., Schultz, C., Sabisch, A., Kastowsky, M., Labischinski, H. New insights into the phase behaviour of a complex anionic amphiphile: Architecture and dynamics of bacterial deep rough mutant lipopolysaccharide membranes as seen by FTIR, X-ray, and molecular modelling techniques. J Molec Struct 214 (1989) 213–246.

Nomura, K., Inaba, T., Morigaki, K., Brandenburg, K., Seydel, U., Kusumoto, S. Interaction of lipopolysaccharide and phospholipid in mixed membranes: solid-state 31P-NMR spectroscopic and microscopic investigations. Biophys J 95 (2008) 1226–1238.

Reichelt, H., Faunce, C.A., Paradies, H.H. The phase diagram of charged colloidal lipid A-diphosphate dispersions. J Phys Chem B 112 (2008) 3290–3293.

Rietschel, E.Th., Brade, L., Schade, U., Seydel, U., Zähringer, U., Brandenburg, K., Helander, I., Holst, O., Kondo, S., Kuhn, H.M., Lindner, B., Röhrscheidt, E., Russa, R., Labischinski, H., Naumann, D., Brade, H. Bacterial Lipopolysaccharides: Relationship of Structure and Conformation to Endotoxic Activity, Seriological Specificity and Biological Function. In Friedmann, H., Klien, T.W., Nakano, M., Nowotny, A. (eds) Endotoxin, Plenum. New York, NY (1990), pp. 81–99.

Rietschel, E.Th., Galanos, C., Tanaka, A., Ruschmann, E., Lüderitz, O., Westphal O. Biological activities of chemically modified endotoxins. Eur J Biochem 22 (1971) 218–224.

Rietschel, E.Th., Seydel, U., Zähringer, U., Schade, U.F., Brade, L., Loppnow, H., Feist, W., Wang, M.-H., Ulmer, A.J., Flad, H.-D., Brandenburg, K., Kirikae, T., Grimmecke, D., Holst, O., Brade, H. Bacterial endotoxin: Molecular relationships between structure and activity. Infect Dis Clin North Am 5 (1991) 753–779.

Risco, C., Carrascosa, J.L., Bosch, M.A. Visualization of lipopolysaccharide aggregates by freeze-fracture and negative staining. J Electron Microsc 42 (1993) 202–204.
Sasaki, H., White, S.H. Aggregation behavior of an ultra-pure lipopolysaccharide that stimulates TLR-4 receptors. Biophys J 95 (2008) 986–993.
Schromm, A.B., Brandenburg, K., Rietschel, E.T., Flad, H.D., Carroll, S.F., Seydel, U., Schromm, A.B. Lipopolysaccharide-binding protein mediates CD14-independent intercalation of lipopolysaccharide into phospholipid membranes. FEBS Lett 399 (1996) 267–271.
Schromm, A.B., Howe, J., Ulmer, A.J., Wiesmüller, K.H., Seyberth, T., Jung, G., Rössle, M., Koch, M.H., Gutsmann, T., Brandenburg, K. Physicochemical and biological analysis of synthetic bacterial lipopeptides: validity of the concept of endotoxic conformation. J Biol Chem 282 (2007) 11030–11037.
Schromm, A.B., Brandenburg, K., Loppnow, H., Moran, A.P., Koch, M.H.J., Rietschel, E.Th., Seydel, U. Biological activities of lipopolysaccharides are determined by the shape of their lipid A portion. Eur J Biochem 267 (2000) 2008–2013.
Schromm, A.B., Brandenburg, K., Loppnow, H., Zähringer, U., Rietschel, E.Th., Carroll, S.F., Koch, M.H.J., Kusumoto, S., Seydel, U., The charge of endotoxin molecules influences their conformation and interleukin-6 inducing capacity. J Immunol 161 (1998) 5464–5471.
Schromm, A.B., Reiling, N., Howe, J., Wiesmuller, K.H., Roessle, M., Brandenburg, K. Influence of serum on the immune recognition of a synthetic lipopeptide mimetic of the 19-kDa lipoprotein from *Mycobacterium tuberculosis*. Innate Immun (2010) (in press).
Seydel, U., Hawkins, L., Schromm, A.B., Heine, H., Scheel, O., Koch, M.H., Brandenburg, K. The generalized endotoxic principle. Eur J Immunol 33 (2003) 1586–1592.
Seydel, U., Koch, M.H.J., Brandenburg, K. Structural polymorphisms of rough mutant lipopolysaccharides Rd to Ra from *Salmonella minnesota*. J Struct Biol 110 (1993) 232–243.
Seydel, U., Oikawa, M., Fukase, K., Kusumoto, S., Brandenburg, K. Intrinsic conformation of lipid A is responsible for agonistic and antagonistic activity. Eur J Biochem 267 (2000) 3032–3039.
Seydel, U., Schromm, A.B., Brade, L., Gronow, S., Andrä, J., Müller, M., Koch, M.H., Fukase, K., Kataoka, M., Hashimoto, M., Kusumoto, S., Brandenburg, K. Physicochemical characterization of carboxymethyl lipid A derivatives in relation to biological activity. FEBS J 272 (2005) 327–340.
Seydel, U., Ulmer, A.J., Uhlig, S., Rietschel, E.Th. Lipopolysaccharide, a membrane-forming and inflammation-inducing bacterial macromolecule. In Zimmer, G. (ed), Membrane Structure in Disease and Drug Therapy. Marcel Dekker Inc, New York, NY (1999), pp. 217–252.
Shands, J.W. The physical structure of bacterial lipopolysaccharides. In Weinbaum, G., Kadis, S., Ajl, S.J. (eds), Microbial Toxins, Vol. IV: Bacterial Endotoxins. Academic, New York, NY (1971), pp. 127–144.
Shnyra, A., Hultenby, K., Lindberg, A.A., Role of the physical state of *Salmonella* lipopolysaccharide in expression of biological and endotoxic properties. Infect Immun 61 (1993) 5351–5360.
Takayama, K., Din, Z.Z., Mukerjee, P., Cooke, P.H., Kirkland, T.N. Physicochemical properties of the lipopolysaccharide unit that activates B lymphocytes. J Biol Chem 265 (1990) 14023–14029.
Teghanemt, A., Prohinar, P., Gioannini, T.L., Weiss, J.P. Transfer of monomeric endotoxin from MD-2 to CD14: characterization and functional consequences. J Biol Chem 282 (2007) 36250–36256.
Tobias, P.S., Soldau, K., Iovine, N.M., Elsbach, P., Weiss, J. Lipopolysaccharide (LPS)-binding proteins BPI and LBP form different types of complexes with LPS. J Biol Chem 272 (1997) 18682–18685.
Toman, R., Garidel, P., Andrä, J., Slaba, K., Hussein, A., Koch, M.H.J., Brandenburg, K. Physicochemical characterization of the endotoxins from *Coxiella burnetii* strain Priscilla in relation to their bioactivities. BMC Biochem 5 (2004) 1.
Zähringer, U., Salvetzki, R., Ulmer, A.J., Rietschel, E.Th., Isolation, chemical analysis and biological investigations of natural lipid A and lipid A-antagonists. J Endotoxin Res 3 (Suppl. 1) (1996) 33.

Chapter 4
The Diversity of the Core Oligosaccharide in Lipopolysaccharides

Alba Silipo and Antonio Molinaro

Abstract Bacterial lipopolysaccharides (LPSs) are the major component of the outer membrane of Gram-negative bacteria. They have a structural role since they contribute to the cellular rigidity by increasing the strength of cell wall and mediating contacts with the external environment that can induce structural changes to allow life in different conditions. Furthermore, the low permeability of the outer membrane acts as a barrier to protect bacteria from host-derived antimicrobial compounds. Lipopolysaccharides are amphiphilic macromolecules generally comprising three defined regions distinguished by their genetics, structures and function: the lipid A, the core oligosaccharide and a polysaccharide portion, the O-chain. In some Gram-negative bacteria LPS can terminate with the core portion to form rough type LPS (R-LPS, LOS). The core oligosaccharide is an often branched and phosphorylated heterooligosaccharide with less than fifteen sugars, more conserved in the inner region, proximal to the lipid A, and often carrying non-stoichiometric substitutions leading to variation and micro-heterogeneity. The core oligosaccharide contributes to the bacterial viability and stability of the outer membrane, can assure the serological specificity and possesses antigenic properties.

Keywords Core oligosaccharide · Endotoxin · Innate immunity · Gram-negative bacteria · Glyco-conjugates

Abbreviations

LPS	lipopolysaccharide
LOS	lipooligosaccharide
OM	outer membrane
Kdo	3-deoxy-D-manno-oct-2-ulopyranosonic acid
Ko	D-glycero-α-D-talo-oct-2-ulopyranosonic acid

A. Molinaro (✉)
Dipartimento di Chimica Organica e Biochimica, Università di Napoli Federico II, Complesso Universitario Monte Santangelo, Via Cintia 4, Napoli I-80126, Italy
e-mail: molinaro@unina.it

X. Wang, P.J. Quinn (eds.), *Endotoxins: Structure, Function and Recognition*, Subcellular Biochemistry 53, DOI 10.1007/978-90-481-9078-2_4,

GlcN	glucosamine
L,D-Hep	D-glycero-D-manno-heptopyranose residue
Ara4N	4-deoxy-4-amino-arabinose
EtN	2-amino-ethanol
P	Phosphate
*PP*EtN	2-aminoethanol diphosphate
*P*EtN	2-aminoethanol phosphate
*P*Cho	2-trimethylaminoethanol phosphate
Ac	Acetyl
Cm	carbamoyl
Pyr	pyruvic acid
R	rough
S	smooth

4.1 Introduction

Both Gram positive and Gram negative bacteria possess a cytoplasmic membrane surrounding the cytosol, which constitutes a physical semi-permeable barrier that regulates the flux of endogenous and exogenous substances in and out the cell. This membrane is in turn enclosed by a layer named peptidoglycan or murein, responsible for the shape and the strength of the cell. Gram negative bacteria possess an additional asymmetric outer membrane (OM) which surrounds a thin layer of peptidoglycan. In these micro-organisms, the external membrane is the first and more immediate line of defence against harsh environment and antimicrobial molecules and makes the cytoplasmic membrane more efficiently protected. The outer membrane is composed of an asymmetric phospholipids bilayer, whose inner leaflet is made of glycerophospholipids while the external leaflet is formed by lipopolysaccharides (LPS) which cover approximately 75% of the outer surface. The remaining space is prevalently filled up by integral membrane proteins like porins, channels for the entrance and the exit of hydrophilic small molecules, and lipoproteins, possessing structural functions. In general, the OM increases the permeability barrier to hydrophobic compounds and higher molecular weight hydrophilic compounds owing to the very low fluidity of the highly ordered structure of the LPS monolayer.

Lipopolysaccharides are heat-stable complex amphiphilic macromolecules indispensable for the bacterial growth, viability and for the correct assembly of the external membrane; they represent a defensive barrier which helps bacteria to resist to antimicrobial compounds and environmental stresses and are involved in many aspects of host–bacterium interactions as recognition, adhesion, colonization, and, in the case of extremophile bacteria, in the survival under harsh conditions (Alexander and Rietschel, 2001; Raetz and Whitfield, 2002; Raetz et al., 2007). LPS are also called endotoxins because they are cell-bound and, once released, play a key role in the pathogenesis of Gram-negative infections, in mechanisms as

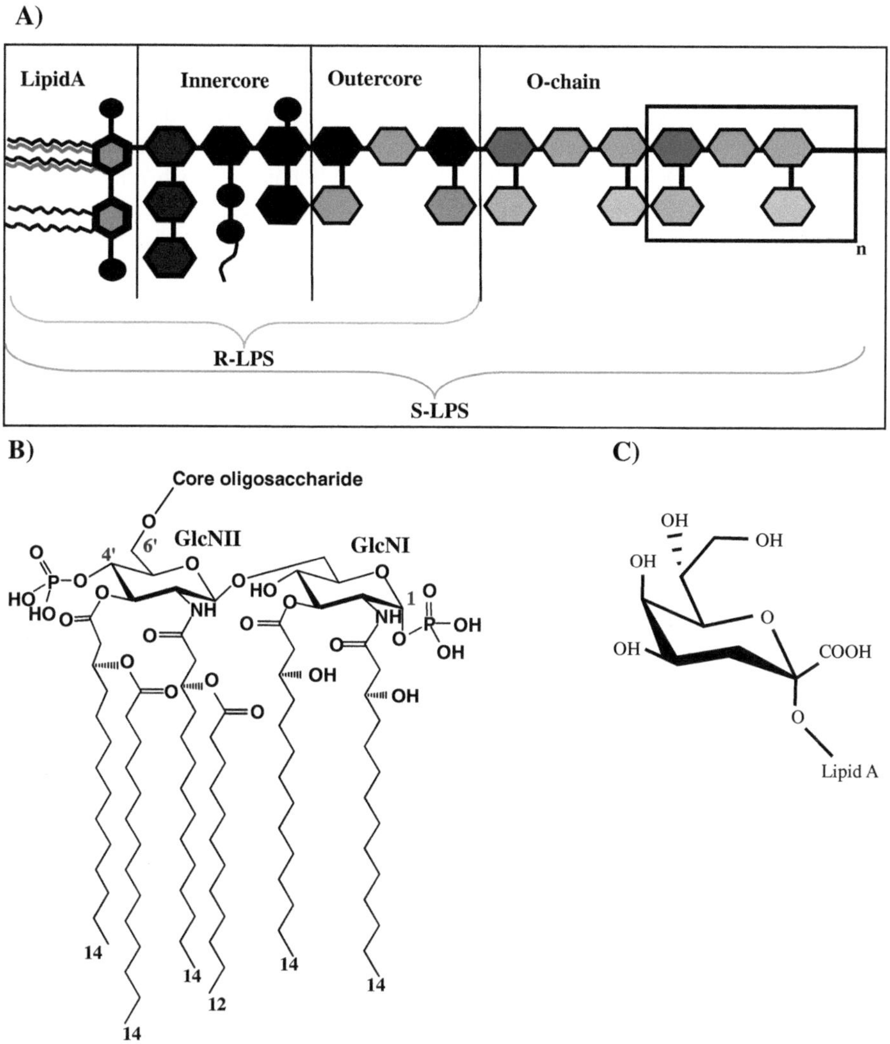

Scheme 4.1 (**a**) Structure of the lipopolysaccharides of Gram-negative bacteria. (**b**) The structure of the lipid A from *E. coli.* (**c**) The structure of Kdo residue (α-3-deoxy-D-manno-oct-2-ulopyranosonic acid). For all structures, where not stated otherwise, sugars are α-D-pyranosides. Residues in bold are present as non-stoichiometric substitutions. Common abbreviations: P, Phosphate, *PP*Etn, 2-aminoethanol diphosphate, *P*EtN, 2-aminoethanol phosphate, *P*Cho, 2-trimethylaminoethanol phosphate; Gly, glycine, Ac, Acetyl, Cm, Carbamoyl, Pyr, Pyruvic Acid

virulence, tolerance for commensal bacteria and symbiosis. In mammalian hosts, they can trigger the activation of both the innate and the adaptative immune system.

Lipopolysaccharides are biosynthesized according to a common structural architecture (Scheme 4.1). They are composed of a hydrophilic hetero-polysaccharide (formed by core oligosaccharide and O-specific polysaccharide or O-chain) covalently linked to a lipophilic domain termed *Lipid A*, which is embedded in the outer leaflet and anchors these macromolecules to the membrane. The LPS leaflet is stabilized by electrostatic interaction of the negatively charged groups present in the lipid A and in the core region (phosphate groups, uronic acids) with divalent cations (Mg^{2+}, Ca^{2+}) which contribute to link the LPS molecules to each other. While a typical glycerophospholipid bilayer is a flexible and fluid system, the LPS layer is a quasi-crystalline and semi-rigid structure with a highly ordered structure and low fluidity. The rigidity of the lipid A-inner core saccharide backbone, the tight packing of the fatty acid residues together with the presence of such strong electrostatic interactions make the LPS layer such a highly structured barrier. The low permeability of the Gram-negative external membrane explains their lower susceptibility to hydrophobic molecules and/or negatively charged antibiotic than Gram-positive microorganism.

The three LPS domains are genetically, biologically and chemically distinct. When present, the O-antigenic polysaccharide confers a smooth appearance to the colony on agar plates and, in this case, lipopolysaccharides are defined of Smooth type or S-LPS. However, some enteric bacteria in particular growth conditions covalently attach the enterobacterial common antigen (ECA, a polysaccharide capsule) or colanic acid (CA, an exopolysaccharide) to the outer core region (Whitfield, 2006; Meredith et al., 2007). Thus, the general architecture of smooth LPS can be defined differently: polysaccharide, core region and lipid A, where for polysaccharide it is meant the O-chain, a capsule such as ECA or an exopolysaccharide as CA. The absence of O-chain gives to the colony a rough aspect and the LPS is then defined of Rough (R)-type or lipooligosaccharide (LOS). LPS not containing O-chain may occur in both wild and laboratory strains possessing mutations in the genes encoding the O-specific polysaccharide biosynthesis or transfer. It has been demonstrated that laboratory deep-rough mutants are able to survive in vitro, and thus the core region and lipid A represent a common structural unit occurring in all LPS important for viability and membrane correct assembly and function. Moreover, in tissues or body fluid such pathogenic bacteria can only survive expressing the polysaccharide O-chain, which protect bacteria from the host environment. Nevertheless, many highly virulent (pathogenic) Gram-negative bacteria posses a R-form LPS: *Neisseria meningitidis*, *Neisseria gonorrhoeae*, *Haemofilus influenzae*, *Bordetella pertussis*, *Campylobacter jejuni*, and several other opportunistic pathogen as *Pseudomonas* and *Burkholderia*. In these species the outer core oligosaccharide is often branched, contributes to the viability and to the membrane function and its variability (like the O-chain in Smooth LPS) assures the serological specificity.

The O-chain is the hydrophilic portion of LPS and renders this macromolecule water soluble and immunogenic (see below). This polysaccharide fraction is more exposed to the selective pressures of the outer environment and to modification induced by external stimuli. It is the most variable portion of LPS and is characterised by a high specificity within a species. The O-chain has different roles, the most important appears to be protective, acting as a defensive barrier and is indispensable in vivo: in tissues or body fluids smooth-type pathogenic bacteria can only persist if they express an O-specific polysaccharide. It has also important functions in the microbial adhesion to the host cells mediated by cell-surface molecules called adhesines (tropism). The O-polysaccharide is composed of up to 50 repeating oligosaccharide units each formed by one to ten sugar residues. This polymer is highly variable and contains different kinds of monosaccharide which can also be very unusual. The O-repeating unit can be homopolymeric or heteropolymeric, linear or branched. The addition of non-carbohydrate substituents like phosphate, amino acids, acetyl or formamide groups, often present in non stoichiometric amount, can complicate the definition of the repeating oligosaccharide.

Lipid A (Scheme 4.1) possesses a rather conservative structure usually consisting of a β-(1→6)-glucosamine disaccharide backbone (GlcNI and GlcNII are the reducing and non reducing unit, see Scheme 4.1) phosphorylated at positions 1 and 4' of GlcNI and GlcNII and acylated with 3-hydroxy fatty acids at positions 2 and 3 of both GlcNs via amide and ester linkages. These primary acyl chains are directly linked to the sugar backbone and are further acylated by their hydroxy groups by secondary acyl moieties. 3-deoxy-D-manno-oct-2-ulopyranosonic acid (Kdo), or derivative of this sugar, is linked to the non reducing glucosamine (GlcNII) of lipid A backbone at position 6' (Scheme 4.1). GlcNs disaccharide carrying a β-(1→4) glycoside linkage is widespread in nature, i.e., in the glycoproteins, chitin and membrane peptidoglycans. Apart from such a general chemical architecture conserved in many bacterial LPS, a number of subtle chemical differences have been identified that are responsible for lipid A variation among bacterial LPS.

The core oligosaccharide of LPS can be built of several different monosaccharides, which can be arranged giving either a linear or a branched architecture, and can be composed of up to fifteen sugars. It is possible to identify an inner and an outer core moiety. The inner core is less variable and its composition is normally characteristic within a *genus* or family; the linkage to the lipid A occurs between position 6' of the GlcNII (Scheme 4.1) and the first residue of the core region, e.g the α-Kdo residue (α-3-deoxy-D-manno-oct-2-ulopyranosonic acid). The Kdo is the only monosaccharide residue always present in the inner core of bacterial lipopolysaccharide and therefore is a chemical hallmark of LPS and a marker of Gram-negative bacteria. In *Acinetobacter haemolyticus* LPS (Holst, 2002) the Kdo residue is replaced by the D-glycero-α-D-talo-oct-2-ulopyranosonic acid (Ko) whereas in LPS of *Shewanella algae* BrY (Vinogradov et al., 2004) is replaced by 8-amino-3,8-dideoxy-α-D-manno-oct-2-ulopyranosonic acid (8-amino-Kdo, Kdo8N).

The main core oligosaccharide chain is usually attached at O-5 of Kdo residue and, in many structures reported so far, the first sugar linked to the Kdo unit has typically a manno configuration such as the L-glycero-D-manno-heptopyranose residue (L,D-Hep) or the less prevalent D,D-Hep which is its biosynthetic precursor. Some genera, as *Francisella, Legionella, Acinetobacter, Chlamydia, Moraxella, Xanthomonas*, *Agrobacterium* and *Rhizobium* produce core oligosaccharides lacking heptoses in which different residues can be attached to O-5 of KdoI[1]: as Man (largely), but also Glc, Gal or GalA (Silipo et al., 2005a; De Castro et al., 2008).

In enterobacterial LPS this first Kdo unit bears at its O-5 position a heptose trisaccharide fragment: α-L,D-Hep*p*III-(1-7)-α-L,D-Hep*p*II-(1-3)-α-L,D-Hep*p*I; Kdo always carries another negatively charged substituent at its O-4, generally a second α-Kdo unit, a phosphate group or a Ko residue (Holst, 2007; Silipo et al., 2004a). Beside lipid A, Ara4N (4-deoxy-4-amino-arabinose) residue is also present in the inner core region of LPS, in *Proteus* and *Serratia* LPSs it is attached to KdoII while in *Burkholderia* LPS, it is always present as terminal non reducing residue of the trisaccharide Ara4N-(1-8)-α-Ko-(2-4)-α-KdoI. In particular, the finding of Ara4N in such chemical arrangement in *Burkholderia* LPS represents a chemio-taxonomic hallmark for this genus, shared by all *Burkholderia* strains, either animal or plant pathogen (Molinaro et al., 2003; Silipo et al., 2005b; De Soyza et al., 2008). The inner core region can be decorated with negatively charged substituents, often present in non stoichiometric amount, like phosphate (P), pyrophosphate (PP), pyrophosphoryl-2-amino-ethanol (PPEtN), phospho-arabinosamine (PAra4N), uronic acids (often Gal*p*A) such as α-galacturonyl phosphate and in a single case a phosphoramide group. Often the anionic groups are present in close proximity and are involved in guaranteeing the correct membrane assembly. Their accumulation in the lipid A-core region is stabilized by divalent cations that connect LPS molecules to each other. These electrostatic interactions contribute to reduce the membrane permeability and to enhance its stability with the formation of a strong, rigid and protective barrier. It has been found that the minimal structural requirement for bacterial growth and viability is the tetra-saccharide formed by Kdo_2-lipid A which is a biosynthetic intermediate acting as the acceptor of the other sugar residues of the core region even though recently it has been shown that, for an *E. coli* mutant, the minimum viable structure is a tetracylated bis-phosporylated lipid A named lipid IVa.

The outer core region is more variable being the part more exposed towards the external pressures. It is normally referred to as the hexoses region and is usually composed of neutral and basic hexoses (Glc*p*, Gal*p*, GlcN*p*, and GalNp). Exceptionally, in *Acinetobacter*, *Rhizobium* and *Agrobacterium* strains the 3-deoxy-D-lyxo-hept-2-ulosaric acid has been found, whereas in *Proteus* and in *Shewanella* LPS, a new kind of glycosydic linkage was discovered that involves an open chain acetal linkage of a GlcN residue that is present as non cyclic carbonyl form (Vinogradov et al., 2003; Vinogradov and Bock, 1999).

The core structure can strongly be influenced by the physiological environment. Under strong stress conditions, LPS can modify its primary structure in order to reinforce the external membrane and to assure the bacteria a best protection. Thus, groups like 2-amino-ethanol (EtN), Ara4N, GlcN, present in the lipid A-core region,

[1]Here and further on, KdoI is meant as the first Kdo of the core region; the branched Kdo residues will be named KdoII and KdoIII. As for Heptose residues, HepI is the Heptose residue directly linked to Kdo; HepII is linked to the first Hep in the main sugar backbone. Branched Hep residues have the highest numbering.

which are positively charged under physiological conditions, can substitute divalent cations in the interaction with the neighborhood phosphate groups. In this way, the entrance of positively charged macromolecules is not allowed. Antibacterial cationic peptides and antibiotic molecules try to penetrate the external membrane substituting the divalent cations present in the outer leaflet (see above the case of *Burkholderia*).

Many highly virulent human pathogen bacteria can make themselves invisible to the host and evade the immune response. The O-antigen of *Helicobacter pylori* (Amano, 2002), the major cause of chronic gastritis and ulcers, expresses Lewis antigenic epitopes, mainly Le^x and Le^y at the non-reducing end unit. This molecular mimicry phenomenon occurs in human pathogen bacteria possessing antigenic determinants chemically indistinguishable from the host's self components. Such bacteria become able to mask themselves to the host immune system, and the microbial products are not recognised as non-self component. The bacterium can thus escape the host immune defence mechanisms, including opsonization and cell damage. The phenomenon of the molecular mimicry has been extensively found also in rough bacteria, as in *N. meningitis* and *H. influenza* (Holst, 2007, 2002; Lerouge and Vanderleyden, 2002), both causative agents of the bacterial meningitis. Their antigenic determinant is constituted by the core oligosaccharide. These invading microorganisms are able to covalently link to their LOS structures sialic acid residues thus mimicking the host cell surface glycoconjugates and avoiding the attack of the MAC on the outer membrane.

4.2 Core Structures

We will discuss below the main classes of core oligosaccharide structures (Holst, 2002, 2007; Holst and Molinaro, 2009) and in addition to these, the structure of the novel of core oligosaccharide published after the year 2007 and not yet included in the earlier reviews.

4.2.1 Enterobacteria

Core oligosaccharide structure from *Enterobacteriaceae* (*Salmonella*, *Klebsiella*, *Escherichia coli*, *Shigella*, *Yersinia*, *Plesiomonas*, *Proteus*, *Hafnia*, *Providencia*) are the first structures fully characterised (Holst, 2002, 2007). In the common structural theme identified in *Enterobacteriaceae* (Holst, 2002, 2007; Olsthoorn et al., 1998) the inner core structure is characterized by a L,D-Hep-(1→7)-L,D-Hep-(1→3)-L,D-Hep-(1→5)-[Kdo-(2→4)]-Kdo sequence that, depending on the *genera*, can be further and differently substituted (Fig. 4.1); usually, core structure are divided into *Salmonella* and non-*Salmonella* group depending on the phosphorylation and glycosylation pattern principally of the Hep residues. In Fig. 4.1 the main and more important core structures from *Salmonella* and non *Salmonella* group have been

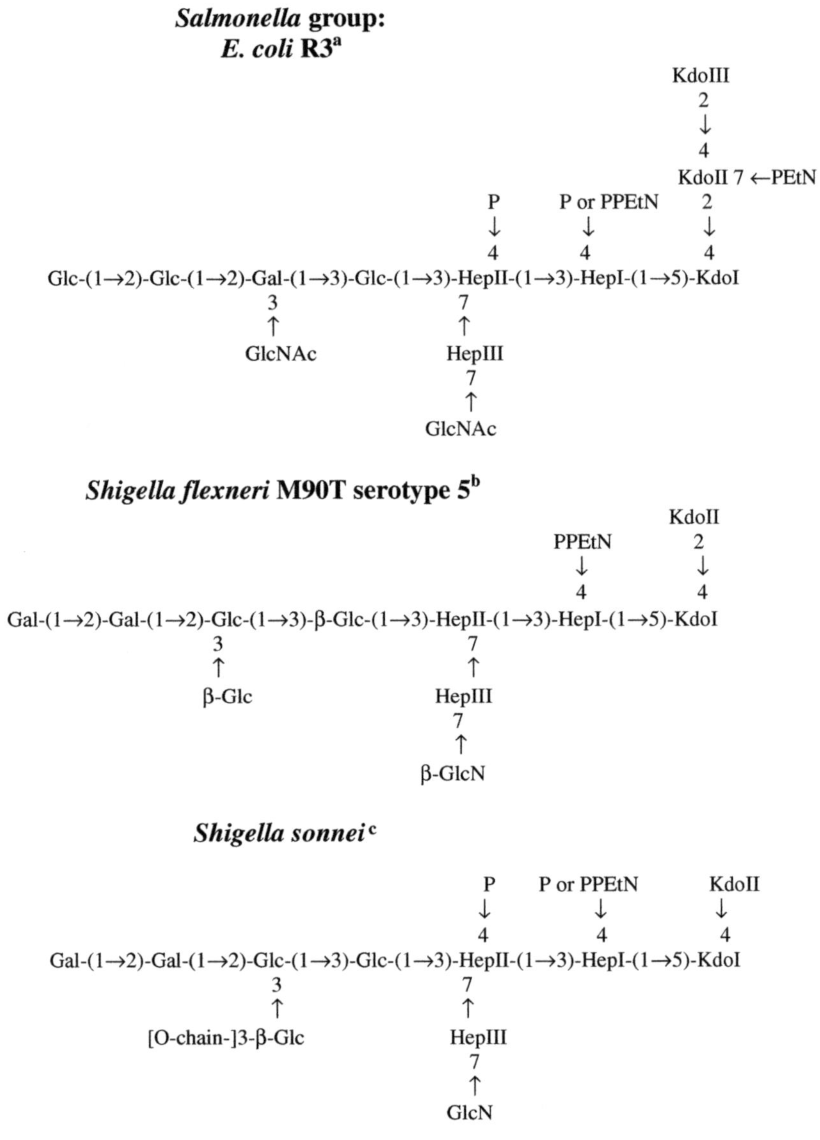

Fig. 4.1 (continued)

***Citrobacter werkmanii* PCM 1548 and 1549**[d]

PPEtN
↓
4
GlcNAc-(1→3)-Gal-(1→4)-Glc-(1→3)-HepII-(1→3)-HepI-(1 →5)-KdoI
2 7
↑ ↑
Gal HepIII
2 7
↑ ↑
Glc β-GalA3PEtN
and
PPEtN
↓
4
Glc-(1→4)-β-Gal-(1→6)-Glc-(1 →2)- Glc-(1 →4)-Glc-(1 →3)-HepII-(1 →3)-HepI-(1 →3)-KdoI
7
↑
HepIII
7
↑
β-GalA3PEtN

Yersinia pestis[e]

KdoII or Ko
β-Glc 2
↓ ↓
4 4
β-GlcNAc-(1→3)-HepII-(1→3)-HepI-(1→5)-KdoI
7
↑
D,D-HepIII or β-Gal

Non Salmonella group:
Plesiomonas shigelloides[f]

KdoII or Ko
β-**Glc** β-Gal 2
↓ ↓ ↓
2 4 4
GlcN-(1→4)-GalA-(1→3)-HepII-(1→3)-HepI-(1→5)-KdoI
7
↑
HepIII
7
↑
GalA
4
↑
GlcN

Fig. 4.1 (continued)

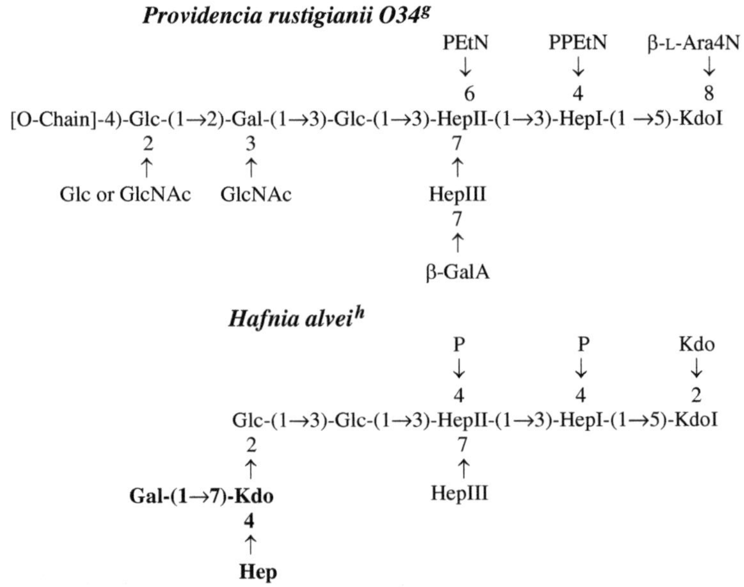

Fig. 4.1 Examples of core region structures of enterobacterial LPS. [a]Heinrichs et al., 1998; [b]Molinaro et al., 2008; [c]Robbins et al., 2009; [d]Kondakova et al., 2009; [e]Holst, 2007; [f]Pieretti et al., 2009; [g]Kondakova et al., 2008b; [h]Lukasiewicz et al., 2009

reported. The core oligosaccharide in *E. coli, Salmonella* (Olsthoorn et al., 1998; Heinrichs et al., 1998; Olsthoorn et al., 2000) and *Shigella* (Robbins et al., 2009; Molinaro et al., 2008) share a common chemical architecture (Fig. 4.1): the first Kdo unit (KdoI) bears at its O-5 position a heptose trisaccharide fragment and at O-4 always a negatively charged substituent, generally a further α-Kdo unit which is in turn substituted by neutral sugar (L-Rha, Gal) or by a third Kdo. Both Hep residues are substituted by phosphate groups (PPEtN) at O-4 and HepII at O-3 is substituted by the first sugar of the outer core, a 3,6 substituted Glc (in the *Salmonella* group).

Bacteria of the genus *Citrobacter* from the family *Enterobacteriaceae* are inhabitants of the intestinal tract and are present in sewage, surface waters, and food contaminated with fecal material. In accordance with close phylogenetic relationship to *Salmonella, Citrobacter* has a so-called *Salmonella*-type enterobacterial LPS core (inner Hep_3Kdo_2 pentasaccharide with both HepI and HepII phosphorylated at O-4 and HepII substituted at O-3 with a Glc residue). Two core structures recently fully characterized (*C. werkmanii* PCM 1548 and PCM 1549) (Kondakova et al., 2009) also contain a GalA residue linked at position 7 of HepIII, which is known to occur in both *Salmonella*-type (*Providencia*, see below) and non-*Salmonella*-type core (*Proteus*, Holst, 2002, 2007).

In species whose core structure belongs to the non *Salmonella* group, e.g. *Yersinia*, *Proteus*, (Holst, 2002, 2007), *Plesiomonas* (Pieretti et al., 2009) both Hep residues are not phosphorylated at O-4, HepI is substituted at O-4 by a β-configured hexose. For what concerns the outer core, HepII at O-3 carries a GlcNAc in *Yersinia* (Holst, 2002, 2007) or a GalA in *Plesiomonas* (Pieretti et al., 2009). The genera *Proteus*, *Providencia*, and *Morganella* are a unique group of Gram-negative bacteria in the family *Enterobacteriaceae*; enterobacteria of the genus *Providencia* are opportunistic pathogens causing diarrhea in travelers and children (Kondakova et al., 2006). Core oligosaccharide from three selected strains of *Providencia* (*P. alcalifaciens O8, P. alcalifaciens O35,* and *P. stuartii O49*) are the first characterized (Kondakova et al., 2006) (Fig. 4.1). Data showed intermediate position of the genus *Providencia* with respect to the core oligosaccharide structures since it shares some features with the so-called *Salmonella* core-type group, which is distinguished by phosphorylation of HepI at position 4 with PPEtN (in the non-*Salmonella* group this position is occupied by a β-Glc) and by the glycosylation of position 3 of HepII by a α–Glc (in the non-*Salmonella* group there is a different sugar). However, the *Providencia* core resembles also the one from *Proteus* belonging to the non-*Salmonella* group, because of the glycosylation of Kdo at position 8 with β-Ara4N, phosphorylation of HepII at position 6 with PEtN, and substitution of HepIII at position 7 with β-GalA. Other lipopolysaccharide from *Providencia* have been isolated and characterised (Kondakova et al., 2007, Kocharova et al., 2008a, b). The one from *Providencia rustigianii* O34, for instance, is decorated with PEtN on HepI and HepII. The former has been reported in the LPS core of *Salmonella*, *E. coli*, and some other enteric bacteria of the so-called *Salmonella* group; HepI with PEtN is the obligatory component of the *Proteus* LPS core from the non-*Salmonella* group. Hence, with respect to the phosphorylation pattern of the inner core, *Providencia* shares peculiar features with both the *Salmonella* and non-*Salmonella* group.

Hafnia alvei is a commensal of the human gastrointestinal tract and not normally pathogenic, but it may cause disease in immunocompromised patients as opportunistic pathogen associated with mixed hospital infections, bacteremia, septicemia, and respiratory diseases. *H. alvei* is often resistant to multiple antibiotics including the Aminopenicillins. LPSs of *H. alvei* are yet another example of enterobacterial endotoxins deviating from the classical scheme of LPS by the presence of an additional Kdo residue in the outer-core region (present in several strains) (Lukasiewicz et al., 2009). The dodecasaccharide from *H. alvei* PCM 32 representing the complete structure of the core contained a →4,5)-Kdo*p* residue in the inner-core region substituted by terminal α-Kdo*p* and the trisaccharide [D-α-D-Hep*p*-(1→4)-[α-D-Gal*p*6OAc-(1→7)]-α–Kdo*p*-(2→] as an integral part of the outer-core OS.

4.2.2 *Pasteurellaceae*

The family of *Pasteurellaceae* (γ-proteobacteria) comprises genera *Haemophilius, Histophilus* and *Pasteurella* and *Histophilus somnus*. Structural studies of LPS from *H. influenzae* have identified a conserved core structure containing the following common inner-core region L-α-D-Hep*p*-(1→2)-[*P*EtN→6]-L-α-D-Hep*p*-(1→3)-[β-D-Glc*p*-(1→4)]-L-α-D-Hep*p*-(1→5)-[*PP*-EtN→4)]-α-Kdo. The inner core region can also carry non-carbohydrate substituents as phosphate (P), 2-aminoethanol phosphate (PEtN), 2-aminoethanol diphosphate (PPEtN), phosphocholine (PCho), Aceyl groups, and glycine (Gly). The outer-core region is much more variable and can mimic host glycolipids. Both the β-D-Glc and the HepIII residues can be further substituted. Structural studies of LPS from various non-typeable, *H. influenzae* strains have revealed that the β-D-Glcp of the inner-core moiety can be glycosylated at both O-4 and O-6 in the same glycoform. The hexose linked to O-4 can be either a β-D-Glc or a β-D-Gal, which can be further extended (Holst, 2007, 2002). The O-6 of Glc has been found substituted by heptose containing extensions [β-D-Gal*p*-(1→4)-D-α-D-Hep*p*-(1→], [β-D-Glc*p*-(1→4)-D-α-D-Hep*p*-(1→] and [β-D-Gal*p*NAc-(1→3)-β-D-Gal*p*-(1→4)-β-D-Gal*p*-(1→6)-L-α-D-Hep*p*-(1→], or truncated versions (Lundström et al., 2008; Schweda et al., 2008). Lipopolysaccharide structures expressed by nontypeable *Haemophilus influenzae* R2846, a strain whose complete genome sequence has been recently reported. A β-D-Glc*p*-(1→4)-D-α-D-Hep*p*-(1→6)-β-D-Glc*p*-(1→4) unit was found linked to the proximal heptose (HepI) of the conserved inner core. O-4 of Glc I was substituted with sialyllacto-*N*-neotetraose [α-Neu5Ac-(2→3)-β-D-Gal*p*-(1→4)-β-D-Glc*p*NAc-(1→3)-β-D-Gal*p*-(1→4)-β-D-Glc*p*-(1→] and the related structure [(*P*EtN→6)-α-D-Gal*p*NAc-(1→6)-β-D-Gal*p*-(1→4)-β-D-Glc*p*NAc-(1→3)-β-D-Gal*p*-(1→4)-β-D-Glc*p*-(1→]. The distal heptose (HepIII) was substituted at O-2 by β-D-Gal; substituents as phosphate, PEtN, PPEtN, PCho, Ac, and Gly were found to substitute the core oligosaccharide. In other recently characterized non typeable *Haemophilus influenzae*, strains 1268 and 1200 (Lundström et al., 2007), HepIII carries globotetraose [β-D-Gal*p*NAc-(1→3)-α-D-Gal*p*-(1→4)-β-D-Gal*p*-(1→4)-β-D-Glc*p*], its truncated versions globoside [α-D-Gal*p*-(1→4)-β-D-Gal*p*-(1→4)-β-D-Glc*p*] and lactose [β-D-Gal*p*-(1→4)-β-D-Glc*p*]; α-D-Glc*p* linked to the middle heptose HepII can be further elongated at O-4 by a β-D-Gal*p* and not by globotriose or truncated analogues as previously found (Mansson et al., 2001) (Fig. 4.2). Previously only either the proximal heptose (HepI) or HepIII were found substituted by these motifs.

The core region from three strains of *Pasteurella multocida* (Harper et al., 2007; St. Michael et al., 2005a, b, c) have been characterised so far (Fig. 4.2). They are all characterised by the simultaneous presence of two glycoforms, of which only one is required for virulence, that shared the same oligosaccharide sequence: L-α-D-Hep*p*-(1→2)-L-α-D-Hep*p*-(1→3)-[β-D-Glc*p*-(1→4)]-[α-D-Glc*p*-(1→6)]-L-α-D-Hep*p*-(1→5)-α-Kdo but differed by substitution at Kdo and HepI. In fact, Kdo could be alternatively substituted at O-4 by a second Kdo unit or by a PEtN residue and, simultaneously, at O-6 of HepI (Fig. 4.2). It has been recently demonstrated the presence of a two acceptor-specific Heptosyl I transferases that add the first heptose residue to the nascent LPS molecule and of both a bifunctional Kdo transferase and a Kdo kinase, which results in the initial assembly of two inner core structures (Harper et al., 2007).

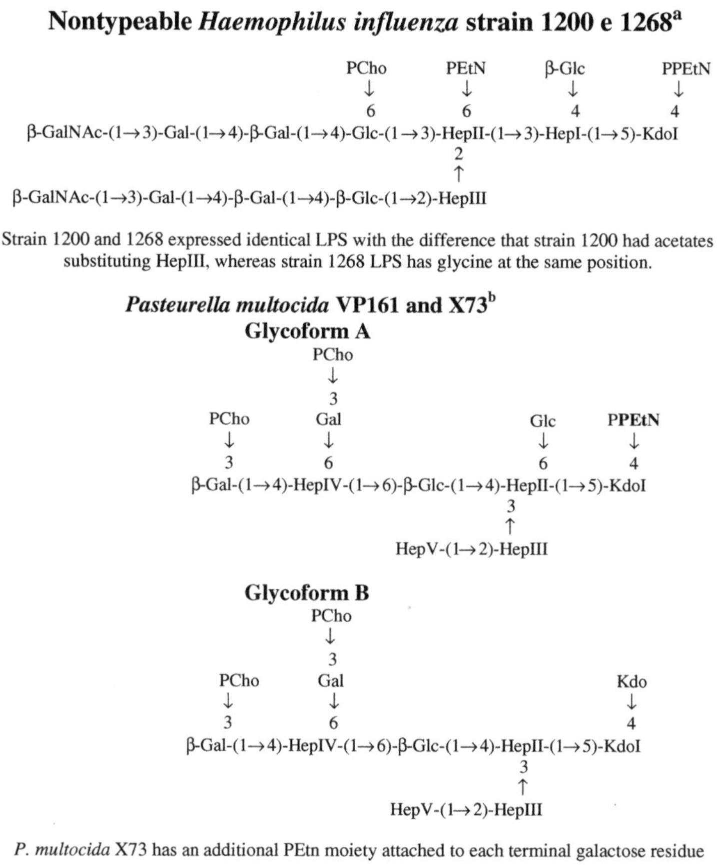

Fig. 4.2 Examples of core regions from *Pasteurellaceae*. [a]Lundström et al., 2007; [b]Harper et al., 2007

4.2.3 Pseudomonas

The core region of *Pseudomonas* is normally highly phosphorylated and appendages such as phosphate, pyrophosphate, PEtN can be found. The heptose region is often substituted by carbohydrate residues. Such substituents are often present in non stoichiometric amounts, in this way contributing to the core structural heterogeneity. In the inner core, a number of chemical peculiarities were found, as the presence of the unique carbamoyl phosphate group substituting O-6 of the second heptose and also the presence of a tri-phosphate group attached to Hep I (Knirel et al., 2006). In few cases, the biosynthetic precursor of L,D-Hep*p*, the D-glycero-D-manno-heptose (D,D-Hep) can be also found in the outer core region (Holst 2002, 2007). Often the Gal*p*N linked at O-3 of HepII residue is amidated by alanine (in few cases, it is present as Gal*p*NAc); two Kdo residues are present in the

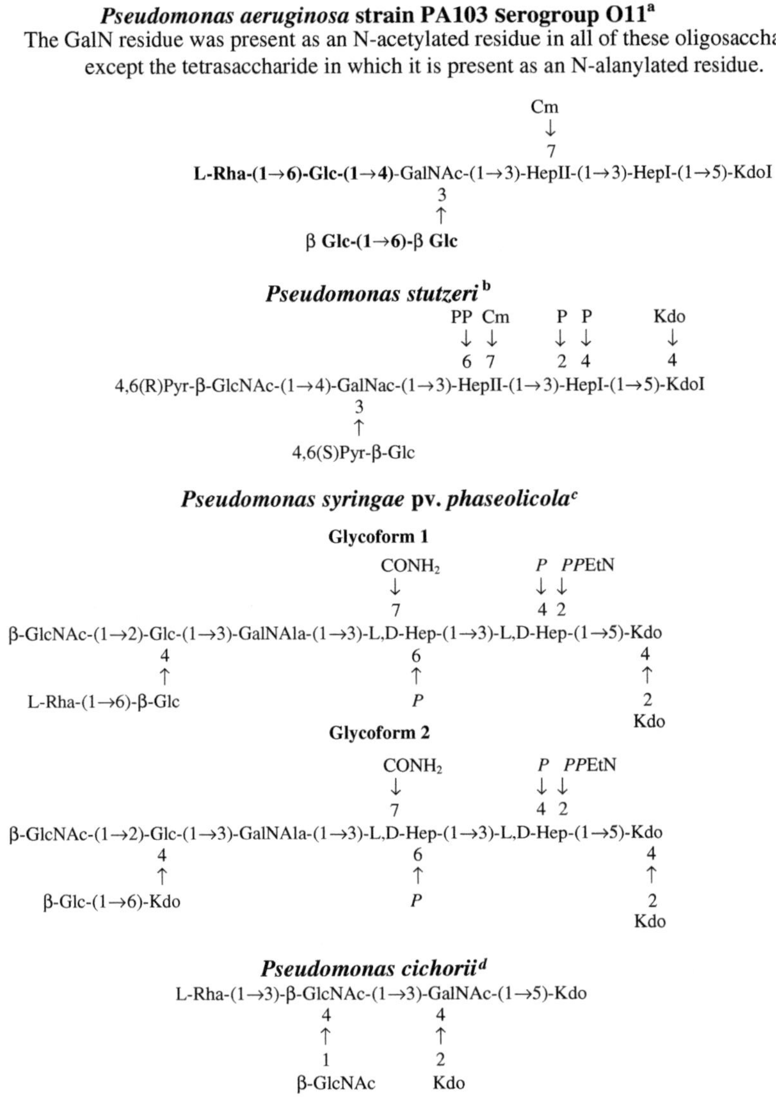

Fig. 4.3 Core regions from *Pseudomonas*. [a]Choudhury et al., 2008; [b]Leone et al., 2004a; [c]Holst, 2007; [d]De Castro et al., 2004

inner core region. Figure 4.3 presents some examples of related structures. The lipopolysaccharide (LPS) of a wbjE mutant of *P. aeruginosa* PA103, a serogroup O11, contained a variety of oligosaccharides, each with two or three phosphate groups present as mono- or pyrophosphates. Similar structures were identified in the core region of *P. stutzeri* OX1 (Leone et al., 2004a, b), no alanine was present but two 4, 6-O-(1-carboxy)-ethylidene residues in the outer core region were found. In contrast, *P. cichorii* LPS core region does not contain any phosphate group, does not contain heptose and a *galacto* configured residue directly binds the Kdo (Fig. 4.3) (De Castro et al., 2004). *P. cichorii* belongs to the RNA group I of *Pseudomonadaceae*, as *P. aeruginosa*, *P. stutzeri* and *P. syringae*.

4.2.4 *Acinetobacter*

The Gram-negative bacterium *Acinetobacter*, belonging to the family of Moraxellaceae, is isolated from soil and water which represent its natural habitats. However, the genus, non pathogenic to healthy individuals, has gained increasing importance as a key source of severe, life-threatening nosocomial infections in immunocompromised patients. *A. baumannii* is among the most frequently isolated species cause of nosocomial pneumonia. The core region of *Acinetobacter* LPS possesses particular structural features since it belongs to the group of heptose-deficient core regions and may contain D-glycero-D-talo-oct-2-ulopyranosonic acid (Ko) which can replace the 3-deoxy-D-manno-oct-2-ulopyranosonic acid (Kdo) residue linking the core region to lipid A. The core region of *A. baumannii* strain NCTC 10303 (Vinogradov et al., 1997a) is devoid of Ko but comprises the tetrasaccharide [α-Kdo-(2→5)-[α-Kdo-(2→4)-]-α-Kdo-(2→5)-α-Kdo-(2→] of which KdoIV is substituted at O-8 by a rhamnan short-chain and KdoIII at O-4 by the disaccharide α-D-Glc*p*NAc-(1→4)-α-D-Glc*p*NA. The core region from *A. baumannii* strain ATCC 19606 (Vinogradov et al., 2002a) contains the branched tetrasaccharide common to several *Acinetobacter* species and composed by three Kdo units and a GlcN: [β-GlcN*p*-(1→7)-α-Kdo-(2→5)-[α-Kdo-(2→4)-]α-Kdo-(2→] and was characterized by a mixture of oligosaccarides a differing in length and for the acetylation of the GalN unit (Fig. 4.4). More recently, the structure of the core oligosaccharide from *Acinetobacter radioresistens* S13, an organic solvent tolerant bacterium, was investigated (Leone et al., 2006); it possesses the branched tetrasaccharide in the inner core region and Glc*p*-rich outer core region.

Acinetobacter lwoffii F78 (Hanuszkiewicz et al., 2008) contains a 4,5,8-tri-substituted Kdo residue that, up to now, was described only in LPS of *Proteus mirabilis, Proteus penneri*, and in *Serratia marcescens*. Furthermore, the first Kdo has been found carrying two ulosonic acid residues [α-Kdo-(2→8)-α-Kdo] in an arrangement which so far has been identified only in *Chlamydial* LPS, as *Chlamydia trachomatis,* that produces a deep-rough LPS characterized by a linear sequence of three Kdo residues: α-Kdo-(2→8)-α-Kdo-(2→4)-α-Kdo-(2→6)-lipid A; whereas in *C. psittaci* a non-stoichiometric fourth Kdo is located at O-4 of the second one. In serological investigations, the anti-chlamydial LPS monoclonal antibody S25-2, which is specific for the epitope α-Kdo-(2→8)-α-Kdo, reacted with *A. lwoffii* F78 LPS. Thus, an LPS was identified outside *Chlamydiaceae* that contains a *Chlamydia*-specific LPS epitope in its core region suggesting an explanation for the cross-reactivity between *Chlamydia* and *Acinetobacter* described earlier.

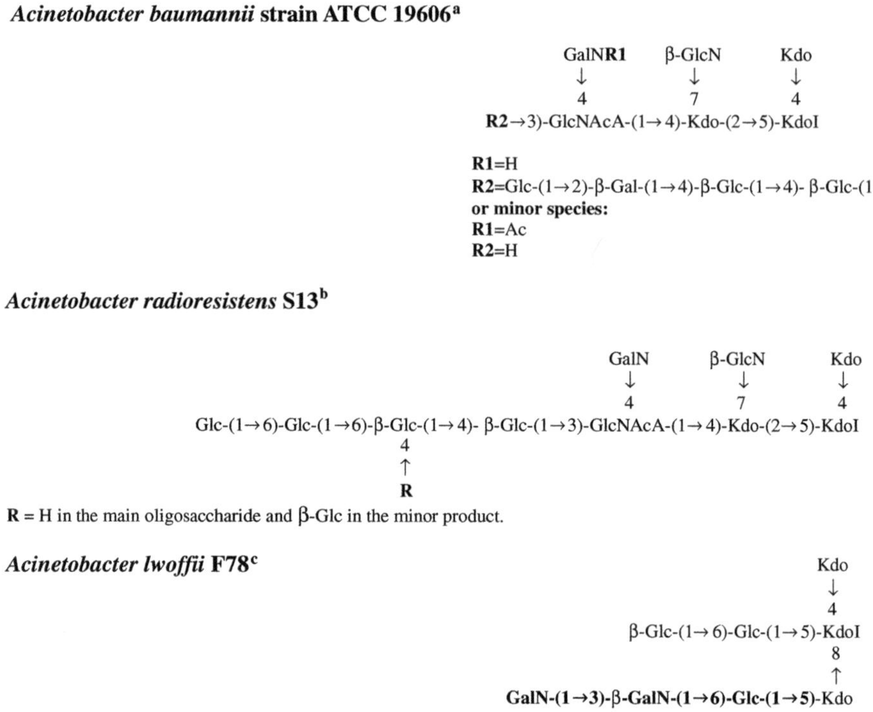

Fig. 4.4 Examples of core regions from *Acinetobacter*. [a]Vinogradov et al., 2002; [b]Leone et al., 2006; [c]Hanuszkiewicz et al., 2008

4.2.5 *Xanthomonas*

In *Xanthomomnas campestris* pv. *campestris* the inner core region (Silipo et al., 2005a) is devoid of heptose and carries anionic substituents located on Kdo and α-mannose residues; the Kdo carries a α-galacturonyl-phosphate substituent at O-4 whereas the adjacent α-mannose is substituted at O-3 by a second α-galacturonyl phosphate or, alternatively, by a phosphoramide group (Fig. 4.5). All of these chemical groups represent a great structural novelty for the core oligosaccharide region of LPSs and were found for the first time in *Xanthomonas*. However, the presence of negatively charged substituents in close proximity to the lipid A-core region is functionally important for intermolecular associations by cross-linking of divalent cations, and it is thought that this electrostatic interaction contributes to enhance the stability of the external bacterial membrane with the formation of a strong, rigid and protective barrier. It has been shown that plant cells recognize both core region and lipid A, and that these moieties induced defense-related gene transcription at different times. Even smaller core oligosaccharides from this LOS were shown to induce oxidative burst in tobacco cells (Kaczyński et al., 2007).

Xanthomonas campestris* pv. *campestris

```
Man-(1→3)-Man-(1→4)-β-Glc-(1→4)-Man-(1→5)-Kdo
                                3         4
                                ↑         ↑
                                R     P←1)-GalA
```

R, phosphoramide or GalA-1*P*.

Fig. 4.5 Core oligosaccharide from *Xanthomonas campestris pv. campestris*. Silipo et al., 2005

4.2.6 Burkholderia

The genus *Burkholderia* represents the RNA group II of *Pseudomonadaceae* and possesses other and significant structural features in its LPS core regions. The common partial structure of the inner core region is L-α-D-Hep*p*-(1→7)-L-α-D-Hep*p*-(1→3)-[β-D-Glc*p*-(1→4)]-L-α-D-Hep*p*-(1→5)-α-Kdo. All the core region identified so far possess a Kdo unit that is in turn substituted at O-5 with a Ko residue. The latter can bear a Ara4N residue at O-8. The Ko monosaccharide is also limited to few bacterial LPS molecules, like *Acinetobacter* (Vinogradov et al., 1997b), *Yersinia* (Vinogradov et al., 2002b) and *Serratia* (Vinogradov et al., 2006). The Ara4N→Ko→Kdo trisaccharide sequence, reported in most *Burkholderia* species, seems unique and distinctive of *Burkholderia* LPSs. Ara4N, bearing a positively charged free amino group, plays a key role in pathogenesis since it reduces the net negatively charged surface on the external membrane rendering it positively charged or in an isoelectric state. In turn, this variation confers resistance to antibiotic compounds and host cationic antimicrobial peptides as demonstrated studying polymyxin B resistant strains. In *B. pyrrocinia* (Silipo et al., 2006), *B. cepacia* (Isshiki et al., 2003) and *B. caryophylli* (Molinaro et al., 2002) the Kdo unit is substituted at O-5 by the pentasaccharide [α-L,D-Hep*p*-(1→7)-[α-L-Rha*p*-(1→2)]-α-L,D-Hep*p*-(1→3)-[β-D-Glc*p*-(1→4)]-α-L,D-Hep*p*-(1→]. The heptose residue of this pentasaccharide portion can be further substituted, in *B. pyrrocinia*, at O-7 by another heptose residue that carries at O-3 a α-D,D-Hep*p*-(1→7)-α-L,D-Hep*p* -(1→ disaccharide giving rise to a heptan pentasaccharide representing the remote region of the outer core. The core region from *B. cenocepacia* LMG 16656, the ET-12 clone type strain (also known as J2315), the most notorious and feared single strain in cystic fibrosis has been elucidated (Fig. 4.6). Moreover, in order to study the biochemical mechanisms of bacterial adaptation, LPS structure from the most representative and problematic group of CF (Cystic Fibrosis) pathogens, the *Burkholderia cepacia* complex (Bcc), namely *Burkholderia multivorans* (genomovar II), *Burkholderia vietnamiensis* (genomovar V), and *Burkholderia cenocepacia* (genomovar V) were respectively isolated from CF patients that underwent lung transplantation. For each clinical isolated, two paired clonal strains were recovered: one pre and the other post transplantation. A review on structural and biological features of LPS from *Burkholderia* has been published recently (De Soyza et al., 2008). Under the structural point of view the core oligosaccharides are characterised for the absence of phosphate groups. The second heptose was differently substituted (2,7- substituted in *B. pyrrocinia* (Silipo et al., 2006), 3,7 substituted in *B. cenocepacia* (Silipo et al., 2007, Ieranò et al., 2010a), 2,3,7 substituted in *B. multivorans* (Ieranò et al., 2008) and *B. vietnamensis* (Ieranò et al., 2009). Recently, genes for the biosynthesis of the core oligosaccharide moiety of the *B. cenocepacia* lipopolysaccharide have been identified (Ortega et al., 2009). The core oligosaccharide region present in *B. cepacia* ATCC 25416 serotype O4 (genomovar I) showed similar components to those of most *Burkholderia* but, differently from many other core structures defined so far, the Ko unit beared a L-α-D-Hep residue (Masoud et al., 2009); moreover, the second heptose in the inner core, HepII, was only substituted at O-3 and a PEtN group was found attached to the last sugar residue of the core (Fig. 4.6).

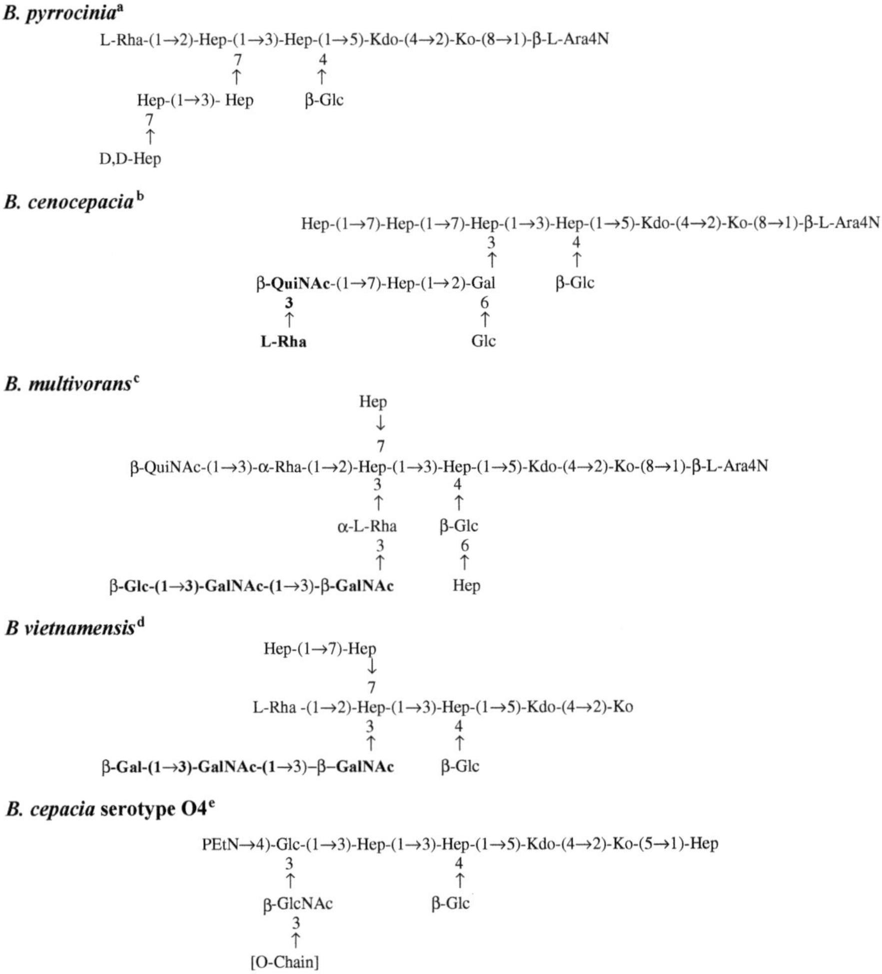

Fig. 4.6 Core oligosaccharide from species belonging to the *Burkholderia cepacia complex*. [a]Silipo et al., 2006; [b]Silipo et al., 2007; Ieranò et al., 2008; [d]Ieranò et al., 2008; [e]Masoud et al., 2009

R. solanacearum Toudk-2

```
                                                                                        β-L-Ara4N
                                                                                            ↑
                                                                                            7
L-Rha-(1→2)- L-Rha-(1→3)- L-Rha-(1→3)-β-GlcNAc-(1→3)- L-Rha-(1→3)-Hep-(1→3)-Hep-(1→5)-Kdo-(4→2)-Ko
                                                                          2                 4
                                                                          ↑                 ↑
                                                                         Rha              β-Glc
```

Fig. 4.7 Core oligosaccharide from *Ralstonia solanacearum Toudk-2,* (Zdorovenko et al., 2008)

4.2.7 Ralstonia

Ralstonia solanacearum is a devastating plant pathogen with a global distribution and an unusually wide host range. In some other respects, the LPS core structure of *R. solanacearum* Toudk-2 (Zdorovenko et al., 2008) resembles that of *B. cepacia* and *B. pyrrocinia*, both bacteria sharing an α-Rha*p*-(1→2)-α-Hep*p*-(1→3)-[β-Glc*p*-(1→4)]-α-Hep*p*-(1→5)-[α-Ko-(2→4)]-Kdo hexasaccharide fragment (Fig. 4.7). This structure similarity is in agreement with a close phylogenetic relationship of these bacteria belonging to the same family *Burkholderia.*

4.2.8 Shewanella

The genus *Shewanella* was created in 1985 to house an abundant group of Gram-negative, facultative anaerobic, readily cultivated γ-Proteobacteria. Bacteria of this genus are mainly associated with severe aquatic habitats (e.g. cold, high pressure and deep-sea, etc.). During the last few years bacteria of the genus *Shewanella* have been under intensive investigation due to their ability to reduce a variety of electron acceptors, including iron, manganese, nitrate, nitrite, thiosulfate, DMSO, trimethylamine N-oxide (TMAO), glycine fumarate and elemental sulfur. Because of their metabolic versatility and wide distribution in a variety of aquatic habitats, *Shewanella*-like organisms are thought to play a significant role in the cycling of organic carbon and other bionutrients. Several core structures of LPS from different *Shewanella* LPS core region have been published so far. The oligosaccharide from *S. oneidensis* MR-1 constitutes the first case in which as first core residue a monosaccharide other than Kdo was identified, namely the 8-amino-3,8-dideoxy-D-*manno*-octulosonic acid (Kdo8N) (Vinogradov et al., 2003). Moreover, a residue of 2-acetamido-2-deoxy-D-galactose in an open-chain form (GalNAc*o*), linked as cyclic acetal to *O*-4 and *O*-6 of D-galactopyranose has also been detected. The structure contains a phosphodiester linkage between the α-D-galactopyranose and D-glycero-D-manno-heptose (D,D-Hep) residues (Fig. 4.8). The occurrence of Kdo8N (Leone et al., 2007a) replacing Kdo has been also detected in other *Shewanella* strains (Leone et al., 2007b). Although the OS structure from *S. putrefaciens* CN32 does not include the Kdo8N, it is still possible to suggest that this sugar residue can be considered as a taxonomic marker for the genus. The oligosaccharide structure from *S. pacifica* KMM 3601, KMM 3605 and KMM 3772 (Leone et al.,

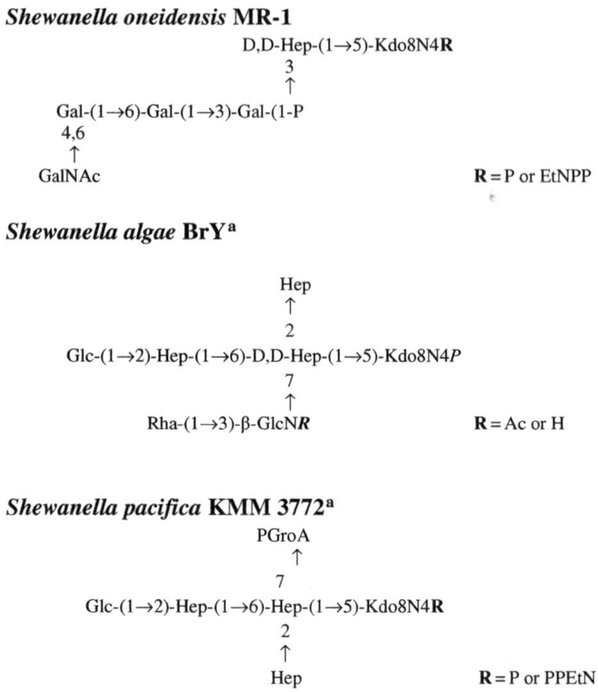

Fig. 4.8 Examples of core regions from *Shewanella*, reviewed in Leone et al., 2008

2007b; Silipo et al., 2005b) is characterized by the presence of glyceric acid, which is attached *via* phosphodiester linkage to the D,D-Hep, contributing to the increasing of the total negative charge of the inner-core region (Fig. 4.8). Although this is a key molecule of the primary metabolism of Gram-negative bacteria, it was never detected before in the core of LPS molecules.

4.2.9 Alteromonadaceae

The core oligosaccharides from LPS from *Pseudoalteromonas* have been described, from *P. carrageenovora* IAM 12662^T (Silipo et al., 2005b) and *P. issachenkonii* KMM 3549^T (Silipo et al., 2004a). The core from *P. carrageenovora* is composed by a mixture of three glycoforms, differing by the length of the sugar chain and the phosphorylation pattern, and is characterized by a strong accumulation of negatively charged groups, as phosphate groups, creating a region with a high charge density (Fig. 4.9). A closely related chemical architecture has been found also in the core region of the LPS from *P. issachenkonii* KMM 3549^T . This halophilic microorganism has bacteriolytic, proteolytic and haemolytic activity and degrades algal polysaccharides, producing a number of glycosyl hydrolases.

The newly defined *Alteromonas* genus comprises few validly described species, namely *A. macleodii*, *A. marina*, *A. stellipolaris*, *A. litorea* and *A. addita*. The structure investigation of the LPS structures from bacteria belonging to this genus has

Fig. 4.9 Examples of core region from *Alteromonadaceae*, reviewed in Leone et al., 2008

***Pseudoalteromonas carrageenovora* IAM 1266**

```
                P        P
                ↓        ↓
                3        4
β-Galp-(1→4)-Hep-(1→5)-Kdo
      4
      ↑
   α-Glcp
```

***Pseudoalteromonas issachenkonii* KMM 3549^T**

```
                          P
                          ↓
                          4
β-Gal-(1→7)-Hep-(1→5)-Kdo
               4
               ↑
Glc-(1→4)-β-Gal
```

***Alteromonas macleodii* ATCC 27126^T [85]**

β-Kdo-(2→7)-Hep-(1→5)-Kdo4P

***Alteromonas addita* KMM 3600^T**

α-Glc-(1-P-3)-α-Hep-(1→5)-α-Kdo4P

only recently begun, and, up to now, only two structures have been given, from *A. macleodii* ATCC 27126^T (Liparoti et al., 2006) and from *A. addita* KMM 3600^T (Fig. 4.9). Interestingly, in both cases bacteria have been found to produce only a R-LPS, provided with an extremely short length oligosaccharide chain with a high negative charge density. Within the core oligosaccharide from *A. macleodii* ATCC 27126^T, the negative charge is inferred, among the others, by the occurrence of Kdo residue that is present as β-configured residue. This feature is rather unusual in polysaccharides in general, being sometimes detected in polysaccharide capsules, but very rare in LPSs. *A. addita*, was first isolated from sea water samples collected at various depth in the Pacific Ocean region of Chazma Bay (Sea of Japan), during a study on free-living microbial colonies in radionuclide contaminated environments. The core showed a base labile substitution of a Glc residue linked *via* phosphodiester bond to a heptose moiety.

4.2.10 Arenibacter Certesii

The recently described genus *Arenibacter* was established to accommodate Gram-negative, strictly aerobic, heterotrophic, dark-orange pigmented, non-motile marine bacteria belonging to the *Cytophaga–Flavobacterium–Bacteroides* phylum. The core region of *Arenibacter certesii* KMM 3941^T (Silipo et al., 2005c) was characterised by a phosphodiester bond connecting two monosaccharides units (Fig. 4.10).

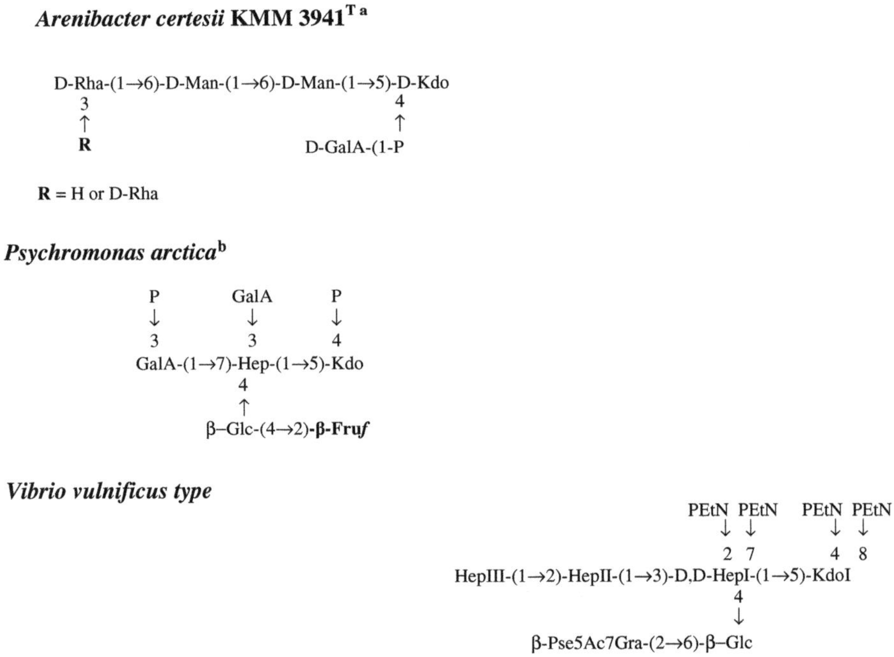

Pse, pseudaminic acid: 5,7-diacylamido-3,5,7,9-tetradeoxynonulosonic acid residue; Gra: glyceric acid.
LPS gave two oligosaccharides, OS1 and OS2; OS2 differed from OS1 by the absence of glyceric acid, acetate, and Pse residues.

Fig. 4.10 Core oligosaccharide from: *Arenibacter certesii* KMM 3941^{T}. [a]Silipo et al., 2005c; *Psychromonas arctica* [b]Corsaro et al., 2008; *Vibrio vulnificus* type strain 27562 [c]Vinogradov et al., 2009

The high number of negative charges in short oligosaccharides present in marine bacteria from *Pseudoalteromonas*, *Shewanella*, *Alteromonas* and *Arenibacter* could be important for maintaining the integrity of the outer membrane exposed to a peculiar external surrounding.

4.2.11 Psychromonas Arctica

The core oligosaccharide from *Psychromonas arctica*, a Gram-negative bacterium isolated in the arctic seawater near Spitzbergen (Svalbard islands, Arctic) has been recently reported (Fig. 4.10) (Corsaro et al., 2008) This biofilm-forming microorganism is classified as an eury-psychrophile, which can actively duplicate in the 0–25 °C temperature range. The core structure arises highly phosphorylated and an interesting feature is the presence of a terminal β-fructofuranose residue, previously found only in the core oligosaccharide of a LPS from *Vibrio cholerae* strains.

4.2.12 *Vibrionaceae*

Vibrio vulnificus is a planktonic marine bacterium that is pathogenic to humans and animals (Vinogradov et al., 2009). Several glycoform composing the core oligosaccharide region were found, some characterized by the presence of PEtN and of a nonulosonic acid, the 5,7-diacylamido-3,5,7,9-tetradeoxynonulosonic acid residue (pseudaminic acid, Pse) which could be substituted at O-7 by a acetyl group or by glyceric acid (Fig. 4.10).

4.2.13 *Rizhobiaceae*

A recent review summarized the structure from *Rizhobiaceae* (De Castro et al., 2008); Figure 4.11 reports the core structures known for *Agrobacterium* and *Rhizobium.* All *Agrobacterium* and *Rhizobium* cores lack of heptoses and phosphate groups and cores share the same residue linked to O-5 of the first Kdo, that is α-D-Man or α-D-Glc; the exception is represented by the core of *Agrobacterium larrymoorei* where the Kdo carries a α-D-Gal. It is interesting to note that this bacterium is not able to infect plants both as pathogenic and symbiontic one. All *Agrobacterium* core, except to that of *A. rubi*, contain only two Kdo units as acid monosaccharides, whereas in the *Rhizobium* core there are additional GalA residues, and, in the case of *Rhizobium loti* NZP2213 and *Rhizobium meliloti* 102F51 additional of 3-deoxy-lyxo-2-heptulosaric acid units, which is also present in the core of *Agrobacterium rubi.*

4.2.14 *Loktanella Rosea*

The core oligosaccharide from *Loktanella rosea* was just identified (Ieranò et al., 2010b). Bacteria belonging to *Loktanella rosea* strain KMM 6003 isolated from seawater samples collected from Sea of Japan are marine microorganisms strictly aerobic and generally halotolerant, with rod-shaped cells. The lipopolysaccharide turned out to be characterized by a novel a trisaccharidic core region exclusively composed of ulosonic sugars and containing a neuraminic acid. It is a short oligosaccharide moiety that at physiological pH is negatively charged. In detail, the first Kdo unit is further substituted at O-4 and O-8 positions by two terminal ulosonic residues: a further Kdo, sitting at O-4 and a Neuraminic acid at position O-8 (Fig. 4.12). Neu5Ac residues have been found in the outer core region, as the case of *Campylobacter jejuni*, *Helicobacter pylori* that can express either Lewis antigens, or resemble structural similarities with glycosphingolipids of the ganglioside group, all attempts to evade host immune response. *Chlamydia trachomatis* produces a deep-rough LPS characterized by a linear sequence of three Kdo residues: α-Kdo-(2→8)-α-Kdo-(2→4)-α-Kdo-(2→6)-lipid A, whereas in *Chlamydia psittaci* a non-stoichiometric fourth Kdo is located at O-4 of the second one. However, in few cases the first Kdo has been found carrying two ulosonic acid residues, as the case

Agrobacterium larrymoorey

```
β-D-Qui3NAcyl-(1→2)-Rha-(1→4)-GalAN-(1→6)-Gal-(1→5)-Kdo
                                          4         4
                                          ↑         ↑
                   β-GlcAN-(1→3)-β-Glc             Kdo
```

Acyl: 3-hydroxy-2,3-dimethyl-5-oxopropylamino

***Agrobacterium tumefaciens* A1, DSM 30150**

```
Man-(1→6)-Man-(1→5)-Kdo
                     4
                     ↑
         β-Gal-(1→8)-Kdo
```

Terminal Mannose and Galactose are non stoichiometric substituents

***Agrobacterium tumefaciens* TT111, DSM 30204**

```
Rha–(1→2)–Rha-(1→2)-Rha-(1→3)-Rha-(1→3)-Man-(1→5)-Kdo
 2                                                  4
 ↑                                                  ↑
Rha                                     β-Gal-(1→8)-Kdo
                                                    4
                                                    ↑
                                       β-GlcN-(1→2)-β-Gal
```

Terminal Rhamnose and Glucosamine are non stoichiometric substituents

***Agrobacterium rubi*[T], DSM 6772**

```
 L-Fuc                          α-D-Dha-(2→3)-GalA-(1→3)-β-D-Glc
 ↓                                                           ↓
 4                                                           4
GlcN-(1→4)-β-GlcA-(1→4)-GalA-(1→3)-β-Glc-(1→6)-Glc-(1→5)-Kdo
 3                                              3R        4
 ↑                                                        ↑
β-Gal-(2←1)-Fuc                                   GalA-(1→7)-Kdo
  3/4R
```

R = Ac

Acetyl groups and GalA on external Kdo are present as non stoichiometric substituents

Fig. 4.11 (continued)

***Agrobacterium radiobacter* Rv3, DSM 30207**

```
Man-(1→2)-Man-(1→3)-Man-(1→3)-Man-(1→5)-Kdo
                    6                  4
                    ↑                  ↑
                   Man       β-Gal-(1→8)-Kdo
```

***Rhizobium etli* CE3**

```
[O-chain]-3-L-Fuc-(1→3)-Man-(1→3)-β-QuiNAc-(1→4)-Kdo      GalA
                                                  ↓         ↓
                                                  6         4
                                                  Gal-(1→6)-Man-(1→5)-Kdo
                                                                      4
                                                                      ↑
                                                        Gal A-(1→4)-Kdo
                                                                    5
                                                                    ↑
                                                                  GalA
```

Fig. 4.11 Examples of core oligosaccharide from *Rizhobiaceae*, reviewed in De castro et al., 2008

***Loktanella rosea*[a]**
α-Kdo-(2→4)-[α–Neu-(2→8)]-α-Kdo

***Halomonas pantelleriensis*[b]**

```
                                                                            P
                                                                            ↑
                                                                            4
[O-chain]-(1→6)-Glc-(1→4)-GlcNAc-(1→4)-β-Glc-(1→3)-Hep-(1→5)-Kdo
                                                    4
                                                    ↑
R= H or Hep                  Glc-(1→7)-D,D-Hep-(1→2)-Hep
                                                      6R
```

Fig. 4.12 Core oligosaccharide from: *Loktanella rosea* [a]Ieranò et al., submitted, and *Halomonas pantelleriensis* [b]Pieretti et al., 2008

of the LPS from *Acinetobacter lwoffii* F78. However a neuraminic acid has never been found directly linked to the Kdo residue and in the inner core region of LPSs.

4.2.15 Halomonas Pantellerensis

Halomonas pantelleriensis is an extremophile, haloalkaliphilic microorganism that requires strictly aerobic conditions for growth (Pieretti et al., 2008); the structure has been reported in Fig. 4.12. The treatment allowed the identification of the linkage between the O-chain and the core structure.

4.3 Conclusion

A selection of the more interesting core structures is discussed in this chapter. The core oligosaccharide region possesses some general architectural principles that generally render the core oligosaccharide of LPS endotoxins a very peculiar if not unique glyco-molecule.

The general principle of a negatively charged core region contributing to the rigidity of the Gram-negative cell wall through intermolecular cationic cross-links is still convincing. The partial structure L,D-Hep-(1→7)/(1→2)-L,D-Hep-(1→3)-L,D-Hep-(1→5)-Kdo, is a common structural theme to a major part of core oligosaccharide regions. Beside a single case, the minimal viable endotoxin structure comprises sugar of the core region. The expression of a (O-specific) polysaccharide in LPS is not a prerequisite for bacterial survival, however, the finding that the polysaccharide portion in S-form LPS may be furnished either by the O-chain or CA or ECA suggests that such LPS form is highly advantageous in many bacteria. The binding of the core region to lipid A occurs always via a Kdo residue [except in *Acinetobacter*, where this Kdo may be replaced in non-stoichiometric amounts by Ko].

Two types of core structures can be found: those containing and those without heptoses. In the first type, L,D-Hep or D,D-Hep alone, or both may be present in a particular core structure. If present, D,D-Hep either decorates the inner core region (e.g. in *Y. enterocolitica*) or is attached to more remote parts of the carbohydrate chain. The D,D-Hep is the genetic precursor of L,D-Hep but however the regulation of the distribution of L,D-Hep and D,D-Hep in the core region is not understood, and it is not known whether L,D-Hep and D,D-Hep are transferred by different transferases.

Acknowledgements Authors wish to thank Prof. M. Parrilli, Prof. R. Lanzetta and Prof. O. Holst for the useful discussions and help through these years.

References

Alexander, C., Rietschel, E.T. Bacterial lipopolysaccharides and innate immunity. J Endotoxin Res 7 (2001) 167–202.

Amano K. An Interesting Property of O-Antigen Polysaccharide in lipopolysaccharide: Cross-talk between Lipopolysaccharides of *Helicobacter pylori* and host cells through lewis blood group antigens. Trends Glycosci Glycotechnol 14 (2002) 105–114.

Choudhury, B., Carlson, R.W., Goldberg, J.B. Characterization of the lipopolysaccharide from a wbjE mutant of the serogroup O11 *Pseudomonas aeruginosa* strain, PA103. Carbohydr Res 343 (2008) 238–248.

Corsaro, M.M., Pieretti, G., Lindner, B., Lanzetta, R., Parrilli, E., Tutino, M.L., Parrilli, M. Highly phosphorylated core oligosaccharide structures from cold-adapted *Psychromonas arctica*, Chem Eur J 14 (2008) 9368–937.

De Castro, C., Molinaro, A., Lanzetta, R., Silipo, A., Parrilli, M. Lipopolysaccharide structures from *Agrobacterium* and *Rhizobiaceae* species. Carbohydr Res 343(12) (2008) 1924–1933.

De Castro, C., Molinaro, A., Nunziata, R., Lanzetta, R., Parrilli, M., Holst, O. A novel core region, lacking heptose and phosphate, of the lipopolysaccharide from the Gram-negative bacterium *Pseudomonas cichorii* (*Pseudomonadaceae* RNA group I). Eur J Org Chem 11 (2004) 2427–2435.

De Soyza, A., Silipo, A., Lanzetta, R., Govan, J.R., Molinaro, A. The chemical and biological features of *Burkholderia cepacia* complex lipopolysaccharides. Innate Immun 14 (2008) 127–144.

Hanuszkiewicz, A., Hübner, G., Vinogradov, E., Lindner, B., Brade, L., Brade, H., Debarry, J., Heine, H., Holst, O. Structural and immunochemical analysis of the lipopolysaccharide from *Acinetobacter lwoffii* F78 located outside *Chlamydiaceae* with a Chlamydia-specific lipopolysaccharide epitope. Chem Eur J 14 (2008) 10251–10258.

Harper, M., Boyce, J.D., Cox, A.D., St. Michael, F., Wilkie, I.W., Blackall, P.J., Adler, B., *Pasteurella multocida* espresses two lipopolysaccharide glycoforms simultaneously, but only a single form is required for virulence: identification of two acceptor-specific heptosyl I transferases. Infect Immun 75 (2007) 3885–3893.

Heinrichs, D.E., Yethon, J.A., Whitfield, C. Molecular basis for structural diversity in the core regions of the lipopolysaccharides of *Escherichia coli* and *Salmonella enterica*. Mol Microbiol 30(2) (1998) 221–232.

Holst, O. Chemical structure of the core region of lipopolysaccharides. an update. Trends Glycosci Glycotechnol 14 (2002) 87–103.

Holst, O., Molinaro, A. 2009. Core oligosaccharide and lipid A components of lipopolysaccharides. In Moran, A., Brennan, P., Holst, O., von Itszstein, M. (eds), Microbial Glycobiology: Structures Relevance and Applications. Elsevier, San Diego, CA, pp. 29–56.

Holst, O., The structures of core regions from enterobacterial lipopolysaccharides – an update. FEMS Microbiol Lett 271 (2007) 3–11.

Ieranò, T., Silipo, A., Sturiale, L., Garozzo, D., Corris, P., Perry, J., Lanzetta, R., Parrilli, M., De Soyza, A., Molinaro, A. Structural elucidation of a novel *B. cenocepacia* ET-12 lipooligosaccharide isolated from a cystic fibrosis patient after lung transplantation. Eur J Org Chem 7 (2010a) 1299–1306.

Ieranò, T., Silipo, A., Nazarenko, E.L., Gorshkova E.L., Ivanova E.P., Garozzo, D., Sturiale, L., Lanzetta, R., Parrilli, M., Molinaro, A. Against the rules: a marine bacterium, Loktanella rosea, possesses a unique lipopolysaccharide. Glycobiology 20(5) (2010b) 586–593.

Ieranò, T., Silipo A., Sturiale, L., Garozzo, D., Brookes, H., Anjam Khan, C.M., Bryant, C., Gould, F.K., Corris, P.A., Lanzetta, R., Parrilli M., De Soyza, A., Molinaro, A. The structure and pro-inflammatory activity of the lipopolysaccharide from *Burkholderia multivorans* and the differences between clonal strains colonizing pre- and post-transplantated lungs. Glycobiology 18(11) (2008) 871–881.

Ieranò, T., Silipo, A., Sturiale, L., Garozzo, D., Bryant, C., Lanzetta, R., Parrilli, M., Aldridge, C., Gould, F.K., Corris, P.A., Khan, C.M., De Soyza, A., Molinaro, A. First structural characterization of *Burkholderia vietnamiensis* lipooligosaccharide from cystic fibrosis-associated lung transplantation strains. Glycobiology 19(11) (2009) 1214–1223.

Isshiki, Y., Zähringer, U., Kawahara, K. Structure of the core oligosaccharide with a characteristic D-glycero-α-D-talo-oct-2-ulosonate-(2→4)-3-deoxy-D-manno-oct-2-ulosonate [α-Ko-(2→4)-Kdo] disaccharide in the lipopolysaccharide of *Burkholderia cepacia*. Carbohydr Res 338 (2003) 2659–2666.

Kaczyński, Z., Braun, S., Lindner, B., Niehaus, K., Holst, O. Investigation of the chemical structure and biological activity of oligosaccharides isolated from rough-type *Xanthomonas campestris* pv. *campestris* B100 lipopolysaccharides. J Endotoxin Res 13 (2007) 101–108.

Knirel, Y.A., Bystrova, O.V., Kocharova, N.A., Zähringer, U., Pier, G.B. Conserved and variable structural features in the lipopolysaccharide of *Pseudomonas aeruginosa*. J Endotoxin Res 12 (2006) 324–3036.

Kocharova, N.A., Kondakova, A.N., Vinogradov, E., Ovchinnikova, O. Lindner B., Shashkov A.S., Rozalski A., Knirel Y.A., full structure of the carbohydrate chain of the lipopolysaccharide of *Providencia rustigianii* O34. Chem Eur J 14 (2008b) 6184–6191.

Kocharova, N.A., Vinogradov, E., Kondakova, A.N., Shashkov, A.S., Rozalski, A., Knirel, Y.A. The full structure of the carbohydrate chain of the lipopolysaccharide of *Providencia alcalifaciens* O19. J Carbohydr Chem 27 (2008a) 320–331.

Kondakova, A.N., Vinogradov, E., Katzenellenbogen, E., Kocharova, N.A., Lindner, B., Knirel, Y.A. Structural studies on the lipopolysaccharide core of bacteria of the genus *Citrobacter*: two different core structures in *Citrobacter* O14 serogroup. J Carbohydr Chem 28 (2009) 298–315.

Kondakova, A.N., Vinogradov, E., Lindner, B., Kocharova, N.A., Rozalski, A., Knirel, Y.A. Elucidation of the lipopolysaccharide core structures of bacteria of the genus *Providencia*. J Carbohydr Chem 25 (2006) 499–520.

Kondakova, A.N., Vinogradov, E., Lindner, B., Kocharova, N.A., Rozalski, A., KnireL, Y.A. Mass-spectrometric studies of *providencia* SR-form lipopolysaccharides and elucidation of the biological repeating unit structure of *Providencia rustigianii* O14-Polysaccharide. J Carbohydr Chem 26 (2007) 497–512.

Leone, S., Izzo, V., Silipo, A., Garozzo, D., Lanzetta, R., Parrilli, M., Molinaro, A., Di Donato, A. A novel type of highly negatively charged lipooligosaccharide from *Pseudomonas stutzeri* OX1 possessing two 4,6-O-(1-carboxy)-ethylidene residues in the outer core region. Eur J Biochem 271 (2004a) 2691–2704.

Leone, S., Izzo, V., Sturiale, L., Garozzo, D., Lanzetta, R., Parrilli, M., Molinaro, A., Di Donato, A. Structure of minor oligosaccharides from the lipopolysaccharide fraction of *Pseudomonas stutzeri* OX1. Carbohydr Res 339 (2004b) 2657–2665.

Leone, S., Molinaro, A., De Castro, C., Baier, A., Nazarenko, E.L., Lanzetta, R., Parrilli, M. Absolute configuration of 8-amino-3,8-dideoxyoct-2-ulosonic acid, the chemical hallmark of lipopolysaccharides of the genus *Shewanella*. J Nat Prod 70 (2007a) 1624–1627.

Leone, S., Molinaro, A., Pessione, E., Mazzoli, R., Giunta, C., Sturiale, L., Garozzo, D., Lanzetta, R., Parrilli, M. Structural elucidation of the core-lipid A backbone from the lipopolysaccharide of *Acinetobacter radioresistens* S13, an organic solvent tolerating Gram-negative bacterium. Carbohydr Res 341 (2006) 582–590.

Leone, S., Silipo, A., Nazarenko, E., Lanzetta, R., Parrilli, M., Molinaro, A. Molecular structure of endotoxins from Gram-negative marine bacteria: an update. Mar Drugs 5(3) (2007b), 85–112.

Lerouge, I., Vanderleyden, J. O-antigen structural variation: mechanisms and possible roles in animal/plant–microbe interactions. FEMS Microbiol Rev 26 (2002) 17–47.

Liparoti, V., Molinaro, A., Sturiale, L., Garozzo, D., Nazarenko, E.L., Gorshkova, R.P., Ivanova, E.P., Shevcenko, L.S., Lanzetta, R., Parrilli, M. Structural analysis of the deep rough lipopolysaccharide from Gram negative bacterium *Alteromonas macleodii* ATCC 27126: the first finding of β-Kdo in the inner core of lipopolysaccharides. Eur J Org Chem (2006) 4710–4716.

Lukasiewicz, J., Niedziel, T., Jachymek, W., Kenne, L., Lugowski, C. Two Kdo-heptose regions identified in *Hafnia alvei* 32 lipopolysaccharide: the complete core structure and serological screening of different *Hafnia* O Serotypes. J Bacteriol 191(2) (2009) 533–544.

Lundström, S.L., Li, J., Deadman, M.E., Hood, D.W., Moxon, E.R., Schweda, E.K. Structural analysis of the lipopolysaccharide from nontypeable *Haemophilus influenzae* strain R2846. Biochemistry 47 (2008) 6025–6038.

Lundström, S.L., Twelkmeyer, B., Sagemark, M.K., Li, J., Richards, J.C., Hood, D.W., Moxon, E.R., Schweda, E.K.H. Novel globoside-like oligosaccharide expression patterns in non-typeable *Haemophilus influenzae* lipopolysaccharide. FEBS J 274 (2007) 4886–4903.

Mansson, M., Bauer, S.H., Hood, D.W., Richards, J.C., Moxon, E.R., Schweda, E.K. A new structural type for Haemophilus influenzae lipopolysaccharide. Structural analysis of the lipopolysaccharide from nontypeable *Haemophilus influenzae* strain 486. Eur J Biochem 268 (2001) 2148–2159.

Masoud, H., Perry, M.B., Brisson, J.R., Uhrin, D., Li, J., Richards, J.C. Structural elucidation of the novel core oligosaccharide from LPS of *Burkholderia cepacia* serogroup O4. Glycobiology 19(5) May (2009) 462–171.

Meredith, T.C., Mamat, U., Kaczynski, Z., Lindner, B., Holst, O., Woodard, R.W. Modification of lipopolysaccharide with colanic acid (M-antigen) repeats in *Escherichia coli*. J Biol Chem 282 (2007) 7790–7798.

Molinaro, A., De Castro, C., Lanzetta, R., Evidente, A., Parrilli, M., Holst, O. Lipopolysaccharides possessing two L-glycero-D-manno-heptopyranosyl-alpha-(1-5)-3-deoxy-D-manno-oct-2-ulopyranosonic acid moieties in the core region. The structure of the core region of the lipopolysaccharides from *Burkholderia caryophylli*. J Biol Chem 277(12) (2002) 10058–10063.

Molinaro, A., Lindner, B., De Castro, C., Nolting, B., Silipo, A., Lanzetta, R., Parrilli, M., Holst, O. The structure of lipid A of the lipopolysacharide from *Burkholderia caryophylli* with a 4-amino-4-deoxy-L-arabinopyranose 1-phosphate residue exclusively in glycosidic linkage. Chem Eur J 9 (2003) 1542–1548.

Molinaro, A., Silipo, A., De Castro, C., Sturiale, L., Nigro, G., Garozzo, D., Bernardini, M.L., Lanzetta, R., Parrilli, M. Full structural characterization of *Shigella flexneri* M90T serotype 5 wild-type R-LPS and its $\Delta galU$ mutant: glycine residue location in the inner core of the lipopolysaccharide. Glycobiology 18(3) (2008) 260–269.

Olsthoorn, M.M., Petersen, B.O., Duus, J., Haverkamp, J., Thomas-Oates, J.E., Bock, K., Holst, O. The structure of the linkage between the O-specific polysaccharide and the core region of the lipopolysaccharide from *Salmonella enterica* serovar Typhimurium revisited. Eur J Biochem 267 (2000) 2014–2027.

Olsthoorn, M.M., Petersen, O., Schlecht, S., Haverkamp, J., Bock, K., Thomas-Oates, J., Holst, O. Identification of a novel core type in *Salmonella* lipopolysaccharide. J Biol Chem 273(7) (1998) 3817–3829.

Ortega, X., Silipo, A., Saldías, M.S., Bates, C.C., Molinaro, A., Valvano, M.A. Biosynthesis and structure of the *Burkholderia cenocepacia* K56-2 lipopolysaccharide core oligosaccharide: truncation of the core oligosaccharide leads to increased binding and sensitivity to polymyxin B. J Biol Chem 284(32) (2009) 21738–21751.

Pieretti, G., Corsaro, M.M., Lanzetta, R., Parrilli, M., Vilches, S., Merino, S., Tomás, J.M. Structural characterization of the core region of the lipopolysaccharide from the haloalkaliphilic *Halomonas pantelleriensis*: identification of the biological O-antigen repeating. Unit Eur J Org Chem (2009) 1365–1371.

Pieretti, G., Corsaro, M.M., Lanzetta, R., Parrilli, M., Nicolaus, B., Gambacorta, A., Lindner, B., Hols, O. Structural characterization of the core region of the lipopolysaccharide from the haloalkaliphilic *Halomonas pantelleriensis*: identification of the biological O-antigen repeating unit. Eur J Org Chem (2008) 721–728.

Raetz, C.R., Reynolds, C.M., Trent, M.S., Bishop, R.E. Lipid A modification systems in Gram-negative bacteria. Annu Rev Biochem 76 (2007) 295–329.

Raetz, C.R., Whitfield, C. Lipopolysaccharide endotoxins. Annu Rev Biochem 71 (2002) 635–700.

Robbins, J.B., Kubler-Kielb, J., Vinogradov, E., Mocca, C., Pozsgay, V., Shiloach, J., Schneerson, R. Synthesis, characterization, and immunogenicity in mice of *Shigella sonnei* O-specific oligosaccharide-core-protein conjugates. PNAS 106 (2009) 7974–7978.

Schweda, E.K.H., Twelkmeyer, B., Li, J. Profiling structural elements of short-chain lipopolysaccharide of non-typeable *Haemophilus influenza*. Innate Immun 14 (2008) 199–211.

Silipo, A., Leone, S., Lanzetta, R., Parrilli, M., Sturiale, L., Garozzo, D., Nazarenko, E.L., Gorshkova, R.P., Ivanova, E.P., Gorshkova, N.M., Molinaro, A. The complete structure of the lipooligosaccharide from the halophilic bacterium *Pseudoalteromonas issachenkonii* KMM 3549T. Carbohydr Res 339 (2004a) 1985–1993.

Silipo, A., Leone, S., Molinaro, A., Sturiale, L., Garozzo, D., Nazarenko, E.L., Gorshkova, R.P., Ivanova, E.P., Lanzetta, R., Parrilli, M. Complete structural elucidation of the novel lipooligosaccharide from the outer membrane of marine bacterium *Shewanella pacifica*. Eur J Org Chem 11 (2005b) 2281–2291.

Silipo, A., Molinaro, A., Comegna, D., Sturiale, L., Cescutti, P., Garozzo, D., Lanzetta, R., Newman, M.A., Parrilli, M. Full structural characterization of the lipooligosaccharide of a *Burkholderia pyrrocina* clinical isolate. Eur J Org Chem 21 (2006) 4874–4883.

Silipo, A., Molinaro, A., Ieranò, T., De Soyza, A., Sturiale, L., Garozzo, D., Aldridge, C., Corris, P.A., Anjam Khan, C.M., Lanzetta, R., Parrilli, M. The complete structure and pro-inflammatory activity of the lipooligosaccharide of the highly epidemic and virulent Gram-negative bacterium *Burkholderia cenocepacia* ET-12 (strain J2315). Chem Eur J 13(12) (2007) 3501–3511.

Silipo, A., Molinaro, A., Sturiale, L., Dow, J.M., Erbs, G., Lanzetta, R., Newman, M.A., Parrilli, M. The elicitation of plant innate immunity by lipooligosaccharide of *Xanthomonas campestris*. J Biol Chem 280 (2005a) 33660–33668.

Silipo, A., Lanzetta, R., Parrilli, M., Sturiale, L., Garozzo, D., Nazarenko, E.L., Gorshkova, R.P., Ivanova, E.P., Molinaro, A. The complete structure of the core carbohydrate backbone from the LPS of marine halophilic bacterium *Pseudoalteromonas carrageenovora* type strain IAM 12662T. Carbohydr Res 340(8) (2005b) 1475–1482.

Silipo, A., Molinaro, A., Nazarenko, E.L., Sturiale, L., Garozzo, D., Gorshkova, R.P., Nedashkovskaya, O.I., Lanzetta, R., Parrilli, M. Structural characterization of the carbohydrate backbone of the lipooligosaccharide of the marine bacterium *Arenibacter certesii* strain KMM 3941(T). Carbohydr Res 2005c, 340, 2540–2549.

St. Michael, F., Li, J., Cox, A.D. Structural analysis of the core oligosaccharide of *Pasteurella multocida* strain X73. Carbohydr Res 340 (2005a) 1253–1257.

St. Michael, F., Li, J., Vinogradov, E., Larocque, S., Harper, M., Cox, A.D. Structural analysis of the lipooligosaccharide of *Pasteurella multocida* strain VP161: identification of both Kdo-P and Kdo-Kdo species in the lipopolysaccharide. Carbohydr Res 340 (2005b) 59–68.

St. Michael, F., Vinogradov, E., Li, J., Cox, A.D. Structural analysis of the lipopolysaccharide of *Pasteurella multocida* strain Pm70 and identification of the putative lipopolysaccharide glycosyltransferases. Glycobiology 15 (2005c) 323–333.

Vinogradov, E., Bock, K. The structure of the core part of *Proteus mirabilis* O27 lipopolysaccharide with a new type of glycosydic linkage. Carbohydr Res 319 (1999) 92–101.

Vinogradov, E., Korenevsky, A., Beveridge, T.J. The structure of the core region of the lipopolysaccharide from *Shewanella algae* BrY, containing 8-amino-3,8-dideoxy-D-manno-oct-2-ulosonic acid. Carbohydr Res 339 (2004) 737–740.

Vinogradov, E., Korenevsky, A., Beveridge, T.J. The structure of the rough-type lipooligosaccharide from *Shewanella oneidensis* MR-1, containing 8-amino-8-deoxy-Kdo and an open-chain form of 2-acetamido-2-deoxy-D-galactose. Carbohydr Res 338 (2003) 1991–1997.

Vinogradov, E., Wilde, C., Anderson, E.M., Nakhamchik, A., Lam, J.S. Rowe-Magnus D.A., Structure of the lipopolysaccharide core of *Vibrio vulnificus* type strain 27562. Carbohydr Res 344 (2009) 484–490.

Vinogradov, E.V., Duus, J., Brade, H., Holst, O. The structure of the carbohydrate backbone of the Lipopolysaccharide from *Acinetobacter baumannii* strain ATCC 19606. Eur J Biochem 269 (2002a) 422–430.

Vinogradov, E.V., Lindner, B., Kocharova, N.A., Senchenkova, S.N., Shashkov, A.S., Knirel, Y.A., Holst, O., Gremyakova, T.A., Shaikhutdinova, R.Z., Anisimov, A.P. The core structure of the lipopolysaccharide from the causative agent of plague, *Yersinia pestis*. Carbohydr Res 337 (2002b) 775–777.

Vinogradov, E.V., Lindner, B., Seltmann, G., Radziejewska-Lebrecht, J., Holst, O. Lipopolysaccharides from *Serratia marcescens* possess one or two 4-amino-4-deoxy-L- arabinopyranose 1-phosphate residues in the lipid A and D-glycero-D-talo-Oct-2-ulopyranosonic acid in the inner core region. Chem Eur J 12 (2006) 6692–6700.

Vinogradov, E.V., Petersen, B.O., Thomas-Oates, J.E., Duus, J., Brade, H., Holst, O. Characterization of a novel branched tetrasaccharide of 3-deoxy-D-manno-oct-2-ulopyranosonic acid. The structure of the carbohydrate backbone of the lipopolysaccharide from *Acinetobacter baumannii* strain NCTC 10303 (ATCC 17904). J Biol Chem 273 (1997a) 28122–28131.

Vinogradov, E.V., Müller-Loennies, S., Petersen, B.O., Meshkov, S., Thomas-Oates, J.E., Holst, O., Brade, H. Structural investigation of the lipopolysaccharide from *Acinetobacter haemolyticus* strain NCTC 10305 (ATCC 17906, DNA group 4). Eur J Bacteriol (1997b), 247, 82–90.

Whitfield, C. Biosynthesis and assembly of capsular polysaccharides in *Escherichia coli*. Annu Rev Biochem 75 (2006) 39–68.

Zdorovenko, E.L., Vinogradov, E., Wydra, K., Lindner, B., Knirel, Y.A. Structure of the oligosaccharide chain of the SR-type lipopolysaccharide of *Ralstonia solanacearum* Toudk-2. Biomacromolecules 9 (2008) 2215–2220.

Chapter 5
Salmonella-Regulated Lipopolysaccharide Modifications

Susan M. Richards, Kristi L. Strandberg, and John S. Gunn

Abstract *Salmonella enterica* are Gram-negative enteric pathogens that cause typhoid fever and gastroenteritis in humans. Many bacteria, including *Salmonella*, use signal transduction cascades such as two-component regulatory systems to detect and respond to stimuli in the local microenvironment. During infection, environmental sensing allows bacteria to regulate gene expression to evade host immune defenses and thrive in vivo. Activation of the *Salmonella* two-component regulatory systems PhoP-PhoQ and PmrA-PmrB and the RcsC-RcsD-RcsB phosphorylay by specific environmental signals in the intestine and within host cells leads to several lipopolysaccharide modifications that promote bacterial survival, cationic antimicrobial peptide resistance and virulence. Many pathogens encode orthologs to *Salmonella* two-component regulatory systems and also modify the lipopolysaccharide to escape killing by the host immune response. However, these organisms often regulate their virulence genes, including those responsible for lipopolysaccharide modification, in ways that differ from *Salmonella*. Further examination of bacterial virulence gene regulation and lipopolysaccharide modifications may lead to improved antimicrobial therapies and vaccines.

Keywords Lipopolysaccharide · *Salmonella* · Two-component regulatory system

Abbreviations

Ara4	4-amino-4-deoxy-L-arabinose (also abbreviated as 4-aminoarabinose)
CAMP	cationic antimicrobial peptide
Hep I	heptose I
Hep II	heptose II

J.S. Gunn (✉)
Department of Molecular Virology, Immunology and Medical Genetics, Center for Microbial Interface Biology, The Ohio State University, 460 W. 12th Avenue, Columbus, OH 43210-1214, USA
e-mail: gunn.43@osu.edu

S.M. Richards, K.L. Strandberg These authors contributed equally to this work

X. Wang, P.J. Quinn (eds.), *Endotoxins: Structure, Function and Recognition*, Subcellular Biochemistry 53, DOI 10.1007/978-90-481-9078-2_5,

L-PG	lysylphosphatidylglycerol
LPS	lipopolysaccharide
NTS	non-typhoidal *Salmonella*
pEtN	phosphoethanolamine
SPI-1	*Salmonella* pathogenicity island-1
SPI-2	*Salmonella* pathogenicity island-2
TLR	Toll-like receptor
TCRS	two-component regulatory system
TTSS	type III secretion system
WT	wild type

5.1 Introduction

5.1.1 Clinical Features and Relevance

Salmonella enterica (*S. enterica*) is a rod-shaped Gram-negative facultative anaerobe and a prominent enteric bacterial pathogen capable of causing food- and water-related diseases. Several serovars of *S. enterica* are associated with human infection. The principle clinical diseases associated with *Salmonella* infection are typhoid fever and gastroenteritis. *S. enterica* serovars Typhi (*S.* Typhi) and Paratyphi (*S.* Paratyphi) are pathogenically exclusive for humans and are known to cause typhoid or enteric fever (Ohl and Miller, 2001). Typhoid fever is a systemic infection characterized by the development of fever, abdominal pain, enterocolitis, and occasionally, a maculopapular rash. The hallmark feature of typhoid fever is the presence of mononuclear cell infiltration and hypertrophy of the intestinal Peyer's patches and mesenteric lymphoidal tissue (Kraus et al., 1999; Ohl and Miller, 2001). There are approximately 20 million cases of typhoid fever worldwide each year, and this acute and often life-threatening infection is responsible for over 200,000 deaths annually (Crump et al., 2004).

Non-typhoidal *Salmonella* (NTS) such as *S. enterica* serovars Typhimurium (*S.* Typhimurium) and Enteriditis (*S.* Enteriditis) are capable of infecting a wide range of hosts, including poultry, cattle, pigs, some strains of inbred mice and humans. In humans, *S.* Typhimurium is known to cause salmonellosis or gastroenteritis, characterized by onset of diarrhea within 12–72 h after infection (Kumar et al., 1982; McCormick et al., 1995; Ohl and Miller, 2001). NTS infections are usually self-limiting and normally do not cause systemic disease. NTS may spread systemically in immunocompromised patients and infected individuals can become carriers (Dutta et al., 2000; Gibbons et al., 2005). NTS serotypes also cause a systemic typhoid-like illness in susceptible mice (Scherer and Miller, 2001; Gibbons et al., 2005). Therefore, infection of susceptible mice with *S.* Typhimurium serves as a laboratory model for typhoid fever (Carter and Collins, 1974). Despite continuous progress in purification techniques, *Salmonella* infections resulting from contaminated food and water remain a problem throughout the world, partially due to the increasing emergence of antibiotic-resistant bacteria (Bhutta et al., 1991). The Centers for Disease Control and Prevention (CDC) estimate that there are approximately 1.4 million annual cases of salmonellosis. Only 40,000 cases are culture-confirmed, and an estimated 400 of these cases are fatal (Centers for Disease Control and Prevention, 2008).

5.1.2 Salmonella Pathogenesis

In order to establish typhoid fever or gastroenteritis, *Salmonella* enters the host through the ingestion of contaminated food or water. Once ingested, *Salmonella* must detect and respond to different host microenvironments, such as the acidic pH of the stomach, bile and innate immune components (Merighi et al., 2005; Groisman and Mouslim, 2006). When typhoidal *Salmonella* (e.g. *S.* Typhi) reach the small intestine, the bacteria must traverse the intestinal barrier to cause infection (Takeuchi, 1967). *Salmonella* invade the intestinal epithelium by preferentially adhering to and entering the microfold cells (M cells), which are specialized epithelial cells present in Peyer's patches of the small intestine (Clark et al., 1994; Jones et al., 1994) (Fig. 5.1). M cells are capable of sampling intestinal contents and transporting acquired antigens to underlying lymphoid cells also present in the Peyer's patch (Brandtzaeg, 1989). *Salmonella* also can invade other intestinal epithelial cells to gain access to deeper tissues (Francis et al., 1992; Scherer and Miller, 2001; Hancock and McPhee, 2005).

Uptake of *Salmonella* by non-phagocytic epithelial cells is facilitated by Type III Secretion Systems (TTSSs). Gram-negative bacteria use these syringe-like complexes to mediate direct translocation of bacterial virulence proteins into the host cell cytoplasm (Hueck, 1998). *Salmonella* Pathogenicity Island 1 (SPI-1) encodes proteins that initiate cytoskeletal rearrangements that lead to membrane ruffling and bacterial internalization (Watson et al., 1995; Galyov et al., 1997) and induce intestinal cell secretion and inflammation.

Once *Salmonella* invade the intestinal barrier, the bacteria encounter host immune cells such as neutrophils, lymphocytes and macrophages, which engulf the bacteria (Fields et al., 1986; McCormick et al., 1993) (Fig. 5.1). *Salmonella* serotypes capable of causing gastroenteritis are killed by host macrophages, and effectively cleared by the immune system. Invasion of M cells and intestinal epithelial cells induces a pro-inflammatory response in the intestinal mucosa (Fig. 5.1a). A combination of inflammatory cytokine production, neutrophil infiltration and fluid and electrolyte secretion by the epithelium results in the diarrhea associated with gastrointestinal enteritis (Fig. 5.1a).

Salmonella serotypes capable of causing typhoid fever localize inside macrophages and dendritic cells and replicate inside these host phagocytic cells during systemic infection (Gahring et al., 1990; Niedergang et al., 2000) (Fig. 5.1b). *Salmonella* Pathogenicity Island 2 (SPI-2) encodes genes required for intracellular replication, including a second TTSS (Shea et al., 1996; Cirillo et al., 1998; Hensel et al., 1998). During typhoid fever, *S.* Typhi breaches the intestinal barrier

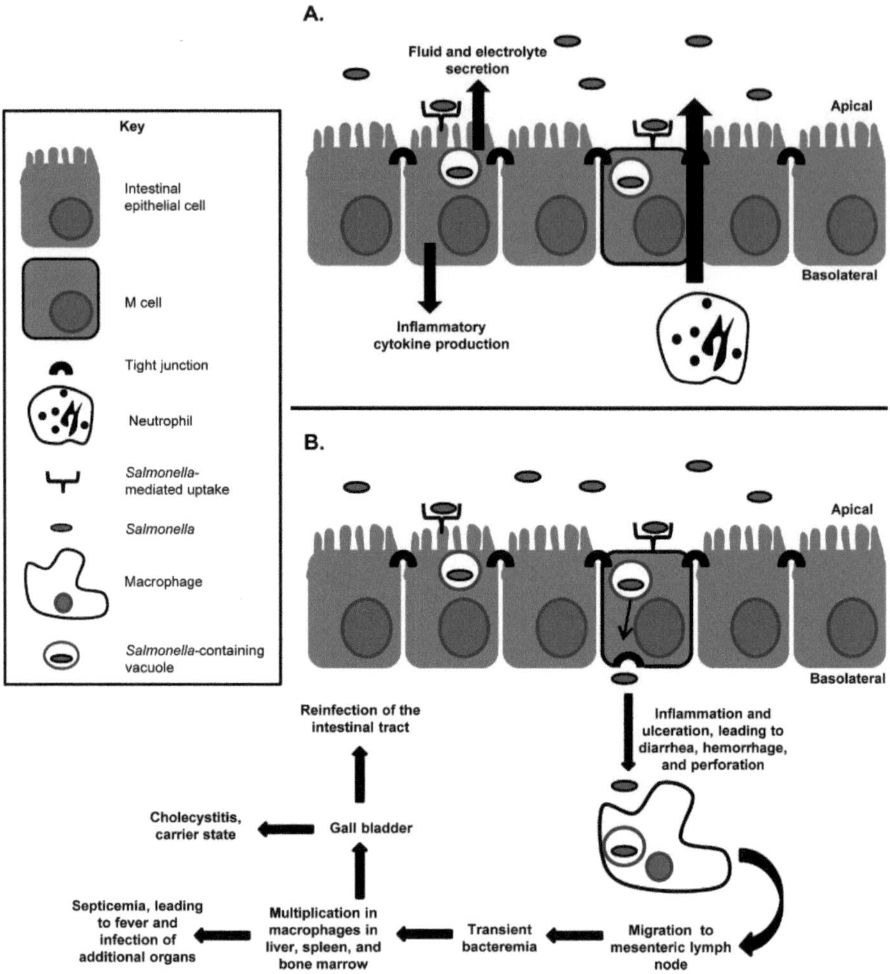

Fig. 5.1 Pathogenesis of *Salmonella* Typhimurium and *Salmonella* Typhi. Upon ingestion, *Salmonella* travel to the small intestine. The bacteria invade Microfold (M) cells and other intestinal epithelial cells through the apical surface to cause infection. (**A**). *Salmonella* Typhimurium remains localized in the small intestine and induces an inflammatory host immune response, resulting in bacterial clearance from immunocompetent individuals. (**B**). *Salmonella* Typhi escapes from intestinal epithelial cells at the basolateral surface of the intestinal epithelium, enters phagocytes, and evades the host innate immune response, resulting in systemic infection

and subsequently invades and replicates inside macrophages present in Peyer's patches. Macrophage uptake protects *Salmonella* from the host humoral immune response and provides access to the circulatory system (Richter-Dahlfors et al., 1997). Internalized *Salmonella* must survive both the nutrient-limiting conditions of the phagosome and the microbicidal environment created by host production of reactive oxygen and nitrogen species, hydrolytic enzymes and cationic antimicrobial peptides (CAMPs). The infected macrophages migrate to mesenteric lymph nodes (Fig. 5.1b). Transient bacteremia is observed in typhoid fever patients. *S.* Typhi then invades and multiplies inside macrophages present in the liver, spleen, and bone marrow (Ohl and Miller, 2001) (Fig. 5.1b). This intracellular replication can lead to septicemia, which results in fever and the infection of additional organs, and colonization of the gallbladder, which results in cholecystitis or a chronic carrier state (Fig. 5.1b). Gallbladder colonization can also lead to reinfection of the intestinal tract (Lai et al., 1992) (Fig. 5.1b).

5.1.3 Host Defenses Against Salmonella Infection

Salmonella in the gut lumen can induce a host inflammatory response mediated by several proteins excreted from the bacterial cytoplasm through *Salmonella* TTSS-1. These virulence factors may play a role in the influx of neutrophils into the intestine and the resulting inflammation that leads to diarrhea and other symptoms (Norris et al., 1998) (Fig. 5.1). Inflammation is an important component of innate immunity that plays a role in recruiting host immune cells to damaged tissue or invading microbes. Interaction with components of the innate immune system initiates a cascade of events that can lead to elimination of the microbe (Singh et al., 2009).

Salmonella can trigger inflammation and stimulate the immune system through many mechanisms, including activation of the Toll-like receptors (TLRs) on host cells. TLR recognition of distinct microbial-associated molecular patterns results in activation of downstream signaling pathways that culminate in expression of proinflammatory cytokines, chemokines, enzymes and peptides (Honda and Takeda, 2009). Macrophages and endothelial cells can detect bacterial lipopolysaccharide (LPS) through TLR4 and the accessory molecule MD2. TLR4/MD2 stimulation activates the host immune system, resulting in a powerful inflammatory response (Poltorak et al., 1998).

Upon host infection or injury, professional phagocytes and epithelial cells in the skin, lungs, intestine and other organs rapidly produce CAMPs, which mediate inflammation and stimulate the host immune system (Jones and Bevins, 1992; Zasloff, 1992). CAMPs are structurally diverse innate immune molecules found in many organisms, from plants and insects to amphibians and mammals. These ubiquitous defense peptides target membranes and have broad-range antimicrobial activity against bacteria, viruses and fungi (Radek and Gallo, 2007). Unsubstituted phosphates present on the lipid A and core regions of LPS give the bacterial surface a negative charge, which is thought to attract CAMPs (Vaara et al., 1979; Vaara, 1981). CAMPs permeabilize bacterial outer and inner membranes, possibly through pore formation or localized solubilization, to cause cell death (Vaara, 1992).

Bacteria sensitive to CAMP-mediated killing exhibit attenuated virulence in the host, while CAMP resistance confers an in vivo survival advantage (Groisman et al., 1992; Gunn et al., 2000; Rosenberger et al., 2004; Tamayo et al., 2005). Since many types of CAMPs are localized along common routes of infection, *Salmonella* and other pathogens have evolved mechanisms to detect and avoid killing by CAMPs (Peschel, 2002; Gunn, 2008; Lai and Gallo, 2009; Otto, 2009). Some bacteria use efflux pumps to prevent peptide entry, while others produce proteases to degrade CAMPs (Peschel, 2002; Lai and Gallo, 2009; Otto, 2009). *S.* Typhimurium environmental sensing activates signal transduction cascades leading to modified gene expression, increased CAMP resistance and evasion of other host immune defenses (Gunn, 2008).

5.1.4 Bacterial Modification of LPS and Other Surface Moieties

LPS comprises most of the outer surface on Gram-negative bacteria and consists of three regions: the lipid A, the core and the O antigen (Fig. 5.2a). LPS is composed of phospholipids and polysaccharides, provides stability to the cell and acts as a permeability barrier. The lipid A portion of LPS is hydrophobic and anchors the glycolipid to the bacterial outer membrane (Fig. 5.2a). The conserved hexa-acylated lipid A of enteric bacteria is known as endotoxin (Raetz, 1990; Raetz and Whitfield, 2002). The O-antigen consists of multiple repeating units of three to six sugars attached to the LPS core region and protrudes from the outer membrane of the cell (Raetz and Whitfield, 2002) (Fig. 5.2a).

In response to environmental stimuli, Gram-negative bacteria can alter the LPS through addition of positively charged molecules to the lipid A and core, alteration of acylation and modification of the chain length of the O-antigen. Both Gram-negative and Gram-positive bacteria can modify surface molecules to resist the action of CAMPs. Although LPS is not a component of the cell wall of Gram-positive bacteria, these bacteria are also known to modify surface molecules in ways that promote immune avoidance. Some Gram-positive bacteria, such as *Staphylococcus aureus*, are able to add D-alanine to teichoic acids present in the cell wall. Strains of *S. aureus* that lack the ability to modify teichoic acids with D-alanine are highly susceptible to killing by human neutrophils (Collins et al., 2002). *S. aureus* is also able to modify anionic phopsholipids present in the bacterial membrane by adding L-lysine, which produces lysylphosphatidylglycerol (L-PG). L-PG adds a positive net charge to the bacterial surface, which contributes to resistance to human defensins and evasion of neutrophil killing (Peschel et al., 2001).

While studies of bacterial surface modification have occurred in both Gram-positive and Gram-negative bacteria, the work on *Salmonella* has pushed the boundries of this field. As detailed below, much of the current research has focused on in vitro stimulation of *Salmonella* two-component regulatory systems (TCRSs) (Gunn and Richards, 2007; Gunn, 2008). In vitro, PhoPQ- and PmrAB-mediated lipid A modifications protect against CAMP killing and partially mask the proinflammatory properties of LPS. These and related regulatory systems also are activated during infection of a host, but the in vivo stimuli detected by these TCRSs remain largely unknown, as are the effects of LPS modification during the course of infection (Gunn, 2001; Gunn, 2008).

A.

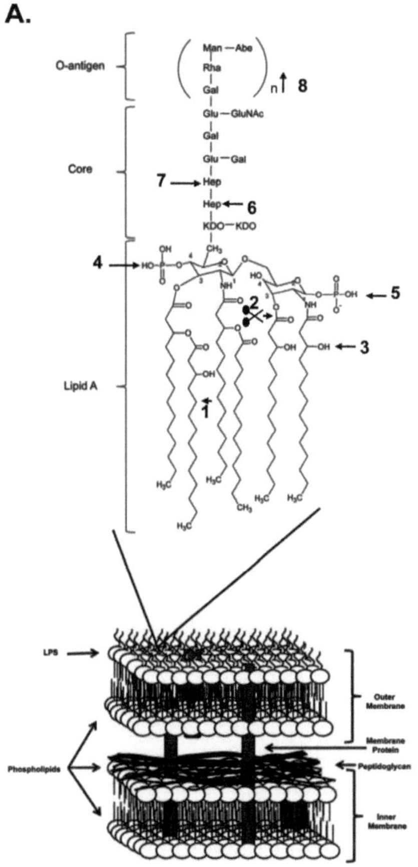

B.

PhoPQ-, PmrAB- and RcsCDB-mediated modifications to *Salmonella enterica* LPS

No. from 1A	Gene	LPS Modification	Structure of Modification
1	*lpxO*	2-hydroxymyristate addition to 3' position of lipid A	←OH
2	*pagL*	Deacylation at 3' position of lipid A	
3	*pagP*	Palmitate chain addition to 2' position of lipid A	
4	*pmrE/ pmrHFIJKL*	Synthesis of aminoarabinose precursor/ Aminoarabinose addition to the 1 and 4' phosphate of lipid A	
5	*pmrC*	Phosphoethanolamine addition to 1 and 4' phosphate of lipid A	(or)
6	*cptA*	Phosphoethanolamine addition to LPS core	
7	*pmrG*	Phosphate addition to heptose	
8	*cld*	O-antigen chain length determinant	

Structure removed
Structure added
n↑ Number increased

Fig. 5.2 Structure of unmodified LPS and *Salmonella*-mediated LPS modifications. (**A**) The structure of unmodified *Salmonella* Typhimurium LPS. (**B**) Activation of the *S.* Typhimurium PhoPQ, PmrAB and RcsCDB regulons results in covalent modification of LPS at the locations specified in the table

5.2 Two-Component Regulatory Systems

Many bacteria regulate gene expression in response to environmental conditions such as nutrient availability and osmolarity. To evade host immune cells and other defense mechanisms encountered in vivo, pathogens such as *S.* Typhimurium must respond to the local environment. These organisms detect environmental signals in part through two-component regulatory systems (TCRSs), which induce signal transduction cascades that regulate bacterial gene expression (Beier and Gross, 2006).

TCRSs usually consist of a sensor histidine protein kinase, often membrane-bound and containing a periplasmic domain, and a cytoplasmic response regulator

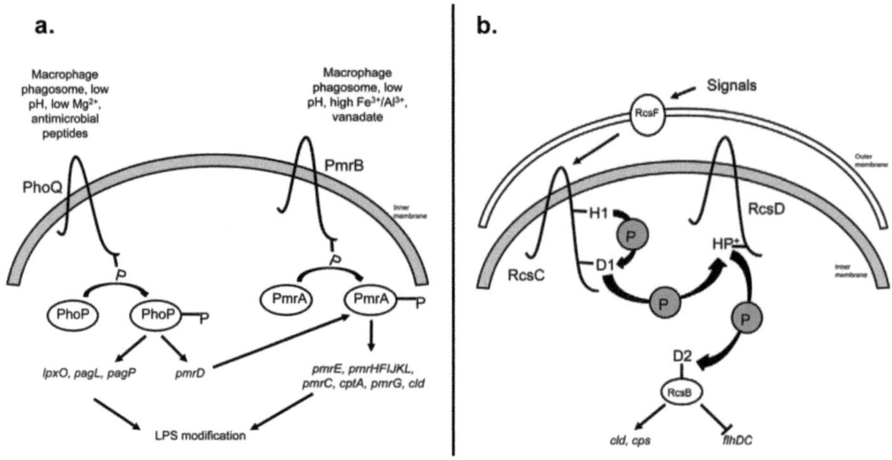

Fig. 5.3 *Salmonella* regulatory system-mediated environmental sensing, gene activation and LPS modification. Specific environmental signals activate the *Salmonella* regulatory systems PhoPQ, PmrAB and RcsCDB. (**a**) In the *Salmonella* two-component regulatory systems (TCRSs) PhoPQ and PmrAB, activation of PhoQ and PmrB results in autophosphorylation and phosphate transfer to PhoP or PmrA, respectively. Phosphorylated PhoP and PmrA activate several genes involved in LPS modification, CAMP resistance and bacterial survival. (**b**) In the RcsCDB phosphorylay, activation of RcsC results in autophosphorylation, phosphotransfer to RcsD and subsequent transfer to RcsC. Phosphorylated RcsC activates *cld*, which mediates O-antigen chain length

protein (Mascher et al., 2006) (Fig. 5.3). Upon sensing of an environmental stimulus, the sensor kinase undergoes a conformational change in the bacterial inner membrane and autophosphorylates a conserved histidine residue. The phosphate is transferred to a conserved aspartate residue on the cytoplasmic response regulator. The phosphorylated response regulator can transfer the phosphate to another cytoplasmic protein in the signal transduction cascade or directly act as a transcription factor to activate or repress genes under the control of the TCRS (Mascher et al., 2006).

Three important regulatory systems among those required for *Salmonella* virulence, LPS modification and CAMP resistance are PhoP-PhoQ (PhoPQ), PmrA-PmrB (PmrAB) and RcsC-RcsD-RcsB (RcsCDB) (Groisman et al., 1989; Miller et al., 1989; Virlogeux et al., 1995; Gunn and Miller, 1996; Guo et al., 1997; Arricau et al., 1998; Gunn et al., 1998) (Fig. 5.3a, b). These signaling cascades are induced in vivo in response to host environmental cues and regulate gene expression to promote *Salmonella* pathogenesis (Alpuche Aranda et al., 1992; Gunn et al., 2000; Merighi et al., 2005; Erickson and Detweiler, 2006). Several other pathogenic bacteria also encode PhoPQ, PmrAB and/or RcsCDB orthologs, illustrating the conserved importance of environmental sensing and LPS modification for bacterial persistence in the host (Oyston et al., 2000; Raetz, 2001; Moskowitz et al., 2004; Monsieurs et al., 2005; Huang et al., 2006; Mohapatra et al., 2007; Zusman et al., 2007; Adams et al., 2009). Below we will detail the functions of these systems, particularly in regard to LPS modification.

5.2.1 PhoP–PhoQ Regulatory System

PhoPQ is a well-studied TCRS in *Salmonella* and other bacteria (Groisman et al., 1989; Miller et al., 1989; Guo et al., 1998; Kawasaki et al., 2005). In this system, PhoQ is the inner membrane-bound sensor kinase and PhoP is the cytoplasmic response regulator. PhoPQ is activated in vitro in response to PhoQ detection of low pH (pH 5.5) or low concentrations (10 micromolar) of divalent cations such as magnesium (Mg^{2+}) or calcium (Ca^{2+}), as well as within the lumen of the small intestine, the macrophage phagolysosome and other host tissues and cell vacuoles (Miller et al., 1989; Alpuche Aranda et al., 1992; Garcia Vescovi et al., 1996; Bearson et al., 1998; Bader et al., 2003; Gibbons et al., 2005; Merighi et al., 2005) (Fig. 5.3a). PhoQ sensing of an environmental activation signal causes a conformational change resulting in autophosphorylation of PhoQ and phosphate transfer to PhoP (Miller et al., 1989; Gunn et al., 1996) (Fig. 5.3a). Signals such as Mg^{2+} and CAMPs interact with the PhoQ periplasmic domain (Bader et al., 2005). Structural analysis of this domain has allowed for the identification of an acidic region that functions in tethering this domain to the phospholipids of the bacterial membrane through the action of cation bridges (Bader et al., 2005; Gunn and Richards, 2007). CAMPs and Mg^{2+} interact with the same acidic regions of the PhoQ periplasmic domain. Although CAMPs and Mg^{2+} both bind to this acidic region, CAMP binding activates PhoQ, while Mg^{2+} (high concentrations of 1-2 millimolar) represses PhoQ (Garcia Vescovi et al., 1996).

Phosphorylated PhoP binds to target gene promoters and regulates transcription of both *pag* (*phoP*-activated genes) and *prg* (*phoP*-repressed genes) (Groisman et al., 1989; Miller et al., 1989) (Fig. 5.3a). The promoters of several PhoP-regulated genes contain a conserved PhoP-binding site, known as a PhoP box ([T/G]TTTAxxxxx[T/G]TTTA) (Kato et al., 1999; Yamamoto et al., 2002; Lejona et al., 2003). Both *S.* Typhimurium and *S.* Typhi PhoPQ directly control expression of over 50 genes, including *phoP* itself, and also indirectly regulate many other genes (Charles et al., 2009).

Activation of *Salmonella phoPQ* and *pag* is required for modification of LPS, intramacrophage survival and wild type (WT) pathogenesis (Miller et al., 1989; Alpuche Aranda et al., 1992; Guo et al., 1997). *Salmonella phoP*- or *phoQ*-null mutants are more susceptible to killing by CAMPs such as polymyxin B and defensins and exhibit attenuated virulence in humans and susceptible mice (Fields et al., 1986; Fields et al., 1989; Miller et al., 1989; Hohmann et al., 1996). Constitutive expression of *phoP* also results in attenuated virulence and decreased survival in macrophages, suggesting that the ability to switch between repression and activation of PhoPQ during infection is important (Miller and Mekalanos, 1990).

Several PhoP-activated genes encode products that mediate *Salmonella* LPS modifications and may promote virulence and CAMP resistance (Gunn and Miller, 1996; Guo et al., 1997; Guo et al., 1998). In addition to being able to directly activate genes associated with LPS modification, PhoPQ is also able to indirectly modify LPS by activating a second TCRS called PmrAB (Gunn and Miller, 1996;

Soncini and Groisman, 1996). Activation of PhoPQ leads to transcription of *pmrD* (Gunn et al., 2000; Kox et al., 2000). Expression of *pmrD* induces PmrAB activation through a phosphorylation-independent mechanism. PmrD binds to and stabilizes PmrA in its phosphorylated form (Kox et al., 2000; Kato and Groisman, 2004). In turn, expression of *pmrA* leads to repression of *pmrD* (Kato et al., 2003).

Upon PhoPQ activation, the aspartyl/asparaginyl beta-hydroxylase LpxO mediates addition of 2-hydroxymyristate to the lipid A (Guo et al., 1997; Gibbons et al., 2000) (Fig. 5.2, number 1). The outer membrane 3-*O*-deacylase PagL catalyzes deacylation at the 3 position of lipid A in the absence of the 4-amino-4-deoxy-L-arabinose (4-aminoarabinose, Ara4N) modification, which can affect *Salmonella* recognition by TLR4 on host cells (Trent et al., 2001a; Kawasaki et al., 2004a, 2005; Tran et al., 2005) (Fig. 5.2, number 2). The outer membrane palmitoyl transferase PagP adds a palmitate chain to lipid A, which reduces membrane fluidity and thereby aids in resistance to CAMP entry into the cell and subsequent killing (Guo et al., 1998) (Fig. 5.2, number 3). Aside from Ara4N and phosphoethanolamine (pEtN; see below), these modifications do not impact *Salmonella* virulence in susceptible mice (Gunn et al., 2000; Tamayo et al., 2005).

5.2.2 PmrA–PmrB Regulatory System

The PmrAB TCRS is encoded on the *Salmonella* genome as part of the three-gene *pmrCAB* operon. The first gene in the operon, *pmrC*, encodes an inner membrane-associated phosphoethanolamine transferase that is induced inside macrophage phagosomes and necessary for addition of pEtN to the lipid A region of LPS (Lee et al., 2004) (Fig. 5.3, number 5). The other two genes express the PmrA–PmrB regulatory system, in which PmrB is the membrane-bound sensor kinase and PmrA is the cytoplasmic transcriptional activator (Roland et al., 1993; Gunn and Miller, 1996; Soncini and Groisman, 1996) (Fig. 5.3a).

PmrAB is activated directly through PmrB sensing of low pH (pH 5.5) or high iron (Fe^{3+}) conditions (100 micromolar). Detection of environmental signals results in PmrB autophosphorylation, phosphate transfer to PmrA, activation of the PmrA regulon and CAMP resistance independent of PhoPQ (Gunn and Miller, 1996) (Fig. 5.3a). As previously mentioned, PmrAB can also be activated indirectly through PhoPQ and PmrD (Gunn et al., 2000; Kox et al., 2000; Wosten et al., 2000).

As high iron concentrations are more common in soil and water than in vivo, the PmrAB operon has been shown to be involved in *Salmonella* survival in non-host niches (Wosten et al., 2000; Chamnongpol et al., 2002). The PhoPQ and PmrAB regulons also are induced by other unknown signals in the intestinal lumen and in macrophages, as PhoP- and PmrA-regulated genes are expressed transiently in the murine intestine before *Salmonella* uptake by host cells (Alpuche Aranda et al., 1992; Gibbons et al., 2005; Merighi et al., 2005). PmrD is repressed by PmrA, highlighting the importance of this small protein in a unique regulatory feedback loop that monitors the amount of activated PmrA in the cell (Kato et al., 2003). In addition, PmrA can autoregulate the *pmrCAB* operon to increase expression of the regulatory system in response to induction signals (Gunn and Miller, 1996). PmrAB also is activated indirectly by another *S.* Typhimurium TCRS known as PreA–PreB (also called QseC–QseB) (Merighi et al., 2006).

PmrAB is a master regulator of LPS structure and modification and also is required for CAMP and Fe^{3+} resistance, as well as WT virulence in susceptible mice (Roland et al., 1993; Gunn et al., 2000; Chamnongpol et al., 2002; Kato et al., 2003; Kawasaki et al., 2005; Tamayo et al., 2005; Merighi et al., 2006). PmrA-mediated covalent addition of Ara4N and pEtN to the lipid A and core oligosaccharide regions of *Salmonella* LPS increases resistance to CAMPs such as polymyxin, CAP-37 and CAP57 (Shafer et al., 1984; Roland et al., 1993; Gunn and Miller, 1996; Gunn et al., 1998; Gunn et al., 2000; Zhou et al., 2001; Lee et al., 2004; Gibbons et al., 2005; Tamayo et al., 2005). *S.* Typhimurium strains with a specific mutation at nucleotide 505 of the *pmrA* operon (within *pmrA*) are constitutively active and resistant to killing by polymxyin and other CAMPs (Roland et al., 1993). These strains produce increased levels of Ara4N and pEtN on the LPS lipid A and core in response to constitutive activation of PmrA ($PmrA^c$) (Vaara et al., 1981; Roland et al., 1993; Helander et al., 1994; Gunn and Miller, 1996). Addition of the positively charged molecules Ara4N and pEtN to LPS reduces the overall negative charge of the bacterial outer membrane, resulting in reduced electrostatic interactions and bacterial repulsion of CAMPs. Therefore, $PmrA^c$ mutants are more resistant than WT *Salmonella* to neutrophil- and CAMP-mediated killing due to decreased CAMP binding to modified LPS (Vaara et al., 1979; Stinavage et al., 1989; Gunn and Miller, 1996).

Synthesis of Ara4N and addition to the lipid A is mediated by the *Salmonella* PmrA-activated genes *pmrE* (also known as *ugd*) and *pmrHFIJKLM* (the *pmrH* operon, also known as the *pbgP* or *arn* operon) (Gunn et al., 2000). Both *pmrE* and the *pmrH* operon are required for modification of the lipid A with Ara4N, resistance to polymyxin and other CAMPs in *S.* Typhimurium and *S.* Typhi (Baker et al., 1999; Gunn et al., 2000) (Fig. 5.2, number 4). These genes also are required for resistance to Fe^{3+} and virulence upon oral infection in BALB/c mice (Gunn and Miller, 1996; Gunn et al., 1998; Gunn et al., 2000; Zhou et al., 2001; Nishino et al., 2006). The PmrA- and RcsB-regulated gene *pmrE* is required for maximum *Salmonella* virulence and encodes a UDP-glucose dehydrogenase involved in synthesis of UDP-glucuronic acid, an Ara4N and colonic acid capsule precursor (Gunn et al., 2000; Yan et al., 2007).

Ara4N synthesis begins with the conversion of UDP-glucose to UDP-glucuronic acid by the UDP-glucose dehydrogenase Ugd. Next, PmrI catalyzes the oxidative decarboxylation of UDP-glucuronic acid (Breazeale et al., 2002; Raetz and Whitfield, 2002). This reaction generates an intermediate termed UDP-4-keto-pyranose. PmrH then cataylzes a transamination (with glutamic acid serving as the amine donor), which generates UDP-Ara4N (Breazeale et al., 2002). PmrF is responsible for transferring Ara4N to undecaprenyl phosphate (Zhou et al., 1999). This lipid carrier mediates translocation of Ara4N to the outer surface of the inner membrane, where PmrK transfers Ara4N to lipid A (Trent et al., 2001b).

The PmrA-activated pEtN phosphotransferases PmrC and CptA are required for addition of pEtN to the 1-phosphate (and 4'-phosphate) of lipid A and the LPS core heptose I (Hep I), respectively (Lee et al., 2004; Tamayo et al., 2005) (Fig. 5.2, numbers 5 and 6, respectively). *Salmonella* polymyxin B resistance is only modestly affected in *pmrC* and *cptA* single mutants, but double mutants show decreased survival in a competition assay against WT *S.* Typhimurium in mice (Tamayo et al., 2005). *S.* Typhimurium *pmrE* or *pmrF* mutants with constitutive activation of *pmrA* cannot add Ara4N to the lipid A but have increased amounts of pEtN on the 1-phosphate and the 4'-phosphate of lipid A (Zhou et al., 2001). In addition, the PmrA-regulated periplasmic phosphatase PmrG removes the phosphate from the heptose II (Hep II) of the LPS core (Nishino et al., 2006) (Fig. 5.2, number 7). This modification further reduces the negative charge of the bacterial outer membrane and promotes Fe^{3+} resistance in soil (Nishino et al., 2006). Furthermore, *ugtL* encodes an inner membrane protein that promotes the formation of monophosphorylated lipid A in *S.* Typhimurium (Shi et al., 2004). This modification was demonstrated, especially in combination with a mutation in *pmrA*, to have a dramatic effect on CAMP resistance.

O-Antigen production in *Salmonella* requires Wzx flippase, Wzy polymerase and Wzz_{st} chain length determinant (also known as *cld*). The wzz_{st} gene is located next to the *ugd* gene in the *Salmonella* genome and downstream of, but transcribed independently from, the genes required for O-antigen biosynthesis. Wzz_{st} mediates production of LPS containing long (L-type) O-antigen with 16–35 subunits and is activated by both PmrA and RcsB at the transcriptional level (see below) (Morona et al., 1995; Raetz and Whitfield, 2002; Delgado et al., 2006). Thus, *Salmonella* LPS likely contains longer chain O-antigen species in vivo, which may promote innate immune resistance.

5.2.3 RcsC–RcsD–RcsB Regulatory System

RcsC–RcsD–RcsB (RcsCDB) is another important *Salmonella* regulatory system involved in LPS modification. The Rcs phosphorelay was named for its role as a regulator of capsule synthesis, since it controls production of several outer membrane proteins and exopolysaccharides, including the colanic acid capsule and Vi antigen, and plays a role in biofilm maturation (Stout and Gottesman, 1990; Arricau et al., 1998; Detweiler et al., 2003; Dominguez-Bernal et al., 2004; Delgado et al., 2006; Erickson and Detweiler, 2006; Huang et al., 2006). The Rcs phosphorelay is found only in *Salmonella* and several other pathogenic *Enterobacteriaceae*.

RcsC and RcsDB are located next to each other in the *Salmonella* genome but are transcribed in opposite directions (Detweiler et al., 2003; Huang et al., 2006). Expression of RcsDB requires activation of the low pH-induced response regulator OmpR (Detweiler et al., 2003; Huang et al., 2006). The outer membrane sensor protein RcsF participates in RcsC activation (Majdalani and Gottesman, 2005; Majdalani et al., 2005) (Fig. 5.3b). RcsCDB is activated by envelope stress, such as high osmolarity, desiccation, low temperature (20°C), high concentrations of zinc and TDP-glucose (a substrate for O-antigen) and exposure to

CAMPs or β-lactam antibiotics (which target peptidoglycan) (Ophir and Gutnick, 1994; Gottesman, 1995; Sledjeski and Gottesman, 1996; Conter et al., 2002; Hagiwara et al., 2003; Mouslim et al., 2003; El-Kazzaz et al., 2004; Kaldalu et al., 2004).

The "modified" hybrid sensor kinase and phospho-transmitter of the Rcs system consists of two proteins, RcsC and RcsD (also known as YojN), which form a heterodimer in the cytoplasmic membrane (Takeda et al., 2001; Majdalani and Gottesman, 2005) (Fig. 5.3b). When RcsC detects an activation signal, it autophosphorylates a conserved histidine residue in its kinase domain and transfers the phosphate group to a conserved aspartate residue in its receiver domain. The intermediate protein RcsD mediates phosphotransfer from the aspartate of RcsC to a conserved histidine on RcsD and then to a conserved aspartate of the cytosolic response regulator RcsB. Phosphorylated RcsB activates target genes of the Rcs regulon (Takeda et al., 2001; Majdalani and Gottesman, 2005; Huang et al., 2006) (Fig. 5.3b).

Upon activation of the Rcs system, RcsB cooperates with the transcription factor RcsA to express the *cps* genes required for colonic acid capsule biosynthesis in *S.* Typhimurium or with TviA to promote Vi antigen production in *S.* Typhi (Gottesman et al., 1985; Mouslim et al., 2003; Erickson and Detweiler, 2006; Huang et al., 2006; Winter et al., 2009). Like PhoPQ and PmrAB, RcsCDB is involved in CAMP resistance and systemic infection in susceptible mice, and it also regulates genes related to chemotaxis, motility, cell division and metabolism (Detweiler et al., 2003; Dominguez-Bernal et al., 2004; Erickson and Detweiler, 2006; Huang et al., 2006). *S.* Typhimurium *rcsC* mutants exhibit impaired systemic infection, and overactivation of the Rcs system also leads to virulence attenuation (Detweiler et al., 2003; Mouslim et al., 2004).

The regulatory protein IgaA (Intracellular growth attenuator-A, also known as YrfF) tightly controls expression of PhoPQ and RcsCDB. IgaA regulation allows *Salmonella* to alter the composition of its surface molecules in response to rapidly changing environmental conditions and to maintain temporal control of host colonization and biofilm development (Cano et al., 2002; Dominguez-Bernal et al., 2004; Tierrez and Garcia-del Portillo, 2004; Huang et al., 2006). IgaA represses the Rcs phosphorelay early during host infection and premature activation results in attenuated virulence. Rcs activation is required at later stages of infection for intracellular growth and full *Salmonella* virulence (Dominguez-Bernal et al., 2004; Huang et al., 2006).

RcsCDB plays a role in LPS biosynthesis and modification by regulating O-antigen production. As mentioned earlier, activation of RcsCDB, PhoPQ and PmrAB promotes expression of wzz_{st} (*cld*), which encodes the chain length determinant of the O-antigen capsule (Morona et al., 1995; Raetz and Whitfield, 2002; Delgado et al., 2006). PmrA and RcsB directly bind to the wzz_{st} promoter to initiate gene transcription. Wzz_{st} mediates production of LPS with increased amounts (16–35 subunits) of L-type O-antigen, which increases *Salmonella* resistance to complement-mediated lysis (Morona et al., 1995; Raetz and Whitfield, 2002; Delgado et al., 2006). In addition, like PhoPQ and PmrAB, RcsCDB participates in activation of *ugd*, which is required for both Ara4N and colanic acid production

(Mouslim and Groisman, 2003; Tierrez and Garcia-del Portillo, 2004). RcsCDB also activates *ydeI*, which encodes a putative periplasmic protein that promotes polymyxin B resistance through means not involving LPS modification (Erickson and Detweiler, 2006).

5.3 Additional Bacteria Capable of Modifying LPS

Many Gram-negative bacteria have evolved mechanisms for LPS modification. Orthologs of the *Salmonella* PmrAB and PhoPQ TCRSs have been shown to regulate LPS modification genes in many different bacterial species, including *Yersinia spp., Pseudomonas aeruginosa* (*P. aeruginosa*) and *Escherichia coli* (*E. coli*) (Oyston et al., 2000; Raetz, 2001; Moskowitz et al., 2004; McPhee et al., 2006). Similar to the effects found in *S.* Typhimurium, lipid A modifications, such as the addition of Ara4N to lipid A, in these bacteria also confer resistance to CAMPs such as polymyxin B.

Some differences exist between *S.* Typhimurium and other Gram-negative bacteria that encode orthologs of PhoPQ, PmrAB and/or RcsCDB. For example, despite the absence of the *pmrD* gene, *Yersinia pestis* (*Y. pestis*) can mediate PhoP-dependent modification of lipid A with Ara4N through transcription of the *pbgP* and *ugd* genes (Winfield et al., 2005). *P. aeruginosa* also lacks the *pmrD* gene, suggesting that PmrA activation in this organism also occurs through a different mechanism than in *S.* Typhimurium (Moskowitz et al., 2004). PhoPQ activates *pmrD* in both *S.* Typhimurium and *E. coli*, but the PmrD proteins produced by these two organisms differ significantly and in *E. coli*, PmrD does not activate PmrA (Monsieurs et al., 2005). Despite differences, both strains are capable of modifying LPS (Monsieurs et al., 2005).

Orthologs of PmrAB have been identified in several bacteria including *Legionella pneumophila* (*L. pneumophila*), *Francisella tularensis* (*F. tularensis*), and *Acinetobacter baumannii* (*A. baumannii*) (Mohapatra et al., 2007; Zusman et al., 2007; Adams et al., 2009). PmrAB and PmrA have been shown to play important roles in intracellular replication in *L. pneumophila* and *F. tularensis,* respectively, but do not appear to affect LPS modification (Mohapatra et al., 2007; Al-Khodor et al., 2009). In *F. tularensis, pmrA* exists as an orphan response regulator gene without a cotranscribed *pmrB*, though unlinked sensor kinases may participate in its regulation (Mohapatra et al., 2007).

5.4 Summary and Significance

Bacteria have the ability to sense and respond to their surroundings. In vivo, *Salmonella* recognition of diverse environmental signals is essential for pathogenesis. TCRSs are primary factors in this process, with PhoPQ, PmrAB and the Rcs system mediating responses in vivo to modify the bacterial cell surface. *Salmonella*

LPS modifications play a role in CAMP resistance and virulence and also influence the host response to infection by altering cell signaling and the release of cytokines and other innate immune factors (Kawasaki et al., 2004b; Lee et al., 2004).

Since LPS is only found in Gram-negative bacteria, it serves as a potential target for development of new antimicrobial therapies. Targeted inhibition of the genes or enzymes involved in LPS modifications could render the bacteria more susceptible to killing by host innate immune responses. In addition, promoters of PhoP- or PmrA-regulated genes that are activated in vivo, such as the *pmrH* operon, may serve as potential promoters to express heterologous antigens in a live-attenuated *Salmonella* vaccine. Further investigation into bacterial TCRSs and LPS modifications will help determine how pathogens develop resistance to and evade killing by host immune defenses and thrive in hostile environments.

Acknowledgements Due to space constrains, the authors regret that all appropriate topics and references could not be covered. Work in the authors' laboratory on this research was supported by The National Institutes of Health Grant AI043521. All chemical structure drawings were created using ACD/ChemSketch 12.0 Freeware.

References

Adams, M., Nickel, G., Bajaksouzian, S., Lavender, H., Murthy, A., Jacobs, M., Bonomo, R. Resistance to colistin in *Acinetobacter baumannii* associated with mutations in the PmrAB two-component system. Antimicro Agents Chemother 53 (2009) 3628–3634.

Al-Khodor, S., Kalachikov, S., Morozova, I., Price, C., Abu Kwaik, Y. The PmrA/PmrB two-component system of *Legionella pneumophila* is a global regulator required for intracellular replication within macrophages and protozoa. Infect Immun 77 (2009) 374–386.

Alpuche Aranda, C., Swanson, J., Loomis, W., Miller, S. *Salmonella* typhimurium activates virulence gene transcription within acidified macrophage phagosomes. Proc Natl Acad Sci USA 89 (1992) 10079–10083.

Arricau, N., Hermant, D., Waxin, H., Ecobichon, C., Duffey, P., Popoff, M. The RcsB-RcsC regulatory system of *Salmonella* typhi differentially modulates the expression of invasion proteins, flagellin and Vi antigen in response to osmolarity. Mol Microbiol 29 (1998) 835–850.

Bader, M., Navarre, W., Shiau, W., Nikaido, H., Frye, J., McClelland, M., Fang, F., Miller, S. Regulation of *Salmonella* typhimurium virulence gene expression by cationic antimicrobial peptides. Mol Microbiol 50 (2003) 219–230.

Bader, M., Sanowar, S., Daley, M., Schneider, A., Cho, U., Xu, W., Klevit, R., Le Moual, H., Miller, S. Recognition of antimicrobial peptides by a bacterial sensor kinase. Cell 122 (2005) 461–472.

Baker, S., Gunn, J., Morona, R. The *Salmonella* typhi melittin resistance gene pqaB affects intracellular growth in PMA-differentiated U937 cells, polymyxin B resistance and lipopolysaccharide. Microbiology 145 (1999) 367–378.

Bearson, B., Wilson, L., Foster, J. A low pH-inducible, PhoPQ-dependent acid tolerance response protects *Salmonella* typhimurium against inorganic acid stress. J Bacteriol 180 (1998) 2409–2417.

Beier, D., Gross, R. Regulation of bacterial virulence by two-component systems. Curr Opin Microbiol 9 (2006) 143–152.

Bhutta, Z., Naqvi, S., Razzaq, R., Farooqui, B. Multidrug-resistant typhoid in children: presentation and clinical features. Rev Infect Dis 13 (1991) 832–836.

Brandtzaeg, P. Overview of the mucosal immune system. Curr Top Microbiol Immunol 146 (1989) 13–25.

Breazeale, S., Ribeiro, A., Raetz, C., Oxidative decarboxylation of UDP-glucuronic acid in extracts of polymyxin-resistant *Escherichia coli*. Origin of lipid a species modified with 4-amino-4-deoxy-L-arabinose. J Biol Chem 277 (2002) 2886–2896.

Cano, D., Dominguez-Bernal, G., Tierrez, A., Garcia-Del Portillo, F., Casadesus, J. Regulation of capsule synthesis and cell motility in *Salmonella* enterica by the essential gene igaA. Genetics 162 (2002) 1513–1523.

Carter, P., Collins, F. The route of enteric infection in normal mice. J Exp Med 139 (1974) 1189–1203.

Centers for Disease Control and Prevention STI. Salmonellosis (2008). http://www.cdc.gov/nczved/dfbmd/disease_listing/salmonellosis_ti.html

Chamnongpol, S., Dodson, W., Cromie, M., Harris, Z., Groisman, E. Fe(III)-mediated cellular toxicity. Mol Microbiol 45 (2002) 711–719.

Charles, R., Harris, J., Chase, M., Lebrun, L., Sheikh, A., LaRocque, R., Longvinenko, T., Rollins, S., Tarique, A., Hohmann, E., Rosenberg, I., Krastins, B., Sarracino, D., Qadri, F., Calderwood, B., Ryan, E. Comparative proteomic analysis of the PhoP regulon in *Salmonella enterica* serovar Typhi versus Typhimurium. PLoS One 4 (2009) e6994.

Cirillo, D., Valdivia, R., Monack, D., Falkow, S. Macrophage-dependent induction of the *Salmonella* pathogenicity island 2 type III secretion system and its role in intracellular survival. Mol Microbiol 30 (1998) 175–188.

Clark, M., Jepson, M., Simmons, N., Hirst, B. Preferential interaction of *Salmonella* typhimurium with mouse Peyer's patch M cells. Res Microbiol 145 (1994) 543–552.

Collins, L., Kristian, S., Weidenmaier, C., Faigle, M., Van Kessel, K., Van Strijp, J., Gotz, F., Neumeister, B., Peschel, A. *Staphylococcus aureus* strains lacking D-alanine modifications of teichoic acids are highly susceptible to human neutrophil killing and are virulence attenuated in mice. J Infect Dis 186 (2002) 214–219.

Conter, A., Sturny, R., Gutierrez, C., Cam, K. The RcsCB His-Asp phosphorelay system is essential to overcome chlorpromazine-induced stress in *Escherichia coli*. J Bacteriol 184 (2002) 2850–2853.

Crump, J., Luby, S., Mintz, E. The global burden of typhoid fever. Bull World Health Organ 82 (2004) 346–353.

Delgado, M., Mouslim, C., Groisman, E. The PmrA/PmrB and RcsC/YojN/RcsB systems control expression of the *Salmonella* O-antigen chain length determinant. Mol Microbiol 60 (2006) 39–50.

Detweiler, C., Monack, D., Brodsky, I., Mathew, H., Falkow, S. virK, somA and rcsC are important for systemic *Salmonella enterica* serovar Typhimurium infection and cationic peptide resistance. Mol Microbiol 48 (2003) 385–400.

Dominguez-Bernal, G., Pucciarelli, M., Ramos-Morales, F., García-Quintanilla, M., Cano, D., Casadesús, J., García-del Portillo, F. Repression of the RcsC-YojN-RcsB phosphorelay by the IgaA protein is a requisite for *Salmonella* virulence. Mol Microbiol 53 (2004) 1437–1449.

Dutta, U., Garg, P., Kumar, R., Tandon, R. Typhoid carriers among patients with gallstones are at increased risk for carcinoma of the gallbladder. Am J Gastroenterol 95 (2000) 784–787.

El-Kazzaz, W., Morita, T., Tagami, H., Inada, T., Aiba, H. Metabolic block at early stages of the glycolytic pathway activates the Rcs phosphorelay system via increased synthesis of dTDP-glucose in *Escherichia coli.* Mol Microbiol 51 (2004) 1117–1128.

Erickson, K., Detweiler, C. The Rcs phosphorelay system is specific to enteric pathogens/commensals and activates ydeI, a gene important for persistent *Salmonella* infection of mice. Mol Microbiol 62 (2006) 883–894.

Fields, P., Groisman, E., Heffron, F. A *Salmonella* locus that controls resistance to microbicidal proteins from phagocytic cells. Science 243 (1989) 1059–1062.

Fields, P., Swanson, R., Haidaris, C., Heffron, F. Mutants of *Salmonella* typhimurium that cannot survive within the macrophage are avirulent. Proc Natl Acad Sci USA 83 (1986) 5189–5193.

Francis, C., Starnbach, M., Falkow, S. Morphological and cytoskeletal changes in epithelial cells occur immediately upon interaction with *Salmonella* typhimurium grown under low-oxygen conditions. Mol Microbiol 6 (1992) 3077–3087.

Gahring, L., Heffron, F., Finlay, B., Falkow, S. Invasion and replication of *Salmonella* typhimurium in animal cells. Infect Immun 58 (1990) 443–448.

Galyov, E., Wood, M., Rosqvist, R., Mullan, P., Watson, P., Hedges, S., Wallis, T. A secreted effector protein of *Salmonella* dublin is translocated into eukaryotic cells and mediates inflammation and fluid secretion in infected ileal mucosa. Mol Microbiol 25 (1997) 903–912.

Garcia Vescovi, E., Soncini, F., Groisman, E. Mg2+ as an extracellular signal: environmental regulation of *Salmonella* virulence. Cell 84 (1996) 165–174.

Gibbons, H., Kalb, S., Cotter, R., Raetz, C. Role of Mg2+ and pH in the modification of *Salmonella* lipid A after endocytosis by macrophage tumour cells. Mol Microbiol 55 (2005) 425–440.

Gibbons, H., Lin, S., Cotter, R., Raetz, C. Oxygen requirement for the biosynthesis of the S-2-hydroxymyristate moiety in *Salmonella* typhimurium lipid A. Function of LpxO, A new Fe2+/alpha-ketoglutarate-dependent dioxygenase homologue. J Biol Chem 275 (2000) 32940–32949.

Gottesman, S. (ed). Regulation of Capsule Synthesis: Modification of the Two-Component Paradigm by an Accessory Unstable Regulator. American Society for Microbiology, Washington DC, (1995).

Gottesman, S., Trisler, P., Torres-Cabassa, A. Regulation of capsular polysaccharide synthesis in *Escherichia coli* K-12: characterization of three regulatory genes. J Bacteriol 162 (1985) 1111–1119.

Groisman, E., Chiao, E., Lipps, C., Heffron, F., *Salmonella* typhimurium phoP virulence gene is a transcriptional regulator. Proc Natl Acad Sci USA 86 (1989) 7077–7081.

Groisman, E., Mouslim, C. Sensing by bacterial regulatory systems in host and non-host environments. Nat Rev Microbiol 4 (2006) 705–709.

Groisman, E., Parra-Lopez, C., Salcedo, M., Lipps, C., Heffron, F. Resistance to host antimicrobial peptides is necessary for *Salmonella* virulence. Proc Natl Acad Sci USA 89 (1992) 11939–11943.

Gunn, J. Bacterial modification of LPS and resistance to antimicrobial peptides. J Endotoxin Res 7 (2001) 57–62.

Gunn, J. The *Salmonella* PmrAB regulon: lipopolysaccharide modifications, antimicrobial peptide resistance and more. Trends Microbiol 16 (2008) 284–290.

Gunn, J., Hohmann, E., Miller, S. Transcriptional regulation of *Salmonella* virulence: a PhoQ periplasmic domain mutation results in increased net phosphotransfer to PhoP. J Bacteriol 178 (1996) 6369–6373.

Gunn, J., Lim, K., Krueger, J., Kim, K., Guo, L., Hackett, M., Miller, S. PmrA-PmrB-regulated genes necessary for 4-aminoarabinose lipid A modification and polymyxin resistance. Mol Microbiol 27 (1998) 1171–1182.

Gunn, J., Miller, S. PhoP-PhoQ activates transcription of pmrAB, encoding a two-component regulatory system involved in *Salmonella* typhimurium antimicrobial peptide resistance. J Bacteriol 178 (1996) 6857–6864.

Gunn, J., Richards, S. Recognition and integration of multiple environmental signals by the bacterial sensor kinase PhoQ. Cell Host Microbe 1 (2007) 163–165.

Gunn, J., Ryan, S., Van Velkinburgh, J., Ernst, R., Miller, S. Genetic and functional analysis of a PmrA-PmrB-regulated locus necessary for lipopolysaccharide modification, antimicrobial peptide resistance, and oral virulence of *Salmonella enterica* serovar typhimurium. Infect Immun 68 (2000) 6139–6146.

Guo, L., Lim, K., Gunn, J., Bainbridge, B., Darveau, R., Hackett, M., Miller, S. Regulation of lipid A modifications by *Salmonella* typhimurium virulence genes phoP-phoQ. Science 276 (1997) 250–253.

Guo, L., Lim, K., Poduje, C., Daniel, M., Gunn, J., Hackett, M., Miller, S. Lipid A acylation and bacterial resistance against vertebrate antimicrobial peptides. Cell 95 (1998) 189–198.

Hagiwara, D., Sugiura, M., Oshima, T., Mori, H., Aiba, H., Yamashino, T., Mizuno, T. Genome-wide analyses revealing a signaling network of the RcsC-YojN-RcsB phosphorelay system in *Escherichia coli*. J Bacteriol 185 (2003) 5735–5746.

Hancock, R., McPhee, J. *Salmonella's* sensor for host defense molecules. Cell 122 (2005) 320–322.

Helander, I., Kilpelainen, I., Vaara, M. Increased substitution of phosphate groups in lipopolysaccharides and lipid A of the polymyxin-resistant pmrA mutants of *Salmonella* typhimurium: a 31P-NMR study. Mol Microbiol 11 (1994) 481–487.

Hensel, M., Shea, J., Waterman, S., Mundy, R., Nikolaus, T., Banks, G., Vazquez-Torres, A., Gleeson, C., Fang, F., Holden, D. Genes encoding putative effector proteins of the type III secretion system of *Salmonella* pathogenicity island 2 are required for bacterial virulence and proliferation in macrophages. Mol Microbiol 30 (1998) 163–174.

Hohmann, E., Oletta, C., Miller, S. Evaluation of a phoP/phoQ-deleted, aroA-deleted live oral *Salmonella* typhi vaccine strain in human volunteers. Vaccine 14 (1996) 19–24.

Honda, K., Takeda, K. Regulatory mechanisms of immune responses to intestinal bacteria. Mucosal Immunol 2 (2009) 187–196.

Huang, Y., Ferrieres, L., Clarke, D. The role of the Rcs phosphorelay in Enterobacteriaceae. Res Microbiol 157 (2006) 206-212.

Hueck, C. Type III protein secretion systems in bacterial pathogens of animals and plants. Microbiol Mol Biol Rev 62 (1998) 379–433.

Jones, B., Ghori, N., Falkow, S. *Salmonella* typhimurium initiates murine infection by penetrating and destroying the specialized epithelial M cells of the Peyer's patches. J Exp Med 180 (1994) 15–23.

Jones, D., Bevins, C. Paneth cells of the human small intestine express an antimicrobial peptide gene. J Biol Chem 267 (1992) 23216–23225.

Kaldalu, N., Mei, R., Lewis, K. Killing by ampicillin and ofloxacin induces overlapping changes in *Escherichia coli* transcription profile. Antimicrob Agents Chemother 48 (2004) 890–896.

Kato, A., Groisman, E. Connecting two-component regulatory systems by a protein that protects a response regulator from dephosphorylation by its cognate sensor. Genes Dev 18 (2004) 2302–2313.

Kato, A., Latifi, T., Groisman, E. Closing the loop: the PmrA/PmrB two-component system negatively controls expression of its posttranscriptional activator PmrD. Proc Natl Acad Sci USA 100 (2003) 4706–4711.

Kato, A., Tanabe, H., Utsumi, R. Molecular characterization of the PhoP-PhoQ two-component system in *Escherichia coli* K-12: identification of extracellular Mg2+-responsive promoters. J Bacteriol 181 (1999) 5516–5520.

Kawasaki, K., Ernst, R., Miller, S. 3-O-deacylation of lipid A by PagL, a PhoP/PhoQ-regulated deacylase of *Salmonella* typhimurium, modulates signaling through Toll-like receptor 4. J Biol Chem 279 (2004a) 20044–20048.

Kawasaki, K., Ernst, R., Miller, S. Deacylation and palmitoylation of lipid A by Salmonellae outer membrane enzymes modulate host signaling through Toll-like receptor 4. J Endotoxin Res 10 (2004b) 439–444.

Kawasaki, K., Ernst, R., Miller, S. Inhibition of *Salmonella enterica* serovar Typhimurium lipopolysaccharide deacylation by aminoarabinose membrane modification. J Bacteriol 187 (2005) 2448–2457.

Kox, L., Wosten, M., Groisman, E. A small protein that mediates the activation of a two-component system by another two-component system. EMBO J 19 (2000) 1861–1872.

Kraus, M., Amatya, B., Kimula, Y. Histopathology of typhoid enteritis: morphologic and immunophenotypic findings. Mod Pathol 12 (1999) 949–955.

Kumar, N., Nostrant, T., Appelman, H. The histopathologic spectrum of acute self-limited colitis (acute infectious-type colitis). Am J Surg Pathol 6 (1982) 523–529.

Lai, C., Chan, R., Cheng, A., Sung, J., Leung, J. Common bile duct stones: a cause of chronic salmonellosis. Am J Gastroenterol 87 (1992) 1198–1199.

Lai, Y., Gallo, R. AMPed up immunity: how antimicrobial peptides have multiple roles in immune defense. Trends Immunol 30 (2009) 131–141.

Lee, H., Hsu, F., Turk, J., Groisman, E. The PmrA-regulated pmrC gene mediates phosphoethanolamine modification of lipid A and polymyxin resistance in *Salmonella enterica.* J Bacteriol 186 (2004) 4124–4133.

Lejona, S., Aguirre, A., Cabeza, M., Garcia Vescovi, E., Soncini, F. Molecular characterization of the Mg2+-responsive PhoP-PhoQ regulon in *Salmonella enterica.* J Bacteriol 185 (2003) 6287–6294.

Majdalani, N., Gottesman, S. The Rcs phosphorelay: a complex signal transduction system. Annu Rev Microbiol 59 (2005) 379–405.

Majdalani, N., Heck, M., Stout, V., Gottesman, S. Role of RcsF in signaling to the Rcs phosphorelay pathway in *Escherichia coli.* J Bacteriol 187 (2005) 6770–6778.

Mascher, T., Helmann, J., Unden, G. Stimulus perception in bacterial signal-transducing histidine kinases. Microbiol Mol Biol Rev 70 (2006) 910–938.

McCormick, B., Colgan, S., Delp-Archer, C., Miller, S., Madara, J. *Salmonella* typhimurium attachment to human intestinal epithelial monolayers: transcellular signalling to subepithelial neutrophils. J Cell Biol 123 (1993) 895–907.

McCormick, B., Miller, S., Carnes, D., Madara, J. Transepithelial signaling to neutrophils by *salmonellae*: a novel virulence mechanism for gastroenteritis. Infect Immun 63 (1995) 2302–2309.

McPhee, J., Bains, M., Winsor, G., Lewenza, S., Kwasnicka, A., Brazas, M., Brinkman, F., Hancock, R. Contribution of the PhoP-PhoQ and PmrA-PmrB two-component regulatory systems to Mg2+-induced gene regulation in *Pseudomonas aeruginosa.* J Bacteriol 188 (2006) 3995–4006.

Merighi, M., Carroll-Portillo, A., Septer, A., Bhatiya, A., Gunn, J. Role of *Salmonella enterica* serovar Typhimurium two-component system PreA/PreB in modulating PmrA-regulated gene transcription. J Bacteriol 188 (2006) 141–149.

Merighi, M., Ellermeier, C., Slauch, J., Gunn, J. Resolvase-in vivo expression technology analysis of the *Salmonella enterica* serovar Typhimurium PhoP and PmrA regulons in BALB/c mice. J Bacteriol 187 (2005) 7407–7416.

Miller, S., Kukral, A., Mekalanos, J. A two-component regulatory system (phoP phoQ) controls *Salmonella* typhimurium virulence. Proc Natl Acad Sci USA 86 (1989) 5054–5058.

Miller, S., Mekalanos, J. Constitutive expression of the phoP regulon attenuates *Salmonella* virulence and survival within macrophages. J Bacteriol 172 (1990) 2485–2490.

Mohapatra, N., Soni, S., Bell, B., Warren, R., Ernst, R., Muszynski, A., Carlson, R., Gunn, J. Identification of an orphan response regulator required for the virulence of *Francisella spp.* and transcription of pathogenicity island genes. Infect Immun 75 (2007) 3305–3314.

Monsieurs, P., De Keersmaecker, S., Navarre, W., Bader, M., De Smet, F., McClelland, M., Fang, F., De Moor, B., Vanderleyden, J., Marchal, K. Comparison of the PhoPQ regulon in *Escherichia coli* and *Salmonella* typhimurium. J Mol Evol 60 (2005) 462–474.

Morona, R., van den Bosch, L., Manning, P. Molecular, genetic, and topological characterization of O-antigen chain length regulation in *Shigella flexneri.* J Bacteriol 177 (1995) 1059–1068.

Moskowitz, S., Ernst, R., Miller, S. PmrAB, a two-component regulatory system of *Pseudomonas aeruginosa* that modulates resistance to cationic antimicrobial peptides and addition of aminoarabinose to lipid A. J Bacteriol 186 (2004) 575–579.

Mouslim, C., Delgado, M., Groisman, E. Activation of the RcsC/YojN/RcsB phosphorelay system attenuates *Salmonella* virulence. Mol Microbiol 54 (2004) 386–395.

Mouslim, C., Groisman, E. Control of the *Salmonella* ugd gene by three two-component regulatory systems. Mol Microbiol 47 (2003) 335–344.

Mouslim, C., Latifi, T., Groisman, E. Signal-dependent requirement for the co-activator protein RcsA in transcription of the RcsB-regulated ugd gene. J Biol Chem 278 (2003) 50588–50595.

Niedergang, F., Sirard, J., Blanc, C., Kraehenbuhl, J. Entry and survival of *Salmonella* typhimurium in dendritic cells and presentation of recombinant antigens do not require macrophage-specific virulence factors, Proc Natl Acad Sci USA 97 (2000) 14650–14655.

Nishino, K., Hsu, F., Turk, J., Cromie, M., Wosten, M., Groisman, E. Identification of the lipopolysaccharide modifications controlled by the *Salmonella* PmrA/PmrB system mediating resistance to Fe(III) and Al(III). Mol Microbiol 61 (2006) 645–654.

Norris, F., Wilson, M., Wallis, T., Galyov, E., Majerus, P. SopB, a protein required for virulence of *Salmonella* dublin, is an inositol phosphate phosphatase. Proc Natl Acad Sci USA 95 (1998) 14057–14059.

Ohl, M., Miller, S. *Salmonella*: a model for bacterial pathogenesis. Annu Rev Med 52 (2001) 259–274.

Ophir, T., Gutnick, D. A Role for Exopolysaccharides in the protection of microorganisms from desiccation. Appl Environ Microbiol 60 (1994) 740–745.

Otto, M. Bacterial sensing of antimicrobial peptides. Contrib Microbiol 16 (2009) 136–149.

Oyston, P., Dorrell, N., Williams, K., Li, S., Green, M., Titball, R., Wren, B. The response regulator PhoP is important for survival under conditions of macrophage-induced stress and virulence in *Yersinia pestis*. Infect Immun 68 (2000) 3419–3425.

Peschel, A. How do bacteria resist human antimicrobial peptides?, Trends Microbiol 10 (2002) 179–186.

Peschel, A., Jack, R., Otto, M., Collins, L., Staubitz, P., Nicholson, G., Kalbacher, H., Nieuwenhuizen, W., Jung, G., Tarkowski, A., van Kessel, K., van Strijp, J. *Staphylococcus aureus* resistance to human defensins and evasion of neutrophil killing via the novel virulence factor MprF is based on modification of membrane lipids with l-lysine. J Exp Med 193 (2001) 1067–1076.

Poltorak, A., He, X., Smirnova, I., Liu, M., Van Huffel, C., Du, X., Birdwell, D., Alejos, E., Silva, M., Galanos, C., Freudenberg, M., Ricciardi-Castagnoli, P., Layton, B., Beutler, B. Defective LPS signaling in C3H/HeJ and C57BL/10ScCr mice: mutations in Tlr4 gene. Science 282 (1998) 2085–2088.

Radek, K., Gallo, R. Antimicrobial peptides: natural effectors of the innate immune system. Semin Immunopathol 29 (2007) 27–43.

Raetz, C. Biochemistry of endotoxins. Annu Rev Biochem 59 (1990) 129–170.

Raetz, C. Regulated covalent modifications of lipid A. J Endotoxin Res 7 (2001) 73–78.

Raetz, C., Whitfield, C. Lipopolysaccharide endotoxins. Annu Rev Biochem 71 (2002) 635–700.

Richter-Dahlfors, A., Buchan, A., Finlay, B. Murine salmonellosis studied by confocal microscopy: *Salmonella* typhimurium resides intracellularly inside macrophages and exerts a cytotoxic effect on phagocytes *in vivo*. J Exp Med 186 (1997) 569–580.

Roland, K., Martin, L., Esther, C., Spitznagel, J. Spontaneous pmrA mutants of *Salmonella* typhimurium LT2 define a new two-component regulatory system with a possible role in virulence. J Bacteriol 175 (1993) 4154–4164.

Rosenberger, C., Gallo, R., Finlay, B. Interplay between antibacterial effectors: a macrophage antimicrobial peptide impairs intracellular *Salmonella* replication. Proc Natl Acad Sci USA 101 (2004) 2422–2427.

Scherer, C., Miller, S. Molecular pathogenesis of *Salmonellae*. In Groisman, E. (ed), Principles of Bacterial Pathogenesis. Academic Press, San Diego, CA (2001), pp. 265–333.

Shafer, W., Martin, L., Spitznagel, J. Cationic antimicrobial proteins isolated from human neutrophil granulocytes in the presence of diisopropyl fluorophosphate. Infect Immun 45 (1984) 29–35.

Shea, J., Hensel, M., Gleeson, C., Holden, D. Identification of a virulence locus encoding a second type III secretion system in *Salmonella* typhimurium. Proc Natl Acad Sci USA 93 (1996) 2593–2597.

Shi, Y., Cromie, M., Hsu, F., Turk, J., Groisman, E. PhoP-regulated *Salmonella* resistance to the antimicrobial peptides magainin 2 and polymyxin B. Mol Microbiol 53 (2004) 229–241.

Singh, V., Singh, K., Amdekar, S., Singh, D., Tripathi, P., Sharma, G., Yadav, H. Innate and specific gut-associated immunity and microbial interference. FEMS Immunol Med Microbiol 55 (2009) 6–12.

Sledjeski, D., Gottesman, S. Osmotic shock induction of capsule synthesis in *Escherichia coli* K-12. J Bacteriol 178 (1996) 1204–1206.

Soncini, F., Groisman, E. Two-component regulatory systems can interact to process multiple environmental signals. J Bacteriol 178 (1996) 6796–6801.

Stinavage, P., Martin, L., Spitznagel, J. O antigen and lipid A phosphoryl groups in resistance of *Salmonella* typhimurium LT-2 to nonoxidative killing in human polymorphonuclear neutrophils. Infect Immun 57 (1989) 3894–3900.

Stout, V., Gottesman, S. RcsB and RcsC: a two-component regulator of capsule synthesis in *Escherichia coli.* J Bacteriol 172 (1990) 659–669.

Takeda, S., Fujisawa, Y., Matsubara, M., Aiba, H., Mizuno, T. A novel feature of the multistep phosphorelay in *Escherichia coli*: a revised model of the RcsC → YojN → RcsB signalling pathway implicated in capsular synthesis and swarming behaviour. Mol Microbiol 40 (2001) 440–450.

Takeuchi, A. Electron microscope studies of experimental *Salmonella* infection. I. Penetration into the intestinal epithelium by *Salmonella* typhimurium, Am J Pathol 50 (1967) 109–136.

Tamayo, R., Choudhury, B., Septer, A., Merighi, M., Carlson, R., Gunn, J., Identification of cptA, a PmrA-regulated locus required for phosphoethanolamine modification of the *Salmonella enterica* serovar typhimurium lipopolysaccharide core. J Bacteriol 187 (2005) 3391–3399.

Tierrez, A., Garcia-del Portillo, F. The *Salmonella* membrane protein IgaA modulates the activity of the RcsC-YojN-RcsB and PhoP-PhoQ regulons. J Bacteriol 186 (2004) 7481–7489.

Tran, A., Lester, M., Stead, C., Raetz, C., Maskell, D., McGrath, S., Cotter, R., Trent, M. Resistance to the antimicrobial peptide polymyxin requires myristoylation of *Escherichia coli* and *Salmonella* typhimurium lipid A. J Biol Chem 280 (2005) 28186–28194.

Trent, M., Pabich, W., Raetz, C., Miller, S. A PhoP/PhoQ-induced Lipase (PagL) that catalyzes 3-O-deacylation of lipid A precursors in membranes of *Salmonella* typhimurium. J Biol Chem 276 (2001a) 9083–9092.

Trent, M., Ribeiro, A., Doerrler, W., Lin, S., Cotter, R., Raetz, C. Accumulation of a polyisoprene-linked amino sugar in polymyxin-resistant *Salmonella* typhimurium and *Escherichia coli*: structural characterization and transfer to lipid A in the periplasm. J Biol Chem 276 (2001b) 43132–43144.

Vaara, M. Increased outer membrane resistance to ethylenediaminetetraacetate and cations in novel lipid A mutants. J Bacteriol 148 (1981) 426–434.

Vaara, M. Agents that increase the permeability of the outer membrane. Microbiol Rev 56 (1992) 395–411.

Vaara, M., Vaara, T., Jensen, M., Helander, I., Nurminen, M., Rietschel, E., Mäkelä, P. Characterization of the lipopolysaccharide from the polymyxin-resistant pmrA mutants of *Salmonella* typhimurium. FEBS Lett 129 (1981) 145–149.

Vaara, M., Vaara, T., Sarvas, M. Decreased binding of polymyxin by polymyxin-resistant mutants of *Salmonella* typhimurium. J Bacteriol 139 (1979) 664–667.

Virlogeux, I., Waxin, H., Ecobichon, C., Popoff, M. Role of the viaB locus in synthesis, transport and expression of *Salmonella* typhi Vi antigen. Microbiology 141 (1995) 3039–3047.

Watson, P., Paulin, S., Bland, A., Jones, P., Wallis, T. Characterization of intestinal invasion by *Salmonella* typhimurium and *Salmonella* dublin and effect of a mutation in the invH gene. Infect Immun 63 (1995) 2743–2754.

Winfield, M., Latifi, T., Groisman, E. Transcriptional regulation of the 4-amino-4-deoxy-L-arabinose biosynthetic genes in *Yersinia pestis*. J Biol Chem 280 (2005) 14765–14772.

Winter, S., Winter, M., Thiennimitr, P., Gerriets, V., Nuccio, S., Rüssmann, H., Bäumler, A. The TviA auxiliary protein renders the *Salmonella enterica* serotype Typhi RcsB regulon responsive to changes in osmolarity. Mol Microbiol 74 (2009) 175–193.

Wosten, M., Kox, L., Chamnongpol, S., Soncini, F., Groisman, E. A signal transduction system that responds to extracellular iron. Cell 103 (2000) 113–125.

Yamamoto, K., Ogasawara, H., Fujita, N., Utsumi, R., Ishihama, A. Novel mode of transcription regulation of divergently overlapping promoters by PhoP, the regulator of two-component system sensing external magnesium availability. Mol Microbiol 45 (2002) 423–438.

Yan, A., Guan, Z., Raetz, C. An undecaprenyl phosphate-aminoarabinose flippase required for polymyxin resistance in *Escherichia coli.* J Biol Chem 282 (2007) 36077–36089.

Zasloff, M. Antibiotic peptides as mediators of innate immunity. Curr Opin Immunol 4 (1992) 3–7.

Zhou, Z., Lin, S., Cotter, R., Raetz, C. Lipid A modifications characteristic of *Salmonella* typhimurium are induced by NH4VO3 in *Escherichia coli* K12. Detection of 4-amino-4-deoxy-L-arabinose, phosphoethanolamine and palmitate. J Biol Chem 274 (1999) 18503–18514.

Zhou, Z., Ribeiro, A., Lin, S., Cotter, R., Miller, S., Raetz, C. Lipid A modifications in polymyxin-resistant *Salmonella* typhimurium: PMRA-dependent 4-amino-4-deoxy-L-arabinose, and phosphoethanolamine incorporation. J Biol Chem 276 (2001) 43111–43121.

Zusman, T., Aloni, G., Halperin, E., Kotzer, H., Degtyar, E., Feldman, M., Segal, G. The response regulator PmrA is a major regulator of the icm/dot type IV secretion system in *Legionella pneumophila* and *Coxiella burnetii.* Mol Microbiol 63 (2007) 1508–1523.

Chapter 6
The Variation of O Antigens in Gram-Negative Bacteria

Lei Wang, Quan Wang, and Peter R. Reeves

Abstract The O antigen, consisting of many repeats of an oligosaccharide unit, is part of the lipopolysaccharide (LPS) in the outer membrane of Gram-negative bacteria. It is on the cell surface and appears to be a major target for both immune system and bacteriophages, and therefore becomes one of the most variable cell constituents. The variability of the O antigen provides the major basis for serotyping schemes of Gram-negative bacteria. The genes responsible for the synthesis of O antigen are usually in a single cluster known as O antigen gene cluster, and their location on the chromosome within a species is generally conserved. Three O antigen biosynthesis pathways including Wzx/Wzy, ABC-transporter and Synthase have been discovered. In this chapter, the traditional and molecular O serotyping schemes are compared, O antigen structures and gene clusters of well-studied species are described, processes for formation and distribution of the variety of O antigens are discussed, and finally, the role of O antigen in bacterial virulence.

Keywords O antigen · Gram-negative bacteria · O antigen diversity · Structure · O antigen gene cluster

Abbreviations

LPS	lipopolysaccharide
RFLP	restriction fragment length polymorphism
Gal	galactose
Gal*f*	galactofuranose
SDS-PAGE	sodium dodecyl sulfate polyacrylamide gel electrophoresis

L. Wang (✉)
TEDA School of Biological Sciences and Biotechnology, Nankai University, 23 Hongda Street, TEDA, Tianjin 300457, China
e-mail: wanglei@nankai.edu.cn

X. Wang, P.J. Quinn (eds.), *Endotoxins: Structure, Function and Recognition*, Subcellular Biochemistry 53, DOI 10.1007/978-90-481-9078-2_6,

Und-P	undecaprenol phosphate
GlcNAc	*N*-acetyl-D-glucosamine
GalNAc	*N*-acetyl-D-galactosamine
TM	transmembrane
IT	initial transferase
CDP-DDH	CDP-3,6-dideoxyhexose
LVS	live vaccine strain
T3SS	type III secretion system

6.1 Introduction

The O antigen (O polysaccharide) is part of the lipopolysaccharide (LPS) in the outer membrane of Gram-negative bacteria. It consists of oligosaccharide repeating units (O units), which usually contain two to eight residues from a broad range of sugars, both common and rare, and their derivatives. The O antigen exhibits variation in the types of sugar present, their arrangement within the O unit and the linkages within and between O units, making it one of the most variable cell constituents. The O antigen is exposed on the cell surface and is highly immunogenic, and also used as a receptor by some bacteriophages, both of which may contribute to maintenance of diversity by intermittent selection against specific O antigen forms. O antigen diversity is thought to be important in allowing the various clones to present variations in surface structures that may of selective advantage in their specific niche. It is also an important virulence factor (Pluschke et al., 1983a; Achtman and Pluschke, 1986), and its variability provides the major basis for serotyping schemes for many Gram-negative bacteria.

Genes for O antigen synthesis are normally found on the chromosome as an O antigen gene cluster, and genetic variation in the gene cluster is the major basis for the diversity of O antigen forms. Genes involved in O antigen synthesis are classified into three main classes: nucleotide sugar synthesis genes; sugar transferase genes and O unit processing genes (Reeves and Wang, 2002). The O antigen present in a strain can change through substitution of the gene cluster by recombination involving the DNA flanking the locus, and as a result related strains may have different O antigens, and members of a serogroup with the same O antigen are commonly not related. Most such events involve donors of the same species, but several cases of intraspecies lateral transfer have been inferred from sequence comparisons.

The variation of O antigens in several species including *E. coli*, *Shigella* and *Salmonella* has been well studied, and several reviews have focused on different aspects of O antigens. This chapter covers most aspects of the O antigens of Gram-negative bacteria including the O serotyping scheme, O antigen structures and gene clusters in different species, the molecular evolution for the formation, redistribution of O antigen diversity and the role of O antigen in bacteria virulence. The focus is on those species where there is sufficient data for the general characteristics of the O antigens of that species to become apparent.

6.2 O Serotyping Schemes

The O antigen is the heat-stable somatic antigen which is part of the LPS, whereas most other antigens used for typing are heat-labile. It is the variation in the O units and the linkages between them that confers specificity to the numerous O antigen forms, and provides the basis for the serotyping schemes for Gram-negative bacteria. Since it was shown between 1945 and 1950 that specific serological types of *E. coli* were associated with epidemics of enteritis in newborn infants in nurseries (Kauffmann, 1947), serotyping has been the most widely used method of identifying strains for epidemiological purposes. O:H serotyping is the "gold standard" in typing of organisms for taxonomy and epidemiology, and is the basic tool used in outbreak investigations and surveillance.

6.2.1 Conventional O Serotyping Methods

There are 2 major surface antigens used for serotyping, first distinguished by Smith and Reagh (1903) and independently by Bayer and Reagh (1904), a heat labile flagellar antigen and a heat stable somatic antigen. In the 1940s, Fritz Kauffmann established the serotyping schemes for *Salmonella* and *E. coli* using these antigens, now known as the H and O antigens respectively, and detailed descriptions of these and several other serotyping schemes, together with the procedures for O antiserum production and O antigen determination, are described by Ewing (Edwards and Ewing, 1972; Ewing, 1986). Since then, more serogroups have been added and more serotyping schemes have been established, and there are currently for example, 174 serogroups of *E. coli* (Stenutz et al., 2006), 34 serogroups of *Shigella* (Liu et al., 2008), 46 serogroups of *Salmonella* (Popoff et al., 2003), and more than 200 serogroups of *Vibrio cholerae* (Shimada et al., 1994) and so on.

Two key steps involved in the traditional methods are antiserum preparation and the agglutination tests. Sera for O determination are produced by immunization of rabbits with cultures that have been heated at 100°C for 2 h. Broth cultures or agar plate suspensions heated at 100°C for 1 h are used as antigens for typing. With these two procedures, bacterial agglutination is a very simple and sensitive method for qualitative O antigen determination.

The development of an O serotyping scheme is complex. Heated O antigen preparations of unknown cultures are tested first for agglutination on slides with droplets of pooled, polyvalent O antisera. If agglutination occurs with one of the pooled antisera, the heated suspensions are then tested for agglutination using the individual O antisera contained in that pool. Next, for positive reactions, the heated suspensions are diluted and titrated with serial dilutions of the relevant O antisera. If a group of isolates do not react well with any of the antisera, then some of them are selected and used to produce new antisera to see if they react well with these isolates, and if so a new serotype is named. If in some cases agglutination does not occur, additional strains are selected to produce more antisera. In this way a serotyping scheme evolves. As an example, Sorensen, U.B. and J. L. Larsen presented a serotyping scheme for *V. anguillarum* based on the detection of O antigens by slide agglutination (Sorensen and Larsen, 1986), and Pedersen, et al extended the serotyping scheme from 16 O serogroups to 23 O serogroups (Pedersen et al., 1999). The schemes vary in the extent to which "untypable" strains are followed up. Some schemes are comprehensive at least for the variation of that species as isolated during routine microbiological analysis. In other cases "untypable" isolates are only followed up if thought to be of medical or veterinary importance, and the number of O antigen forms not reported is not known.

Traditional serotyping, which detects surface antigen expression, retains its important position by virtue of accuracy, specificity and stability. However, conventional serotyping by agglutination of somatic antigens using the polyclonal antisera is time consuming, expensive, and available only in a small number of reference laboratories. Recently, new antibody methods have been investigated for serotyping, including ELISA (Trautmann et al., 1996; Correia Barbosa et al., 2000) and antibody microarray which allow parallel analysis of multiple antigens (Anjum et al., 2006).

6.2.2 Molecular Typing Methods

There are several molecular typing schemes in use for various species, most of which are not related to traditional serotyping schemes. However molecular techniques can be applied to the genes that determine the serogroups, and in this way the relationships of the earlier and current data are clear, and there is potential to replace serotyping with molecular techniques, while retaining the ability to identify the many well–known strains characterized by their serogroup, Genotyping methods are generic and often easily applied and in addition are independent of expression which offers advantages for molecular "serotyping". Molecular approaches that can be used to support or replace the classical serotyping method include restriction fragment length polymorphism (RFLP), gene-specific PCR and microarrays.

6.2.2.1 PCR-RFLP

Coimbra et al. proposed a comprehensive approach for molecular serotype determination by using long-range PCR to amplify *E. coli* O antigen gene clusters, followed by enzymatic restriction (PCR-RFLP). They used *Mbo*II on 148 serogroups and the method proved to be effective for rapid characterization of the known *E. coli* O serotypes and for the detection of new O serotypes (Coimbra et al., 2000). The approach has been extended to *Klebsiella* and *Chlamydia trachomatis* (Choi et al., 2001; Brisse et al., 2004) and may be generally applicable.

6.2.2.2 Gene-Specific PCR

Genes for sugar transferases, and *wzx* and *wzy* genes, are often specific to an individual O antigen gene cluster, and with very few exceptions an O antigen can be defined by presence of a combination of 3 or 4 genes, and in many cases by *wzy* alone, These genes have the potential to be used in PCR-based assays for rapid identification and detection of relevant strains (Wang and Reeves, 1998). Many PCR based typing methods targeting O antigen specific genes have proved to be reliable, rapid and sensitive for detecting isolates of *E. coli*, *Shigella* and *Salmonella* from clinical, food and environmental samples (DebRoy et al., 2004; Feng et al., 2004; Han et al., 2007).

PCR-based methods are also ideal for rapid detection of organisms at low concentrations due to the high sensitivity and specificity. However, the necessary O antigen gene sequences are often not known, and if several strains are present, the results may not be easily interpreted when serogroup identification relies on a combination of genes.

6.2.2.3 DNA Microarray

There are however difficulties in running multiplex PCR, and it is difficult to differentiate PCR products of approximately the same size in a multiplex PCR or PCR-RFLP gel. However DNA microarrays do not have these problems, and have been used to serotype pathogenic strains and proved to be rapid, reliable and sensitive. The approach involves the immobilization of numerous oligonucleotide DNA probes on a solid support to which fluorescence-labeled amplified target DNA is hybridized, and is a powerful tool for the detection of pathogens by virtue of high throughput, speed and sensitivity.

The *wzy* gene is the most commonly used target for an oligo DNA microarray as it is very variable as indicated by sequence comparisons and hybridization results, showing strong correlation with classical agglutination. However a given *wzy* gene may be shared by several serogroups, and *wzx* and glycosyl transferase genes can also be used for discrimination. Other genes and particularly the sugar pathway genes often show cross hybridization signals with other unrelated serotypes. DNA microarrays targeting O-serotype-specific genes to detect *Shigella*, *E. coli* and *Salmonella* etc have been reported, and can be used as an alternative to the traditional serotyping procedure (Li et al., 2006; Wen et al., 2006; Ballmer et al., 2007; Wang et al., 2007).

DNA microarrays have many advantages. The whole operation is easier and more efficient than conventional serological methods or multiplex PCR. In our experience the results are repeatable and give good identification. A drawback is that it takes time to generate the necessary probes, and of course one needs to have the necessary equipment. Also it still takes about 12 h after growing the bacteria, whereas serology can give an answer more quickly, provided that all necessary sera are present. An offsetting advantage is that one can include in one microarray probes for other genes of interest such as antibiotic resistance genes, or genes for various stages of serotyping, such as H and O antigen genes. There is the possibility for high-throughput use, but probably the price has to come down first.

6.3 O antigen Diversity in Gram-Negative Bacteria

An understanding of O antigen diversity is critical for developing insights into the evolution of Gram-negative bacteria and for vaccine development strategies. This section is focused on the structural diversity of the O units in the families that have been more extensively studied so far, such as *Enterobacteriaceae, Pseudomonadaceae*, *Vibrionaceae*, etc.

6.3.1 Enterobacteriaceae

6.3.1.1 *Escherichia* and *Shigella*

E. coli includes both commensal and pathogenic forma, while *Shigella* is a well-known human pathogen that causes diseases such as diarrhea and bacillary dysentery (shigellosis). *E. coli* and *Shigella* have long been known to be closely related, and most *Shigella* serotypes fall into three clusters within *E. coli* (Pupo et al., 2000), but are formally treated as 4 species of a separate genus. The current *E. coli* typing scheme comprises O antigens 1 to 181, but O31, O47, O67, O72, O93, O94 and O122 have been removed, some having been shown to be the same as others, and others were reclassified into other genera. A few of the serogroups are divided into subgroups. Some subgroups have also been removed. For example, the *E. coli* subgroup 73-1 has been typed as *E. coli* O73:K-:H33 and subgroups 62D1 was suggested to belong to the genus *Erwinia herbicola* (Stenutz et al., 2006). The genus *Shigella* is divided to 4 species – *S. flexneri, S. dysenteriae, S. boydii* and *S. sonnei*. Although 46 *Shigella* serotypes were recognized, there are only 34 basic O antigen structures involved, mostly because almost all *S. flexneri* O antigens have the same basic structure as discussed below. Also 18 of the 34 O antigens are also found in traditional *E. coli*. Three are very similar to *E. coli* O antigens and 13 are unique to *Shigella* strains (Liu et al., 2008). The total for traditional *E. coli* and *Shigella* is thus 190 (174+16).

About 96 of the *E. coli* O antigen structures have been determined, in some cases including variants, and all are included in the ECODAB database (Stenutz et al., 2006). The structures of all the *Shigella* O antigen have been identified (Liu et al., 2008). Two to seven sugar residues are usually found in the O units and the topology of the O units may be linear, branched or double branched. The most common topology is four sugars in the backbone being linear and a single terminal residue in the side-chain. 3- and 5-residue backbones are common, whereas 2- and 6-residue backbones are present in only a few cases (Varki et al., 1999). While most sugars in the O units of *E. coli* and *Shigella* are hexoses, a number of unusual sugars such as pentoses, deoxyhexoses, lactyl substituted hexoses, heptoses and nonuloses, are also found. The most unusual sugar is a higher acidic diamino sugar, a derivative of 5,7-diamino-3,5,7,9-tetradeoxy-L-glycero-L-manno-non-2-ulosonic (pseudaminic) acid, found in *S. boydii* type 7. Most sugars exist in the pyranose form, but D-galactose and D-fucose are in the furanose form in some O antigens and the pentose D-ribose is always in the furanose form. Some sugars in the sidechains are present only at non-terminal positions, e.g. *N*-acetyl-D-glucosamine, whereas others are only found as terminal residues in the O unit, e.g. colitose.

6.3.1.2 *Salmonella*

The genus *Salmonella* consists of two species, *S. bongori* and *S. enterica,* the latter being divided into six subspecies: *entericae*, *salamae*, *arizonae*, *diarizonae*, *houtenae*, *indica,* also known as subspecies 1, 2, 3a, 3b, 4 and 6 respectively. Strains of *Salmonella* fall into 46 O serogroups, and structures are known for 37 of them. The topology of the O units may be linear, branched or double branched, and there are generally three to five sugar residues. The most common sugars are D-mannose, D-galactose, L-rhamnose and D-glycerol. Sugars such as 3,6-dideoxyhexoses, 6-deoxyaminosugar, 4-acetamido-4,6-dideoxy-D-rhamnose (*N*-acetylperosamine) are also found (Knirel and Kochetkov, 1994).

6.3.1.3 *Yersinia*

The genus *Yersinia* consists of eleven species, three of which, *Y. pestis*, *Y. pseudotuberculosis* and *Y. enterocolitica,* can be human pathogens (Miller et al., 1988). There are about 35 recognized *Yersinia* O antigens based on serology. The strains of *Y. pseudotuberculosis* were divided into seven serogroups, and several of them were subdivided, but as serogroups were determined largely by immunodominant sidebranch residues, the subgroups would generally be treated as groups in other species, and on that basis there are now 15 serogroups. *Y. pestis* lacks O specific chains (Kenne and Lindberg, 1983; Minka and Bruneteau, 1998), but genetically it is a variant of *Y. pseudotuberculosis* and has a defective 1b gene cluster. *Y. enterocolitica*, *Y. frideriksenii, Y. intermedia*, and *Y. kristensenii*, have a shared serological classification.

Many of the *Yersinia* O antigen chemical structures have been described, including 15 structures for 9 *Y. pseudotuberculosis* serogroups (Knirel and Kochetkov, 1994; Kondakova et al., 2008a, b; Cunneen et al., 2009; De Castro et al., 2009; Kondakova et al., 2009a, b, c, d; De Castro et al., 2010). The O antigens of *Yersinia* are either homopolysaccharides or branched heteropolysaccharides. The backbones of the O units usually have two to four sugar residues, and for *Y. pseudotuberculosis*, they usually carry 4 or 5 sugar residues (Skurnik and Bengoechea, 2003).

The O antigens contain mainly neutral sugars, of which the most common are D-glycerol, D-glucose, D-mannose and L-rhamnose. Two sugars, yersiniose A (4C-[(R)-1-hydroxyethl]-3,6-dideoxy-D-*xylo*-hexose) and yersiniose B (4C-[(S)-1-hydroxyethyl]-3,6-dideoxy-D-*xylo*hexose), occur rarely in nature, otherwise only reported as a component in *Legionella* and *Burkholderia* O antigens (Sonesson and Jantzen, 1992; Knirel and Kochetkov, 1994; Mattos et al., 2005).

6.3.1.4 *Citrobacter*

The genus *Citrobacter* is closely related to *Salmonella* and *E. coli.* 43 O serogroups of *Citrobacter* have been identified (Sedlak and Slajsova, 1966a, b), and more than 25 structures have been determined. The O units range from 1 to 7 sugars. There are three homopolymer O antigens, with either 4-deoxy-D-arabino-hexose, 4-acetamido-4-deoxy-D-rhamnose (*N*-acetyl-D-perosamine) or 3-deoxy-3-(L-glyceroylamino)-D-fucose. The O antigens contain mainly neutral sugars, of which the most common are D-glucose, D-mannose and D-rhamnose. Two sugars, 4-deoxy-D-arabino-hexose and 3,6-dideoxy-D-xylo-hexose (abequose), are uncommon in nature (Knirel et al., 2002).

6.3.1.5 *Klebsiella*

Organisms of the *Klebsiella* group, in particular *K. pneumoniae* and *K. oxytoca*, are frequently isolated nosocomial pathogens (Jarvis et al., 1985). There are only 12 recognized *Klebsiella* O serogroups. 13 O antigen structures of *K. pneumoniae* (serogoups O1, O3 to O12, and subgroups O2a, and O2a,c) have been determined. There are usually two to five sugar residues in the O unit, and the topology is linear, except for serogroup O9, which has a D-Gal sidebranch. Some serogroups – O1, O2a,c and O8, have two types of galactan with different structures. One type (→3)-α-D-Gal-(1→3)-β-D-Gal-(1→) is common to serogroups O1 and O8, and (→3)-α-D-Gal-(1→3)-β-D-Gal*f*-(1→) is common to serogroups O1, O2a and O2a,c, and also present in an O-acetylated form in serogroup O8. The neutral sugars D-mannose, D-galactose, L-rhamnose, and D-ribofuranose are characteristic for the O antigens. None of the polysaccharides include acid components (Knirel and Kochetkov, 1994).

6.3.1.6 *Serratia*

Serratia is a widely distributed saprophytic genus, the best-known species is *S. marcescens*, present in the environment including in foods and can be pathogenic. 26 O serotypes have been identified (Traub, 1985; Gaston and Pitt, 1989; Traub, 1991), and all of the O antigen structures have been determined. The O units are linear or single branched. They are neutral and built up of common sugars, such as D-glucose, D-galactose, D-ribose, L-rhamnose, *N*-acetyl-D-glucosamine, and *N*-acetyl-D-galactosamine. Most of them have disaccharide O units which are sometimes O-acetylated. In some cases serological cross reactions between *S. marcescens* serogroups are associated with structural microheterogeneity of the O antigens (Knirel and Kochetkov, 1994).

6.3.1.7 *Hafnia*

Only a single species, *H. alvei* has been designated in the genus (Sakazaki and Tamura, 1992), and 39 O serogroups have been described (Baturo and Raginskaya, 1978). Chemical studies on LPS isolated from 33 strains of *H. alvei* started in 1988, and SDS-PAGE analysis of the LPSs and their sugar composition was described (Romanowska et al., 1988). Further analysis of 24 O antigens showed an unusual richness of sugar components as well as a variety of non-sugar side groups. D-glucose, D-galactose, D-glucosamine, D-galactosamine, L-fucose, L-rhamnose, *N*-aceyl-D-fucosamine and *N*-acetyl-D-quinovosamine are often present in the O antigens. Among the 24 O antigens examined, 11 contain phosphodiester linkages, which is a teichoic acid-like characteristic. Generally, glycerol participates in these linkages. It is interesting to note that fucosamine as well as quinovosamine may have amino groups situated at position 2, 3 or 4 in *H. alvei* O antigens. O-acetyl groups appear as the side groups in many O antigens, and incomplete substitution can result in the microheterogeneity often observed in *H. alvei* O antigens (Romanowska, 2000).

6.3.1.8 *Proteus*

Currently, there are 75 O serogroups identified in *Proteus* (Penner and Hennessy, 1980; Larsson, 1984; Zych et al., 2000; Sidorczyk et al., 2002), and most of the O antigen structures have been described. They are generally acidic due to the presence of uronic acids and various non-carbohydrate acidic components, including phosphate groups (Knirel et al., 1993; Rozalski et al., 1997). Besides the typical sugar constituents widespread in nature, like hexoses, hexosoamines, and uronic acids, they also contain 6-deoxyamino sugars including L-fucosamine, L-quinovosamine, D-quinovoso-3-amine, and D-fucoso-3-amine. Various non-sugar constituents such as amino acids (L- and D-alanine, L-serine, L-threonine, and L-lysine) can be attached to the carboxyl group of uronic acids, and other unusual acidic components like (*R*)- and (*S*)-lactic acid ethers and (*R*)-hydroxy butyryl, pyruvic, and phosphate groups are also found (Rozalski et al., 1997).

6.3.2 *Pseudomonadaceae*

Pseudomonas is a well studied genus of *Pseudomonadaceae*. Some species such as *P. aeruginosa* are pathogens for humans and animals, and some such as *P. syringae* are plant pathogens.

6.3.2.1 *P. aeruginosa*

At least 20 distinct O antigen serogroups of *P. aeruginosa* have been identified, and 13 O antigen structures have been studied (Spiers et al., 2000; Raymond et al., 2002). Several of the strains have 2 or more O antigen forms with related structures varying in properties such as sidegroup substituents or one of the linkages between sugars. The O units usually have three to four sugar residues, and the topology is linear. Typical sugars within include D-glucose, D-fucose, D-quinovose, D-galactose, D-mannose and L-rhamnose (Knirel, 1990; Bystrova et al., 2006).

6.3.2.2 *P. syringae*

There are 10 serogroups identified in *P. syringae*. Typically, the O antigens have a linear main chain of L- or D- rhamnose with tri- or tetra-saccharide repeating units, and most have a side branch of D-rhamnose, D-fucofuranose, *N*-acetyl-D-glucosamine or *N*-acetyl-D-fucoso-3-amine (Knirel and Kochetkov, 1994). As for *P. aeruginoas*, structural heterogeneity has been reported in some cases with for example two types of pentasaccharide repeating units in serogrop O9, differing in the position of sidebranch residues (Zdorovenko et al., 2003).

6.3.3 Vibrionaceae

The genus *Vibrio* in the family of *Vibrionaceae* includes many species, of which *V. cholerae*, *V. vulnificus*, *V. parahaemolyticus* are well known human pathogens. Around 200 serogroups of *V. cholerae* have been identified, and 14 have been identified in *V. parahaemolyticus*.

About 33 *Vibrio* O antigen structures have been determined. One to six sugar residues are found in the O units and the topology of the O units may be linear, single branched, double branched or treble branched. Most are heteropolysaccharides. The most common sugars in the O unit are hexosamines, while pentoses, deoxyhexoses, and heptoses are also found. Typical sugar components are L-rhamnose, L-rhamnose-4-amine, di-*N*-acetyl-bacillosamine, *N*-acetyl-D-quinovosamine, *N*-acetyl-D-galactosamine, D-galactose, D-galacturonic acid, *N*-acetyl-L-fucosamine, L-fucose, *N*-acetyl-2-amino-2-deoxy-D-mannopyranose uronic acid, D-ManNAc3NAmN, *N*-acetyl-D-glucosamine, D-GlcNAc3NAN, and D-glucuronic acid. The most unusual sugar is a higher acidic di-amino sugar, a derivative of 5-acetamidino-7-acetamido-3,5,7,9-tetradeoxy-L-glycero-L-manno-nonulosonic acid.

6.4 General Properties of O Antigen Gene Clusters

More than 230 O antigen gene clusters have been reported, the majority being from *E. coli*, *S. enterica* and *Shigella* spp. For all of them the GC content of the cluster is lower than that of the genome average, and this is usually taken to indicate it was laterally transferred from a species with a low GC, but it does seem surprising that the bias is almost always in the same direction.

6.4.1 O Antigen Biosynthesis

Three major classes of genes – nucleotide sugar synthesis genes, sugar transferase genes, and O unit processing genes are commonly found in O antigen gene clusters. Nucleotide sugar synthesis genes for sugars specific to the O antigen are usually in the O antigen gene cluster, but genes for sugars also found in other structures or used in the metabolism of that species are usually at other loci. Glycosyl transferase genes are responsible for the specific linkages in the O antigen, and result in the very high level of diversity.

O antigen synthesis is initiated by transfer of a sugar phosphate from an NDP-sugar to undecaprenol phosphate (Und-P) (Reeves and Wang, 2002). In most *E. coli* and *Shiglla*, and a high proportion of *Salmonella*, WecA, encoded in the gene cluster for enterobacterial common antigen, is the first transferase responsible for the transfer of GlcNAc-P or GalNAc-P from UDP-GlcNAc or UDP-GalNAc respectively to UndP. In some *S. enterica*, such as groups B and D1, WbaP, located in the O antigen gene cluster, is the initial transferase (IT) to transfer Gal-P. WbpL is the IT in *P. aeruginose*, which can transfer a variety of *N*-acetyl sugars (Reeves and Cunneen, 2009).

Three different pathways, the Wzx/Wzy, ABC transporter, and the synthase pathways are described for the assembly and processing of O antigens (Fig. 6.1). Most O antigens appear to be synthesized by the Wzx/Wzy pathway. In this case, sugars and other constituents are transferred sequentially to the first sugar to form an O unit, which is then translocated by Wzx across the inner membrane, to put the O unit on the periplasmic face, where it is polymerized by Wzy to generate the polymer, with the number of O units in the final O antigen being regulated by Wzz. Wzx and Wzy are hydrophobic proteins with about 12 predicted transmembrane (TM) segments, and a long cytoplasmic loop between TM segments is usually found in Wzy.

The ABC transporter pathway differs from the Wzx/Wzy pathway in that the complete O antigen chain is synthesized on the cytoplasmic face of the inner membrane, and translocation of the UndPP-O-antigen is carried out by ABC transporter proteins Wzm (carries out the export process) and Wzt (the ATP-binding component). The polysaccharides involved are usually homopolymers or have only 2 constituent sugars. This pathway is rare among O antigens, except for all of the known O antigens of *K .pneumoniae* and *V. cholerae*, and most of the *Y. enterocolitica* O antigens (Rick et al., 1994; Skurnik, 2003; Chatterjee and Chaudhuri, 2004). This pathway has also been described in *E. coli* O8, O9, O9a, O52 and O99, the A-band O antigen of *P. aeruginosa,* and in *L. pneumophila* O1 (Rick et al., 1994; Kido et al., 1995; Feng et al., 2004; Cazalet et al., 2008; Kintz and Goldberg, 2008). In most cases the role of the ABC transporter is inferred from the presence of *wzm* and

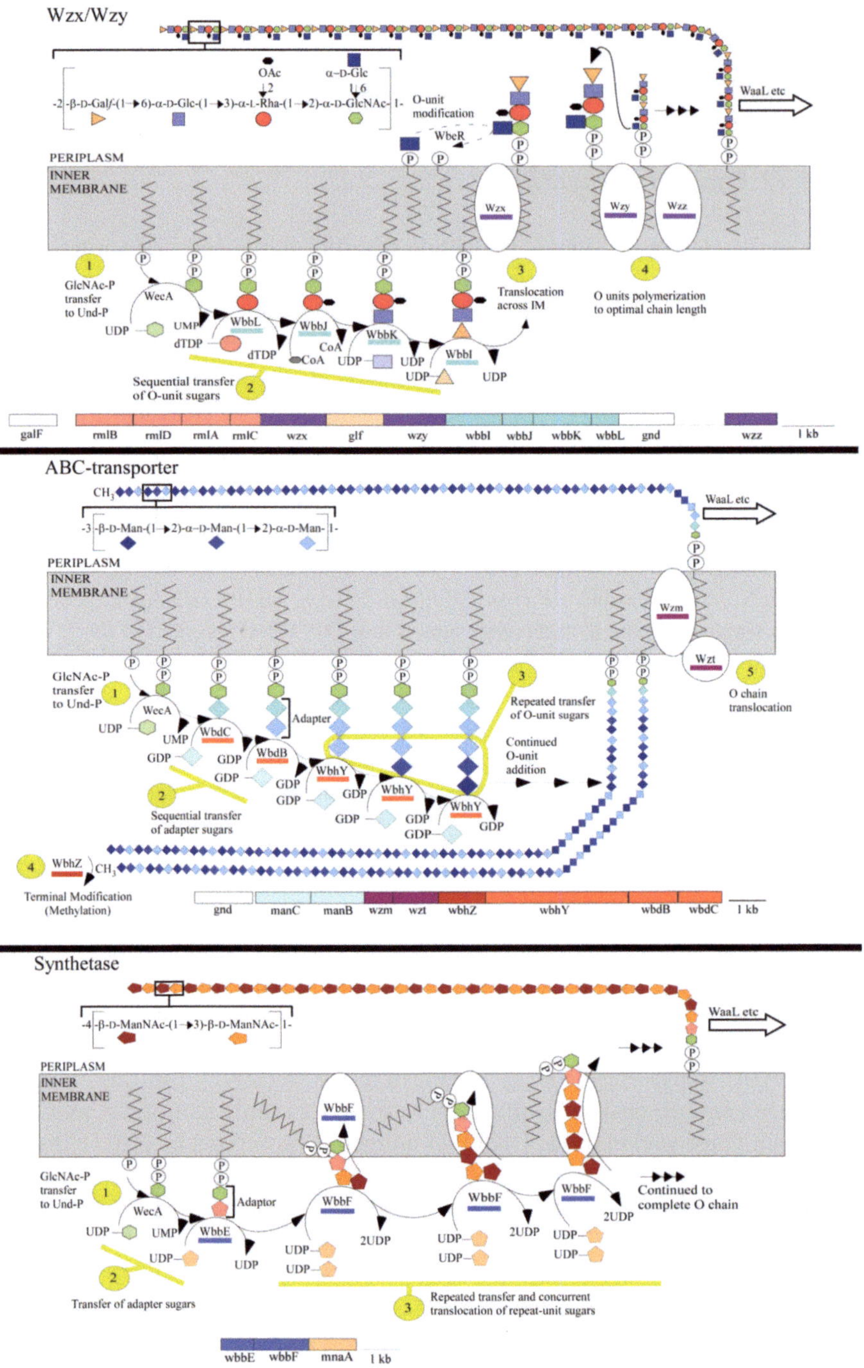

Fig. 6.1 (continued)

wzt genes and absence of *wzx* and *wzy* genes in the gene cluster. The best studied are those of the O8 and O9a O antigens of *E. coli.*

The synthase pathway is the simplest, with a single integral membrane protein responsible both for sequential addition of the sugars, and the concurrent extrusion of the nascent polymer across the cell membrane. The synthase pathway is also very rare with *S. enterica* O54 the only described case for O antigens (Keenleyside et al., 1994). The repeating units in the synthase pathway are generally homopolymers or have two sugars. For all three pathways, the complete O antigen chain is then transferred by the ligase, WaaL, to lipid A-core to make a complete LPS molecule (Whitfield, 1995; Valvano, 2003; Kaniuk et al., 2004).

6.4.2 Examples of O Antigen Gene Clusters

6.4.2.1 *E. coli, S. enterica* and *Shigella* spp

Thus far 148 *E. coli*, 15 *S. enterica* and 34 *Shigella* O antigen gene clusters have been reported. In these species, the O antigen gene cluster generally lies between the *galF* and *gnd* genes, and most of them have the Wzx/Wzy pathway (Liu et al., 2008). A conserved 39-bp JUMPStart sequence, which is required for the regulation of downstream genes, is located in the intergenic region between *galF* and the O antigen gene cluster (Hobbs and Reeves, 1994; Marolda and Valvano, 1998).

Although the O antigen gene clusters of most *E. coli* serotypes are located between *galF* and *gnd*, there are also some exceptions to these "rules". The *E. coli* O8, O9 and O9a gene clusters are between *gnd* and the *his* operon. They are identical to the *K. pneumoniae* O5 and O3 O antigens, respectively, whose O antigen gene clusters are also between *gnd* and the *his* operon, and all have the ABC transporter pathway (Rick et al., 1994; Sugiyama et al., 1994, 1998).

In *S. sonnei*, which is different from the other *Shigella* strains in having a major deletion in the O antigen gene cluster between *galF* and *gnd*, the functional

Fig. 6.1 (continued) O antigen biosynthesis pathways. Top panel: *E. coli* O16 (Wzx/Wzy pathway), centre panel: *E. coli* O8 (ABC transporter pathway), and bottom panel: *S. enterica* O54 (Synthase pathway). Synthesis in all three pathways is initiated by transfer of GlcNAc-P onto UndP. The pathway steps are numbered sequentially and the respective gene clusters and repeating units are indicated. Each sugar is represented by a different symbol. The boxes are for the respective biosynthesis genes, and the bars are for transferases and processing proteins that are encoded by genes residing within the O antigen gene cluster. O unit modification in *E. coli* O16 occurs after repeat unit synthesis and is indicated by a dashed line. The ligation of the O antigen from UndPP to lipid-A-core by WaaL and the final export of the LPS molecule to the outer membrane, both common to all three pathways are not shown. Not drawn to scale. From Reeves and Cunneen (2009), copyright Elsevier

gene cluster is on a plasmid. An identical O antigen is also found in *Plesiomonas shigelloides*. The high level of sequence similarity establishes that the *S. sonnei* O antigen gene cluster is derived from *P. shigelloides* (Shepherd et al., 2000).

For *Salmonella* the convention has been adopted of defining an O antigen by the major O antigen gene cluster, and treating variation due to genes outside of that as variation within the serogroup. This has lead to some revision of the typing scheme as the genetic data becomes available. For example serogroups E2 and E3 were removed when it was found that the differences from E1 were due to presence of genes on prophages. It is important to know that this is not the case for some other genera and so the numbers of O antigens reported for different species are not entirely comparable.

6.4.2.2 *Enterobacter Sakazakii*

E. sakazakii is an opportunistic pathogen that has been associated with food-borne illness in neonates. Only two major serotypes – O1 and O2 have been identified. The two O antigen gene clusters are located between the *galF* and *gnd* genes, and both have the Wzx/Wzy pathway (Mullane et al., 2008).

6.4.2.3 *P. aeruginosa*

P. aeruginosa produces two different forms of O polysaccharide called A-band and B-band O antigens. A-band O antigen, also known as common antigen, is a homopolymer composed of D-rhamnose and found in almost all isolates. B-band O antigen is an immunodominant protective antigen, responsible for serogroup specificity (Kintz and Goldberg, 2008).

O antigen gene cluster sequences have been reported for 10 of the *P. aeruginosa* B-band O antigens. The B-band O antigen gene clusters are between *himD* and *tyrB*, and all include the *wzx* and *wzy* genes.

The genes for A-band synthesis are also located as a gene cluster on the chromosome. The genes involved in D-Rha synthesis (*rmd*, *gmd*, and *wbpW*) are positioned at the beginning of the cluster, followed by those coding for an ABC transport system (*wzm* and *wzt*) and those involved in the assembly of the D-rhamnan O unit (*wbpX*, *wbpY*, and *wbpZ*) (Kintz and Goldberg, 2008).

6.4.2.4 *V. cholerae*

Four *V. cholerae* O antigen gene clusters (O1, O139, O37 and O22) have been reported, all located between the *gmhD* and *rjg* genes, and all have the ABC transporter pathway (Chatterjee and Chaudhuri, 2004).

In *V. cholerae* O1, the O antigen gene cluster consists of five major regions: (1) perosamine biosynthesis genes (*manC*, *manB*, *gmd* and *wbeE*); (2) O antigen transport genes (*wbeG*, *wzm* and *wzt*); (3) tetronate biosynthesis genes (*wbeK*, *wbeL*, *wbeM*, *wbeN* and *wbeO*); (4) O antigen modification genes (*wbeT*); and (5) 3 additional genes (*wbeU*, *wbeV*, and *wbeW*) (Chatterjee and Chaudhuri, 2004).

Molecular analysis of an O139 Bengal isolate showed that a 22-kbp DNA segment encoding the O1 antigen (the *wbe* region) had been replaced by a 35-kb DNA segment (the *wbf* region) specifying the new O139 serogroup antigen. The region is complex and has a number of redundancies, and some genes are responsible for both O antigen and capsular synthesis. Unusually, the O139 only contains only *wzm* as N assembly gene without the presence of *wzt* (Blokesch and Schoolnik, 2007).

Among the different non-O1, non-O139 strains, an O37 strain has played an important role in localized cholera outbreaks (Kamal, 1971). Twenty-three ORFs were identified between genes *gmhD* and *rjg*, in the O37 strain, and many of the ORFs (ORF-1 to ORF-13 and ORF-18) encoded enzymes involved in polysaccharide biosynthesis. A 1549-bp promoter region separates *gmhD* and ORF-1, and this region contains a putative promoter and ops elements (an 8-bp subsequence within JUMPStart which has been demonstrated to be an elongation factor for O antigen transcription). There is an IS element in the interval between the region that is unique to O37 and the right junction. The region downstream of the IS element has three ORFs almost identical to the O1 O antigen gene cluster: *wbeV*, *galE* and *wbeW*, followed by *rjg* (Hobbs and Reeves, 1994; Bailey et al., 1997; Chatterjee and Chaudhuri, 2004).

The O antigen gene cluster of O22 is 35.9-kb long and closely related to that of O139.

6.4.2.5 *Yersinia*

Yersinia spp have two loci for O antigen gene clusters. Gene clusters between the *hemH* and *gsk* genes have *wzx* and *wzy* as the assembly genes, and those at the other site upstream of *galF* have ABC transporter pathway genes. An exception is the O antigen gene cluster of *Y. kristensenii* O11, located at a novel chromosomal locus between *aroA* and *cmk* where it is flanked by remnant *galF* and *gnd* genes. The sequence similarities of the *galF* and *gnd* genes suggest a lateral transfer of this O antigen gene cluster from an *E. coli*-like donor (Cunneen and Reeves, 2007). 20 *Yersinia* gene clusters have been reported (Zhang et al., 1996; Pacinelli et al., 2002; Bogdanovich et al., 2003; Cunneen and Reeves, 2007; Cunneen et al., 2009).

In *Y. enterocolitica* O3 and O9, the major O antigen is a homopolymer, proposed to have the ABC transporter pathway, but, the locations of the O antigen clusters are presently unknown. These two serogroups, also have an "outer core" with a gene cluster between *hemH* and *gsk* (Skurnik, 2003). The outer core gene clusters have a *wzx* gene and an IT gene, and the outer core corresponds to a single O unit of an O antigen, so the location between *hemH* and *gsk* is appropriate. The O antigen gene cluster of *Y. enterocolitica* serotype O8 locates between *hemH* and *gsk*, and contains 18 genes including *wzx*, and *wzy* (Zhang et al., 1996; Skurnik and Bengoechea, 2003).

Sequence analyses of the *Y. pseudotuberculosis* O antigen gene clusters between *hemH* and *gsk* revealed a pattern of modules for different backbone structures and for different sidebranch sugars, with indications of homologous recombination events during evolution of the diversity (Skurnik et al., 2000; Pacinelli et al., 2002;

Skurnik and Bengoechea, 2003). All serotypes except O7, O9 and O10 were shown to carry the genes involved in CDP-3,6-dideoxyhexose (CDP-DDH) synthesis in the beginning of the O antigen gene cluster. The O antigen gene clusters also contain the genes involved in other NDP-sugar biosynthesis, the genes encoding respective glycosyltransferases and *wzx*, *wzy*, *wzz* (Cunneen et al., 2009).

6.4.2.6 *Legionella Pneumophila*

L. pneumophila is the causative agent of Legionnaires' disease, a pneumonia with sometimes fatal progression. There are 15 serogroups of *L. pneumophila*, but the only gene cluster reported is that for *L. pneumophila* O1. The genes involved are dispersed in a 30 kb cluster for the LPS biosynthesis. There are 30 ORFs in the gene locus, most of which are still not clearly identified. Genes involved in O antigen synthesis include sugar biosynthesis and transferase genes (*mnaA*, *neuB*, *neuA*, *wecA*) and O antigen translocation genes (*wzt*, *wzm*). Southern blot analysis indicated the entire gene locus to be present in *L. pneumophila* O1 strains, whereas only parts of the region hybridized to DNA from *L. pneumophila* O2 to O15 strains (Luneberg et al., 2000; Cazalet et al., 2008).

6.4.2.7 *C. freundii*

Only 1 *C. freundii* (F90) O antigen gene cluster has been reported, which is also located between *galF* and *gnd* genes as in *E. coli*. *C. freundii* F90 shares the same O antigen structure with *E. coli* O157 and *S. enterica* O30. Its O antigen gene cluster is 14.2 kb long and 11 open reading frames were found. The three O antigen gene clusters in *C. freundii* F90, *E. coli* O157 and *S. enterica* O30 have the same transcriptional direction, the same genes and the same organization, and the only difference is that a *N*-acetyltransferase gene, *wbdR*, at the 3'end of the *E. coli* O157 gene cluster is absent in the other two clusters. The levels of divergence indicated that the O antigen gene cluster had been present in the common ancestor for these species and diverged with the many other shared genes (Samuel et al., 2004).

6.5 Formation and Distribution of O Antigen Gene Clusters

The diversity of O antigens within species appears to be dynamic, because related species generally have different sets of O antigens, as discussed for several genera in section 3. To account for this there must be an ongoing gain and loss of O antigens as species evolve. Related strains also often have different O antigens and this will usually be due to recombination within the species. New O antigen gene clusters can arise in many ways, and various mechanism for change of O antigen are discussed in this section.

Some O antigens are now known to be derived from others by mutation. For example a single amino acid substitution in the mannosyltansferase WbdA is responsible for the change of *E. coli* O9 O antigen to O9a (Kido and Kobayashi,

2000). A second example of O antigen variation caused by nucleotide substitutions involves *E. coli* O107 and O117. They differ in substitution of only a few amino acid residues, but the glycosyltranferases have different specificities resulting in the presence of different sugars as the third residue in the O107 and O117 structures (Wang et al., 2009).

However in most cases change is not due simply to mutation. As most O antigen gene clusters in a species are at the same locus on the chromosome, transfer of O antigen gene clusters between strains can occur by homologous recombination in the flanking genes, and this will substitute one O antigen for another. O antigen gene clusters also transfer between closely related species by homologous recombination in flanking DNA, as for intraspecies transfer. This is much less common but can lead to gain of an additional O antigen for a species. IS elements have also been proposed to play important roles in mediating intraspecies and interspecies O antigen gene transfer, and were also noted to play a role in the diversification of O antigen forms by gene inactivation. Phages were also found to play an important role by carrying genes responsible for the O antigen modification, for example in serotype-conversion in bacteria such as *S. flexneri*.

6.5.1 Lateral Transfer of O Antigen Genes by Homologous Recombination

The O antigen present in a strain can change by recombination involving the DNA flanking the locus. In general most forms of the O antigen in a species are encoded by a gene cluster at the same locus, such as O antigen genes in *E. coli*, *Shigella*, and *Salmonella* etc clustered between *galF* and *gnd* on the chromosome. Intraspecies O antigen gene cluster transfer by homologous recombination in the flanking genes usually substitutes one O antigen for another. The movement of O antigen gene clusters by homologous recombination has been documented in *S. enterica* (Selander et al., 1991), and recombination sites involved in this have been observed in the adjacent *gnd* gene (Thampapillai et al., 1994). Lateral transfer mediated by homologous recombination can involve genes within in the O antigen gene cluster if they are present in a group of antigens (Li and Reeves, 2000). In another example comparison of the O antigen gene clusters of *E. coli* O86 and O127 showed that five genes downstream of *manB* and upstream of *gnd* were transferred by recombination between O86 and O127, with the recombination sites in the 3' end of the *manB* gene and 5' end of the *gnd* gene (Feng et al., 2005).

O antigen genes can also transfer between closely related species by homologous recombination in flanking DNA as for intraspecies transfer. For example, the O antigen gene clusters of *E. coli* O8 and O9a map between *gnd* and *his* instead of between *galF* and *gnd*, which is atypical in *E. coli*, and are identical to O5 and O3 of *K. pneumoniae*, in which the homologues also map between *gnd* and *his*. It seems clear that the *E. coli* O8 and O9a gene clusters were transferred from *K. pneumoniae* O5 and O3, respectively (Sugiyama et al., 1997, 1998).

6.5.2 Insertion and Deletion of O Antigen Genes Mediated by IS Elements

IS elements are known to play important roles in evolution of bacterial genome (Bennett, 2004). Insertion of IS elements can result in the inactivation of genes, and the combination of two or more IS elements can also result in the mobilization of large segments of DNA and lead to generation of new O antigen forms. Another example of the role of IS elements in evolution of O antigens is given by the O antigen gene cluster of *S. enterica* D2, which is a chimera of the *S. enterica* D1 and E1 O antigen gene clusters. The 5' end of the gene cluster is the same as that of D1, and the 3' end the same as that of E1, and at the junction of the two components there is an H repeat type of IS element which was proposed to have mediated an intraspecific recombination event to generate the D2 O antigen gene cluster (Xiang et al., 1994). In *V. cholerae* O1, the entry of a methyl tranferase gene, *wbeT*, into the gene cluster, was also proposed to have been mediated an IS element adjacent to the *wbeT* gene, which converts the Inaba to the Ogawa form of the *V. cholerae* O1 (Manning et al., 1993). A mechanism for the role of IS elements found at one end of an inserted section of DNA was given by Xiang et al (1994). IS elements are found in seven of the *Shigella* O antigen gene clusters, and some of them seem to have been directly involved in formation of the new O antigen forms (Liu et al., 2008). For example in *S. boydii* type 6, a ribosyltransferase gene, *wbaM,* is interrupted by an IS element, generating the only difference between the O antigen gene clusters of *S. boydii* type 6 and 10, and accounts for the structural difference (Senchenkova et al., 2005). Similarly the *E. coli* O40 O antigen gene cluster differs from that of *S. dysenteriae* type 9 in having an IS element inserted into the 5'end of a pyruvyltransferase gene, *wffR*, resulting in the lack of the pyruvate acetal, again the only difference between the O antigen structures of the 2 O antigens (Liu et al., 2008).

IS elements have also been proposed to mediate transfer of O antigen genes between species. The case of O antigen gene cluster in *Y. kristensenii* O11 is a good example. As discussed above this gene cluster is not located at the usual *Yersinia* O antigen gene cluster locus, and is also flanked by remnant *galF* and *gnd* genes suggesting lateral transfer of this O antigen gene cluster from an *E. coli*-like donor. A mechaism is suggested by the presence of remnant IS sequences flanking the *galF* and *gnd* remnants and these were proposed to have mediated the lateral O antigen gene cluster transfer (Cunneen and Reeves, 2007).

6.5.3 Role of Plasmids in Evolution of O Antigen gene Clusters

Plasmids are found to play an important role in transferring O antigen clusters or O antigen genes to generate new O antigen forms. A good example of O antigen gene transfer mediated by a plasmid has been reported for *S. sonnei.* The O antigen gene cluster of *S. sonnei* is on a plasmid instead of on the chromosome between *galF* and *gnd*. It also had a *wzz* gene as the first gene in the O antigen gene cluster, which is usually located outside of the O antigen gene cluster and between *gnd* and *ugd* in *E. coli* (Bastin et al., 1993; Franco et al., 1998). The location of the gene cluster on a plasmid and the atypical location of *wzz* gene suggest a relatively recent transfer from another species. There has also been a large deletion in *sonnei* in the region where the O antigen gene cluster is usually located (Lai et al., 1998). Thus it seems clear that the *S. sonnei* O antigen appeared in the *E. coli/Shigella* species by interspecies transfer of an O antigen gene cluster on a plasmid and loss of function of the preexisting chromosomal gene cluster.

It has been found that the *P. shigelloides* O17 has the same O antigen structure and high level of O antigen gene sequence similarity with those of *S. sonnei*, making *P. shigelloides* O17 a potentianl donor of the O antigen gene cluster *S. sonnei* (Rauss et al., 1970; Taylor et al., 1993; Shepherd et al., 2000). The *S. enterica* O54 gene cluster, like that of *S. sonnei* is on a plasmid, but no potential donor has been identified (Keenleyside and Whitefield, 1996).

Our final example is the O antigen of *S. dysenteriae* type 1, that differs from that of *E. coli* O148 by having a galactose residue in place of a glucose residue. The O antigen gene clusters of the two strains have the same organization and high level of DNA identity, except that in *S. dysenteriae* type 1, a glucosyltransferase gene – *wbbG* is interrupted by a deletion, and a galactosyltransferase gene – *wbbP*, located on a plasmid, is responsible for the transfer of galactose to make a major antigenic epitope of the *S. dysenteriae* type 1 O antigen. It seems clear that the O antigen gene cluster of *S. dysenteriae* type 1 is derived from that of *E. coli* O148 by loss of the proposed glusosyltransferase gene *wbbG* and gain of a plasmid-borne galactosyltransferase gene *wbbP*, although the potential donor is not clear (Feng et al., 2007).

6.5.4 O Antigen Modification Mediated by Phages

O antigen modification can create antigenic variation, which enhances the survival of the bacteria. Serotype-converting bacteriophages have been found to play an important role in this. The association between O antigen modification and temperate bacteriophages in *S. flexneri* has been known for many years. All serotypes except serotype 6 of *S. flexneri* share a common polysaccharide backbone, and the addition of glucosyl and/or O-acetyl groups to different sugars in the backbone results in the serotype variation among *S. flexneri* (Fig. 6.2). Six different serotype specific glucosyl transferases carried by phages --GtrI, II, IV, V, X, Ic have been characterized which, together with a single O-acetyl transferase gene also on a phage, account for conversion of a serotype Y to the other serotypes (Stagg et al., 2009). Each of the six bacteriophages involved in the glucosylation of the O antigen backbone have three genes for the process. Two of them, *gtrA* and *gtrB* are highly conserved, and are involved in the transfer of glucose from UDP-glucose to UndP and translocating UndP-glucose to face the periplasmic space. The third gene, catalyzing the linkage of a glucosyl group to the O antigen backbone, is serotype specific (Allison and Verma, 2000). The *O*-acetyltranferase gene, *oac*, is responsible

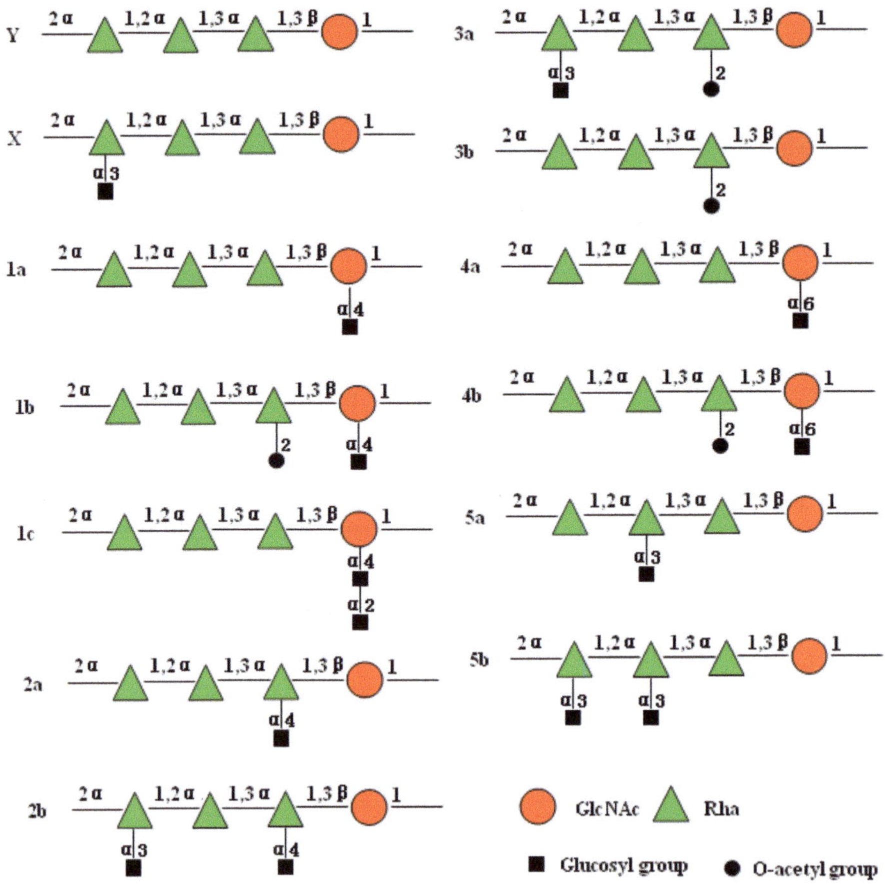

Fig. 6.2 O antigens from different *S. flexneri* serotypes. The O antigen of serotype Y is basic for all serotypes. Other serotypes differ form serotype Y by the addition of either glucosyl or O-acetyl groups to different sugars in the O unit via different linkages indicated. Abbreviations: GlcNac, *N*-acetylglucosamine; Rha, rhamnose. The data is from Alison and Verma, 2000; Stagg, Tang et al. 2009

for the O-acetyl modification of the O antigen backbone (Clark et al., 1991; Verma et al., 1991).

A group of *E. coli* strains including *E. coli* O17, O44, O73, O77 and O106 were also found to share a common four-sugar O unit backbone, and vary in presence of sidebranch glucose residues. They possess almost identical O antigen gene clusters with the same set of genes. The *E. coli* O77 antigen does not have any substitutions, the other O antigens in this group differ by the addition of one or two glucose side branches at various positions of the backbone. Three genes for glucosyl modification of O antigen backbone were found in a prophage on the chromosome of the O44 strain, as for *S. flexneri,* and are proposed to be responsible for the O44 modification

(Wang et al., 2007). It appears that this group of *E. coli* resembles the *Shigella flexneri* group.

Strains belonging to these *E. coli* O serogroups have been reported to be pathogenic, and as for *S. flexneri,* O antigen modification was proposed to play an important role in enhancing survival and pathogenicity (Allison and Verma, 2000; West et al., 2005; Wang et al., 2007).

6.6 O antigen and Virulence

The O antigen plays a critical role in the interactions between bacteria and the host, and consequently contributes to the gain and maintenance of pathogen virulence, synthesis and expression of O antigens is generally considered to be of importance for the virulence of many pathogens.

6.6.1 The Complete Loss of O Antigen Leads to Severe Attenuation of Virulence

O antigen is known to be an important factor in the ability of many bacteria to evade complement. This has been shown for several species and one example is given. Inactivation of *wbtA* (the first gene of the predicted O antigen gene cluster) *F. tularensis* live vaccine strain (LVS) led to the complete loss of O antigen, and strong attenuation of virulence of the mutant was noted on challenging mice via the sensitive intraperitoneal route compared to that of parental LVS (Sebastian et al., 2007). Also when exposed to human serum the *wbtA* mutant was totally eliminated, whereas the parental LVS survived, revealing an extreme sensitivity to serum. Likewise while the LVS strain multiplied efficiently in murine macrophages during the 24-h period of the assay, the replication of the *wbtA* mutant was totally abolished (Sebastian et al., 2007).

6.6.2 The O Antigen Differences Can Account for Differences in the Nature of Pathogenicity

More interesting is that a change in O antigen can affect virulence. For example the O18 antigen is known to play a key role in the virulence of the archetypal *E. coli* O18:K1:H7 meningitis clone. It was found that O1:K1 and O18:K1 bacteria differ in their potential to cause newborn meningitis and to multiply in the bloodstream of infected newborn rats. In in vitro experiments, O18:K1 bacteria were resistant to the bactericidal activity of adult rat serum, whereas O1:K1 bacteria were killed by the classical complement pathway of such sera. It was also shown that the O18 polysaccharide is involved in resistance to the classical complement pathway in guinea pig serum, and the crucial role of O18 in sustained high-level bacteremia, which is necessary for blood–brain barrier penetration (Pluschke et al., 1983a, b; Achtman and Pluschke, 1986).

Similar studies have been carried out to explore the virulence of avian pathogenic *E. coli* O78:K80:H9. They constructed a mutant lacking antigen O78 and two derivative strains supplemented with antigen O1 or O26. They found that the loss of O78 was associated with lower pathogenicity and that substitution by O1 or O26 did not fully restore the initial virulence (Mellata et al., 2003).

6.6.3 The Effect of the Chemical Composition and Structure of an O Side Branch on Virulence

Most bacteria can change the structure of their outer membrane or modify their antigenic determinants in accordance with the host reaction, thereby increasing their virulence, and penetration into the host organs.

West et al showed that the loss of glycosylation of a *S. flexneri* O5a strain led to a change in the organization of the O antigen form and a doubling of the distance that the LPS projected from the bilayer outer membrane. This was attributed to an ability of the glycosylated O antigen to form a compact helical structure as predicted by modeling. The effect of losing glycosylation was interference by the more extended O antigen with the operation of the type III secretion system (T3SS) used by *Shigella* to inject proteins into host cells, limiting bacterial invasion and the usual vigorous inflammatory response. The change in O antigen had little effect on total amount of LPS or protection against innate immunity (West et al., 2005).

6.6.4 The O Antigen Chain Lengths is Important for the Full Virulence

In another pathogen, *Y. enterocolitica* O8, LPS structure was also shown to be important as both rough mutants and semi-rough mutants (with one O unit) were severely impaired in their ability to colonize mice in contrast to the wild-type strain, but having a single O unit had less effect on colonisation than total absence of O antigen. However, the mere presence of the O antigen is not sufficient to confer full virulence, therefore, the length and the proper distribution of the O units is important (Bengoechea et al., 2004).

In another study panel of *S. enterica* Typhimurium variants with different O antigen lengths were constructed by substitution of the original *wzz* gene with *wzz* genes from other strains or species. There was a trend toward decreased bacterial ingestion as the length of O antigen increased, with the rough mutant taken up most efficiently compared to other strains (Murray et al., 2006).

6.7 Conclusions

O serotyping schemes have been established for a range of species of Gram-negative bacteria, and in many cases the O antigen structures and gene clusters have been well studied. The best-known are *E. coli/Shigella* and *Salmonella*, and for others that we discuss there is sufficient data to show up the variation between species. In some species, the O antigen is usually linear but in other species mostly branched. There is also variation in charge with O antigens in some species generally neutral while in others they are acidic due to specific components in the individual O antigen. Some species have mostly common sugars whereas for other species a number of the antigens include rare sugars, and in such cases variations in the pathway genes are found in the gene clusters. *Y. pseudotuberculosis* is an extreme example of this with most O antigens having a dideoxyhexose. Some sugars are only found in certain species, such as yersiniose in only 3 species. All O antigens require an initial sugar transferase to start O unit synthesis, and the gene encoding the IT can be present within or outside of the O antigen gene cluster, and again this is generally consistent within a species. Of the three pathways, the Wzx/Wzy pathway is used for most of the O antigens, and the ABC transporter pathway for only a few, but where found may be common for that species. The synthase pathway has only been reported in *S. enterica* O54. The O antigen genes usually cluster in one conserved locus on the chromosome, but sometimes there are 2 major loci, such as in *Yersinia,* where it appears that one is for O antigens using the Wzx/Wzy pathway and the other for those with the ABC transporter pathway. New O antigens have now been shown to be derived from others in several ways such as homologous recombination, IS mediated events, plasmid mediated events or by phage carrying modification genes. It is also well established that O antigen has a significant role in bacterial virulence.

References

Achtman, M., Pluschke, G. Clonal analysis of descent and virulence among selected *Escherichia coli.* Annu Rev Microbiol 40 (1986) 185–210.

Allison, G.E., Verma, N.K. Serotype-converting bacteriophages and O-antigen modification in *Shigella flexneri.* Trends Microbiol 8 (2000) 17–23.

Anjum, M.F., Tucker, J.D., Sprigings, K.A., Woodward, M.J., Ehricht, R. Use of miniaturized protein arrays for *Escherichia coli* O serotyping. Clin Vaccine Immunol 13 (2006) 561–567.

Bailey, M.J.A., Hughes, C., Koronakis, V. RfaH and the *ops* element, components of a novel system controlling bacterial transcription elongation. Mol Microbiol 26 (1997) 845–851.

Ballmer, K., Korczak, B.M., Kuhnert, P., Slickers, P., Ehricht, R., Hachler, H. Fast DNA serotyping of *Escherichia coli* by use of an oligonucleotide microarray. J Clin Microbiol 45 (2007) 370–379.

Bastin, D.A., Brown, P.K., Haase, A., Stevenson, G., Reeves, P.R. Repeat unit polysaccharides of bacteria: a model for polymerisation resembling that of ribosomes and fatty acid synthetase, with a novel mechanism for determining chain length. Mol Microbiol 7 (1993) 725–734.

Baturo, A.P., Raginskaya, V.P. Antigenic scheme for the Hafniae. Int J Syst Bacteriol 28 (1978) 126–127.

Bengoechea, J.A., Najdenski, H., Skurnik, M. Lipopolysaccharide O antigen status of *Yersinia enterocolitica* O:8 is essential for virulence and absence of O antigen affects the expression of other *Yersinia* virulence factors. Mol Microbiol 52 (2004) 451–469.

Bennett, P.M. Genome plasticity: insertion sequence elements, transposons and integrons, and DNA rearrangement. Methods Mol Biol 266 (2004) 71–113.

Blokesch, M., Schoolnik, G.K. Serogroup conversion of *Vibrio cholerae* in aquatic reservoirs. PLoS Pathog 3 (2007) e81.

Bogdanovich, T., Carniel, E., Fukushima, H., Skurnik, M. Use of O-antigen gene cluster-specific PCRs for the identification and O-genotyping of *Yersinia pseudotuberculosis* and *Yersinia pestis*. J Clin Microbiol 41 (2003) 5103–5112.

Brisse, S., Issenhuth-Jeanjean, S., Grimont, P.A. Molecular serotyping of *Klebsiella* species isolates by restriction of the amplified capsular antigen gene cluster. J Clin Microbiol 42 (2004) 3388–3398.

Bystrova, O.V., Knirel, Y.A., Lindner, B., Kocharova, N.A., Kondakova, A.N., Zahringer, U., Pier, G.B. Structures of the core oligosaccharide and O-units in the R- and SR-type lipopolysaccharides of reference strains of *Pseudomonas aeruginosa* O-serogroups. FEMS Immunol Med Microbiol 46 (2006) 85–99.

Cazalet, C., Jarraud, S., Ghavi-Helm, Y., Kunst, F., Glaser, P., Etienne, J. Buchrieser, C., Multigenome analysis identifies a worldwide distributed epidemic *Legionella pneumophila* clone that emerged within a highly diverse species. Genome Res 18 (2008) 431–441.

Chatterjee, S.N., Chaudhuri, K. Lipopolysaccharides of *Vibrio cholerae* II. Genetics of biosynthesis. Biochim Biophys Acta 1690 (2004) 93–109.

Choi, T.Y., Kim, D.A., Seo, Y.H. Evaluation of serotyping using monoclonal antibodies and PCR-RFLP for *Chlamydia trachomatis* serotype identification. J Korean Med Sci 16 (2001) 15–19.

Clark, C.A., Beltrame, J., Manning, P.A. The *oac* gene encoding a lipopolysaccharide O-antigen acetylase maps adjacent to the integrase-encoding gene on the genome of *Shigella flexneri* bacteriophage Sf6. Gene 107 (1991) 43–52.

Coimbra, R.S., Grimont, F., Lenormand, P., Burguiere, P., Beutin, L., Grimont, P.A. Identification of *Escherichia coli* O-serogroups by restriction of the amplified O-antigen gene cluster (rfb-RFLP). Res Microbiol 151 (2000) 639–654.

Correia Barbosa, S.F., Alkmin, M.G., Landgraf, I.M. Detecting polysaccharide antigen of *Neisseria meningitidis* group C in cerebrospinal fluid by dot-ELISA assay. Braz J Infect Dis 4 (2000) 144–150.

Cunneen, M.M., De Castro, C., Kenyon, J., Parrilli, M., Reeves, P.R., Molinaro, A., Holst, O., Skurnik, M. The O-specific polysaccharide structure and biosynthetic gene cluster of *Yersinia pseudotuberculosis* serotype O:11. Carbohydr Res 344 (2009) 1533–1540.

Cunneen, M.M., Reeves, P.R. The *Yersinia kristensenii* O11 O-antigen gene cluster was acquired by lateral gene transfer and incorporated at a novel chromosomal locus. Mol Biol Evol 24 (2007) 1355–1365.

De Castro, C., Kenyon, J., Cunneen, M.M., Reeves, P.R., Molinaro, A., Holst, O., Skurnik, M. Genetic characterization and structural analysis of the O-specific polysaccharide of *Yersinia pseudotuberculosis* serotype O:1c. Innate Immun In press (2010).

De Castro, C., Skurnik, M., Molinaro, A., Holst, O. Characterization of the specific O-polysaccharide structure and biosynthetic gene cluster of *Yersinia pseudotuberculosis* serotype O:15. Innate Immun 15 (2009) 351–359.

DebRoy, C., Roberts, E., Kundrat, J., Davis, M.A., Briggs, C.E., Fratamico, P.M. Detection of *Escherichia coli* serogroups O26 and O113 by PCR amplification of the *wzx* and *wzy* genes. Appl Environ Microbiol 70 (2004) 1830–1832.

Edwards, P.R., Ewing, W.H. Identification of Enterobacteriaceae. Burgess Publishing Co., Minneapolis (1972).

Ewing, W.H. Edwards and Ewing's Identification of the *Enterobacteriaceae*., Elsevier Science Publishers, Amsterdam, The Netherlands (1986).

Kauffmann, F. The Serology of the Coli Group, J Immunol 57 (1947) 71–100.

Feng, L., Han, W., Wang, Q., Bastin, D.A., Wang, L. Characterization of *Escherichia coli* O86 O-antigen gene cluster and identification of O86-specific genes. Vet Microbiol 106 (2005) 241–248.

Feng, L., Perepelov, A.V., Zhao, G., Shevelev, S.D., Wang, Q., Senchenkova, S.N., Shashkov, A.S., Geng, Y., Reeves, P.R., Knirel, Y.A., Wang, L. Structural and genetic evidence that the *Escherichia coli* O148 O antigen is the precursor of the *Shigella dysenteriae* type 1 O antigen and identification of a glucosyltransferase gene. Microbiology 153 (2007) 139–147.

Feng, L., Senchenkova, S.N., Yang, J., Shashkov, A.S., Tao, J., Guo, H., Cheng, J., Ren, Y., Knirel, Y.A., Reeves, P.R., Wang, L. Synthesis of the heteropolysaccharide O antigen of *Escherichia coli* O52 requires an ABC transporter: structural and genetic evidence. J Bacteriol 186 (2004) 4510–4519.

Feng, L., Wang, W., Tao, J., Guo, H., Krause, G., Beutin, L., Wang, L. Identification of *Escherichia coli* O114 O-antigen gene cluster and development of an O114 serogroup-specific PCR assay. J Clin Microbiol 42 (2004) 3799–3804.

Franco, V.A., Liu, D., Reeves, P.R. The Wzz (Cld) protein in *Escherichia coli*: amino acid sequence variation determines O antigen chain length specificity. J Bacteriol 180 (1998) 2670–2675.

Gaston, M.A., Pitt, T.L. Improved O-serotyping method for *Serratia marcescens.* J Clin Microbiol 27 (1989) 2702–2705.

Han, W., Liu, B., Cao, B., Beutin, L., Kruger, U., Liu, H., Li, Y., Liu, Y., Feng, L., Wang, L. DNA microarray-based identification of serogroups and virulence gene patterns of *Escherichia coli* isolates associated with porcine postweaning diarrhea and edema disease. Appl Environ Microbiol 73 (2007) 4082–4088.

Hobbs, M., Reeves, P. The JUMPstart sequence: a 39 bp element common to several polysaccharide gene clusters. Mol Microbiol 12 (1994) 855–856.

Jarvis, V.R., Munn, V.P., Highsmith, A.K., Culver, D.H., Hughes, J.M. The epidemiology of nosocomial infection caused by *Klebsiella pneumoniae*. Infect Control 6 (1985) 68–74.

Kamal, A.M. Outbreak of gastroenteritis by non- agglutinable (NAG) vibrios in the Republic of the Sudan. J Egypt Pub Health Assoc 46 (1971) 125–174.

Kaniuk, N.A., Vinogradov, E., Whitfield, C. Investigation of the structural requirements in the lipopolysaccharide core acceptor for ligation of O antigens in the genus *Salmonella*: WaaL "ligase" is not the sole determinant of acceptor specificity. J Biol Chem 279 (2004) 36470–36480.

Keenleyside, W.J., Perry, M., Maclean, L., Poppe, C., Whitfield, C. A plasmid-encoded *rfb*O:54 gene cluster is required for biosynthesis of the O:54 antigen in *Salmonella enterica* serovar Borreze. Mol Microbiol 11 (1994) 437–448.

Keenleyside, W.J., Whitefield, C. A novel pathway for O-polysaccharide biosynthesis in *Salmonella enterica* serovar Borreze. J Biol Chem 271 (1996) 28581–28592.

Kenne, L., Lindberg, B. Bacterial polysaccharides. In Aspinall, G.O. (eds), The Polysaccharides, 2, Harcourt Brace Jovanovich, New York, NY (1983), pp. 287–363.

Kido, N., Kobayashi, H. A single amino acid substitution in a mannosyltransferase, WbdA, converts the *Escherichia coli* O9 polysaccharide into O9a: generation of a new O-serotype group. J Bacteriol 182 (2000) 2567–2573.

Kido, N., Torgov, V.I., Sugiyama, T., Uchiya, K., Sugihara, H., Komatsu, T., Kato, N., Jann, K. Expression of the O9 polysaccharide of *Escherichia coli*: sequencing of the *E. coli* O9 *rfb* gene cluster, characterization of mannosyl transferases, and evidence for an ATP-binding cassette transport system. J Bacteriol 177 (1995) 2178–2187.

Kintz, E., Goldberg, J.B. Regulation of lipopolysaccharide O antigen expression in *Pseudomonas aeruginosa.* Future Microbiol 3 (2008) 191–203.

Knirel, Y.A. Polysaccharide antigens of *Pseudomonas aeruginosa.* Crit Rev Microbiol 17 (1990) 273–304.

Knirel, Y.A., Kocharova, N.A., Bystrova, O.A., Katzenellenbogen, E., Gamian, A. Structures and Serology of the O-Specific Polysaccharides of Bacteria of the Genus *Citrobacter.* Arch Immunol Ther Exp 50 (2002) 379–391.

Knirel, Y.A., Kochetkov, N.K. The structure of lipopolysaccharides of Gram-negative bacteria. III. The strcucture of O antigens: a review. Biochemistry (Moscow) 59 (1994) 1325–1383.

Knirel, Y.A., Vinogradov, E.V., Shashkov, A.S., Sidorczyk, Z., Rozalski, A., Radziejewska-Lebrecht, J., Kaca, W. Structural study of O-specific polysaccharides of *Proteus*. J Carbohydr Chem 12 (1993) 379–414.

Kondakova, A.N., Bystrova, O.V., Shaikhutdinova, R.Z., Ivanov, S.A., Dentovskaya, S.V., Shashkov, A.S., Knirel, Y.A., Anisimov, A.P. Reinvestigation of the O-antigens of *Yersinia pseudotuberculosis*: revision of the O2c and confirmation of the O3 antigen structures. Carbohydr Res 343 (2008a) 2486–2488.

Kondakova, A.N., Bystrova, O.V., Shaikhutdinova, R.Z., Ivanov, S.A., Dentovskaya, S.V., Shashkov, A.S., Knirel, Y.A., Anisimov, A.P. Structure of the O-antigen of *Yersinia pseudotuberculosis* O:4a revised. Carbohydr Res 344 (2009a) 531–534.

Kondakova, A.N., Bystrova, O.V., Shaikhutdinova, R.Z., Ivanov, S.A., Dentovskaya, S.V., Shashkov, A.S., Knirel, Y.A., Anisimov, A.P. Structure of the O-antigen of *Yersinia pseudotuberculosis* O:4b. Carbohydr Res 344 (2009b) 152–154.

Kondakova, A.N., Bystrova, O.V., Shaikhutdinova, R.Z., Ivanov, S.A., Dentovskaya, S.V., Shashkov, A.S., Knirel, Y.A., Anisimov, A.P. Structure of the O-polysaccharide of *Yersinia pseudotuberculosis* O:2b. Carbohydr Res 344 (2009c) 405–407.

Kondakova, A.N., Ho, N., Bystrova, O.V., Shashkov, A.S., Lindner, B., Creuzenet, C., Knirel, Y.A. Structural studies of the O-antigens of *Yersinia pseudotuberculosis* O:2a and mutants thereof with impaired 6-deoxy-D-manno-heptose biosynthesis pathway. Carbohydr Res 343 (2008b) 1383–1389.

Kondakova, A.N., Shaikhutdinova, R.Z., Ivanov, S.A., Dentovskaya, S.V., Shashkov, A.S., Anisimov, A.P., Knirel, Y.A. Revision of the O-polysaccharide structure of *Yersinia pseudotuberculosis* O:1b. Carbohydr Res 344 (2009d) 2421–2423.

Lai, V., Wang, L., Reeves, P.R. *Escherichia coli* clone Sonnei (*Shigella sonnei*) had a chromosomal O-antigen gene cluster prior to gaining its current plasmid-borne O-antigen genes. J Bacteriol 180 (1998) 2983–2986.

Larsson, P. Serology of *Proteus mirabilis* and *Proteus vulgaris*. Methods Microbiol 14 (1984) 187–214.

Li, Q., Reeves, P.R. Genetic variation of dTDP-L-rhamnose pathway genes in *Salmonella enterica*. Microbiology 146 (2000) 2291–2307.

Li, Y., Liu, D., Cao, B., Han, W., Liu, Y., Liu, F., Guo, X., Bastin, D.A., Feng, L., Wang, L. Development of a serotype-specific DNA microarray for identification of some *Shigella* and pathogenic *Escherichia coli* strains. J Clin Microbiol 44 (2006) 4376–4383.

Liu, B., Knirel, Y.A., Feng, L., Perepelov, A.V., Senchenkova, S.N., Wang, Q., Reeves, P.R., Wang, L. Structure and genetics of *Shigella* O antigens. FEMS Microbiol Rev 32 (2008) 627–653.

Luneberg, E., Zetzmann, N., Alber, D., Knirel, Y.A., Kooistra, O., Zahringer, U., Frosch, M. Cloning and functional characterization of a 30 kb gene locus required for lipopolysaccharide biosynthesis in *Legionella pneumophila*. Int J Med Microbiol 290 (2000) 37–49.

Manning, P.A., Stroeher, U.H., Morona, R. Molecular basis for O-antigen biosynthesis in *Vibrio cholerae* O1: Ogawa-Inaba switching. In Wachsmuth, I.K., Blake, P., Olsvik, O. (eds), *Vibrio Cholerae* and Cholera. American Society for Microbiology, Washington, DC (1993), pp. 77–94.

Marolda, C.L., Valvano, M.A. Promoter region of the *Escherichia coli* O7-specific lipopolysaccharide gene cluster: structural and functional characterization of an upstream untranslated mRNA sequence. J Bacteriol 180 (1998) 3070–3079.

Mattos, K.A., Todeschini, A.R., Heise, N., Jones, C., Previato, J.O., Mendonca-Previato, L. Nitrogen-fixing bacterium *Burkholderia brasiliensis* produces a novel yersiniose A-containing O-polysaccharide. Glycobiology 15 (2005) 313–321.

Mellata, M., Dho-Moulin, M., Dozois, C.M., Curtiss, R., 3rd, Brown, P.K., Arne, P., Bree, A., Desautels, C., Fairbrother, J.M. Role of virulence factors in resistance of avian pathogenic *Escherichia coli* to serum and in pathogenicity. Infect Immun 71 (2003) 536–540.

Miller, V.L., Finlay, B.B., Falkow, S. Factors essential for the penetration of mammalian cells by *Yersinia*. Curr Top Microbiol Immunol 138 (1988) 15–39.

Minka, S., Bruneteau, M. Isolation and chemical characterization of type R lipopolysaccharides of a hypovirulent strain of *Yersinia pestis*. Can J Microbiol 44 (1998) 477–481.

Mullane, N., P. O'Gaora, Nally, J.E., Iversen, C., Whyte, P., Wall, P.G., Fanning, S. Molecular analysis of the Enterobacter sakazakii O-antigen gene locus. Appl Environ Microbiol 74 (2008) 3783–3794.

Murray, G.L., Attridge, S.R., Morona, R. Altering the length of the lipopolysaccharide O antigen has an impact on the interaction of *Salmonella enterica* serovar Typhimurium with macrophages and complement. J Bacteriol 188 (2006) 2735–2739.

Pacinelli, E., Wang, L., Reeves, P.R. Relationship of *Yersinia pseudotuberculosis* O antigens IA, IIA, and IVB: the IIA gene cluster was derived from that of IVB. Infect Immun 70 (2002) 3271–3276.

Pedersen, K., Grisez, L., R. van Houdt, Tiainen, T., Ollevier, F., Larsen, J.L. Extended serotyping scheme for *Vibrio anguillarum* with the definition and characterization of seven provisional O-serogroups. Curr Microbiol 38 (1999) 183–189.

Penner, J.L., Hennessy, J.N. Separate O-grouping schemes for serotyping clinical isolates of *Proteus vulgaris* and *Proteus mirabilis*. J Clin Microbiol 12 (1980) 304–309.

Pluschke, G., Mayden, J., Achtman, M., Levine, R.P. Role of the capsule and the O-antigen in resistance of O18:K1 *Escherichia coli* to complement-mediated killing. J Bacteriol 42 (1983a) 907–913.

Pluschke, G., Mercer, A., Kusecek, B., Pohl, A., Achtman, M. Induction of bacteremia in newborn rats by *Escherichia coli* K1 is correlated with only certain O (lipopolysaccharide) antigen types. Infect Immun 39 (1983b) 599–608.

Popoff, M.Y., Bockemuhl, J., Gheesling, L.L. Supplement 2001 (no. 45) to the Kauffmann-White scheme. Res Microbiol 154 (2003) 173–174.

Pupo, G.M., Lan, R., Reeves, P.R. Multiple independent origins of *Shigella* clones of *Escherichia coli* and convergent evolution of many of their characteristics. Proc Natl Acad Sci USA 97 (2000) 10567–10572.

Rauss, K., Kontrohr, T., Vertenyi, A., Szendrei, L. Serological and chemical studies of *Sh. sonnei, Pseudomonas shigelloides* and C27 strains. Acta Microbiol Acad Sci Hung 17 (1970) 157–166.

Raymond, C.K., Sims, E.H., Kas, A., Spencer, D.H., Kutyavin, T.V., Ivey, R.G., Zhou, Y., Kaul, R., Clendenning, J.B., Olson, M.V. Genetic variation at the O-antigen biosynthetic locus in *Pseudomonas aeruginosa*. J Bacteriol 184 (2002) 3614–3622.

Reeves, P.P., Wang, L. Genomic organization of LPS-specific loci. Curr Top Microbiol Immunol 264 (2002) 109–135.

Reeves, P.R., Cunneen, M.M. Biosynthesis of O-antigen chains and assembly. In Moran, A., Brennan, P., Holst, O., Itzstein, M.v. (eds), Microbial Glycobiology: Structures, Relevance and Applications, Academic, New York, NY (2009).

Reeves, P.R., Wang, L. Genomic organization of LPS-specific loci. Curr Top Microbiol Immunol 264 (2002) 109–135.

Rick, P.D., Hubbard, G.L., Barr, K. Role of the rfe gene in the synthesis of the O8 Antigen in *Escherichia coli* K-12. J Bacteriol 176 (1994) 2877–2884.

Romanowska, A., Katzenellenbogen, E., Kulakowska, M., Gamian, A., Witkowska, D., Mulczyk, M., Romanowska, E. *Hafnia alvei* lipopolysaccharides: isolation, sugar composition and SDS-PAGE analysis. FEMS Microbiol Immunol 1 (1988) 151–155.

Romanowska, E. Immunochemical aspects of *Hafnia alvei* O antigens. FEMS Immunol Med Microbiol 27 (2000) 219–225.

Rozalski, A., Sidorczyk, Z., Kotelko, K. Potential virulence factors of *Proteus* bacilli. Microbiol Mol Biol Rev 61 (1997) 65–89.

Sakazaki, R., Tamura, K. The genus *Hafnia*. In Balows, A. (ed), The Prokaryotes, Springer, New York, NY (1992), pp. 2816–2821.

Samuel, G., Hogbin, J.P., Wang, L., Reeves, P.R. Relationships of the *Escherichia coli* O157, O111, and O55 O-antigen gene clusters with those of *Salmonella enterica* and *Citrobacter freundii*, which express identical O antigens. J Bacteriol 186 (2004) 6536–6543.

Scheutz, F., Cheasty, T., Woodward, D., Smith, H.R. Designation of O174 and O175 to temporary O groups OX3 and OX7, and six new *E. coli* O groups that include Verocytotoxin-producing *E. coli* (VTEC): O176, O177, O178, O179, O180 and O181. Apmis 112 (2004) 569–584.

Sebastian, S., Dillon, S.T., Lynch, J.G., Blalock, L.T., Balon, E., Lee, K.T., Comstock, L.E., Conlan, J.W., Rubin, E.J., Tzianabos, A.O., Kasper, D.L. A defined O-antigen polysaccharide mutant of *Francisella tularensis* live vaccine strain has attenuated virulence while retaining its protective capacity. Infect Immun 75 (2007) 2591–2602.

Sedlak, J., Slajsova, M. Antigen structure and antigenic relationships of the species *Citrobacter.* Zentralbl Bakteriol 200 (1966a) 369–374.

Sedlak, J., Slajsova, M. On the antigenic relationships of certain *Citrobacter* and *Hafnia* cultures. J Gen Microbiol 43 (1966b) 151–158.

Selander, R.K., Beltran, P., Smith, N.H. Evolutionary genetics of *Salmonella.* In Selander, R.K., Clark, A.G., Whittam, T.S. (eds), Evolution at the molecular level. Sinauer Associates, Sunderland, Massecheusetts (1991), pp. 25–27.

Senchenkova, S.N., Feng, L., Yang, J., Shashkov, A.S., Cheng, J., Liu, D., Knirel, Y.A., Reeves, P.R., Jin, Q., Ye, Q., Wang, L. Structural and genetic characterization of the *Shigella boydii* Type 10 and Type 6 O antigens. J Bacteriol 187 (2005) 2551–2554.

Shepherd, J.G., Wang, L., Reeves, P.R. Comparison of O-antigen gene clusters of *Escherichia coli* (*Shigella*) sonnei and *Plesiomonas shigelloides* O17: Sonnei gained its current plasmid-borne O-antigen genes from *P. shigelloides* in a recent event. Infect Immun 68 (2000) 6056–6061.

Shimada, T., Arakawa, E., Itoh, K., Okitsu, T., Matsushima, A., Asai, Y., Yamai, S., Nakazato, T., Nair, G.B., Albert, M.J., Takeda, Y. Extended serotyping scheme for *Vibrio cholerae.* Curr Microbiol 28 (1994) 175–178.

Sidorczyk, Z., Zych, K., Kolodziejska, K., Drzewiecka, D., Zabłotni, A. Progress in serological classification of further strains from genus *Protesu* and determination of epitopes and new serogroup. 2nd German-Polish-Russian Meeting an Bacterial Carbohydrates, Moscow, Russia, (2002).

Skurnik, M. Molecular genetics, biochemistry and biological role of *Yersinia* lipopolysaccharide. In Skurnik, M., Bengoechea, J.A., Granfors, K. (eds), The Genus *Yersinia*: Entering the Functional Genomics Era., Kluwer Plenum, New York, NY (2003), pp. 187–197.

Skurnik, M., Bengoechea, J.A. The biosynthesis and biological role of lipopolysaccharide O-antigens of pathogenic *Yersiniae.* Carbohydr Res 338 (2003) 2521–2529.

Skurnik, M., Peippo, A., Ervela, E. Characterization of the O-antigen gene clusters of *Yersinia pseudotuberculosis* and the cryptic O-antigen gene cluster of *Yersinia pestis* shows that the plague bacillus is most closely related to and has evolved from *Y. pseudotuberculosis* serotype O: 1b. Mol Microbiol 37 (2000) 316–330.

Sonesson, A., Jantzen, E. The branched-chain octose yersiniose A is a lipopolysaccharide constituent of *Legionella micdadei* and *Legionella macehernii.* J Microbiol Methods 15 (1992) 241–248.

Sorensen, U.B., Larsen, J.L. Serotyping of *Vibrio anguillarum.* Appl Environ Microbiol 51 (1986) 593–597.

Spiers, A.J., Buckling, A., Rainey, P.B. The causes of *Pseudomonas* diversity. Microbiology 146 (Pt 10) (2000) 2345–2350.

Stagg, R.M., Tang, S.S., Carlin, N.I., Talukder, K.A., Cam, P.D., Verma, N.K. A novel glucosyltransferase involved in O-antigen modification of *Shigella flexneri* serotype 1c. J Bacteriol 191 (2009) 6612–6617.

Stenutz, R., Weintraub, A., Widmalm, G. The structures of *Escherichia coli* O-polysaccharide antigens. FEMS Microbiol Rev 30 (2006) 382–403.

Sugiyama, T., Kido, N., Kato, Y., Koide, N., Yoshida, T., Yokochi, T. Evolutionary relationship among rfb gene clusters synthesizing mannose homopolymer as O-specific polysaccharides in *Escherichia coli* and *Klebsiella.* Gene 198 (1997) 111–113.

Sugiyama, T., Kido, N., Kato, Y., Koide, N., Yoshida, T., Yokochi, T. Generation of *Escherichia coli* O9a serotype, a subtype of *E. coli* O9, by transfer of the *wb∗* gene cluster of *Klebsiella* O3 into *E. coli* via recombination. J Bacteriol 180 (1998) 2775–2778.

Sugiyama, T., Kido, N., Komatsu, T., Ohta, M., Jann, K., Jann, B., Saeki, A., Kato, N. Genetic analysis of *Escherichia coli* O9 *rfb:* identification and DNA sequence of phosphomannomutase and GDP-mannose pyrophosphorylase genes. Microbiology 140 (1994) 59–71.

Taylor, D.N., Trofa, A.C., Sadoff, J., Chu, C., Bryla, D., Shiloach, J., Cohen, D., Ashkenazi, S., Lerman, Y., Egan, W., Schneerson, R., Robbins, J.B. Synthesis, characterization, and clinical evaluation of conjugate vaccines composed of the O-specific polysaccharides of *Shigella dysenteriae* type 1, *Shigella flexneri* type 2a, and *Shigella sonnei* (*Plesiomonas shigelloides*) bound to bacterial toxoids. Infect Immun 61 (1993) 3678–3687.

Thampapillai, G., Lan, R., Reeves, P.R. Molecular evolution in the *gnd* locus of *Salmonella enterica.* Mol Biol Evol 11 (1994) 813–828.

Traub, W.H. Serotyping of *Sermtia marcescens*: identification of a new O-antigen (024). Zentralbl Bakteriol Mihbiol HygA 259 (1985) 485–488.

Traub, W.H. Serotyping of *Serratia marcescens*: detection of two new O-antigens (025 and 026). Int J Med Microbiol 275 (1991) 495–499.

Trautmann, M., Cross, A.S., Reich, G., Held, H., Podschun, R., Marre, R., Evaluation of a competitive ELISA method for the determination of *Klebsiella* O antigens. J Med Microbiol 44 (1996) 44–51.

Valvano, M.A. Export of O-specific lipopolysaccharide. Front Biosci 8 (2003) s452–471.

Varki, A., Cummings, R., Esko, J. Essentials of Glycobiology, Cold Spring Harbor Laboratory Press, Cold Spring Harbor (NY) (1999).

Verma, N.K., Brandt, J.B., Verma, K.J., Lindberg, A.A. Molecular characterisation of the O-acetyl transferase gene of converting phage Sf6 that adds group antigen 6 to *Shigella flexneri.* Mol Microbiol 5 (1991) 71–75.

Wang, L., Reeves, P.R. Organization of *Escherichia coli* O157 O antigen gene cluster and identification of its specific genes. Infect Immun 66 (1998) 3545–3551.

Wang, Q., Perepelov, A.V., Feng, L., Knirel, Y.A., Li, Y., Wang, L. Genetic and structural analyses of *Escherichia coli* O107 and O117 O-antigens. FEMS Immunol Med Microbiol 55 (2009) 47–54.

Wang, Q., Wang, M., Kong, F., Gilbert, G.L., Cao, B., Wang, L., Feng, L. Development of a DNA microarray to identify the *Streptococcus pneumoniae* serotypes contained in the 23-valent pneumococcal polysaccharide vaccine and closely related serotypes. J Microbiol Methods 68 (2007) 128–136.

Wang, W., Perepelov, A.V., Feng, L., Shevelev, S.D., Wang, Q., Senchenkova, S.N., Han, W., Li, Y., Shashkov, A.S., Knirel, Y.A., Reeves, P.R., Wang, L. A group of *Escherichia coli* and *Salmonella enterica* O antigens sharing a common backbone structure. Microbiology 153 (2007) 2159–2167.

Wen, L., Wang, Q., Li, Y., Kong, F., Gilbert, G.L., Cao, B., Wang, L., Feng, L. Use of a serotype-specific DNA microarray for identification of group B *Streptococcus* (*Streptococcus agalactiae*). J Clin Microbiol 44 (2006) 1447–1452.

West, N.P., Sansonetti, P., Mounier, J., Exley, R.M., Parsot, C., Guadagnini, S., Prevost, M.C., Prochnicka-Chalufour, A., Delepierre, M., Tanguy, M., Tang, C.M. Optimization of virulence functions through glucosylation of *Shigella* LPS. Science 307 (2005) 1313–1317.

Whitfield, C. Biosynthesis of lipopolysaccharide O antigens. Trends Microbiol 3 (1995) 178–185.

Xiang, S.H., Hobbs, M., Reeves, P.R. Molecular analysis of the *rfb* gene cluster of a group D2 *Salmonella enterica* strain: evidence for its origin from an insertion sequence-mediated recombination event between group E and D1 strains. J Bacteriol 176 (1994) 4357–4365.

Zdorovenko, E.L., Zatonskii, G.V., Kocharova, N.A., Shashkov, A.S., Knirel, Y.A., Ovod, V.V. Structure of the O polysaccharides and serological classification of *Pseudomonas syringae* pv. porri from genomospecies 4. Eur J Biochem 270 (2003) 20–27.
Zhang, L., Toivanen, P., Skurnik, M. The gene cluster directing O-antigen biosynthesis in *Yersinia enterocolitica* serotype O8: idnetification of the genes for mannose and galactose biosynthesis and the gene for the O-antigen polymerase. Microbiology 142 (1996) 277–288.
Zych, K., Kowalczyk, M., Knirel, Y.A., Sidorczyk, Z. New serogroups of the genus *Proteus* consisting of *Proteus penneri* strains only. Determination of some LPS epitopes responsible for specificity. Adv Exp Med Biol 485 (2000) 339–344.

Chapter 7
Regulators of TLR4 Signaling by Endotoxins

Anne F. McGettrick and Luke A.J. O'Neill

Abstract The stimulation of TLR4 by LPS activates two distinct signaling pathways leading to the expression of diverse inflammatory genes. Intensive studies over the past decade have revealed the components involved in these signaling pathways, however, more recently the focus has shifted somewhat towards the components that regulate these pathways. Several regulatory mechanisms, including localisation of components, splice variants and inhibitory molecules will be discussed in this review.

Keywords Chaperones · Inhibitors of TLR4 signaling · LPS · MyD88 · TRIF

Abbreviations

PRRs	pathogen recognition receptors
PAMPS	pathogen-associated molecular patterns
IFNs	interferons
TLRs	Toll-like receptors
LPS	lipopolysacchairde
TIR	toll-IL-1 receptor
MYD88	myeloid differentiation primary response gene (88)
Mal	MyD88-adaptor like
TRIF	TIR domain-containing adapter inducing IFN-beta
TRAM	TRIF-related adaptor molecule
Mal	MyD88-adaptor like
TAG	TRAM adaptor with GOLD domain

A.F. McGettrick (✉)
School of Biochemistry and Immunology, Trinity College Dublin, Dublin 2, Ireland
e-mail: mcgettra@tcd.ie

X. Wang, P.J. Quinn (eds.), *Endotoxins: Structure, Function and Recognition*,
Subcellular Biochemistry 53, DOI 10.1007/978-90-481-9078-2_7,

7.1 Introduction

Early innate immune responses to foreign pathogens are critical in enabling the host to effectively eliminate the pathogen. Pathogen recognition receptors (PRRs) can recognise a broad range of pathogens and discriminate between non-self and self (Medzhitov and Janeway, 2002). Toll-like receptors (TLR) are one of the main families of PRRs. They sense a variety of pathogen-associated molecular patterns (PAMPS) (Akira et al., 2006) and the detection of these PAMPS initiates a signal cascade that results in the activation of transcription factor such as NF-κB and IRF3 leading to the production of proinflammatory cytokines and type 1 interferons (IFNs). The first Toll protein was discovered in Drosphilia and was found to play a role in dorsal–ventral patterning during embryogenesis (Anderson et al., 1985; Morisato and Anderson, 1995). It was later discovered to also have a role in antifungal responses in the Drosphilia (Lemaitre et al. 1996; Cherry and Silverman, 2006). The role of Toll-like receptors in innate immunity was first discovered when the C3H/HeJ mice, which have a defective response to bacterial endotoxin, were found to have a missense mutation in the Toll homolog, TLR4 (Poltorak et al., 1998). Since this discovery at least 11 human TLRs and murine 13 TLRs have been described (Kawai and Akira, 2006) and they all play a role in the innate immune response, recognising different PAMPS ranging from lipids to lipopeptides, to proteins and nucleic acid.

Upon recognition of their PAMPS, TLRs form homo- or heterdimers in order to signal. TLR2 heterodimerises with TLR1 or TLR6 to recognise a wide variety of lipopetides present on gram negative bacteria, fungi, parasites and viruses. TLR3 recognises viral double stranded RNA. The gram-negative bacterial product lipopolysacchairde (LPS) is sensed by TLR4 while TLR5 recognises bacterial flaggellin. TLR7/8 recognises guanosine- or uridine-rich single-stranded RNA from viruses and TLR9 recognises unmethylated CpG motifs in viral DNA. Wang et al. (2006) has recently shown that TLR7, 8 and 9 can interact with each other adding another layer of complexity to their signalling. TLR10 is thought to heterodimerise with TLR2, however a ligand for this heterodimer has yet to be discovered (Akira et al., 2006).

TLRs are type 1 membrane proteins which are characterised by an extracellular domain of leucine rich repeats and an intracellular toll-IL-1 receptor (TIR) domain. Stimulation of the TLRs results in the recruitment of TIR domain-containing cytoplasmic proteins, via a TIR–TIR interaction, resulting in the activation of downstream signaling pathways. There are five known TIR domain-containing adaptors: MyD88, Mal, TRIF, TRAM and SARM (Kenny and O'Neill, 2008). Different TLRs utilise different adaptors to induce the production of a variety of pro-inflammatory cytokines. MyD88 is the universal adaptor, utilised by all of the TLRs, with the exception of TLR3. MyD88 was first described as a myeloid differentiation primary response gene that was rapidly induced upon IL-6-stimulated differentiation of M1 myeloleukaemic cells into macrophages (Lord et al., 1990). Later it was discovered that MyD88 played a key role in IL-1R signaling (Muzio et al., 1997; Adachi et al., 1998) and subsequently in TLR signaling (Kawai et al., 1999). The generation of MyD88-deficient mice led to the realisation that other adaptors may be involved in signaling from certain TLRs (Kawai et al., 1999; Akira et al., 2000). As expected, MyD88-deficient mice fail to respond to TLR5, 7 and 9 ligands (Adachi et al., 1998) suggesting that MyD88 is the sole adaptor utilized by TLR5, 7, 8 and 9. The response to TLR3 was completely normal and in response to TLR4 there was normal induction of IFNβ and delayed activation of NF-κB. This indicated that other adaptors were utilized by these pathways. MyD88-deficient mice also fail to respond to TLR2, however, it was later discovered that the surfaces of TLR2, TLR4 and MyD88 are largely electropositive leaving MyD88 unable to bind these receptors. The bridging adaptor, Mal, is largely electronegative allowing it to bridge MyD88 to TLR2 and 4 (Dunne et al., 2003). TLR3 utilises the adaptor TRIF to activate the NF-κB pathway and the IRF3 pathway. While TLR4 utilises Mal and MyD88 to activate the NF-κB pathway, it utilises TRIF and TRAM to activate the type 1 IFN pathway (Akira et al., 2006; McGettrick and O'Neill, 2004). This review will concentrate on the mechanism of regulation of these TLR4 signaling pathways. The importance of regulating the signaling pathways emanating from TLR4 became apparent when it was discovered that TLR4 was the receptor involved in LPS induced sepsis. Much research has been done on the different modes of regulation of TLR4 and a very complex story involving many different mechanisms is emerging. This review will examine many of the proteins involved in the regulation of TLR4.

7.2 TLR4 Signaling Pathway

Several proteins, including LPS binding protein (LBP), CD14 and MD2, are required to allow TLR4 bind LPS. LBP, a serum glycoprotein, is a lipid transferase which promotes the transfer of LPS from bacterial outer membranes to CD14 (Tobias et al., 1995; Wright et al., 1990). CD14 is a glycoprotein that exists in two forms, membrane bound or soluble. CD14 presents LPS to its signaling receptor TLR4. MD2 is a small glycoprotein which again has two forms, membrane bound and secreted. The membrane bound form associates with TLR4 in the ER while the secreted form is secreted as both monomers and dimers. MD2 presents LPS as a monomeric LPS-MD-2 complex that directly and potently activates TLR4 (Shimazu et al., 1999). Only the monomers can bind TLR4, thus the dimers may be act as negative regulators binding LPS and preventing it from interacting with TLR4 (Teghanemt et al., 2008). The levels of LBP, CD14 and MD2 available determine the ability of TLR4 to recognise LPS. Low concentrations of LBP enhance LPS responses while high concentrations in plasma actually inhibit LPS responses (Zweigner et al., 2001). While soluble CD14 is known to promote LPS responses under certain conditions it can also inhibit LPS responses by diverting LPS away from membrane bound CD14 towards lipidproteins (Kitchens et al., 2001). In vivo experiments in mice have shown that injection of soluble CD14 can in fact rescue the mice from lethal doses of LPS (Haziot et al., 1995). Upon recognising LPS, TLR4 homodimerises, initiating the binding of its adaptor proteins. The subsequent pathways are described in Fig. 7.1.

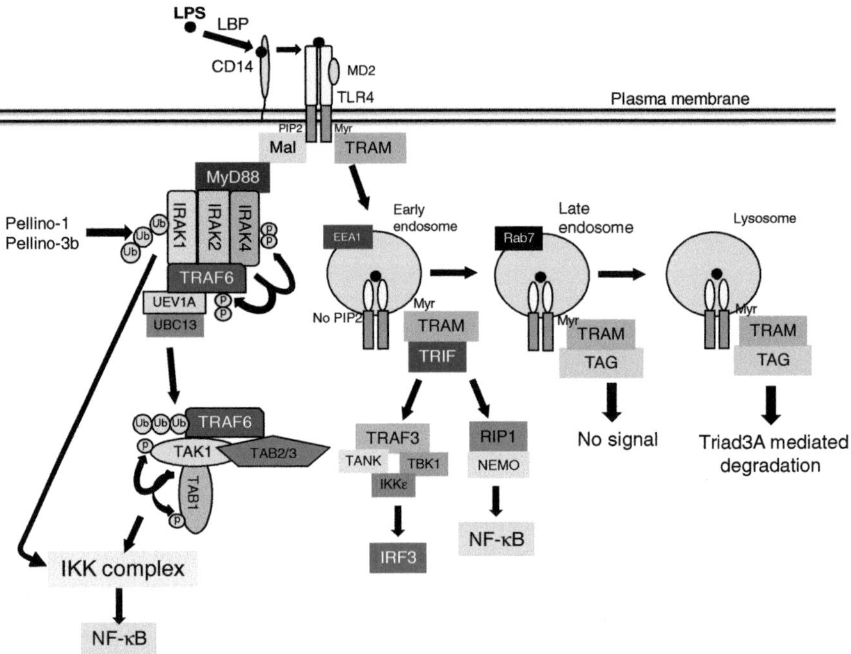

Fig. 7.1 Signaling from TLR4. LBP, CD14 and MD2 assist LPS to interact with TLR4. This induces the binding of Mal and MyD88 to TLR4 at the plasma membrane resulting in the activation and phosphorylation of IRAK4 and TRAK6. TRAF6 interacts with UEV1A and UBC13 to promote its ubiquitination leading to the activation of TAK1. TAK1 then interacts with TAB1, TAB2 and TAB3 which activates the IKK complex resulting in the activation of the NF-κB pathway. Pellino-1 and -3b promote IRAK-1 ubiquitination which promotes NF-κB activation via an interaction with NEMO. TLR4 then moves into the endosome in a dynamin and clathrin dependent manner causing the displacement of Mal and allowing for the binding of TRAM and TRIF. The N-terminal of TRIF recruits TRAF3 which in turn recruits TANK, TBK1 and IKKε resulting in the activation of IRF3. The C-terminal of TRIF interacts with RIP1 which interacts with NEMO leading to the activation of the IKK complex. The binding of TAG to TRAM in the late endosome results in the dissociation of TRIF from TRAM, terminating the signal. Ubiquitinated TLR4 then moves in the lysosome where it is degraded

7.2.1 MyD88-Dependent Pathway

The downstream pathway from MyD88 has been extensively studied. As mentioned above, in the case of TLR4, Mal is required to bridge MyD88 and TLR4. MyD88 possesses a death domain which allows it to interact with the death domain of the IRAKs. Similar to MyD88 knockout cells, IRAK-4 knockout macrophages show severe impairment in the production of proinflammatory cytokines in response to LPS (Suzuki et al., 2002). IRAK-4 then recruits IRAK-1 to the complex, leading to its phosphorylation and activation. IRAK-4 and IRAK-1 dissociate from the TLR complex and interact with tumour necrosis factor (TNF) receptor-associated factor-6 (TRAF-6), an E3 ubiquitin ligase. Interestingly, IRAK-1 knockout macrophages only show partial reduction to the levels of cytokine production in response to LPS, suggesting that other molecules are involved in this stage of the pathway. Recent research has indicated a central role for IRAK-2 in LPS signaling (Keating et al., 2007). IRAK-2 can associate to MyD88 and Mal (Muzio et al., 1997; Fitzgerald et al., 2001) and knockdown of IRAK-2 expression suppressed TLR4 signaling to NF-κB in human cells. Expression of IRAK-2, but not IRAK-1, led to TRAF6 ubiquitination, an event critical for NF-κB activation. TRAF6 interacts with ubiquitin-conjugating enzyme 13 (UBC13) and ubiquitin-conjugating enzyme E2 variant 1 isoform A (UEV1A) to promote the synthesis of lysine 63-linked polyubiquitin chains, which activate TAK1 (Chen, 2005). TAK1, in combination with TAB1, TAB2 and TAB3 activates kinases upstream of p38 and JNK and the IKK complex (Sato et al., 2005), which contains NF-κB essential modulator (NEMO) and IKK2, which phosphorylates IκB prompting its degradation, thereby freeing NF-κB to translocate to the nucleus and activate the expression of genes encoding proinflammatory cytokines. The story is somewhat more complex than that, however, as IRAK-1 also undergoes lysine 63-linked ubiquitination and Pellino-1 and Pellino-3b have been implicated in this process (Butler et al., 2007; Ordureau et al., 2008). These proteins are reported to be E3 ligases and their activity is greatly enhanced upon phosphorylation by IRAK-1 or IRAK-4 (Schauvliege et al., 2007) suggesting that the initial role of IRAK-1 and IRAK-4 may be to activate the Pellinos allowing them to polyubiquitinate IRAK-1. This recruits NEMO to IRAK-1 thus providing a second, TRAF6-independent pathway to activate the IKK complex (Conze et al., 2008).

As mentioned above, Mal is the bridging adaptor between TLR4 and MyD88 in the activation of NF-κB (Yamamoto et al., 2002; Horng et al., 2002). Mal plays a vital role in the innate immune response as was demonstrated by the discovery of a SNP in human Mal gene which confers resistant to the infectious diseases malaria, tuberculosis (TB) and pneunococcal pneumonia (Khor et al., 2007). The SNP results in a change in the amino acid at position 180 from a serine to a leucine. Khor et al. (2007) measured the incidence of the S180L variant in several populations and observed that being heterozygous halved the risk of developing malaria, TB and pneunococcal pneumonia. The S180L variant cannot reconstitute TLR2 signaling in Mal-deficient cells and this probably stems from the fact that molecular modelling suggests that the S180L variant cannot bind TLR2. The advantage to being heterozygous may therefore be that the presence of one wild-type allele produces a sufficient host defence while not allowing for an overactive immune response which may occur in wild-type homozygotes. Leucine homozygotes cannot mount any immune response, leaving them very susceptible to disease progression. A novel role for Mal in NF-κB activation has recently emerged. It directly interacts with TRAF6 via a TRAF6-binding motif, localising TRAF6 to the plasma membrane (Verstak et al., 2009). Disruption of the TRAF6-binding motif of MAL inhibited TLR-mediated NF-κB activation (Ye et al., 2002). Verstak et al. (2009) discovered that the interaction of MAL and TRAF6 mediates Ser phosphorylation of the p65 subunit of NF-κB and thus controls transcriptional activation but not nuclear translocation of NF-κB. These studies suggest that Mal's role as a bridging adaptor between TLR4 and MyD88 is not its only role in TLR4 signaling.

7.2.2 *MyD88-Independent Pathway*

TLR4 also activates a MyD88-independeny pathway that results in the induction of type 1 IFN. TRAM acts as a bridging adaptor bringing TLR4 and TRIF together. The N-terminal and C-terminal regions of TRIF have distinct functions. The C-terminal region of TRIF contains a Rip homotypic interaction motif (RHIM) which interacts with and activates RIP1 (receptor-interacting protein 1). RIP1 is a serine/threonine kinase that interacts with NEMO leading to the activation of the IKK complex (Poyet et al., 2000). The absence of RIP1 abolishes TRIF-dependent NF-κB activation (Meylan et al., 2004). RIP1 is not however, involved in IRF3 activation by LPS. The N-terminal region of TRIF appears to be involved in the activation of the IRF3 pathway. The N-terminal recruits TRAF3 which can associate with TANK, TBK1 and IKKϵ which phosphorylate IRF3 leading to the induction of the IFNβ pathway (Sharma et al., 2003; Fitzgerald et al., 2003). As expected, TRAF3-deficient cells could not induce type 1 IFN (Hacker et al., 2006; Oganesyan et al., 2006). TRIF can bind directly to TLR3 but TRAM is required to bridge TRIF to TLR4. TRAM-deficient mice showed normal responses to the ligands for TLR2, 3, 7, 9 and also IL-1β. The activation of the MyD88-dependent pathway by LPS was unaffected in the TRAM-deficient mice but the activation of the MyD88-independent pathway was abolished in response to TLR4 (Yamamoto et al., 2003).

7.3 Localisation of TLR4 and its Adaptor Molecules

The localisation of TLR4 in resting cells and upon activation by LPS appears to act as a form of regulation, by preventing overactivation, and prolonged activation of the receptor. In resting cells TLR4 is located in the Golgi and at the plasma membrane of human monocytes and cycles rapidly between the two membranes (Husebye et al., 2006; Latz et al., 2002). Initially it was thought that LPS travelled to the Golgi to activate TLR4 but, although LPS does traffic to the Golgi alongside TLR4, this is not essential for TLR4 signaling as inhibition of this trafficking as no effect on LPS signaling. MD2 is essential for the translocation of TLR4 from the Golgi to the plasma membrane (Nagai et al., 2002) and as mentioned above for the activation of TLR4 by LPS (Shimazu et al., 1999). It now appears that MD2/CD14/TLR4 cycles between the Golgi and the plasma membrane until engaged at the surface by LPS (Thieblemont and Wright, 1999; Latz et al., 2002). Several chaperones are known to be involved in the shuttling of TLR4 to the plasma membrane and these will be discussed in more detail later. In resting cells, Mal and MyD88 do not colocalise with TLR4 thus providing another barrier to unregulated activation of the pathway. MyD88 is present in condense particles scattered throughout the cytosol with no clear localisation to markers for the ER, Golgi, endosomes or lysosomes in resting cells (Jaunin et al., 1998; Honda et al., 2005; Kagan and Medzhitov, 2006). Mal contains a PtdIns(4,5)P_2 binding domain and localises to PtdIns(4,5)P_2 rich lipid rafts in resting cells. Upon LPS stimulation TLR4 moves to the lipid rafts, as does MyD88, colocalising them with Mal. Tanimura et al. (2008) observed that the recruitment of LPS to TLR4 induced the translocation of TLR4 from the plasma membrane to the endosome. During endocytosis it is known that PtdIns(4,5)P_2 concentrations drop (Botelho et al., 2000). This results in the dissociation of Mal from TLR4 allowing for the association of TRAM and TRIF, resulting in the activation of the IFN pathway in the endosome (Husebye et al., 2006). Once again, TRAM and TRIF do not colocalise prior to LPS signaling. TRIF is diffusely expressed in the cytosol in resting cells (Honda et al. 2004). TRIF relocates into lipid rafts in the plasma membrane (Wong et al., 2009) and early endosomes upon LPS stimulation where it colocalises with TRAM (Tanimura et al., 2008). TRAM is myristoylated and this localises to the plasma membrane (Rowe et al., 2006) and in rab5+ early endosomes (Kagan et al., 2008), where is partially colocalises with CD14 but not with TLR4 in resting cells (Tanimura et al., 2008). TRAM translocates from the plasma membrane to early endosomes where it colocalises with TLR4 upon LPS stimulation (Palsson-McDermott et al., 2009).

TLR4 is known to be ubiquitinated upon LPS stimulation by the E3 ubiquitin-protein ligase Triad3A. Triad3A is a RING finger protein that enhances ubiquitination and promotes TLR4 degradation leading to decreased signaling from the NF-κB pathway (Chuang and Ulevitch, 2004). The effect of Triad3A on signals downstream of the NF-κB pathway was measured but the effect of Triad3A on IRF3 signaling has yet to be investigated. Ubiquitination, however, was shown to have no effect on the endocytosis of the TLR4 (Husebye et al. 2006). TLR4 is also tyrosine phosphorylated (Husebye et al., 2006) and the combination of ubiquitination and tyrosine phosphorylation often targets transmembrane proteins to the lysosome for degradation. Indeed, the inhibition of the maturation of endosomes to lysosomes prevented the degradation of TLR4 following LPS stimulation. Another protein involved in this process is Hrs. TLR4, mainly in its ubiquitinated form, associates with Hrs, a protein located in the clathrin-coated limiting membrane of early/sorting endosome which targets ubiquitinated proteins to the lysosome for degradation.

7.4 Chaperones

As discussed above, LPS responsiveness is in part controlled by the amount of TLR4 present on the cell surface. The maturation of TLR4 in the ER and Golgi and the trafficking of TLR4 from the Golgi to the plasma membrane are regulated by several chaperone proteins, including gp96 and PRAT4A. The gp96 is an endoplasmic reticulum paralog of the heat shock protein 90 (HSP90) and chaperones multiple protein substrates. The gp96 associates with immature TLR4 in the ER. Calcium concentration in the ER is critical for this association suggesting the gp96 and TLR4 only interact under certain conditions in the ER (Randow and Seed, 2001). Macrophages lacking gp96 failed to respond to TLR4 ligands once again demonstrating the importance of this chaperone in TLR4 signaling (Yang et al., 2007). While gp96 appears to be a more general chaperone, protein associated with TLR4 A (PRAT4A) appears to be more dedicated to the TLRs as the expression of several other cell surface receptors was unaffected in the PRAT4A$^{-/-}$ cells. Like gp96, PRAT4A is an ER-resident protein. In immunocytes from PRAT4A$^{-/-}$ mice there was impaired expression of TLR2 and 4 on the cell surface and impaired cytokine production to several TLR ligands. PRAT4A$^{-/-}$ bone marrow chimeric mice were resistant to LPS induced sepsis. PRAT4A mRNA expression was down-regulated upon TLR stimulation, while gp96 mRNA levels were unaffected. (Takahashi et al., 2007).

7.5 Negative Regulators of TLR4 Signaling

Several proteins act as negative regulators of the TLR pathways. They are outlined below and in Table 7.1.

7.5.1 Cell Surface Receptors

Several cell surface receptors act as negative regulators of the TLR4 pathway, for example, ST2L, RP105 and SIGIRR. RP105 contains an extracellular LRR with homology to TLR4 and a short cytoplasmic domain (Miyake et al., 1995). RP105 associates with MD1, a MD2 homolog. RP105/MD1 binds directly to TLR4/MD2 preventing LPS from binding (Miyake et al., 1998). RP105-deficient mice are hyper-responsive to LPS but respond normally to CpG DNA indicating that RP105 is a specific inhibitor of the TLR4 pathway (Divanovic et al., 2005). ST2L is a

Table 7.1 Negative regulators of the TLR pathways. Several proteins act as negative regulators of the TLR pathways. This table lists these regulators and describes their targets and functions

Negative regulators	Target and function
ST2L	Sequester MyD88 and Mal away from TLR4
RP105/MD1	RP105/MD1 binds TLR4/MD2 preventing LPS from binding
SIGIRR	SIGIRR sequesters IRAKs and TRAF6 preventing IL-1R and certain TLRs from signaling
MyD88 s	This splice variant cannot recruit IRAK-4 thus inhibiting downstream signaling
TAG	This splice variant of TRAM disrupts the TRAM-TRIF interaction thus inhibiting TLR4-dependent IFN production
IRAK-1c	This splice variant lacks kinase activity and sequesters MyD88, IRAK-2 and TRAF6 away from IRAK-1
IRAK-M	IRAK-M lacks kinase activity and prevents dissociation of IRAK-1 and IRAK-4 from MyD88
TRAF1	TRAF1 inhibits TRIF-mediated NF-κB and IRF3 activation
TRAF4	TRAF4 inhibits NF-κB activation probably via interactions with TRAF6 and TRIF
TANK	TANK negatively regulates the TLR NF-κB pathway
RIP3	RIP3 sequesters RIP1 away from TRIF
A20	A20 cleaves ubiqutinated TRAF6 limiting the activation of NF-κB and interacts with TBK1 and IKKε to inhibit IFN production
SOCS-1	The exact role of SOCS-1 is still under debate
Rab7b	Rab7b promotes the degradation of TLR4

transmembrane protein which is a homolog of the IL-1 receptor. It contains an extracellular Ig-like domain and an intracellular TIR domain. ST2L sequesters MyD88 and MAL preventing them from binding TLR4. ST2-deficient macrophages showed enhanced production of proinflammatory cytokines upon LPS stimulation (Brint et al., 2004). SIGIRR is also a member of the IL-1R family. It contains one extracellular immunoglobulin domain and an intracellular TIR domain but it cannot bind IL-1 or enhance IL-1-dependent signaling. It is in fact a negative regulator of signaling from IL-1R and certain TLRs (Thomassen, et al., 1999). SIGIRR-deficient mice show enhanced cytokine production in response to LPS or IL-1 (Wald et al., 2003). It sequesters IRAKs and TRAF6 away from the receptor thus blocking signaling.

7.5.2 Splice Variants

Splice variants play an important role in the regulation of signaling pathways and several splice variants have been implicated in the regulation of the TLR4 pathway. A splice variant of TLR4 itself has been described. In mouse, an alternatively spliced mouse TLR4 contains an additional exon between the second and third exon of the reported mTLR4 gene. This exon contains an in-frame stop codon producing a 122 amino acid protein. In Chinese hamster ovary (CHO)-K1 cells, this protein was partly secreted and was therefore named soluble mTLR4 (smTLR4). In a mouse macrophage cell line, the overexpression of smTLR4 resulted in the inhibition of LPS-mediated TNF-α production and NF-κB activation. LPS also increased the mRNA for smTLR4 in mouse macrophages. smTLR4 may therefore act as a feedback mechanism to inhibit excessive LPS responses (Iwami et al., 2000). Four splice variants of TLR4 in humans have been described but their significance has yet to be elucidated (Jaresova et al., 2007).

MyD88s is a splice variant of MyD88 that is upregulated in monocytes following continuous stimulation with LPS (Burns et al., 2003). MyD88s lacks the intermediate domain between the TIR domain and death domain, which includes a C-terminal helix. This helix has recently been shown to be critical for the interaction of MyD88 and IRAK-4 (Mendoza-barbera, et al., 2009), MyD88s therefore cannot recruit IRAK-4 thus inhibiting downstream signaling. It only inhibits the MyD88-dependent pathway activated by TLR4. TAG, a splice variant of TRAM, inhibits the MyD88-independent pathway. TAG contains a Golgi dynamics (GOLD) domain upstream of the TIR domain (Palsson-McDermott et al., 2009). The GOLD domain is found in several proteins involved in Golgi dynamics and secretion as well as localisation of proteins to membranous vesicles (Anantharaman and Aravind, 2002). Overexpression studies and specific knockdown of TAG revealed that TAG specifically inhibited the IRF3 pathway activated by LPS. TAG moves to late endosomes upon LPS stimulation where it encounters TLR4 and TRAM. TAG disrupts the TRIF-TRAM complex inhibiting type 1 IFN production as well as promoting TLR4 degradation.

Several other splice variants of downstream signaling proteins also act to inhibit these pathways. Examples include, IRAK-1c, a splice variant of IRAK-1. It lacks a region encoded by exon 11 of the IRAK-1 gene. IRAK-1c, alongside IRAK-1, is expressed in most tissues. Interestingly, IRAK-1c is the predominant form of IRAK-1 expressed in the brain. Unlike IRAK-1, IRAK-1c lacks kinase activity and cannot be phosphorylated by IRAK-4. IRAK-1c can however interact with IRAK-2, MyD88, and TRAF6, thus sequestering them away from IRAK-1 and inhibiting NF-κB activation upon LPS stimulation (Rao et al., 2005). In mice, several splice variants of IRAK-2, were found. No evidence of similar alternative splicing was found for the human IRAK-2 gene. When overexpressed, IRAK-2a and IRAK-2b potentiated NF-κB activation by LPS while IRAK-2c was shown to be inhibitory. LPS induced the expression of IRAK-2c suggesting a negative feedback loop (Hardy and O'Neill, 2004).

7.5.3 Inhibitory Molecules

Several inhibitory proteins have been discovered that inhibit at different points on the TLR4 signaling pathways.

7.5.3.1 IRAK-M

IRAK-M, a member of the IRAK family, has no kinase activity and its expression is restricted to monocytes/macrophages, while other IRAKs are ubiquitously expressed. IRAK-M has been shown to prevent dissociation of IRAK-1 and IRAK-4 from MyD88 thus blocking the formation of IRAK-TRAF6 complexes. IRAK-$M^{-/-}$ macrophages show increased NF-κB activity and increased expression of various inflammatory cytokines upon stimulation with several TLR ligands. Endotoxin tolerance was significantly reduced in IRAK-$M^{-/-}$ cells. This indicated that IRAK-M acts as an inhibitor of cytokine production by TLRs (Kobayashi et al., 2002). It is now however thought that IRAKM in fact acts to inhibit the NF-κB alternative pathway which involves NFκB-inducing kinase (NIK)-mediated IKKα-dependent p100 cleavage and nuclear translocation of RelB/p52, rather than the classical NF-κB pathway (Su et al., 2009). Su et al. examined the involvement of IRAK-M in LPS-TLR4-mediated signaling and found that levels of IL-6 and GM-CSF were similar in $IRAKM^{-/-}$ BMDMs and wild-type BMDMs following LPS stimulation, indicating that IRAKM does not in fact play a role in regulating TLR4 NF-κB signaling. However, TRAF3 renders NIK stable and active (Qing et al. 2005; He et al., 2007) thus IRAK-M may play a role in the regulation of the TRIF-TRAM pathway which activates TRAF3 and this will need to be addressed in future.

7.5.3.2 TRAF1 and TRAF4

A yeast two hybrid identified an interaction between TRAF1 and TRIF. Overexpression of TRAF1 inhibits TRIF-mediated NF-κB activation and IFN production suggesting that it is an inhibitor of the TRIF-dependent pathway. Overexpression of TRIF results in the caspase-dependent cleavage of TRAF1. The cleaved C-terminal fragment of TRAF1 has been shown to inhibit NF-κB activation by TNF and Fas ligand (Leo et al., 2001; Henkler et al., 2003). However, it is the N-terminal portion of TRAF1 that is responsible for the inhibition of NF-κB activation and IFN production by the TRIF pathway (Su et al., 2006) suggesting a novel method of inhibition. The overexpression of TRAF4, another member of the TRAF family, inhibits NF-κB activation and this may be due to its interaction with TRAF6 and TRIF. TRAF4 mRNA expression was upregulated in RPMI 8226 cells following LPS stimulation suggesting that TRAF4 participates a feedback loop to regulate TLR4 signaling (Takeshita et al., 2005).

7.5.3.3 TANK

TANK (also known as I-TRAF) is a TRAF-binding protein (Cheng and Baltimore, 1996; Rothe et al., 1996), interacting with TRAF1, TRAF2, TRAF3, TRAF5 and TRAF6 as well as IKKϵ and TBK1 (Pomerantz, and Baltimore, 1999; Nomura, et al., 2000). TANK has initially thought to act as a positive regulator of NF-κB activation. It was also reported that TANK functions as an adaptor that bridges TRAF3 and TBK1-IKKϵ suggesting that TANK is required for the production of type I interferon in response to TLR stimulation (Guo and Cheng, 2007). However, $TANK^{-/-}$ mice demonstrated that TANK is not involved in the IFN pathway and is in fact a negative regulator of the TLR NF-κB pathway. Macrophages and B cells from $TANK^{-/-}$ mice had increased NF-κB activation in response to TLR ligands. Polyubiquitination of TRAF6 was upregulated in $TANK^{-/-}$ macrophages, which indicates that TANK suppresses TRAF ubiquitination leading to suppression of TLR signaling (Kawagoe et al., 2009).

7.5.3.4 RIP3

As mentioned above RIP1 binds to the RHIM domain of TRIF to activate the TRIF-dependent NF- κB. RIP3 is another serine/threonine kinase that promotes the phosphorylation of RIP1 resulting in a reduction in NF-κB activation by RIP1. RIP3-induced inhibition was specific for TRIF, as NF-κB was not substantially inhibited when triggered by overexpression of MyD88 (Meylan et al., 2004). It is likely that RIP3 competitively binds to TRIF preventing RIP1 from binding.

7.5.3.5 A20

A20 is a Zinc finger protein containing a deubiquitinase domain and it is induced in response to LPS. Overexpression of A20 abolishes TLR4 activation of NF-κB and AP-1, and the induction of the chemokine IL-8 (O'Reilly and Moynagh, 2003). A20 deficient mice are hypersensitive to endotoxin shock, strongly suggests that A20 is involved in negative feedback regulation of TLR signaling. A20 interacts with TRAF6 (Heyninck and Beyaert, 1999). As mentioned above

TRAF6 undergoes K63-linked ubiquitination which is vital for NF-κB activation (Deng et al., 2000; Chen, 2005). A20 cleaves ubiquitinated TRAF6 and it was thought that this was the mechanism by which A20 limited the activation of NF-κB. However, several studies have found that A20 mutants lacking the ubiquitin-modifying functions still possess significant NF-κB inhibitory activity in many types of cells (Song et al., 1996; Evans et al., 2004; Li et al., 2008). A20 is localized to an endocytic membrane compartment that is in association with the lysosome. The lysosomal association of A20 requires its carboxy terminal zinc finger domain but not its deubiquitinase domain. A20 mutants defective in membrane association show reduced inhibition of NF-κB activity. These findings suggest the involvement of a lysosome-associated mechanism in A20-dependent termination of NF-κB signaling. Saitoh et al. (2005) also showed a role for A20 in the inhibition of the IRF3 pathway. A20 interacts with TBK1 and IKKϵ, and inhibits TLR3 or virus-induced IRF3 dimerization and ISRE-dependent transcriptional activation. It is likely that it works in the same way for TLR4-induced IRF3 activation.

7.5.3.6 SOCS-1

The role of SOCS-1 in the regulation of TLR4 signaling is still under debate. Several groups have published seemingly contradictory results and it remains to be seen if SOCS-1 plays a direct or indirect role in TLR regulation. Initially, it was thought that autocrine IFN-β caused induction of SOCS following LPS stimulation (Crespo et al., 2000). Baetz et al. (2004) however showed direct induction of SOCS by TLR ligands as intermediate protein synthesis was not necessary and DCs from IFNAR$^{-/-}$ mice could still induce SOCS-1 production following LPS stimulation. SOCS-1$^{-/-}$ mice were unable to mount LPS tolerance and SOCS-1–deficient macrophages produced increased amounts of inflammatory cytokines such as IL-6 and TNF in response to LPS (Kinjyo et al., 2002; Nakagawa et al., 2002). However, Dalpke et al. (2001) showed, following overexpression studies in RAW264.7 macrophages, no role for SOCS-1 in direct TLR regulation. An additional study reported that SOCS-1 interacted directly with the p65 subunit of the NF-κB complex (Ryo et al., 2003) and a more recent study suggests that SOCS-1 might be found within the nucleus where it could interact with NF-κB (Maine et al., 2007). Mansell et al. (2006) found that, like in earlier studies, the p38 phosphorylation and IκBα and p42/p44 phosphorylation was unaltered in SOCS-1–deficient macrophages in response to LPS. They did however describe a direct role for SOCS-1 in TLR4 signaling. SOCS-1 binds Mal, resulting in the polyubiquitination of Mal, leading to its degradation. As mentioned above Mal associates with TRAF6 to promote phosphorylation of p65 and the transactivation of NF-κB. SOCS-1 may regulate this pathway, rather than the canonical NF-κB pathway mediated by MyD88 or the p38 pathway. Prele et al. (2008) demonstrated that SOCS-1 had no effect on the activation or on the DNA binding capacity of NF-κB following LPS stimulation. However, SOCS-1 did regulate the IFN-dependent pathways in LPS-activated cells as evidenced by reduced IFN-β production and STAT1 phosphorylation (Mansell et al., 2006). Further studies will be needed to determine the exact role of SOCS-1 in TLR4 regulation.

7.5.3.7 Rab7b

Rab7b is a lysosome-associated small GTPase that is involved in trafficking and lysosomal degradation of several receptors. Rab7b is localized in late endosome where it colocalizes with TLR4 upon LPS stimulation and can decrease the protein level of TLR4. Rab7b negatively regulates NF-κB and IRF3 signaling pathways in macrophages by promoting the lysosomal degradation of TLR4 and decreasing the plasma membrane TLR4 expression level (Wang et al., 2007).

7.6 Conclusion

TLR4 plays a vital role in our response to gram negative bacteria. The overactivation the TLR4 signaling pathway can result in the development of sepsis and therefore needs to be tightly regulated. The highly complex nature of the TLR4 signaling pathway has emerged over the past decade and new research continues to unravel the components involved in the pathway. More recently the complexity of the regulation of this pathway is being discovered. Regulation involves the localisation of components, the concentration of components, splice variants and inhibitors at every level of the pathway. Every new regulator found gives scientist a better understanding of the cells natural regulatory mechanism and gives them a better chance of developing drugs to specifically inhibit the TLR4 pathway when required.

References

Adachi, O., Kawai, T., Takeda, K., Matsumoto, M., Tsutsui, H., Sakagami, M., Nakanishi, K., et al. Targeted disruption of the MyD88 gene results in loss of IL-1- and IL-18-mediated function. Immunity 9 (1998) 143–150.

Akira, S., Hoshino, K., Kaisho, T. The role of Toll-like receptors and MyD88 in innate immune responses. J Endotoxin Res, 6 (2000) 383–387.

Akira, S., Uematsu, S., Takeuchi, O. Pathogen recognition and innate immunity. Cell 124 (2006) 783–801.

Anantharaman, V., Aravind, L. The GOLD domain, a novel protein module involved in Golgi function and secretion. Genome Biol 3 (2002) research0023.

Anderson, K.V., Bokla, L., Nusslein-Volhard, C. Establishment of dorsal-ventral polarity in the *Drosophila embryo*: the induction of polarity by the Toll gene product. Cell 42 (1985) 791–798.

Baetz, A., Frey, M., Heeg, K., Dalpke, A.H. Suppressor of cytokine signaling (SOCS) proteins indirectly regulate toll-like receptor signaling in innate immune cells. J Biol Chem 279 (2004) 54708–54715.

Botelho, R.J., Teruel, M., Dierckman, R., Anderson, R., Wells, A., York, J.D., Meyer, T., et al. Localized biphasic changes in phosphatidylinositol-4,5-bisphosphate at sites of phagocytosis. J Cell Biol 151 (2000) 1353–1368.

Brint, E.K., Xu, D., Liu, H., Dunne, A., McKenzie, A.N., O'Neill, L.A., Liew, F.Y. ST2 is an inhibitor of interleukin 1 receptor and Toll-like receptor 4 signaling and maintains endotoxin tolerance. Nat Immunol 5 (2004) 373–379.

Burns, K., Janssens, S., Brissoni, B., Olivos, N., Beyaert, R., Tschopp, J. Inhibition of interleukin 1 receptor/Toll-like receptor signaling through the alternatively spliced, short form of MyD88 is due to its failure to recruit IRAK-4. J Exp Med, 197 (2003) 263–268.

Butler, M.P., Hanly, J. A., Moynagh, P.N. Kinase-active interleukin-1 receptor-associated kinases promote polyubiquitination and degradation of the Pellino family: direct evidence for PELLINO proteins being ubiquitin-protein isopeptide ligases. J Biol Chem 282 (2007) 29729–29737.

Chen, Z.J. Ubiquitin signalling in the NF-kappaB pathway. Nat Cell Biol 7 (2005) 758–765.

Cheng, G., Baltimore, D. TANK, a co-inducer with TRAF2 of TNF- and CD 40L-mediated NF-kappaB activation. Genes Dev 10 (1996) 963–973.

Cherry, S., Silverman, N. Host-pathogen interactions in drosophila: new tricks from an old friend. Nat Immunol 7 (2006) 911–917.

Chuang, T.H., Ulevitch, R.J. Triad3A, an E3 ubiquitin-protein ligase regulating Toll-like receptors. Nat Immunol 5 (2004) 495–502.

Conze, D.B., Wu, C.J., Thomas, J.A., Landstrom, A., Ashwell, J.D. Lys63-linked polyubiquitination of IRAK-1 is required for interleukin-1 receptor- and toll-like receptor-mediated NF-kappaB activation. Mol Cell Biol 28 (2008) 3538–3547.

Crespo, A., Filla, M.B., Russell, S.W. Murphy, W.J., Indirect induction of suppressor of cytokine signalling-1 in macrophages stimulated with bacterial lipopolysaccharide: partial role of autocrine/paracrine interferon-alpha/beta. Biochem J 349 (2000) 99–104.

Dalpke, A.H., Opper, S., Zimmermann, S., Heeg, K. Suppressors of cytokine signaling (SOCS)-1 and SOCS-3 are induced by CpG-DNA and modulate cytokine responses in APCs. J Immunol 166 (2001) 7082–7089.

Deng, L., Wang, C., Spencer, E., Yang, L., Braun, A., You, J., Slaughter, C., et al., Activation of the IkappaB kinase complex by TRAF6 requires a dimeric ubiquitin-conjugating enzyme complex and a unique polyubiquitin chain. Cell 103 (2000) 351–361.

Divanovic, S., Trompette, A., Atabani, S.F., Madan, R., Golenbock, D.T., Visintin, A., Finberg, R.W., et al., Negative regulation of Toll-like receptor 4 signaling by the Toll-like receptor homolog RP105. Nat Immunol 6 (2005) 571–578.

Dunne, A., Ejdeback, M., Ludidi, P.L., O'Neill, L.A., Gay, N.J. Structural complementarity of Toll/interleukin-1 receptor domains in Toll-like receptors and the adaptors Mal and MyD88. J Biol Chem 278 (2003) 41443–41451.

Evans, P.C., Ovaa, H., Hamon, M., Kilshaw, P.J., Hamm, S., Bauer, S., Ploegh, H.L., Smith, T.S. Zinc-finger protein A20, a regulator of inflammation and cell survival, has de-ubiquitinating activity. Biochem J 378 (2004) 727–734.

Fitzgerald, K.A., McWhirter, S.M., Faia, K.L., Rowe, D.C., Latz, E., Golenbock, D.T., Coyle, A.J., et al., IKKepsilon and TBK1 are essential components of the IRF3 signaling pathway. Nat Immunol 4 (2003) 491–496.

Fitzgerald, K.A., Palsson-McDermott, E.M., Bowie, A.G., Jefferies, C.A., Mansell, A.S., Brady, G., Brint, E., et al., Mal (MyD88-adapter-like) is required for Toll-like receptor-4 signal transduction. Nature 413 (2001) 78–83.

Guo, B., Cheng, G. Modulation of the interferon antiviral response by the TBK1/IKKi adaptor protein TANK. J Biol Chem 282 (2007) 11817–11826.

Hacker, H., Redecke, V., Blagoev, B., Kratchmarova, I., Hsu, L.C., Wang, G.G., Kamps, M.P., et al., Specificity in Toll-like receptor signalling through distinct effector functions of TRAF3 and TRAF6. Nature 439 (2006) 204–207.

Hardy, M.P., O'Neill, L.A. The murine IRAK2 gene encodes four alternatively spliced isoforms, two of which are inhibitory. J Biol Chem 279 (2004) 27699–27708.

Haziot, A., Rong, G.W., Lin, X.Y., Silver, J., Goyert, S.M. Recombinant soluble CD14 prevents mortality in mice treated with endotoxin (lipopolysaccharide). J Immunol 154 (1995) 6529–6532.

He, J.Q., Saha, S.K., Kang, J.R., Zarnegar, B., Cheng, G. Specificity of TRAF3 in its negative regulation of the noncanonical NF-kappa B pathway. J Biol Chem 282 (2007) 3688–3694.

Henkler, F., Baumann, B., Fotin-Mleczek, M., Weingartner, M., Schwenzer, R., Peters, N., Graness, A., et al., Caspase-mediated cleavage converts the tumor necrosis factor (TNF) receptor-associated factor (TRAF)-1 from a selective modulator of TNF receptor signaling to a general inhibitor of NF-kappaB activation. J Biol Chem 278 (2003) 29216–29230.

Heyninck, K., Beyaert, R. The cytokine-inducible zinc finger protein A20 inhibits IL-1-induced NF-kappaB activation at the level of TRAF6. FEBS Lett 442 (1999) 147–150.

Honda, K., Ohba, Y., Yanai, H., Negishi, H., Mizutani, T., Takaoka, A., Taya, C., et al., Spatiotemporal regulation of MyD88-IRF-7 signalling for robust type-I interferon induction. Nature 434 (2005) 1035–1040.

Honda, K., Yanai, H., Mizutani, T., Negishi, H., Shimada, N., Suzuki, N., Ohba, Y., et al., Role of a transductional-transcriptional processor complex involving MyD88 and IRF-7 in Toll-like receptor signaling. Proc Natl Acad Sci USA 101 (2004) 15416–15421.

Horng, T., Barton, G.M., Flavell, R.A., Medzhitov, R., The adaptor molecule TIRAP provides signalling specificity for Toll-like receptors. Nature 420 (2002) 329–333.

Husebye, H., Halaas, O., Stenmark, H., Tunheim, G., Sandanger, O., Bogen, B., Brech, A., et al., Endocytic pathways regulate Toll-like receptor 4 signaling and link innate and adaptive immunity. Embo J 25 (2006) 683–692.

Iwami, K.I., Matsuguchi, T., Masuda, A., Kikuchi, T., Musikacharoen, T., Yoshikai, Y. Cutting edge: naturally occurring soluble form of mouse Toll-like receptor 4 inhibits lipopolysaccharide signaling. J Immunol 165 (2000) 6682–6686.

Jaresova, I., Rozkova, D., Spisek, R., Janda, A., Brazova, J., Sediva, A., Kinetics of Toll-like receptor-4 splice variants expression in lipopolysaccharide-stimulated antigen presenting cells of healthy donors and patients with cystic fibrosis. Microbes Infect 9 (2007) 1359–1367.

Jaunin, F., Burns, K., Tschopp, J., Martin, T.E., Fakan, S. Ultrastructural distribution of the death-domain-containing MyD88 protein in HeLa cells. Exp Cell Res 1998, 243 (1998) 67–75.

Kagan, J.C., Medzhitov, R. Phosphoinositide-mediated adaptor recruitment controls Toll-like receptor signaling. Cell 2006, 125 (2006) 943–955.

Kagan, J.C., Su, T., Horng, T., Chow, A., Akira, S., Medzhitov R. TRAM couples endocytosis of Toll-like receptor 4 to the induction of interferon-beta. Nat Immunol 2008, 9 (2008) 361–368.

Kawagoe, T., Takeuchi, O., Takabatake, Y., Kato, H., Isaka, Y., Tsujimura, T., Akira, S. TANK is a negative regulator of Toll-like receptor signaling and is critical for the prevention of autoimmune nephritis. Nat Immunol 10 (2009) 965–972.

Kawai, T., Adachi, O., Ogawa, T., Takeda, K., Akira, S. Unresponsiveness of MyD88-deficient mice to endotoxin. Immunity 11 (1999) 115–122.

Kawai, T., Akira, S. TLR signaling. Cell Death Differ 13 (2006) 816–825.

Keating, S.E., Maloney, G.M., Moran, E.M., Bowie, A.G. IRAK-2 participates in multiple toll-like receptor signaling pathways to NFkappaB via activation of TRAF6 ubiquitination. J Biol Chem 282 (2007) 33435–33443.

Kenny, E.F., O'Neill, L.A. Signalling adaptors used by Toll-like receptors: an update. Cytokine 43 (2008) 342–349.

Khor, C.C., Chapman, S.J., Vannberg, F.O., Dunne, A., Murphy, C., Ling, E.Y., Frodsham, A.J., et al., A Mal functional variant is associated with protection against invasive pneumococcal disease, bacteremia, malaria and tuberculosis. Nat Genet 39 (2007) 523–528.

Kinjyo, I., Hanada, T., Inagaki-Ohara, K., Mori, H., Aki, D., Ohishi, M., Yoshida, H., Kubo, M., Yoshimura, A. SOCS1/JAB is a negative regulator of LPS-induced macrophage activation. Immunity 17 (2002) 583.

Kitchens, R.L., Thompson, P.A., Viriyakosol, S., O'Keefe, G.E., Munford, R.S. Plasma CD14 decreases monocyte responses to LPS by transferring cell-bound LPS to plasma lipoproteins. J Clin Invest 108 (2001) 485-493.

Kobayashi, K., Hernandez, L.D., Galan, J.E., Janeway, C.A., Jr., Medzhitov, R., Flavell, R.A. IRAK-M is a negative regulator of Toll-like receptor signaling. Cell 110 (2002) 191–202.

Latz, E., Visintin, A., Lien, E., Fitzgerald, K.A., Monks, B.G., Kurt-Jones, E.A., Golenbock, D.T., Espevik, T. Lipopolysaccharide rapidly traffics to and from the Golgi apparatus with the toll-like receptor 4-MD-2-CD14 complex in a process that is distinct from the initiation of signal transduction. J Biol Chem 277 (2002) 47834–47843.

Lemaitre, B., Nicolas, E., Michaut, L., Reichhart, J.M., Hoffmann, J.A. The dorsoventral regulatory gene cassette spatzle/Toll/cactus controls the potent antifungal response in Drosophila adults. Cell 86 (1996) 973–983.

Leo, E., Deveraux, Q.L., Buchholtz, C., Welsh, K., Matsuzawa, S., Stennicke, H.R., Salvesen, G.S., Reed, J.C. TRAF1 is a substrate of caspases activated during tumor necrosis factor receptor-alpha-induced apoptosis. J Biol Chem 276 (2001) 8087–8093.

Li, L., Hailey, D.W., Soetandyo, N., Li, W., Lippincott-Schwartz, J., Shu, H.B., Ye, Y. Localization of A20 to a lysosome-associated compartment and its role in NFkappaB signaling. Biochim Biophys Acta 1783 (2008) 1140–1149.

Lord, K.A., Hoffman-Liebermann, B., Liebermann, D.A. Nucleotide sequence and expression of a cDNA encoding MyD88, a novel myeloid differentiation primary response gene induced by IL6. Oncogene 5 (1990) 1095–1097.

Maine, G.N., Mao, X., Komarck, C.M., Burstein, E. COMMD1 promotes the ubiquitination of NF-kappaB subunits through a cullin-containing ubiquitin ligase. Embo J 26 (2007) 436–447.

Mansell, A., Smith, R., Doyle, S.L., Gray, P., Fenner, J.E., Crack, P.J., Nicholson, S.E., et al., Suppressor of cytokine signaling 1 negatively regulates Toll-like receptor signaling by mediating Mal degradation. Nat Immunol 7 (2006) 148–155.

McGettrick, A.F., O'Neill, L.A. The expanding family of MyD88-like adaptors in Toll-like receptor signal transduction. Mol Immunol 41 (2004) 577–582.

Medzhitov, R., Janeway, C.A., Jr. Decoding the patterns of self and nonself by the innate immune system. Science 296 (2002) 298–300.

Mendoza-Barbera, E., Corral-Rodriguez, M.A., Soares-Schanoski, A., Velarde, M., Macieira, S., Messerschmidt, A., Lopez-Collazo, E., et al. Contribution of globular death domains and unstructured linkers to MyD88.IRAK-4 heterodimer formation: an explanation for the antagonistic activity of MyD88 s. Biochem Biophys Res Commun 380 (2009) 183–187.

Meylan, E., Burns, K., Hofmann, K., Blancheteau, V., Martinon, F., Kelliher, M., Tschopp, J. RIP1 is an essential mediator of Toll-like receptor 3-induced NF-kappa B activation. Nat Immunol 2004, 5 (2004) 503–507.

Miyake, K., Shimazu, R., Kondo, J., Niki, T., Akashi, S., Ogata, H., Yamashita, Y., et al. Mouse MD-1, a molecule that is physically associated with RP105 and positively regulates its expression. J Immunol 1998, 161 (1998) 1348–1353.

Miyake, K., Yamashita, Y., Ogata, M., Sudo, T., Kimoto, M. RP105, a novel B cell surface molecule implicated in B cell activation, is a member of the leucine-rich repeat protein family. J Immunol 154 (1995) 3333–3340.

Morisato, D., Anderson, K.V. Signaling pathways that establish the dorsal-ventral pattern of the *Drosophila embryo.* Annu Rev Genet 1995, 29:371–399.

Muzio, M., Ni, J., Feng, P., Dixit, V.M. IRAK (Pelle) family member IRAK-2 and MyD88 as proximal mediators of IL-1 signaling. Science 278 (1997) 1612–1615.

Nagai, Y., Akashi, S., Nagafuku, M., Ogata, M., Iwakura, Y., Akira, S., Kitamura, T., et al. Essential role of MD-2 in LPS responsiveness and TLR4 distribution. Nat Immunol 3 (2002) 667–672.

Nakagawa, R., Naka, T., Tsutsui, H., Fujimoto, M., Kimura, A., Abe, T., Seki, E., Sato, S., Takeuchi, O., Takeda, K., Akira, S., Yamanishi, K., Kawase, I., Nakanishi, K., Kishimoto, T. SOCS-1 participates in negative regulation of LPS responses. Immunity 17 (2002) 677.

Nomura, F., Kawai, T., Nakanishi, K., Akira, S. NF-kappaB activation through IKK-i-dependent I-TRAF/TANK phosphorylation. Genes Cells 5 (2000) 191–202.

Oganesyan, G., Saha, S.K., Guo, B., He, J.Q., Shahangian, A., Zarnegar, B., Perry, A., et al. Critical role of TRAF3 in the Toll-like receptor-dependent and -independent antiviral response. Nature 439 (2006) 208–211.

Ordureau, A., Smith, H., Windheim, M., Peggie, M., Carrick, E., Morrice, N., Cohen, P. The IRAK-catalysed activation of the E3 ligase function of Pellino isoforms induces the Lys63-linked polyubiquitination of IRAK1. Biochem J 409 (2008) 43–52.

O'Reilly, S.M., Moynagh, P.N. Regulation of Toll-like receptor 4 signalling by A20 zinc finger protein. Biochem Biophys Res Commun 303 (2003) 586–593.

Palsson-McDermott, E.M., Doyle, S.L., McGettrick, A.F., Hardy, M., Husebye, H., Banahan, K., Gong, M., et al. TAG, a splice variant of the adaptor TRAM, negatively regulates the adaptor MyD88-independent TLR4 pathway. Nat Immunol 10 (2009) 579–586.

Poltorak, A., He, X., Smirnova, I., Liu, M.Y., Van Huffel, C., Du, X., Birdwell, D., et al. Defective LPS signaling in C3H/HeJ and C57BL/10ScCr mice: mutations in Tlr4 gene. Science 282 (1998) 2085–2088.

Pomerantz, J.L., Baltimore, D. NF-kappaB activation by a signaling complex containing TRAF2, TANK and TBK1, a novel IKK-related kinase. Embo J 18 (1999) 6694–6704.

Poyet, J.L., Srinivasula, S.M., Lin, J.H., Fernandes-Alnemri, T., Yamaoka, S., Tsichlis, P.N., Alnemri, E.S. Activation of the Ikappa B kinases by RIP via IKKgamma /NEMO-mediated oligomerization. J Biol Chem 275 (2000) 37966–37977.

Prele, C.M., Woodward, E.A., Bisley, J., Keith-Magee, A., Nicholson, S.E., Hart, P.H. SOCS1 regulates the IFN but not NFkappaB pathway in TLR-stimulated human monocytes and macrophages. J Immunol 181 (2008) 8018–8026.

Qing, G., Qu, Z., Xiao, G. Stabilization of basally translated NF-kappaB-inducing kinase (NIK) protein functions as a molecular switch of processing of NF-kappaB2 p100. J Biol Chem 280 (2005) 40578–40582.

Randow, F., Seed, B. Endoplasmic reticulum chaperone gp96 is required for innate immunity but not cell viability. Nat Cell Biol 3 (2001) 891-896.

Rao, N., Nguyen, S., Ngo, K., Fung-Leung, W.P. A novel splice variant of interleukin-1 receptor (IL-1R)-associated kinase 1 plays a negative regulatory role in Toll/IL-1R-induced inflammatory signaling. Mol Cell Biol 25 (2005) 6521–6532.

Rothe, M., Xiong, J., Shu, H.B., Williamson, K., Goddard, A., Goeddel, D.V. I-TRAF is a novel TRAF-interacting protein that regulates TRAF-mediated signal transduction. Proc Natl Acad Sci USA 93 (1996) 8241–8246.

Rowe, D.C., McGettrick, A.F., Latz, E., Monks, B.G., Gay, N.J., Yamamoto, M., Akira, S., et al. The myristoylation of TRIF-related adaptor molecule is essential for Toll-like receptor 4 signal transduction. Proc Natl Acad Sci USA 103 (2006) 6299–6304.

Ryo, A., Suizu, F., Yoshida, Y., Perrem, K., Liou, Y.C., Wulf, G., Rottapel, R., et al. Regulation of NF-kappaB signaling by Pin1-dependent prolyl isomerization and ubiquitin-mediated proteolysis of p65/RelA. Mol Cell 12 (2003) 1413–1426.

Saitoh, T., Yamamoto, M., Miyagishi, M., Taira, K., Nakanishi, M., Fujita, T., Akira, S., et al., A20 is a negative regulator of IFN regulatory factor 3 signaling, J Immunol 174 (2005) 1507–1512.

Sato, S., Sanjo, H., Takeda, K., Ninomiya-Tsuji, J., Yamamoto, M., Kawai, T., Matsumoto, K., et al. Essential function for the kinase TAK1 in innate and adaptive immune responses. Nat Immunol 6 (2005) 1087–1095.

Schauvliege, R., Janssens, S., Beyaert, R. Pellino proteins: novel players in TLR and IL-1R signalling. J Cell Mol Med 11 (2007) 453–461.

Sharma, S., tenOever, B.R., Grandvaux, N., Zhou, G.P., Lin, R., Hiscott, J. Triggering the interferon antiviral response through an IKK-related pathway. Science 300 (2003) 1148–1151.

Shimazu, R., Akashi, S., Ogata, H., Nagai, Y., Fukudome, K., Miyake, K., Kimoto, M. MD-2, a molecule that confers lipopolysaccharide responsiveness on Toll-like receptor 4. J Exp Med 189 (1999) 1777–1782.

Song, H.Y., Rothe, M., Goeddel, D.V. The tumor necrosis factor-inducible zinc finger protein A20 interacts with TRAF1/TRAF2 and inhibits NF-kappaB activation. Proc Natl Acad Sci USA 93 (1996) 6721–6725.

Su, J., Zhang, T., Tyson, J., Li, L. The Interleukin-1 Receptor-Associated Kinase M Selectively Inhibits the Alternative, Instead of the Classical NFkappaB Pathway. J Innate Immun 1 (2009) 164–174.

Su, X., Li, S., Meng, M., Qian, W., Xie, W., Chen, D., Zhai, Z. et al., TNF receptor-associated factor-1 (TRAF1) negatively regulates Toll/IL-1 receptor domain-containing adaptor inducing IFN-beta (TRIF)-mediated signaling. Eur J Immunol 36 (2006) 199–206.

Suzuki, N., Suzuki, S., Duncan, G.S., Millar, D.G., Wada, T., Mirtsos, C., Takada, H.. et al. Severe impairment of interleukin-1 and Toll-like receptor signalling in mice lacking IRAK-4. Nature 416 (2002) 750–756.

Takahashi, K., Shibata, T., Akashi-Takamura, S., Kiyokawa, T., Wakabayashi, Y., Tanimura, N., Kobayashi, T., et al. A protein associated with Toll-like receptor (TLR) 4 (PRAT4A) is required for TLR-dependent immune responses. J Exp Med 204 (2007) 2963–2976.

Takeshita, F., Ishii, K.J., Kobiyama, K., Kojima, Y., Coban, C., Sasaki, S., Ishii, N., et al. TRAF4 acts as a silencer in TLR-mediated signaling through the association with TRAF6 and TRIF. Eur J Immunol 35 (2005) 2477–2485.

Tanimura, N., Saitoh, S., Matsumoto, F., Akashi-Takamura, S., Miyake, K. Roles for LPS-dependent interaction and relocation of TLR4 and TRAM in TRIF-signaling. Biochem Biophys Res Commun 368 (2008) 94–99.

Teghanemt, A., Widstrom, R.L., Gioannini, T.L., Weiss, J.P.. Isolation of monomeric and dimeric secreted MD-2. Endotoxin.sCD14 and Toll-like receptor 4 ectodomain selectively react with the monomeric form of secreted MD-2. J Biol Chem. 283 (2008) 21881–21889.

Thieblemont, N., Wright, S.D.. Transport of bacterial lipopolysaccharide to the golgi apparatus, J Exp Med. 190 (1999) 523–534.

Thomassen, E., Renshaw, B.R., Sims, J.E.. Identification and characterization of SIGIRR, a molecule representing a novel subtype of the IL-1R superfamily. Cytokine. 11 (1999) 389–399.

Tobias, P.S., Soldau, K., Gegner, J.A., Mintz, D., Ulevitch, R.J. Lipopolysaccharide binding protein-mediated complexation of lipopolysaccharide with soluble CD14. J Biol Chem. 270 (1995) 10482–10488.

Verstak, B., Nagpal, K., Bottomley, S.P., Golenbock, D.T., Hertzog, P.J., Mansell, A. MyD88 adapter-like (Mal)/TIRAP interaction with TRAF6 is critical for TLR2- and TLR4-mediated NF-kappaB proinflammatory responses. J Biol Chem 284 (2009) 24192–24203.

Wald, D., Qin, J., Zhao, Z., Qian, Y., Naramura, M., Tian, L., Towne, J., et al. SIGIRR, a negative regulator of Toll-like receptor-interleukin 1 receptor signaling, Nat Immunol 4 (2003) 920–927.

Wang, J., Shao, Y., Bennett, T.A., Shankar, R.A., Wightman, P.D., Reddy, L.G. The functional effects of physical interactions among Toll-like receptors 7, 8, and 9. J Biol Chem. 281 (2006) 37427–37434.

Wang, Y., Chen, T., Han, C., He, D., Liu, H., An, H., Cai, Z., et al. Lysosome-associated small Rab GTPase Rab7b negatively regulates TLR4 signaling in macrophages by promoting lysosomal degradation of TLR4. Blood 110 (2007) 962–971.

Wong, S.W., Kwon, M.J., Choi, A.M., Kim, H.P., Nakahira, K., Hwang, D.H. Fatty acids modulate Toll-like Receptor 4 activation through regulation of receptor dimerization and recruitment into lipid rafts in a reactive oxygen species-dependent manner. J Biol Chem 284 (2009) 27384–27392.

Wright, S.D., Ramos, R.A., Tobias, P.S., Ulevitch, R.J., Mathison, J.C. CD14, a receptor for complexes of lipopolysaccharide (LPS) and LPS binding protein. Science 249 (1990) 1431–1433.

Yamamoto, M., Sato, S., Hemmi, H., Sanjo, H., Uematsu, S., Kaisho, T., Hoshino, K., et al. Essential role for TIRAP in activation of the signalling cascade shared by TLR2 and TLR4. Nature 420 (2002) 324–329.

Yamamoto, M., Sato, S., Hemmi, H., Uematsu, S., Hoshino, K., Kaisho, T., Takeuchi, O., et al. TRAM is specifically involved in the Toll-like receptor 4-mediated MyD88-independent signaling pathway. Nat Immunol 4 (2003) 1144–1150.

Yang, Y., Liu, B., Dai, J., Srivastava, P.K., Zammit, D.J., Lefrancois, L., Li, Z. Heat shock protein gp96 is a master chaperone for toll-like receptors and is important in the innate function of macrophages. Immunity 26 (2007) 215–226.

Ye, H., Arron, J.R., Lamothe, B., Cirilli, M., Kobayashi, T., Shevde, N.K., Segal, D., et al. Distinct molecular mechanism for initiating TRAF6 signalling. Nature 418 (2002) 443–447.

Zweigner, J., Gramm, H.J., Singer, O.C., Wegscheider, K., Schumann, R.R. High concentrations of lipopolysaccharide-binding protein in serum of patients with severe sepsis or septic shock inhibit the lipopolysaccharide response in human monocytes. Blood 2001, 98 (2001) 3800–3808.

Chapter 8
Membrane Partitioning: Is Location Everything When It Comes to Endotoxin Recognition?

Martha Triantafilou and Kathy Triantafilou

Abstract Lipid rafts are envisaged as islands of highly ordered saturated lipids and cholesterol that are laterally mobile in the plane of the plasma membrane. Lipid rafts are thought to provide a means to explain the spatial segregation of certain signalling pathways emanating from the cell surface. They seem to provide the necessary microenvironment in order for certain specialised signalling events to take place- such as the innate immune recognition. The innate immune system seems to employ germ-lined encoded receptors, called pattern recognition receptors (PRRs) in order to "sense" pathogens. One family of such receptors are the Toll like receptors (TLRs), which are the central "sensing" apparatus of the innate immune system. In recent years, it has become apparent that TLRs are recruited into membrane microdomains in response to ligands and these constitute signalling platforms, which transducer singals that lead to innate immune activation. In this chapter will review all past and current literature concerning recruitment of TLRs into lipid rafts and how this membrane compartmentalization is crucial for innate immune responses.

Keywords Lipopolysaccharide · Endotoxin · Toll-like receptors · Innate immunity · Lipid rafts

Abbreviations

LBP	lipopolysaccharide binding protein
LPS	lipopolysaccharide
LTA	lipoteichoic acid
MHC	major histocompatibility complex

K. Triantafilou (✉)
Infection and Immunity Group, School of Life Sciences, University of Sussex,
Falmer, Brighton, BN1 9QG, UK
e-mail: K.Triantafilou@sussex.ac.uk

X. Wang, P.J. Quinn (eds.), *Endotoxins: Structure, Function and Recognition*,
Subcellular Biochemistry 53, DOI 10.1007/978-90-481-9078-2_8,

NF-kB	nuclear factor kappa B
PRRs	pattern recognition receptor
TLR	Toll like receptor

8.1 Introduction

Over the last couple of decades, evidence has accumulated for organisation of the plasma membrane into lipid-based microdomains or lipid rafts. Lipid rafts are envisaged as islands of highly ordered saturated lipids and cholesterol that are laterally mobile in the plane of a more fluid disordered bilayer of largely unsaturated lipids (Pralle et al., 2000; Simons and Ikonen, 1997). The hallmark of the lipid raft hypothesis are the spontaneous partitioning of lipids and proteins in discrete membrane domains, a behaviour based on their physico-chemical characteristics and the possibility to recover these microdomains and their associated protein machinery as detergent-resistant entities using biochemical flotation experiments. Microdomains appear as small dynamic structures that can aggregate into larger platforms in response to various stimuli (Harder et al., 1998).

Lipid rafts are thought to provide a means to explain the spatial segregation of certain signalling pathways emanating from the cell surface. They seem to provide the necessary microenvironment in order for certain specialised signalling events to take place. Recent studies have shown the importance of lipid raft formation in the acquired immune response. Major Histocompatibility Complex (MHC)-restricted T-cell activation seems to be facilitated by lipid raft formation (Anderson et al., 2000). Furthermore, we have recently found that mediators of the innate immune response also concentrate in lipid rafts in order to facilitate signal transduction (Triantafilou et al., 2002, 2004d), thus suggesting that both the acquired and innate immune system utilise membrane partitioning as means of activation against invading pathogens. Crucial receptors for both innate and acquired immunity, seem to oligomerize in non-random membrane structures, bringing together their signalling machinery. Thus accumulation of receptors within these "floating islands" on the cell membrane seems to bring together intracellularly all the adaptor molecules that are necessary for signalling. In this chapter, we will investigate further the mechanisms of innate immune recognition and review past and current literature that lead us to believe that membrane partitioning and lipid rafts play a central role in innate immune activation.

8.2 The Innate Immune System

The function of the innate immune system is thought to be the recognition of invading pathogens, the activation of inflammation to control the pathogen, and the subsequent activation of the acquired immune response. It constitutes the most archaic part of our immune defences and has survived through years of evolution.

As part of its mechanism of activation, the innate immune system employs germlined encoded receptors, called pattern recognition receptors (PRRs) in order to "sense" pathogens. These PRRs recognise a restricted collection of microbial signatures, able to sense different types of microbial pathogens ranging from bacteria and viruses to fungi and spirochetes. One such family of PRRs is the Toll like receptor (TLR) family.

8.2.1 The Toll Like Receptor Family

The TLR family of proteins is an integral part of the human innate immune system (Akira, 2001; Medzhitov and Janeway, 2002). TLRs are expressed on immune cells and are able to distinguish a great variety of microbial ligands, such as cell wall components like lipopolysaccharide (LPS) from Gram-negative bacteria and lipoteichoic acid from Gram-positive bacteria, bacterial flagellin, CpG DNA, as well as viral DNA or single stranded RNA (Takeda et al., 2003).

This family of at least ten encoded receptors is able to "sense" microbial signatures and trigger activation leading to pro-inflammatory cytokine secretion. TLR4 was found to recognise bacterial LPS or endotoxin (Poltorak et al., 1998; Qureshi et al., 1999), TLR2 was found to recognise lipoteichoic acid (LTA) and peptidoglycan (Takeuchi et al., 1999), TLR3 was able to sense double stranded viral RNA (Alexopoulou et al., 2001), TLR5 was found to recognise bacterial flagellin (Hayashi et al., 2001), TLR7 (Lund et al., 2004) and TLR8 (Heil et al., 2004) to sense single stranded viral RNA, whereas TLR9 to recognise bacterial CpG DNA (Hemni et al., 2000). In addition, TLR2 was found to recognise different motifs including several components of Gram-positive bacteria such as peptidoglycan (Yoshimura et al., 1999), LTA (Schwandner et al., 1999), lipoarabinomanan (Means et al., 1999), lipoproteins (Takeuchi et al., 2002), as well as different LPS from certain Gram-negative bacteria (Werts et al., 2001), yeast (Underhill et al., 1999), spirochete and fungi (Gantner et al., 2003; Heine and Lien, 2003) through its unique ability to hetero-dimerize with TLRs 1 and 6 (Ozinsky et al., 2000). Studies using di-acylated and tri-acylated lipoproteins have revealed that diacylated lipoproteins require TLR2/6 heterodimers for activation, whereas tri-acylated lipoproteins induce activation of the innate immune system independently of TLR6 and mainly through TLR2/TLR1 heterodimers (Alexopoulou et al., 2002; Buwitt-Beckmann et al., 2005; Morr et al., 2002; Takeuchi et al., 2001; Takeuchi et al., 2002; Triantafilou et al., 2006).

All identified TLRs are type I transmembrane proteins, whose intracellular domains contain regions homologous to the intracellular domains of IL-1R and are referred to as TIR domains (Takeda et al., 2003). These intracellular domains are able to trigger signalling pathways known to activate the nuclear factor kappa B (NF-kB) (Medzhitov et al., 1998; O'Neill, 2000), which in turn leads to the secretion of pro-inflammatory cytokines such as TNF-α, IL-6 and IL-8. The membrane distribution of TLRs as well as their intracellular trafficking has only now beginning to be investigated. Most TLRs (TLR1, TLR2, TLR4, TLR5, TLR6) seem to activate cells by engaging their ligands on the cell surface, whereas TLR3, TLR7, TLR8 and TLR9 seem to trigger signalling intracellularly. These TLRs have been shown to reside in the ER and to recognise their ligands once they have been endocytosed (Heil et al., 2003; Nishiya and DeFranco, 2004).

8.2.2 Innate Immune Recognition of Bacterial Endotoxin or Lipopolysaccharide

TLR4 is the most studied TLR, mainly because of its involvement with sepsis and septic shock. Sepsis is a paradoxical and complex disorder that results from an over-reaction of our innate immune system to bacterial infections. The mechanisms that are designed to protect the host against infection by bacterial pathogens, either Gram-negative or Gram-positive, can lead to oversecretion of cytokines and fatal sepsis syndrome. It is now widely accepted that the over-reaction of the host occurs at the level of the innate immune system and is directly linked to the recognition of bacterial cell wall components, such as LPS from Gram-negative bacteria or LTA from Gram-positive bacteria. Thus the recognition of bacterial products by the innate immune system seems to be detrimental for the host.

In the last quarter of a century, great leaps forward in our understanding of the molecular events that lead to the innate recognition of pathogens have occurred. One of the seminal discoveries has been the identification of a serum protein, LBP, which binds LPS or LTA and delivers it to its cellular targets (Tobias et al., 1986). Probably the most important discovery has been that the main family of receptors employed by the innate immune system are the TLRs.

As far as sepsis and bacterial recognition is concerned, TLR4 seems to be the central sensor of Gram-negative bacterial products (Poltorak et al., 2000; Qureshi, Lariviere et al., 1999), whereas TLR2 seems to be the key receptor in activating the immune system against Gram-positive bacteria (Schwandner et al., 1999). In addition to the involvement of TLRs, other accessory molecules seem to be involved. CD14 is believed to act as a transfer molecule for both Gram-negative and Gram-positive bacteria (Gupta et al., 1996; Wright et al., 1990). In the case of LPS recognition, it has been further shown that a soluble molecule, MD-2 is involved (Nagai et al., 2002), as well as activation clusters involving several other receptors (Heine et al., 2003; Pfeiffer et al., 2001; Triantafilou et al., 2001). In the case of LTA recognition, TLR2 seems to form receptor clusters as well, comprising of at least CD14, TLR2, TLR6 and CD36 (Hoebe et al., 2005). Thus we are moving away from the single-receptor model of activation, and a more complex picture is emerging. The mechanism that leads to activation seems to involve the careful interplay of several receptor molecules as well as serum proteins. Therefore such a complex orchestration of events requires non-random membrane architecture specifically geared to bring receptor molecules together and trigger activation within the lipid bilayer – and lipid rafts or membrane microdomains seem to provide this platform.

8.2.3 Protein–Protein Interactions in Innate Immunity: PRRs Are Part of Multi-component Sensor Apparatuses

Although it is now widely known that PRRs employed by the innate immune system have the ability to bind and recognise conserved products of pathogens that are unique to the invading microorganisms but not to the host, it is becoming increasingly apparent that the model of a single PRR recognising foreign antigen is an oversimplified one. With the discovery of the Toll-like receptors as the main signal transducing molecules of the innate immune system, an onslaught of research has shown that PRRs are part of multi-component sensor apparatuses.

TLRs have been shown to function as homo- or heterodimers and to even form functional interactions with non-TLR molecules. Many of these interactions are highly stable, whereas others are transient, forming dynamic associations in response to specific stimuli. Whether homotypic, heterotypic, stable or transient, these different protein combinations generate considerable functional diversity for the innate immune system by triggering distinct signalling cascades leading to cellular activation. There are a number of examples that suggest that TLR associations are required for cellular activation. TLR4 seems to form a complex with at least two other molecules, CD14 and MD2, in order to recognise bacterial LPS (Visintin et al., 2003). In addition, it seems to associate with a Toll-like receptor homologue RP105, which acts as a negative regulator of TLR4 responses (Divanovic et al., 2005). TLR2 has been found to heterodimerize with TLR1 or TLR6 for recognition of yeast components (Ozinsky, et al., 2000), and to associate with TLR1 for the recognition of bacterial lipoproteins. In addition, TLR2 has been shown to also interact with scavenger receptors in order to recognise lipoproteins (Hoebe et al., 2005) and most recently it was shown that TLR2 associates with CXCR4, which acts as a negative regulator of TLR2 responses (Hajishengallis et al., 2008).

Functional associations of TLRs with non-TLR molecules have also been demonstrated, for example TLR2 association with dectin-1 is required for macrophage and dendritic cell activation by β-glucan-containing particles. More recently, functional interactions of TLR2 and CD36 have been shown to be involved in the recognition of diacylglycerides (Hoebe et al., 2005). TLR4 seems to be the best example of TLRs associating with non-TLR molecules. As it has already been mentioned, TLR4 has been shown to form at least a tri-molecular complex with CD14 and MD2 in order to recognise bacterial LPS (Poltorak et al., 2000). The possibility that additional receptor components such as heat shock proteins (Byrd et al., 1999; Triantafilou et al., 2001), CXCR4 (Triantafilou et al., 2001), or CD55 (Heine et al., 2003) have been suggested to be part of this activation cluster, possibly acting as additional LPS transfer molecules. Furthermore, it has been demonstrated that different "shapes" of LPS induce the formation of different activation clusters, involving the association of TLR4 with a variety of molecules mentioned above, which seems to determine LPS responses (Triantafilou et al., 2004b).

Recent structural studies have shed some light into TLR associations, supporting the hypothesis of cluster formation, since all TLRs that have been crystallised have been found to be in a dimer formation, thus the hypothesis has been put forward that dimerisation or clustering might be a common feature of the TLRs and might be essential for signalling.

Structural studies of TLRs have been an attractive area of research since structural information is crucial in understanding receptor function. In 2005, the crystal structure of TLR3 was the first one to be revealed (Choe et al., 2005). It was surprising, that although the structure did not have a ligand, TLR3 was crystallised as a dimer. In 2007 and 2008, three structures of TLR-ligand complexes were revealed, TLR1-TLR2-lipoprotein, TLR4-MD-2-Eritoran, and TLR3- dsRNA (Jin et al., 2007; Kim et al., 2007; Liu et al., 2008). The ectodomains were found to form dimers, which were strikingly similar in shape. Prior to the publication of the crystal structures, Gay et al. (Gay and Gangloff, 2008) suggested a possible model of activation, where dimerization was ligand induced. These observations have suggested the hypothesis that dimerization of the ectodomains forces the intracellular TIR domains to dimerize, and this initiates signalling by recruiting the intracellular adaptor molecules, such as MyD88, MAL, TRIF and TRAM in order to initiate signalling. The structures of the TIR domains of TLR1, TLR2 and TLR10 have been revealed (Nyman et al., 2008). Interestingly, the TIR domain of TLR10 was shown to be involved in a homodimeric interaction. However, it is not certain whether the structure seen in the crystal corresponds to a physiologically relevant dimer of TLR10 TIR domains because they have been found to exist as monomers in solution. Interestingly, Motshwene et al. (2009) have recently suggested that MyD88 interacts with IRAK4 in an 8:4 ratio in solution, suggesting that maybe there is higher oligomer formation.

In order for such higher oligomers to be formed and in order to have such a well orchestrated accumulation of receptors and signalling machinery membrane partitioning seems to be crucial for the formation of these "TLR multi-component sensor apparatuses".

8.2.4 Is TLR4 Recruited in Membrane Microdomains Upon Ligand Engagement?

As already mentioned, TLR4 is the most studied TLR and it is no surprise that it was the first one to be shown to be recruited to lipid rafts upon stimulation by bacterial LPS (Triantafilou et al., 2002). Within these membrane microdomains it was shown that TLR4 formed clusters with non-TLR molecules that tailored the immune response against the particular pathogen (Humphries et al., 2005; Triantafilou et al., 2001; Triantafilou et al., 2004a; Triantafilou and Triantafilou, 2003, 2005).

It was subsequently shown that this accumulation in lipid rafts also influenced its internalization and targeting. TLR4 was found to accumulate in lipid rafts, to internalize in a lipid-raft dependent manner and to be targeted to the Golgi apparatus (Latz et al., 2002). This intracellular targeting was shown to be independent of signalling, thus suggesting that accumulation in lipid rafts only facilitated ligand recognition and signalling that was initiated at the cell surface and not in the intracellular compartments where TLR4 was targeted to (Latz et al., 2002).

More recently it had been proposed that the molecular mechanism for signalling by the TLRs must involve a series of protein conformational changes initiated by dimerization of their extracellular domains (Gay et al. 2006). It was suggested that this receptor–receptor association of the extracellular domains forced the association of the cytoplasmic domains as well. Motshwene et al. (2009) recently proved this experimentally, demonstrating that the death domains of human MyD88, one of the adaptor proteins used by all bit one of the TLRs, and IRAK4 assemble into closed complexes with stoichiometries of 7:4 and 8:4, which they called the Myddosome. The ability to form 7:4 or even 8:4 stochiometries suggests a mechanism by which clusters of activated receptors concentrate in lipid rafts and their intracellular machinery clusters as well, forming a signalling platform that seems to be crucial for TLR activation.

8.2.5 Does Membrane-Partitioning Play a Major Role in Protein Uptake and Intracellular Routing?

Lipid rafts were originally proposed as an explanation for a non-random membrane architecture and their function was originally thought to be linked with membrane trafficking. However, rafts proved to be able to influence organization of membrane receptors and bioactivity as well as membrane trafficking.

It is now emerging that this membrane partitioning might play a major role in protein uptake and intracellular routing. It is becoming more apparent that this differential sorting on the cell surface might pre-dispose the intracellular fate of a given molecule. Since the discovery of clathrin coated pits by Roth and Porter (1964), as specialised sites for the selective recruitment of specialised molecules that are internalised into eukaryotic cells, clathrin-independent endocytic pathways have now emerged. Endocytic pathways that do not rely on the formation of clathrin coated pits include the earliest identified pathways such as phagocytosis, macropinosis, and caveolae. Caveolae defined as small, uncoated invaginations in the plasma membrane containing the plasma protein caveolin-1 has recently been shown to be able to bind cholesterol and to be resistant to detergent extraction (Sargiacomo et al., 1993) and this has led to the suggestion that caveolae might constitute a type of lipid raft (Harder and Simon, 1997). Lipid rafts are increasingly becoming linked with clathrin-independent endocytosis, since nearly all molecules that are known to be internalised independently of clathrin are found in biochemically-defined rafts (Nichols and Lippincott-Schwartz, 2001). It has been suggested that raft-components might be taken up preferentially by clathrin-independent endocytosis. There are likely to be several types of clathrin-independent endocytosis. The extent to which these different pathways require lipid rafts to operate, or are somehow selective for lipid rafts is currently the subject of intensive investigation. Recently, Nichols et al., have described a rapid lipid-raft-dependent targeting from the cell surface to the Golgi apparatus (Nichols et al., 2001). In addition, a new clathrin-independent mechanism has been described that can lead to delivery of receptor molecules from the plasma membrane to caveolin-1 containing endosomes, termed "caveosomes" (Pelkmans and Helenius, 2002). With the emergence of these new clathrin-independent uptake mechanisms the idea that different types of endocytosis have markedly different functions is beginning to become apparent. Ultimately we have to speculate that sorting at the plasma membrane might pre-dispose the intracellular route that a molecule might take. If that is the case, then where are the raft-associated molecules, such as TLR4, targeted to? And most importantly why?

This intracellular targeting seems to be independent of signalling. TLR2 has also been found to reside in lipid rafts after stimulation by Gram-positive bacterial products and to be similarly targeted to the Golgi apparatus (Triantafilou et al., 2004c). The question that remains is whether lipid raft-association is common for all TLRs expressed at the cell surface? If this is the case, do they all follow the same intracellular route? Do different signalling cascades require differential targeting of TLRs and their ligands?

In the case of the ER-resident TLRs, very little evidence of their trafficking upon stimulation exists. To date only TLR9 has been found to translocate from the ER to lysosomes in response to its ligand, CpG DNA (Latz et al., 2004). Based on the findings for TLR9, a hypothesis has been put forward that ER-resident TLRs might become accessible to endosomal and lysosomal compartments after the ER fuses with sites of microbial entry. If this is the case, then it would seem that ER membrane fusion might be critical for microbial recognition by ER-resident TLRs.

8.2.6 Concluding Remarks

Cell membranes display a tremendous complexity of lipids and proteins designed to perform the functions cells require. In order to co-ordinate these functions, the membrane seems to be able to segregate it constituents. In this way, the membrane is able to compartmentalize, segregate receptors as well as their signalling machinery and create oligomeric signalling platforms in order to transducer signals. Once the required function has subsided, these segregated islands are involved in internalization and membrane trafficking, thus bringing the whole function to a close. The innate and acquired immune systems seem to utilise this membrane partitioning for their functions. In this chapter, we have extensively looked at the use of this membrane partitioning by the innate immune system and most particular by the TLRs. The molecular mechanism involved in LPS recognition and TLR signalling in general, utilises a series of protein–lipid as well as protein–protein interactions. The plasma membrane seems to be heterogeneous and to coalesce to more stable membrane-ordered assemblies upon activation by ligands. This partitioning of the membrane and the assembly of more stable raft platforms in the functionalized state must be initiated by raft-resident proteins, which form protein–lipid as well as protein–protein interactions. The TLRs associate with the raft-resident proteins and are recruited to these "floating islands" forming higher oligomers, both extracellularly as well as inside the cell, concentrating their signalling machinery which finally leads to a functional, focused and co-ordinated activation of the innate immune system.

References

Akira, S. Toll-like receptors and innate immunity. AdvImmunol 78 (2001) 1–56.

Alexopoulou, L., Holt, A.C., Medzhitov, R., Flavell, R.A. Recognition of double-stranded RNA and activation of NF-kappaB by Toll-like receptor 3. Nature 413 (2001) 732–738.

Alexopoulou, L., Thomas, V., Schnare, M., Lobet, Y., Anguita, J., Schoen, R.T., Medzhitov, R., Fikrig, E., Flavell, R.A. Hyporesponsiveness to vaccination with *Borrelia burgdorferi* OspA in humans and in TLR1- and TLR2-deficient mice. Nat Immunol 8 (2002) 878–884.

Anderson, H.A., Hiltbold, E.M., Roche, P.A. Concentration of MHC class II molecules in lipid rafts facilitates antigen presentation. Nat Immunol 1 (2000) 156–162.

Buwitt-Beckmann, U., Heine, H., Wiesmuller, K.H., Jung, G., Brock, R., Akira, S., Ulmer, A.J. Toll-like receptor 6-independent signaling by diacylated lipopeptides. Eur J Immunol 35 (2005) 282–289.

Byrd, C.A., Bornmann, W., Erdjument-Bromage, H., Tempst, P., Pavletich, N., Rosen, N., Nathan, C.F., Ding, A. Heat shock 90 mediates macrophage activation by Taxol and bacterial lipopolysaccharide. Proc Natl Acad Sci USA 96 (1999) 5645–5650.

Choe, J., Kelker, M.S., Wilson, I.A. Crystal structure of human toll-like receptor 3 (TLR3) ectodomain. Science 309 (2005) 581–585.

Divanovic, S., Trompette, A., Atabani, S.F., Madan, R., Golenbock, D.T., Visintin, A., Finberg, R.W., Tarakhovsky, A., Vogel, S.N., Belkaid, Y., Kurt-Jones, E.A., Karp, C.L. Negative regulation of Toll-like receptor 4 signalling by the Toll-like receptor homolog RP105. Nat Immunol 6 (2005) 571–578.

Gantner, B.N., Simmons, R.M., Canavera, S.J., Akira, S., Underhill, D.M. Collaborative induction of inflammatory responses by dectin-1 and Toll-like receptor 2. J Exp Med 197 (2003) 1107–1117.

Gay, N.J. Gangloff, M., Structure of toll-like receptors. Handb Exp Pharmacol (2008) 181–200.

Gay, N.J., Gangloff, M., Weber, A.N. Toll-like receptors as molecular switches. Nat Rev Immunol 6(9) (2006) 693–698.

Gupta, D., Kirkland, T.N., Viriyakosol, S., Dziarski, R. CD14 is a cell-activating receptor for bacterial peptidoglycan. J Biol Chem 271 (1996) 23310–23316.

Hajishengallis, G., Wang, M., Liang, S., Triantafilou, M., Triantafilou, K. Pathogen induction of CXCR4/TLR2 cross-talk impairs host defense function. Proc Natl Acad Sci USA 105 (2008) 13532–13537.

Harder, T., Scheiffele, P., Verkade, P., Simons, K. Lipid domain structure of the plasma membrane revealed by patching of membrane components. J Cell Biol 141 (1998) 929–942.

Harder, T., Simon, K. Caveolae, DIGs and the dynamics of spingolipid-cholesterol microdomains. Curr Opin Cell Biol 9 (1997) 534–542.

Hayashi, F., Smith, K.D., Ozinsky, A., Hawn, T.R., Yi, E.C., Goodlett, D.R., Eng, J.K., Akira, S., Underhill, D.M., Aderem, A. The innate immune response to bacterial flagellin is mediated by Toll-like receptor 5. Nature 410 (2001) 1099–1103.

Heil, F., Ahmad-Nejad, P., Hemmi, H., Hochrein, H., Ampenberger, F., Gellert, T., Dietrich, H., Lipford, G., Takeda, K., Akira, S., Wagner, H., Bauer, S. The Toll-like receptor 7 (TLR7)-specific stimulus loxoribine uncovers a strong relationship within the TLR7, 8 and 9 subfamily. Eur J Immunol 33 (2003) 2987–2997.

Heil, F., Hemmi, H., Hochrein, H., Ampenberger, F., Kirschning, C., Akira, S., Lipford, G., Wagner, H., Bauer, S. Species-specific recognition of single-stranded RNA via Toll-like receptor 7 and 8. Science 303 (2004) 1526–1529.

Heine, H., El-Samalouti, V.T., Notzel, C., Pfeiffer, A., Lentschat, A., Kusumoto, S., Schmitz, G., Hamann, L., Ulmer, A.J. CD55/decay accelerating factor is part of the lipopolysaccharide-induced receptor complex. Eur J Immunol 33 (2003) 1399–1408.

Heine, H., Lien, E. Toll-like receptors and their function in innate and adaptive immunity. Int Arch Allergy Immunol 130 (2003) 180–192.

Hemni, H., Takeuchi, O., Kawai, T., Kaisho, T., Sato, S., Sanjo, H., Matsumoto, M., Hoshino, K., Wagner, H., Takeda, K., Akira, S. A Toll-like receptor recognises bacterial DNA. Nature 408 (2000) 740–745.

Hoebe, K., Georgel, P., Rutschmann, S., Du, X., Mudd, S., Crozat, K., Sovath, S., Shamel, L., Hartung, T., Zahringer, U., Beutler, B. CD36 is a sensor of diacylglycerides. Nature 433 (2005) 523–527.

Humphries, H.E., Triantafilou, M., Makepeace, B.L., Heckels, J.E., Triantafilou, K., Christodoulides, M. Activation of human meningeal cells is modulated by lipopolysaccharide (LPS) and non-LPS components of *Neisseria meningitidis* and is independent of Toll-like receptor (TLR)4 and TLR2 signalling. Cell Microbiol 7 (2005) 415–430.

Jin, M.S., Kim, S.E., Heo, J.Y., Lee, M.E., Kim, H.M., Paik, S.G., Lee, H., Lee, J.O. Crystal structure of the TLR1-TLR2 heterodimer induced by binding of a tri-acylated lipopeptide. Cell 130 (2007) 1071–1082.

Kim, H.M., Park, B.S., Kim, J.I., Kim, S.E., Lee, J., Oh, S.C., Enkhbayar, P., Matsushima, N., Lee, H., Yoo, O.J., Lee, J.O. Crystal structure of the TLR4-MD-2 complex with bound endotoxin antagonist Eritoran. Cell 130 (2007) 906–917.

Latz, E., Schoenemeyer, A., Visintin, A., Fitzgerald, K.A., Monks, B.G., Knetter, C.F., Lien, E., Nilsen, N.J., Espevik, T., Golenbock, D.T. TLR9 signals after translocating from the ER to CpG DNA in the lysosome. Nat Immunol 5 (2004) 190–198.

Latz, E., Visintin, A., Lien, E., Fitzgerald, K., Monks, B., Kurt-Jones, E., Golenbock, D.T., Espevik, T. LPS rapidly transfers to and from the Golgi apparatus with the TLR4/MD-2/CD14 complex in a process that is distinct from the initiation of signal transduction. J Biol Chem 277 (2002) 47834–47843.

Liu, L., Botos, I., Wang, Y., Leonard, J.N., Shiloach, J., Segal, D.M. Davies, D.R., Structural basis of toll-like receptor 3 signaling with double-stranded RNA. Science 320 (2008) 379–381.

Lund, J.M., Alexopoulou, L., Sato, A., Karow, M., Adams, N.C., Gale, N.W., Iwasaki, A., Flavell, R.A. Recognition of single-stranded RNA viruses by Toll-like receptor 7. Proc Natl Acad Sci USA 101 (2004) 5598–5603.

Means, T.K., Lien, E., Yoshimura, A., Wang, Y., Golenbock, D.T., Fenton, M.J. The CD14 ligands Lipoarabinomannan and lipopolysaccharide differ in their requirement for Toll-like receptors. J Immunol 163 (1999) 6748–6755.

Medzhitov, R., Janeway, C.A. Decoding the patterns of self and nonself by the innate immune system. Science 296 (2002) 298–300.

Medzhitov, R., Preston-Hurlburt, P., Kopp, E., Stadlen, A., Chen, C., Ghosh, S., Janeway, C.A. MyD88 is an adaptor protein in the hToll/IL-1 receptor family signaling pathways. Mol Cell 2 (1998) 253–258.

Morr, M., Takeuchi, O., Akashi, S., Simon, M.M., Muhlradt, P.F. Differential recognition of structural details of bacterial lipopeptides by Toll-like receptors. Eur J Immunol 32 (2002) 3337–3347.

Motshwene, P.G., Moncrieffe, M.C., Grossmann, J.G., Kao, C., Ayaluru, M., Sandercock, A.M., Robinson, C.V., Latz, E., Gay, N.J. An oligomeric signaling platform formed by the Toll-like receptor signal transducers MyD88 and IRAK-4. J Biol Chem 284(37) (2009) 25404–25411.

Nagai, Y., Akashi, S., Nagafuku, M., Ogata, M., Iwakura, Y., Akira, S., Kitamura, T., Kosugi, A., Kimoto, M., Miyake, K. Essential role of MD-2 in LPS responsiveness and TLR4 distribution. Nat Immunol 3 (2002) 667–672.

Nichols, B.J., Kenworthy, A.K., Polishchuk, R.S., Lodge, R., Roberts, T.H., Hirschberg, K., Lippincott-Schwartz, J. Rapid Cycling of lipid raft markers between the cell surface and golgi complex. J Cell Biol 153 (2001) 529–541.

Nichols, B.J., Lippincott-Schwartz, J. Endocytosis without clathrin coated pits. Trends Cell Biol 11 (2001) 406–412.
Nishiya, T., DeFranco, A.L. Ligand-regulated chimeric receptor approach reveals distinctive subcellular localisation and signaling properties of the Toll-like receptors. J Biol Chem 279 (2004) 19008–19017.
Nyman, T., Stenmark, P., Flodin, S., Johansson, I., Hammarstrom, M., Nordlund, P. The crystal structure of the human toll-like receptor 10 cytoplasmic domain reveals a putative signaling dimer. J Biol Chem 283 (2008) 11861–11865.
O'Neill, L., The Toll/interleukin-1 receptor domain: a molecular switch for inflammation and host defense. Biochem Soc Trans 28 (2000) 557–563.
Ozinsky, A., Underhill, D.M., Fontenot, J.D., Hajjar, A.M., Smith, K.D., Wilson, C.B., Schroeder, L., Aderem, A. The repertoire for pattern recognition of pathogens by the innate immune system is defined by cooperation between Toll-like receptors. Proc Natl Acad Sci USA 97 (2000) 13766–13771.
Pelkmans, L., Helenius, A. Endocytosis via caveolae. Traffic 3 (2002) 311–320.
Pfeiffer, A., Bottcher, A., Orso, E., Kapinsky, M., Nagy, P., Bodnar, A., Spreitzer, I., Liebisch, G., Drobnik, W., Gempel, K., Horn, M., Holmer, S., Hartung, T., Multhoff, G., Schutz, G., Schindler, H., Ulmer, A.J., Heine, H., Stelter, F., Schutt, C., Rothe, G., Szollosi, J., Damjanovitch, S., Schmitz, G. Lipopolysaccharide and ceramide docking to CD14 provokes ligand-specific receptor clustering in rafts. Eur J Immunol 31 (2001) 3153–3164.
Poltorak, A., He, X.L., Smirnova, I., Liu, M.Y., VanHuffel, C., Du, X., Birdwell, D., Alejos, E., Silva, M., Galanos, C., Freudenberg, M., RicciardiCastagnoli, P., Layton, B., Beutrel, B. Defective LPS signaling in C3H/Hej and C57BL/10ScCr mice: Mutations in TLR4 gene. Science 282 (1998) 2085–2088.
Poltorak, A., Ricciardi-Castagnoli, P., Citterio, S., Beutler, B. Physical contact between lipopolysaccharide and Toll-like receptor 4 revealed by genetic complementation. Proc Natl Acad Sci USA 97 (2000) 2163–2167.
Pralle, A., Keller, P., Florin, E.L., Simons, K., Horber, J.K.H. Sphingolipid-cholesterol rafts diffuse as small entities in the plasma membrane of mammalian cells. J Cell Biol 148 (2000) 997–1007.
Qureshi, S.T., Lariviere, L., Leveque, G., Clermont, S., Moore, K.J., Gros, P., Malo, D. Endotoxin-tolerant mice have mutations in toll-like receptor 4 (TLR4). J Exp Med 189 (1999) 615–625.
Roth, T.F., Porter, K.R. Yolk protein uptake in the oocyte of the mosquito *Aedes aegypti*. J Cell Biol 20 (1964) 313–330.
Sargiacomo, M., Sudol, M., Tang, A., Lisanti, M.P. Signal transducing molecules and GPI-linked proteins form a caveolin-rich insoluble complex in MDCK cells. J Cell Biol 122 (1993) 789–807.
Schwandner, R., Dziarski, R., Wesche, H., Rothe, M., Kirschning, C.J. Peptidoglycan- and Lipoteichoic acid-induced cell activation is mediated by Toll-like receptor 2. J Biol Chem 274 (1999) 17406–17409.
Simons, K., Ikonen, E. Functional rafts in membranes. Nature 387 (1997) 569–570.
Takeda, K., Kaisho, T., Akira, S. Toll-like receptors. Ann Rev Immunol 21 (2003) 335–376.
Takeuchi, O., Hoshino, K., Kawai, T., Sanjo, H., Takada, H., Ogawa, T., Takeda, K., Akira, S. Differential roles of TLR2 and TLR4 in recognition of gram-negative and gram-positive bacterial cell wall components. Immunity 11 (1999) 443–451.
Takeuchi, O., Kawai, T., Muhlradt, P.F., Morr, M., Randolf, J.D., Zychlinsky, A., Takeda, K., Akira, S. Discrimination of bacterial lipoproteins by Toll-like receptor 6. Int Immunol 13 (2001) 933–940.
Takeuchi, O., Sato, S., Horiuchi, T., Hoshino, K., Takeda, K., Dong, Z., Modlin, R.L., Akira, S. Cutting edge: role of Toll-like receptor 1 in mediating immune response to microbial lipoproteins. J Immunol 169 (2002) 10–14.
Tobias, P.S., Soldau, K., Ulevitch, R.J. Isolation of a lipopolysaccharide-binding acute phase reactant from rabbit serum. J Exp Med 164 (1986) 785–793.

Triantafilou, K., Triantafilou, M., Dedrick, R.L. A CD14-independent LPS receptor cluster. Nat Immunol 4 (2001) 338–345.

Triantafilou, M., Brandenburg, K., Kusumoto, S., Fukase, K., Mackie, A., Seydel, U., Triantafilou, K. Combinational clustering of receptors following stimulation by bacterial products determines lipopolysaccharide responses. Biochem J 381, (2004a) 527–536.

Triantafilou, M., Brandenburg, K., Kusumoto, S., Fukase, K., Mackie, A., Seydel, U., Triantafilou, K. Combinational clustering of receptors following stimulation by bacterial products determines LPS responses. Biochem J 381 (2004b) 527–536.

Triantafilou, M., Gamper, F.G., Haston, R.M., Mouratis, M.A., Morath, S., Hartung, T., Triantafilou, K. Membrane sorting of toll-like receptor (TLR)-2/6 and TLR2/1 heterodimers at the cell surface determines heterotypic associations with cd36 and intracellular targeting. J Biol Chem 281 (2006) 31002–31011.

Triantafilou, M., Manukyan, M., Mackie, A., Morath, S., Hartung, T., Heine, H., Triantafilou, K. Lipoteichoic acid and Toll-like receptor 2 internalization and targeting to the Golgi are lipid-raft dependent. J Biol Chem 279 (2004c) 40882–40889.

Triantafilou, M., Miyake, K., Golenbock, D., Triantafilou, K. Mediators of the innate immune recognition of bacteria concentrate in lipid rafts and facilitate lipopolysaccharide-induced cell activation. J Cell Sci 115 (2002) 2603–2611.

Triantafilou, M., Morath, S., Mackie, A., Hartung, T., Triantafilou, K. Lateral diffusion of Toll-like receptors reveals that they are transiently confined within lipid rafts on the plasma membrane. J Cell Sci 117 (2004d) 4007–4014.

Triantafilou, M., Triantafilou, K. Receptor cluster formation during activation by bacterial products. J Endotoxin Res 9 (2003) 331–335.

Triantafilou, M., Triantafilou, K. The dynamics of LPS recognition: complex orchestration of multiple receptors. J Endotoxin Res 11 (2005) 5–11.

Underhill, D.M., Ozinsky, A., Hajjar, A.M., Stevens, A., Wilson, C.B., Bassetti, M., Aderem, A. The Toll-like receptor 2 is recruited to macrophage phagosomes and discriminates between pathogens. Nature 401 (1999) 811–815.

Visintin, A., Latz, E., Monks, B.G., Espevik, T., Golenbock, D.T. Lysines 128 and 132 enable lipopolysaccharide binding to MD2, leading to Toll-like receptor 4-aggregation and signal transduction. J Biol Chem 278 (2003) 48313–48320.

Werts, C., Tapping, R.I., Mathison, J.C., Chuang, T.H., Kravchenko, V., Saint Girons, I., Haake, D.A., Godowski, P.J., Hayashi, F., Ozinsky, A., Underhill, D.M., Kirschning, C.J., Wagner, H., Aderem, A., Tobias, P.S., Ulevitch, R.J. Leptospiral lipopolysaccharide activates cells through a TLR-2-dependent mechanism. Nat Immunol 2 (2001) 346–352.

Wright, S.D., Ramos, R.A., Tobias, P.S., Ulevitch, R.J., Mathison, J.C. CD14, a receptor for complexes of lipopolysaccharide (LPS) and LPS binding-protein. Science 249 (1990), 1431–1433.

Yoshimura, A., Lien, E., Ingalls, R.R., Tuomanen, E., Dziarski, R., Golenbock, D. Cutting edge: recognition of Gram-positive bacterial cell wall components by the innate immune system occurs via Toll-like receptor 2. J Immunol 163 (1999) 1–5.

Part II
Infection, Treatment and Immunity

Chapter 9
Endotoxin Detection – from *Limulus* Amebocyte Lysate to Recombinant Factor C

Jeak Ling Ding and Bow Ho

Abstract Gram negative bacterial endotoxin is a biological pyrogen that causes fever when introduced intravenously. The endotoxin, also known as lipopolysaccharide (LPS), is found in the outer membrane of Gram-negative bacteria. During Gram-negative sepsis, endotoxin stimulates host macrophages to release inflammatory cytokines. However, excessive inflammation causes multiple organ failure and death. Endotoxins, which are ubiquitous pathogenic molecules, are a bane to the pharmaceutical industry and healthcare community. Thus early and sensitive detection of endotoxin is crucial to prevent endotoxaemia. The limulus amebocyte lysate (LAL) has been widely used for ~30 years for the detection of endotoxin in the quality assurance of injectable drugs and medical devices. The LAL constitutes a cascade of serine proteases which are triggered by trace levels of endotoxin, culminating in a gel clot at the end of the reaction. The Factor C, which normally exists as a zymogen, is the primer of this coagulation cascade. In vivo, Factor C is the perfect biosensor, which alerts the horseshoe crab of the presence of a Gram-negative invader. The hemostatic end-point entraps the invader, killing it and limiting further infection. However, as an in vitro endotoxin detection tool, variations in the sensitivity and specificity of LAL to endotoxin, and the dwindling supply of horseshoe crabs are posing increasing challenges to the biotechnology industry. This has necessitated the innovation of an alternative test for endotoxin. Thus, Factor C became the obvious, albeit tricky target for the recombinant technology effort. This chapter documents the *backwater of mining* the natural blood lysate of the endangered species to the monumental effort of genetic engineering, to produce recombinant Factor C (rFC). The rFC is a 132 kDa molecule, which was produced as a proenzyme inducible by the presence of trace levels of endotoxin. The rFC forms the basis of the "PyroGene" kit, which is a novel micro-enzymatic endotoxin diagnostic assay for high-throughput screens of endotoxin. Using the rFC, Lonza Inc. has

J.L. Ding (✉)
Department of Biological Sciences, Faculty of Science, National University of Singapore, 14 Science Drive 4, Singapore 117543
e-mail: dbsdjl@nus.edu.sg

X. Wang, P.J. Quinn (eds.), *Endotoxins: Structure, Function and Recognition*, Subcellular Biochemistry 53, DOI 10.1007/978-90-481-9078-2_9,

spawned the "PyroSense" which serves as checkpoints of the biotechnology production line. Thus, from cloning to commercial applications, the rFC has initiated a new era in endotoxin-testing for the quality assurance of biomedical products and for the healthcare industry, whilst sparing the endangered horseshoe crabs.

Keywords Endotoxin · Limulus amoebocyte lysate test · LAL · Molecular cloning · Pyrogene assay · Recombinant Factor C

Abbreviations

BPI	bactericidal/permeability-increasing protein
HDL	high density lipoprotein
GlcNAc	N-acetylglucosamine
GNB	Gram negative bacteria
KDO	2-keto-3-deoxy-D-mano-octonate
LAL	limulus amoebocyte lysate
LALF	limulus anti-LPS factor
LBP	LPS-binding protein
LDL	low density lipoprotein
LPS	lipopolysaccharide
rFC	recombinant Factor C
TNF-α	tumor necrosis factor alpha

9.1 General Introduction

Bacterial infections can be traced to exogenous and endogenous sources (Lent et al., 2001; Van Leeuwen et al., 1994; Lemaire et al., 1997). Gram negative bacteria (GNB) are ubiquitous and amongst the most challenging pathogens to the human host (Breithaupt, 1999; Zasloff, 2002; Boneca, 2005). Infection by GNB is the leading cause of sepsis (Bone, 1996; McCormick et al., 2001). It can cause excessive release of inflammatory cytokines, which can lead to multiple organ failure and death. The rising acquisition of multiple antibiotic resistance by superbugs is worsened by the lag in discovering and/or innovating new and more powerful antibiotics including antimicrobial peptides (Gradishar et al., 1995; Hancock and Chapple, 1999). There is a great urgency to circumvent the threat of new and more deadly species, which are capable of growing rapidly, even under challenging environmental conditions. The sporadic spread of Gram negative infection is exacerbated by the bioactive endotoxin (lipopolysaccharide, LPS) found on the outer membrane of the bacteria. The LPS and its juxtaposing anionic microbial phospholipids form the protective armour surrounding the bacterium. Upon death either by natural turnover events or killed by antibiotics during its invasion of the host, the Gram negative bacterium sheds its outer membrane releasing LPS. Thus, LPS is ubiquitous in nature, occurring in water, soil and the human gut which hosts commensals. Intravenous introduction of the LPS commences a string of medical problems instigated by the bioactive pharmacophore, the lipid A moiety of the LPS molecule. Thus, early and highly sensitive detection of LPS is paramount to preventing and/or arresting the cascade of ill effects caused by LPS.

9.2 Gram-Negative Bacterial Membrane – A Wall of Fire

The Gram-negative bacterial membrane is a very well studied outer cell wall of a microbe. It represents the metabolite-based armour at the frontline of defense of the bacteria. Absent in eukaryotic hosts, and uniquely displayed on the outer membrane of the bacteria, these metabolites are collectively known as pathogen-associated molecular patterns (PAMPs). One such example is the LPS. For its grave ability to provoke pyrogenic action in an infected host, the GNB outer membrane indeed befits the description, "a wall of fire", as it is also needed as a fortress to protect the bacterium. Whether it is intact on the bacterium or released as cell wall fragments or as free LPS molecules in a mammalian host, the LPS wreaks inflammatory havoc leading to septic shock, and potentially, death to the invaded host-victim if not controlled on time. Despite extensive research on Gram negative septicaemia and efforts to develop antibiotics, infection by GNB is still a leading cause of sepsis, accounting for 45–60% of sepsis caused by bacterial infection (Bone, 1996; McCormick et al., 2001). The LPS stimulates the host's macrophages to release inflammatory cytokines, causing inflammation, which alerts the host of pathogen invasion. However, persistent exposure to LPS and excessive inflammation causes septic shock. Subsequently, multiple organ failure ensues and becomes the main clinical problem and cause of mortality (Brady and Otto, 2001). The patient can be rapidly killed by septic shock even before the bacteria could cause any direct harm (Ruiter et al., 1981). A multicentre observational cohort study had projected an estimated 751,000 cases of sepsis per annum in the United States alone (Chaby, 1999; Angus and Wax, 2001), which is comparable to that attributed to AIDS. In a small and developed country like Singapore with a population of 5 million, septicaemia is the 10th principal cause of death, with approximately one hundred casualties each year (http://www/moh.gov.sg/corp/publications/statistics/principal.do).

9.3 Lipopolysaccharide: A Mediator of Septic Shock – Pathophysiological Properties

The cascade of pathological outcomes and potential mortality triggered by LPS has drawn much research attention on the chemical structure-activity of the LPS molecule in order to help understand the molecular biology of Gram negative septicaemia. The LPS is also referred to as endotoxin because of its pyrogenic properties (fever causing) in human and other mammalian hosts. Introduced in the 19th century, this terminology, pyrogen, describes a component of GNB responsible for the pathophysiological phenomena associated with infection by GNB. Synthesized by bacteria as diverse as those responsible for cholera, whooping cough, plague and nitrogen fixation, a community's fear of endotoxin continues to attract wide interest by virtue of its role in bacterial infection and sepsis. Anchored on the outer cell wall of the GNB (Ulmer et al., 2002, Brandenburg and Wiese, 2004), the LPS is an essential component of virtually all GNBs.

Unlike other toxic bacterial products that confer survival advantages but are otherwise non-essential, the LPS is presumably essential in all GNBs. The inability of GNB mutants to synthesise the minimally required LPS structure during the assembly of the outer membrane is a lethal event leading to the non-proliferation of the bacterium (Raetz, 1993). In contrast to most bacterial toxins, which directly injure eukaryotic cells by disrupting vital host cell functions, the LPS exerts its harmful effects more obnoxiously and indirectly by eliciting an exaggerated response of the host immune system (Corriveau and Danner, 1993; Karima et al., 1999). The host's innate immune response to LPS is beneficial in the event of a GNB infection as it creates a hostile environment for the invading bacteria. In the event that the innate immune system over-reacts, as in the case of overwhelming bacteraemia, the overproduction and systemic release of potent host-derived mediators such as proinflammatory cytokines, nitric oxide and eicosanoids may initiate a cascade of events which culminate in pyrogenic effect, shock, organ failure and death (Karima et al., 1999). The effects exerted by LPS are further complicated in vivo as LPS also stimulates the release of anti-inflammatory mediators such as transforming growth factor β, TGF-β (Dinarello, 1991). Thus, the phenomenon of endotoxaemia may actually represent conditions where there are imbalances between the proinflammatory and anti-inflammatory effects of LPS. The mammalian response to LPS is a complex and highly regulated process. It relies on both the humoral and cell membrane-bound recognition receptors, for example, TLR4 and CD14, which specifically interact with the lipid A moiety of LPS (Wright, 1991).

9.4 The Structure of LPS

Research on the LPS structure has focused on those derived from the enterobacteria (Takayama et al., 1983; Rietschel et al., 1996). Generally, the LPS is a structurally heterogenous, extremely resilient, indomitable and ubiquitous chemical molecule. The Gram-negative bacterium has been endowed with such a uniquely thermostable shield, which is also fairly insensitive to pH changes. Destruction of the endotoxin requires baking at 200°C for two hours or more, or shorter durations in the presence of high concentrations of acids or bases, for example in citric acid, pH 1.0 for 3 min. The envelope of a single *E. coli* is estimated to contain 2×10^6 LPS molecules, constituting about 20 femtograms (Minabe et al., 1994). LPS molecules are of great compositional and structural diversity, and yet, they are constructed according to a common architectural principle (Fig. 9.1). The LPS has a tripartite structure (Fig. 9.1) comprising three covalently linked domains: the O-specific chain, the core oligosaccharide and lipid A.

9.4.1 The O-Specific Chain

The O-specific chain is made up of a chain of repeating oligosaccharides of 3–8 units, which are specific to bacterial strains (Westphal et al., 1983; Raetz, 1990).

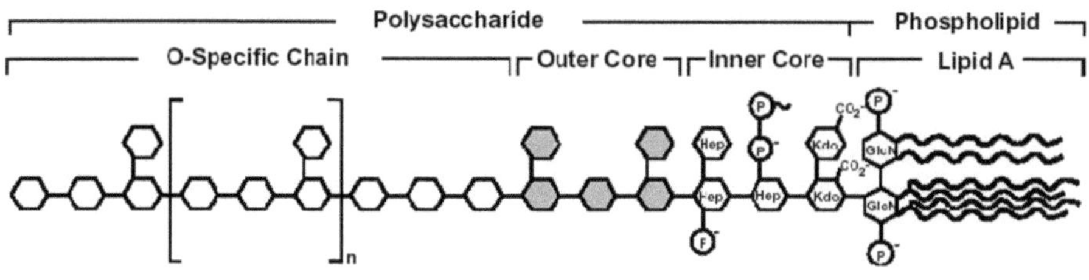

LPS

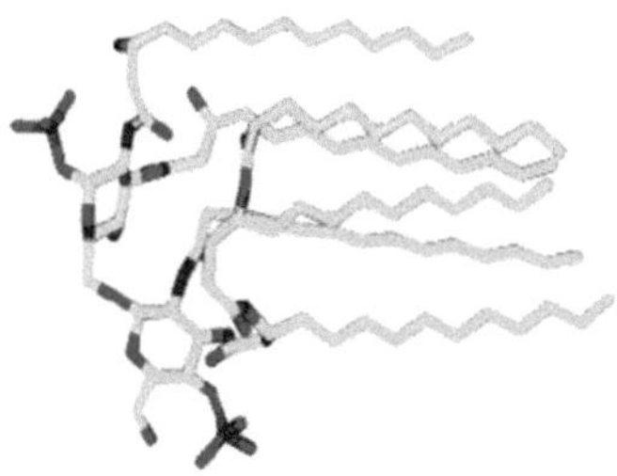

LipidA

Fig. 9.1 The structure of lipopolysaccharide, LPS. LPS consists of an O-specific antigen, a core oligosaccharide and the lipid A moiety. The core oligosaccharide, which varies from one bacterial species to another, is made up of outer and inner sugar regions. Lipid A virtually always includes two glucosamine residues modified by phosphates and a variable number of fatty acid chains (Frecer et al., 2000). The LPS structure was kindly contributed by Professor Helmut Brade (Research Center Borstel, Leibniz-Center for Medicine and Biosciences, Medical and Biochemical Mikrobiology Parkallee 22, D-23845 Borstel, Germany)

This domain determines the serological identity of the respective bacterium (Petsch and Anspach, 2000). Therefore, considerable structural diversity is noted amongst the O-specific antigen chain structures of LPS from different GNBs. Furthermore, the O-chain polysaccharides are the major immune-reactive domain of the LPS molecule, hence, this domain is often referred to as the O-specific antigen. It is responsible for defining the serotypic specificity of an individual GNB strain. Although the O-specific chain of LPS confers survival advantages, such as prevention of serum-mediated lysis of the bacterium, it is reported to be unessential for bacterial growth or survival (Raetz et al., 1988). This is reasonably logical since it would be consistent with the manner in which the GNBs are able to adapt to different modes of culture (agar plate or LB broth), and survive against challenges of the changing environment.

9.4.2 The Core Oligosaccharide Domain

The core oligosaccharide is a complex, non-repeating oligosaccharide that contains a unique inner sugar, KDO (2-keto-3-deoxy-D-mano-octonate)-heptose region which links the core oligosaccharide to the lipid A moiety. The core oligosaccharide is more conserved than the highly variable O-specific chain (Brade et al., 1988).

It is made up primarily of glucose, galactose and N-acetylglucosamine (GlcNAc) (Jansson et al., 1981; Rick, 1987). The hexose molecules in the outer core are more variable in structure than the inner core. The KDO sugar is linked directly to the lipid A moiety (Rietschel and Brade, 1992). Since it is buried within the LPS molecule, it was believed to be sterically-hindered from interaction with the host immune system (Giglioti and Shenep, 1985; Pollack et al., 1989; Heumann et al., 1991). It has been established that at least part of the core oligosaccharide, viz, the KDO-containing part is essential for bacterial viability (Raetz et al., 1991).

9.4.3 The Lipid A

The lipid A which is the minimum structure of LPS capable of sustaining bacterial growth and survival, is the most conserved moiety of the LPS molecules derived from diverse strains of GNB. The lipid A moiety, attached to two or three KDO residues (Lynn, 1998), acts as the membrane anchor of the LPS molecule. Lipid A is considered to be the bioactive centre of the LPS as virtually all LPS-induced biological responses in the host cells are lipid A-dependent. The synthetic lipid A (Galanos et al., 1992) was shown to display full endotoxic activity compared to the free lipid A cleaved from the native LPS molecule (Takayama et al., 1983), thus strongly suggesting the pathophysiological significance of the lipid A moiety. The lipid A is composed of a phosphorylated β1,6-linked D-glucosamine disaccharide (Frecer et al., 2000a, b) that carries variable numbers of asymmetrically placed amide or ester-linked acyl chains. This structure is the minimal requirement for the cytokine inducing capacity of lipid A (Rietschel and Brade, 1992). The unique structure of lipid A most likely reflects its important roles in the outer membrane assembly and functions, and it ensures resistance to phospholipases.

The structures of GNB lipid A such as those of *Escherichia coli* and *Salmonella typhimurium* have been elucidated (Takayama et al., 1983). The classical *sn*-1,2-diacylglycerol moiety of membrane phospholipids is replaced by a 2,3-diacylglucosamine unit in the lipid A moiety (Raetz, 1990). The acyl chains that are attached to the glucosamine backbone of lipid A differ from those attached to the membrane phospholipids in that they are 2–6 carbon atoms shorter and contain an (R)-3-hydroxyl substituent. The unique structure of lipid A presumably reflects its specific roles in the outer membrane assembly and function, and it ensures resistance to phosholipases. The lipid A of *E. coli* and *S. typhimurium* are β(1,6)-linked disaccharides of D-glucosamine that are acylated with (R)-3-hydroxytetradecanoic acid at positions 2, 2', 3, and 3', and phosphorylated at positions 1 and 4'. The two (R)-3-hydroxyl-acyl groups at positions 2' and 3' of the nonreducing glucosamine are further esterified with dodecanoic acid and tetradecanoic acid. The envelope of a single *E. coli* cell contains approximately 2×10^6 lipid A residues and approximately 2×10^7 glycerophospholipids, quantities consistent with the existence of one monolayer of lipid A and three monolayers of glycerophospholipids.

Since the structural determination of lipid A has focused on those of *E. coli* and *S. typhimurium* and that similar lipid A structures are reported to be present in the Enterobacteriaceae family, their lipid A structures are taken to be the prototype of lipid A structures, viz, "usual" lipid A (Mayer and Weckesser, 1984). However, intrinsic heterogeneity and species- and strain- specific modifications of lipid A do occur. Variations in the phosphorylation pattern, acylation pattern and the fatty acid chain length of lipid A (Kawata et al., 1999) in other GNB species as opposed to those of *E. coli* and *S. typhimurium* lipid A could lead to a change in the endotoxic potency of the molecule. Many of the variant, "unusual" lipid A were observed to be non-endotoxic, even though the exact prerequisites of endotoxic activity remain unclear. For the full expression of endotoxic activities, the structural requirements of lipid A allow for only small variations from those of *E. coli* or *S. typhimurium*, viz, a β(1,6)-linked D-glucosamine disaccharide carrying two negatively charged phosphates and six saturated fatty acids in a defined asymmetrical 4/2 distribution.

9.5 Plasma LPS-Binding Proteins Protect and Provoke Septic Shock – The Achilles Heel?

As a pathogen-associated molecular pattern (PAMP) molecule of the GNB, the LPS is a key virulence factor. During infection, the host's innate immune defenses immediately respond to the GNB invasion via LPS-induced signal transduction pathways, finally resulting in inflammation and septic shock (Fig. 9.2). Upon infection, the bacteria release LPS into the bloodstream to trigger the innate immune system, which initiates a series of defenses against the invasive GNB via specific recognition mechanisms (Ng et al., 2007; Jiang et al., 2007; Le Saux et al., 2008; Zhang et al., 2009; Jiang et al., 2009).

Certain circulating proteins in the plasma of mammals appear to interact with LPS in ways that can inhibit or enhance the immunological response to LPS (Corriveau and Danner, 1993). Plasma lipopolyproteins such as low density lipoprotein (LDL) and high density lipoprotein (HDL), which are involved in the transport and metabolic regulation of triglycerides and cholesterol, can bind LPS, reduce its toxicity and promote its clearance from circulation through the hepatobiliary system (Harris et al., 1993; Read et al., 1993; Feingold et al., 1995; Chaby, 1999). Non-lipoprotein factors, such as LPS binding protein (LBP) and soluble CD14 (sCD14), discussed later in this section, appear to be important in facilitating LPS-lipoprotein interaction. In this regard, plasma lipoproteins form part of the innate immune defense system against LPS.

The LBP, a 60 kDa glycoprotein of hepatic origin acts as an opsonin by enhancing the interaction of LPS with phagocytes. LBP binds to LPS via the lipid A moiety (Schumann et al., 1990). The LBP-LPS complex subsequently interacts with CD14 (Wright et al., 1990), a 55 kDa glycophosphatidylinositol (GPI)-linked receptor protein that is found on the surface of macrophages, monocytes and neutrophils (Ziegler-Heitbrock and Ulevitch, 1993; Kim et al., 2005).

The LBP facilitates the binding of LPS or GNB to phagocytes (Tobias et al., 1988; Corriveau and Danner, 1993; Yu and Wright, 1996). The enhanced interaction between LPS and phagocytes causes a marked increase in cell activation. Addition of small amounts of LBP to cultured macrophages increases by 100-fold the ability of LPS to induce the production of tumor necrosis factor alpha (TNF-α), a potent

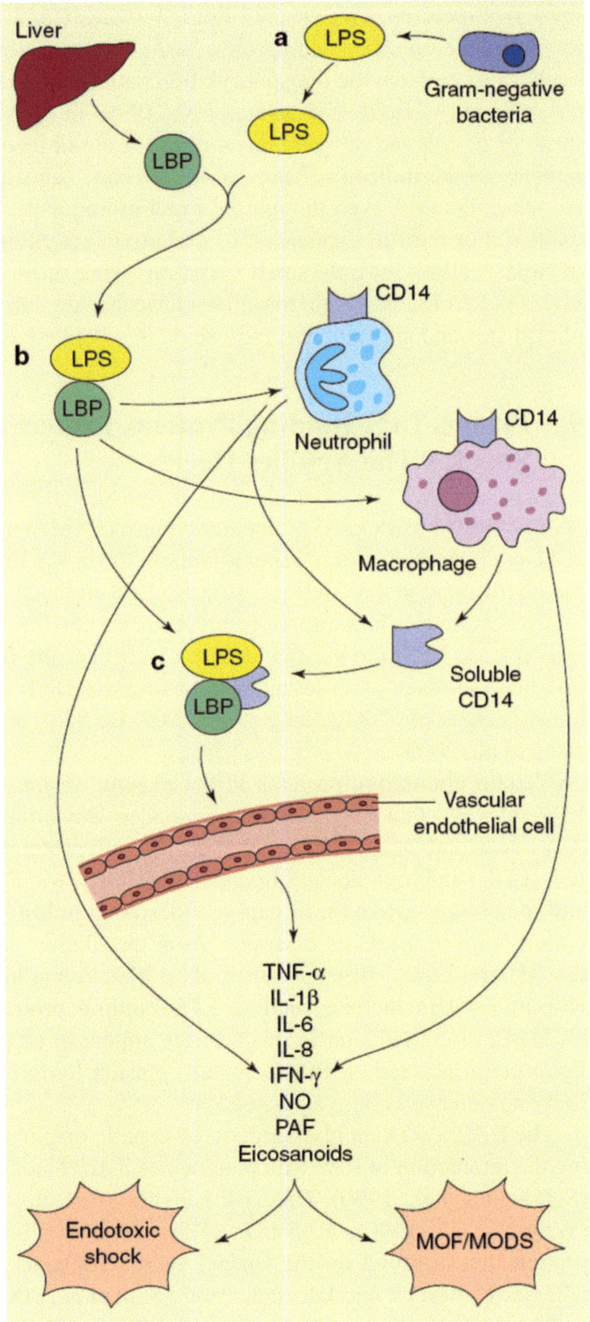

Fig. 9.2 The host cellular activation by lipopolysaccharide. (**a**) In the plasma, LPS is released from the Gram negative bacteria. (**b**) LPS-binding protein (LBP) transfers LPS to CD14 and facilitates the interactions of LPS with CD14 expressed on the surface of monocytes/macrophages or neutrophils. Endothelial cells and some other types of cells do not express CD14. (**c**) LPS stimulates these cells by binding soluble CD14. IFN-γ, interferon γ; IL-1, interleukin 1; MOF/MODS, multiple organ failure/multiple organ dysfunction syndrome; NO, nitric oxide; PAF, platelet-activating factor; TNF-α, tumor necrosis factor. Adapted from (Karima et al., 1999), with permission from Elsevier, Copyright Clearance Centre

proinflammatory cytokine. This phenomenon has been observed with different types and sources of LPS (Schumann et al., 1990). In acute phase response, the plasma LBP level can escalate 1000-fold, from a basal concentration of 0.5 g/ml to 50 g/ml, within 24 h, leading to a surge in cell activation. Besides LBP, septin (the product of a proteolytic cascade distinct from the coagulation and complement systems in the human plasma) binds LPS and mediates LPS recognition by phagocytes in a manner similar to LBP (Wright et al., 1992).

Bactericidal/permeability-increasing protein (BPI) is a 55 kDa inducible cationic protein of neutrophil origin that binds the lipid A moiety and neutralizes a variety of LPS. BPI is a boomerang-shaped molecule, comprising two domains located at the amino and carboxy termini that are structurally similar. These domains are hydrophobic and they bind lipid A (Beamer et al., 1997). Although BPI is thought to be primarily involved in the non-oxidative killing of ingested bacteria, evidence suggesting localization of BPI on the surface of neutrophils indicate that BPI is able to detoxify LPS in circulation (Weersink et al., 1992).

In the horseshoe crab, the LPS-binding proteins include the Limulus anti-LPS factor (LALF), which is a small basic protein of 101 amino acids that not only binds LPS but elicits a strong antibacterial effect on GNB (Morita et al., 1985b). Based on the sequence similarity and the crystal structure of LALF, Hoess et al. (1993) suggested that an exposed amphipathic loop on LALF, distinguished by an alternating series of positively charged and hydrophobic residues, represents an LPS-binding motif.

Another LPS sensor is the Toll-like receptor 4 (TLR4)-MD-2 complex, which is localized on the immune-responsive cell membrane (Yang et al., 2000). Protein kinases, such as p38 and JNK (Sweet and Hume, 1996) are triggered by LPS induction, leading to the activation of several transcription factors such as NF-κB, which in turn activates transcription of genes coding for numerous proinflammatory cytokines, tissue factors, adhesion molecules and inducible nitric oxide synthase. The overproduction of these potent mediators initiates a series of pathophysiological events that culminates in clinical manifestations of sepsis. Amongst the proinflammatory cytokines, TNF-α plays a critical role in the inflammatory response and is often regarded as a hallmark of LPS-induced inflammation (Tang et al., 2005). Following LPS challenge, the immunological cascade encompassing CD14, TLR4, MAPK and NF-κB (Guha and Mackman, 2001; Akira et al., 2006) is swiftly activated to sensitize the host to an LPS-induced uncontrolled acute inflammatory response that can result in septic shock, multiple organ failure/deaths, MOF/MODS (Fig. 9.2) (Li et al., 2006).

The LPS also interacts with the humoral immune system. It activates both the complement cascade which further fuels the inflammatory response, and the coagulation cascade which leads to disseminated intravascular coagulation that quickly depletes the clotting components in the blood leading to haemorrhage (Glauser et al., 1991). Therefore, at the initial stage of Gram negative infection, the binding of the LPS-receptors to LPS is required to trigger an inflammatory response, and yet, an immune over-reaction leads to septic shock and further downstream casualties. Therefore, LPS-binding proteins are analogous to an "Achilles heel", protecting and yet provoking inflammatory responses. Thus, recent studies are directed towards the intervention of this early step with potential LPS-binding drugs to compete against binding of LPS-effectors, and to attenuate the consequential damage to the host. However, an even more upstream step is the detection of LPS in a parenteral product, which may circumvent the need to intervene LPS-intoxication. This might indeed be the best option towards prevention rather than resolution-after-contamination/infection, since the LPS molecule is extremely indomitable.

9.6 Overcoming the LPS Problem – The Horseshoe Crab, a Creature Small and Great

LPS is the best-studied biological pyrogen. Upon intravenous introduction, it causes fever/pyrogenic action. Owing to its ubiquity, the LPS has been a bane to the pharmaceutical and medical industries since parenteral preparations contaminated with trace levels of LPS (in the picogram levels), can elicit dramatic effects in the patients. Thus, reliable endotoxin diagnostics and therapeutics are urgently sought.

The horseshoe crab hemolymph contains mainly one type of blood cells called amoebocytes (Fig. 9.3), which are extremely sensitive to LPS. During a Gram negative infection, the amoebocytes release granular components into the plasma

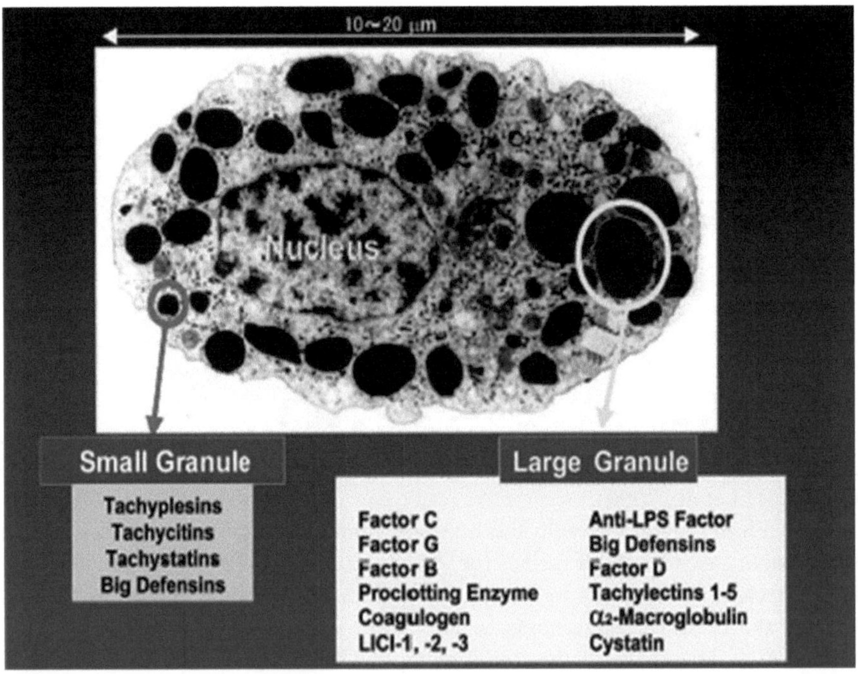

Fig. 9.3 The amoebocytes of horseshoe crab. Some small antimicrobial molecules like antimicrobial peptides that have been identified are in small granules, while Factor C and other defense molecules are localized in the large granules. Adapted from (Iwanaga and Lee, 2005), with permission from Editor-in-Chief of the Journal of Biochemistry and Molecular Biology

to participate in self-defense via blood coagulation (Iwanaga and Lee, 2005), which incapacitates the invading microbe. The amoebocytes contain two kinds of secretory granules, the large and small granules. Studies on these granules suggest that coagulation factors such as Factor C, are localized in the large granules whereas the antimicrobial peptides such as tachyplesin, are contained exclusively in the small granules (Iwanaga, 2002).

In the past decade, the molecular mechanisms of the coagulation cascade have been established (Navas et al., 1990; Ho et al., 1993; Ding et al., 1993a; Iwanaga et al., 1994; Ding et al., 1995). LPS from the GNB induces the amoebocytes to degranulate, thus initiating the blood coagulation cascade (Fig. 9.4), which is an important defense mechanism used by horseshoe crabs to kill and trap the invading GNB (Armstrong and Rickles, 1982; Ding et al., 1993b). This cascade is based on three serine protease zymogens – Factor C, Factor B, proclotting enzyme and one clottable protein, coagulogen (Muta and Iwanaga, 1996). Factor C, at the first step of the coagulation pathway, is sensitive to LPS. Interestingly, no homologues of Factor C have been found in the mammals, although the C-terminal serine protease domain has substantial similarities (36.7%) with human α-thrombin (Li et al., 2007). Thus, Factor C is a unique LPS-binding protein found only in the horseshoe crab.

As a "living fossil" which has survived for several hundred million years, the evolutionary success of the horseshoe crab attests to its strong innate immune defense ability, which it uses to thrive in microbiologically harsh habitats with abundant variety of disease-causing pathogens.

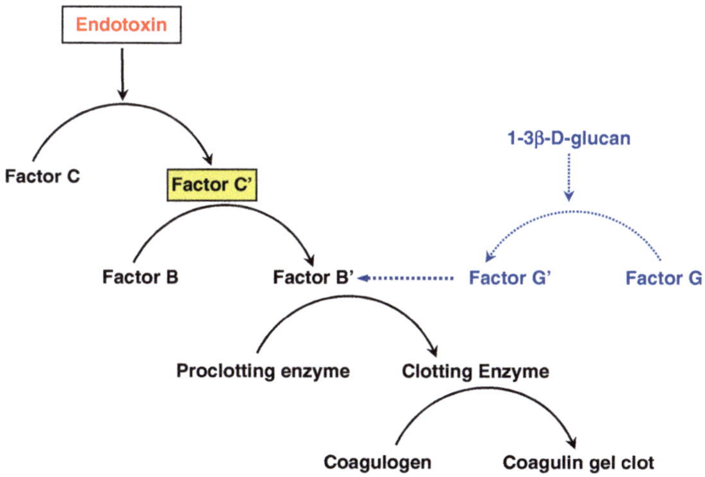

Fig. 9.4 The coagulation cascade in the horseshoe crab amoebocyte lysate. In the presence of LPS, the Factor C serine protease zymogen is autocatalytically activated to an active form, Factor C', which activates the proenzyme Factor B into Factor B'. This in turn activates proclotting to active clotting enzyme. Clotting enzyme then converts coagulogen into a coagulin gel clot, which traps the invading bacteria

The American horseshoe crab, *Limulus polyphemus* amoebocyte lysate (LAL) is well known and its coagulation cascade has been fully characterized and shown to defend the animal against the invasion of Gram-negative bacteria (Armstrong and Rickles, 1982). Owing to its extreme sensitivity to the bacterial endotoxin, LAL has been widely marketed for decades as a tool for detecting LPS in pharmaceuticals, parenterals and surgical implants, water and food (Levin et al., 1970; Novitsky, 1994). The LAL can detect femtogramme levels of LPS (Ho, 1983).

9.7 Drawbacks with LAL

Although the LAL test has been introduced since the 1970s, as a replacement for the rabbit pyrogen test for the quality assurance of parenterals and medical devices, it has suffered many severe setbacks. First, batch-to-batch and seasonal variations in LAL preparations cause differential sensitivity to LPS. Second, the lack of specificity for endotoxin is compounded by fungal contaminant, 1–3 β-D glucan (Fig. 9.4), which switches on the alternate coagulation pathway (Iwanaga et al., 1985), resulting in false positive test for pyrogen. Third, besides providing their newly laid eggs an energy boosting meal for migratory birds, the biomedical and economic importance of the horseshoe crab blood has caused the population of the American *Limulus polyphemus* to drop alarmingly. We describe below, some of the problems associated with the LAL test.

9.7.1 Differential Endotoxin Reactivities and Lack of Specificity

Despite years of intense efforts made with various methods of extraction of the amoebocyte lysate, the traditional clot assay and other quantitative assays were plagued with variations in sensitivity and specificity for LPS (Jorgensen and Smith, 1973). The specificity of the LAL assay for endotoxin is interfered by proteins and cofactors such as thrombin, thromboplastin, and certain synthetic polynucleotides, which can all give false positive results. On the other hand, peptidoglycan from the Gram positive bacteria, exotoxins from group A Streptococci (Brunson and Watson, 1976), simple polysaccharides including yeast mannans and bacterial dextrans and dithiols can also activate LAL to give a false positive or false negative result in some batches of LAL. As shown in the coagulation pathways (Fig. 9.4), the main pathway is triggered by endotoxin-sensitive Factor C, which activates intermediate serine proteases in the coagulation event to cause gelation in vivo, or to cleave a chromogenic or fluorogenic substrate in an in vitro reaction. However, an alternative cascade driven by the fungal toxin 1–3 β-D-glucan activates Factor G, which joins the main pathway and ultimately also leads to coagulin formation in vivo, and/or hydrolysis of the synthetic substrate in an in vitro assay, causing a false positive result. Thus, fungal contaminant will give a false positive result, which may be interpreted as endotoxin-positive. This explains the phenomenon of differential LAL reactivity in the absence of pyrogenicity, which has been widely observed and presents a recurring problem to endotoxin testing.

9.7.2 Problems with Sample and Specimen Preparations

Since LAL contains a series of coagulation enzymes, the pH, temperature and ionic strength have a critical influence over the precise biochemical reactions to ensure an appropriate end-point, coagulin formation. Components of the test sample can interfere with any step in the coagulation cascade, thereby affecting the final result. EDTA was found to inhibit endotoxin-induced LAL reaction (Morita et al., 1985), thus indicating the importance of cationic balance in this reaction. The LPS monomers, which contain both hydrophilic sugar groups and hydrophobic lipid A portion, have a propensity to form heteropolymers /micelles of different sizes, with the lipid A portion hidden from the aqueous environment although it is the lipid A that is the endotoxic moiety of the LPS molecule. Attempts are being made to overcome this problem; BioWhittaker Inc. (now Lonza Inc.) and BioDTech Inc. have produced dispersing agents, which increase the amount of detectable endotoxin in inhibitory samples.

9.7.3 LAL Production Endangers the Horseshoe Crab

Despite all the advancements made with new hardware and software and novel design methodologies, the biotechnology and medical industries still rely heavily on the horseshoe crab blood lysate to pass quality assurance of injectables and medical devices. Over-harvesting of the horseshoe crab by the Biotechnology Industry and the food chain-induced loss of the horseshoe crab spawned eggs, due to feeding by migratory birds, have dramatically dwindled the horseshoe crab population into potential extinction (Widener and Barlow, 1999). Furthermore, due to urbanization, the Japanese horseshoe crab was pronounced endangered since 1973 (Sekiguchi and Nakamura, 1979) and much effort continues in re-spawning and conserving the species (Sekiguchi et al., 1982, 1988).

These drawbacks have prompted many researchers to look into long term alternatives like: (i) farming of horseshoe crabs for the procurement of blood (Kropach, 1979), (ii) tissue culture of amoebocytes from which LAL originates (Pearson and Woodland, 1979) and (iii) genetic engineering of lysate proteins such as Factor C (Muta et al., 1991; Ding et al., 1995), the first serine protease of the coagulation cascade which is enzymatically-activated by LPS. It was necessary to face the challenge of producing a genetically-engineered LAL to replace the conventional LAL test; one which is more reliable, sensitive and specific for LPS and without relying on harvesting blood from the endangered horseshoe crab.

9.8 Factor C: A Horseshoe Crab Serine Protease with Multiple High Affinity LPS-Binding Sites – LPS Detection and Prevention Strategies – Towards Non-LAL Based LPS Detection

After 2–3 decades of conventional pyrogen testing using the blood extract of the horseshoe crab, and the problems associated with LAL, hereon, we describe a breakthrough in genetic engineering (US Patent No: 5,712,144) and molecular expression of recombinant Factor C, which yields enzymatically active rFC that is activated by trace levels of LPS, with a remarkable sensitivity of 0.001 EU/ml. Thus, rFC serves as a novel, perpectual, environmentally-friendly and standardised source of "LAL" for sensitive and specific detection of LPS. Using rFC, a novel micro-assay (US Patent No: 6,645,724B1) was developed for high throughput screens of pyrogen in pharmaceuticals and parenterals. A US Biotech Company that is a major producer of LAL is commercialising the rFC as a novel endotoxin diagnostic that sets a new standard for pyrogen testing. Furthermore, being capable of binding both free and bound LPS / lipid A (the biologically potent moiety of LPS) with high affinity, the rFC has other potential applications, such as the removal of LPS from contaminated samples (US Patent No: 6,645,724), as well as for the development of novel endotoxin therapeutics and antimicrobials.

The initial discovery by Bang (1956) that Gram negative bacterial endotoxin causes limulus blood to clot, followed by the formulation of the first LAL (Levin et al., 1970), have led to the FDA-approval (Sullivan and Watson, 1978) and commercialization (Novitsky, 1984) of LAL for testing endotoxin. Since the mid-1970s, LAL has been used widely for the quality assurance of many products approved by FDA. Therefore, LAL has superceded the US Pharmacopioea (USA)-approved rabbit pyrogen test. The latter is time consuming, expensive and often subjective. Over the years, LAL-based quantitation of endotoxin became somewhat possible with the design of new methods such as chromogenic, colorimetric and turbidometric assays (Iwanaga et al., 1978), although still haunted by problems with either the sample to be tested or the LAL preparation itself. Being at the initial step of the coagulation cascade, Factor C functions as a very sensitive and specific biosensor of LPS, capable of detecting picogram to nanogram levels of LPS (Ho, 1983), hence it was a prudent step to genetically engineer Factor C to replace the LAL.

9.9 Genetic Engineering and Production of Recombinant Factor C (rFC) – Necessity Spawns Innovation: Cloning and Subcloning the Factor C cDNA into Bacterial, Yeast, Insect and Mammalian Cells

Factor C is the LPS-sensitive intracellular serine protease zymogen that initiates the coagulation cascade system. It exists as a single- and a double-chain form (Ding et al., 1993). While the catalytic site of the molecule is found in the light chain, the

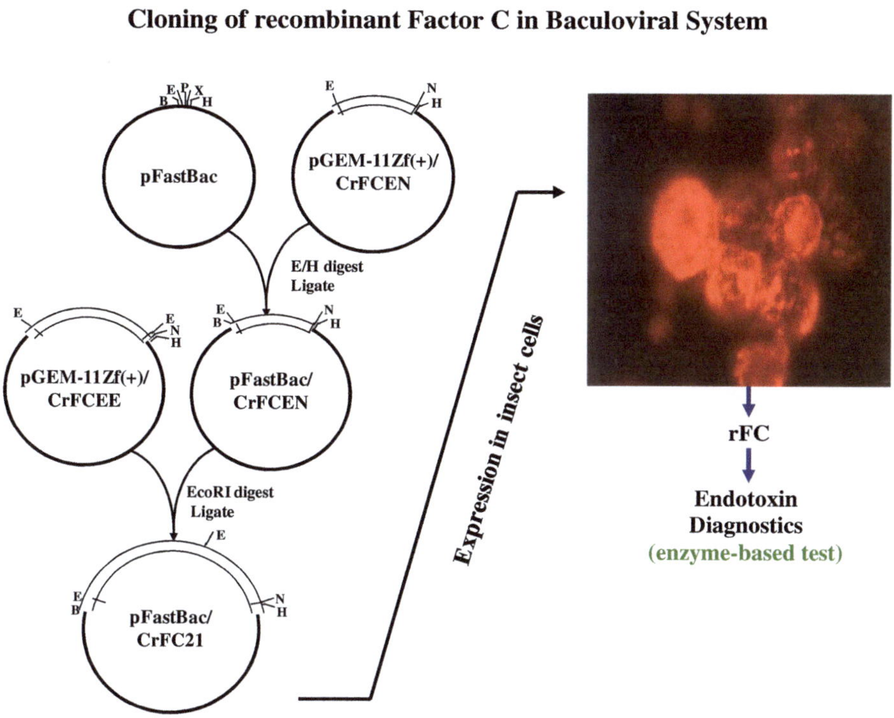

Fig. 9.5 Molecular cloning and expression of Factor C in the baculoviral system and expression in insect cells produced an rFC with remarkable sensitivity of 0.001 EU/mL endotoxin. (B: *BamHI*, E: *EcoRI*, X: *XhoI*, H: *HindIII*, P: *PstI*, N: *NcoI*)

endotoxin-binding site is located at the NH_2-terminal heavy chain region (Nakamura et al., 1988). The Japanese horseshoe crab (*Tachypleus tridentatus*) Factor C had been cloned in two separate and overlapping partial fragments (Muta et al., 1991). Ding et al. (1995) cloned the full length Factor C, originally in the *E. coli* and then recloned and expressed the recombinant Factor C (rFC) in different hosts, including yeast, *Saccharomyces cerevisiae* (Ding et al., 1997; Pui et al., 1997), *Pichia pastoris* and mammalian cell line, Cos-1 (Roopashree et al., 1996, 1997) and insect cells using the baculoviral system (Ding and Ho, 2001) (Fig. 9.5).

9.10 Development of a Quantitative Endotoxin Assay

Based on rFC, a modern and yet simple, rapid, specific and sensitive diagnostic test for endotoxin has been developed (Ding and Ho, 2001). rFC is a proenzyme until it encounters trace levels of endotoxin where it unequivocally exhibits full enzymatic activity, hence, acting as a very sensitive and specific biosensor for endotoxin. The resulting activated rFC acts as a catalyst to hydrolyse a synthetic substrate to form a quantifiable product, which measures the level of endotoxin. A fluorimetric assay

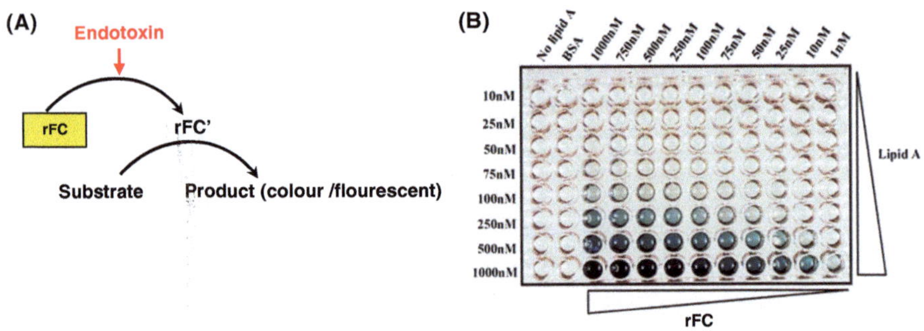

Fig. 9.6 Recombinant Factor C (rFC) is activated by endotoxin into rFC'. (**a**) A single-step activation of rFC which hydrolyses a substrate to yield either a colored (colorimetric assay) or fluorescent (fluorimetric assay) product, hence, (**b**) reporting on the presence and quantity of endotoxin present in samples (a colorimetric assay, in this case)

for endotoxin uses rFC zymogen, which, on activation by endotoxin, hydrolyses a fluorigenic substrate such as Boc-Val-Pro-Arg-MCA (Boc, butoxy-carbonyl; MCA, 7-amido-4-methylcoumarin). The fluorimetric product is measured at an excitation of 380 nm and an emission of 40 nm. Raising the amount of rFC in the assay was found to further increase the sensitivity of the detection of endotoxin from 0.005 EU/ml to 0.001 EU/ml (Ding and Ho, 2001). A comparison of rFC with commercial LAL, under the same assay conditions showed rFC to have lower background reading and a more sensitive response to endotoxin (Ding and Ho, 2001).

A colorimetric assay was also established using Boc-Val-Pro-Arg-pNA (pNA, p-nitroanilide) as chromogenic substrate that is hydrolysed by endotoxin-activated rFC to yield a measurable colorimetric product. A further demonstration that rFC is capable of detecting endotoxin (Ding and Ho – US patent filed no: 6,645,724B1) is depicted in Fig. 9.6. Compared to the coagulation cascade that occurs in the LAL, which is present in the conventional LAL-assay, the rFC affords a rapid and high-throughput assay in one single step. This reaction also obviates any potential interference of the specificity for LPS detection since no other blood proteins are present in the rFC.

9.11 Commercialization of the Endotoxin Detection Kit – The Route to PyroGene and Pyrosense

A microfluorimetric or microcolorimetric assay integrated into the rFC assay has allowed high-throughput screens of LPS in multiple samples. Currently, the rFC is used in an LPS-assay to sensitively and specifically detect LPS (Ding and Ho, 2001). The rFC has been incorporated into the PyroGene kit, which was launched in 2004 (Cambrex Inc., USA). Recently, Lonza Inc. (USA) applied rFC in PyroSense, which has potentials for continuous online monitoring of endotoxin in water and other fluids used for large-scale production of biomedical products.

The endotoxin test has a large market in drug companies that use LAL to detect endotoxin contamination in injectable products. For quality assurance and validation, every pharmaceutical company uses it for process monitoring. Furthermore, medical device firms require this test to ensure that catheters, pacemakers and other implantable devices are endotoxin-free. In gene therapy, where isolated plasmid DNA are used as therapeutics, international regulatory agencies are obliged to set stringent guidelines with regard to endotoxin levels (Levy et al., 2000). With this great impact on the future therapeutic and diagnostic approaches, and the growing need for the removal of endotoxins, the advent of rFC-based PyroGene test is timely to provide a sensitive, rapid and quantitative diagnostic for endotoxin. The rFC serves as a perpetual source of a genetically engineered biosensor for LPS. Finally, the horseshoe crab species should not need to be harvested for biomedical consumption. Towards an "environmentally-friendly science," the genetic engineering feat has contributed not only to the conservation of this "living fossil", but also to a more accurate, more sensitive and reliable, and less variable endotoxin test for the human healthcare industry.

Acknowledgements We thank the BMRC, Agency for Science and Research (A∗STAR, Singapore), Ministry of Education (AcRF, Tier 2) and the National University of Singapore (FRC, NUS) for financial support.

References

Akira, S., Uematsu, S., Takeuchi, O. Pathogen recognition and innate immunity. Cell 124 (2006) 783–801.

Angus, D.C., Wax, R.S. Epidemiology of sepsis: an update. Crit Care Med 29 (2001) S109–16.

Armstrong, P.B., Rickles, F.R. Endotoxin-induced degranulation of the Limulus amoebocyte. Exp Cell Res 140 (1982) 15–24.

Bang, F.B. A bacterial disease of *Limulus polyphemus*. Bull Johns Hopkins Hosp 98 (1956) 325–351.

Beamer, R.J. Carroll, S.F., Eisenberg, D. Crystal structure of human BPI and two bound phospholipids at 2.4 angstrom resolution. Science 276 (1997) 1861–1864.

Boneca, I.G. The role of peptidoglycan in pathogenesis. Curr Opin Microbiol 8 (2005) 46–53.

Bone, R.C. The sepsis syndrome. efinition and general approach to management. Clin Chest Med 17 (1996) 175–181.

Brade, H., Brade, I., Schade, U., Zabringer, U., Holst, O., Kuhn, H.M., Rozalski, A., Rohrscheidt, E., Rietschel, T.T. Structure, endotoxicity, immunogenecity and antigenicity of bacterial lipopolysaccharides (endotoxins, O-antigens). Progr Clin Biol Res 272 (1988) 17–45.

Brandenburg, K., Wiese, A. Endotoxins: relationships between structure, function, and activity. Curr Top Med Chem 4 (2004) 1127–1146.

Brady, C.A., Otto, C.M. Systemic inflammatory response syndrome, sepsis, and multiple organ dysfunction. Vet Clin North Am Small Anim Pract 31 (2001) 1147–1162, v–vi.

Breithaupt, H. The new antibiotics. Nat Biotechnol 17 (1999) 1165–1169.

Brunson, K.W., Watson, D.W. Limulus amoebocyte lysate reaction with Streptococcal pyrogenic exotoxin. Infect Immun 14 (1976) 1256–1258.

Chaby, R. Strategies for the control of LPS-mediated pathophysiological disorders. Drug Discov Today 4 (1999) 209–221.

Corriveau, C.C., Danner, R.L. Endotoxin as a therapeutic target in septic shock. Infect Agents Dis 2 (1993) 35–43.

Ding, J.L., Ho, B. New era in pyrogen testing. Trends in Biotechnol (2001) 277–281.

Ding, J.L., Chai, C., Pui, A.W.M., Ho, B. Expression of full length and deletion homologues of Carcinoscorpius rotundicauda Factor C in Saccharomyces cerevisiae: immunoreactivity and endotoxin binding. J Endotoxin Res 4 (1997) 33–43.

Ding, J.L., Navas, M.A.A., Ho, B. Two forms of Factor C from the amoebocytes of *Carcinoscorpius rotundicauda*: Purification and characterisation. Biochim et Biophys Acta 1202 (1993a) 149–156.

Ding, J.L., Navas, M.A.A., Ho, B. Molecular cloning and sequence analysis of Factor C cDNA from the Singapore horseshoe crab, *Carcinoscorpius rotundicauda*. Mol Mar Biol Biotechnol 4 (1995) 90–103.

Ding, J.L., Sababathy, Tr.K. Ho, B. Morphological changes in *C. rotundicauda* amoebocytes and *E. coli* during their interaction. Cytobios 75 (1993b) 21–32.

Dinarello, C.A. The proinflammatory cytokines interleukin-1 and tumour necrosis factor and treatment of the spetic shock syndrome. J Infect Dis 167 (1991) 1177–1184.

Feingold, K.R., Shigenaga, J.K., Rapp, J.H., Funk, J.L., Moser, A.H. Role for circulating lipoproteins in protection from endotoxin toxicity. Infect Immun 63 (1995) 2041–2046.

Frecer, V., Ho, B., Ding, J.L. Molecular dynamics study on lipid A from *Escherichia coli*: insights into its mechanism of biological action. Biochim Biophys Acta 1466 (2000a) 87–104.

Frecer, V., Ho, B., Ding, J.L. Interpretation of biological activity data of bacterial endotoxins by simple molecular models of mechanism of action. Eur J Biochem 267 (2000b) 837–52.

Galanos, C., Freudenberg, M.A, Katschinski, T., Salmao, R., Mossmann, H., Kumazawa, Y., Tumor necrosis factor and host response to endotoxin. In Ryan, J.L., Morrison, D.C. (eds), Immunopharmacology and Pathophysiology Vol. II, CRC Press, Boca Raton, FL (1992). pp. 75–104.

Giglioti, F., Shenep, J.L. Failure of monoclonal antibodies to core glycolipid to bind intact strains of *Escherichia coli*. J Infect Dis 151 (1985) 1005–1011.

Gradishar, W.J., Vogelzang, N.J., Kilton, L.J., Leibach, S.J., Rademaker, A.W., French, S., Benson, A.B., 3rd. A phase II clinical trial of echinomycin in metastatic soft tissue sarcoma. An Illinois Cancer Center Study. Invest New Drugs 13 (1995) 171–174.

Glauser, M.P., Zanetti, G., Baumgartner, J.D., Cohen, J. Septic shock: pathogenesis. Lancet 338 (1991) 732–736.

Guha, M., Mackman, N. LPS induction of gene expression in human monocytes. Cell Signal 13 (2001) 85–94.

Hancock, R.E., Chapple, D.S. Peptide antibiotics. Antimicro Ag Chemother 43 (1999) 1317–1323.

Harris, H.W., Grunfeld, C., Feingold, K.R., Read, T.E., Kane, J.P., Jones, A.L., Eichbaum, E.E., Bland, G.F., Rapp, J.H. Chylomicrons alter the fate of endotoxin, decreasing tumour necrosis factor release and preventing death. J Clin Invest 91 (1993) 1028–1034.

Heumann, D., Baumgartner, J.D., Jacot-Guillarmod, H., Glauser, M.P. Antibodies to core lipopolysaccharide determinants: absence of cross-reactiveity with heterologous lipopolysaccharide. J Infect Dis 163 (1991) 762–768.

Ho, B. An improved Limulus gelation assay. Microbios Letts 24 (1983) 81–84.

Ho, B., Kim, J.C., Ding, J.L. Electrophoretic analysis of endotoxin-activatied gelation reaction of *Carcinoscorpius rotundicauda* amoebocyte lysate. Biochem Mol Biol Intl 29 (1993) 687–694.

Hoess, A., Watson, S., Siber, G.R., Liddington, R. Crystal structure of an endotoxin-neutralizing protein from the horseshoe crab, Limulus anti-LPS factor, at 1.5 A resolution. EMBO J 12 (1993) 3351–3356.

Jansson, P.E., Lindberg, A.A., Lindberg, B., Wollin, R. Structural studies on the hexose region of the core in lipopolysaccharides from Enterobacteriaceae. Eur J Biochem 115 (1981) 571–577.

Iwanaga, S., Lee, B.L. Recent advances in the innate immunity of invertebrate animals. J Biochem Mol Biol 38 (2005) 128–150.

Iwanaga, S. The molecular basis of innate immunity in the horseshoe crab. Curr Opin Immunol 14 (2002), 87–95.

Iwanaga, S., Muta, T., Shigenaga, T., Miura, Y., Seki, N., Saito, T.S.K., Kawabata, S. Role of Hemocyte-derived granular components in invertebrate defense. NY Acad Sci 712 (1994) 102–116.

Iwanaga, S., Morita, T., Harada, T., Nakamura, S., Niwa, M., Takada, K., Kimura, T., Sakakibaru, S. Chromogenic substrates for horseshoe crab clotting enzyme. Its application for the assay of bacterial endotoxins. Haemostasis 7 (1978) 183–188.

Iwanaga, S., Morita, T., Miyata, T., Nakamura, T. Haemolymph coagulation in Limulus. In Leive L. et al. (eds). Microbiology, American Society of Microbiology, Washington (1985). pp. 29–32.

Jiang, N., Tan, N.S., Ho, B. Respiratory proteins generate ROS as an antimicrobial strategy. Nat Immunol 8 (2007) 1114–1122.

Jiang, N., Thangamani, S., Chor, C.F., Wang, S., Winarsih, I., Du, R., Sivaraman, J., Ho, B., Ding, J.L. A novel serine protease inhibitor acts as an immunomodulatory switch while maintaining homeostasis. J Innate Immunity 1 (2009) 465–479.

Jorgensen, J.H., Smith, R.F. Preparation, sensitivity and specificity of Limulus lysate for endotoxin assay. Appl Microbiol 26 (1973) 43–48.

Karima, R., Matsumoto, S., Higashi, H., Matsushima, K. The molecular pathogenesis of endotoxic shock and organ failure. Mol Med Today 5(3) (1999) 123–132.

Kawata, T., Bristol, J.R., Rossignol, D.P., Rose, J.R., Kobayashi, S., Yokohama, H., Ishibashi, A., Christ, W.J., Katayama, K., Yamatsu, I., Kishi, Y. E5531, a synthetic non-toxic lipid A derivative blocks the immunobiological activities of lipopolysaccharides. Br J Pharmacol 127 (1999) 853–862.

Kim, J.I., Lee, C.J., Jin, M.S., Lee, C.H., Paik, S.G., Lee, H., Lee, J.O. Crystal structure of CD14 and its implications for lipopolysaccharide signaling. J Biol Chem 280 (2005), 11347–11351.

Kropach, C. Observation on the potential of amoebocyte lysate. In Cohen, E., Levin, J., Bang, F.B. (eds), Progress in Clinical and Biological Research. Vol. 29, Alan, R. Liss, NY (1979) pp. 103–106.

Lemaire, L.C., van Lanschot, J.J., Stoutenbeek, C.P., van Deventer, S.J., Wells, C.L., Gouma, D.J. Bacterial translocation in multiple organ failure: cause or epiphenomenon still unproven. Br J Surg 84 (1997) 1340–1350.

Lent, M., Hirshberg, A., Margolis, G. Systemic toxins. Signs, symptoms and management of patients in septic shock. J Exp Med 26 (2001) 54–65; quiz 66–67.

Le Saux, A., Ng, P.M.L., Koh, J.Y.Y., Low, D.H.P., Leong, G.E.L., Ho, B., Ding, J.L. The macromolecular assembly of pathogen recognition receptors is impelled by serine proteases, via their complement control protein (CCP) modules. J Mol Biol 377 (2008) 902–913.

Levin, J., Tomasulo, P.A., Oser, RS. Detection of endotoxin in human blood and demonstration of an inhibitor. J Lab Clin Med 75 (1970) 903–911.

Levy, M.S., O'Kennedy, R.D., Ayazi-Shamlou, P., Dunnill, P. Biochemical engineering approaches to the challenges of producing pure plasmid DNA. Trends Biotechnol 18 (2000) 296–305.

Li, P., Sun, M., Ho, B., Ding, J.L. The specificity of Sushi peptides for endotoxin and anionic phospholipids: potential application of POPG as an adjuvant for anti-LPS strategies. Biochem Soc Trans 34 (2006) 270–272.

Li, P., Ho, B., Ding, J.L. Recombinant Factor C competes against LBP to bind lipopolysaccharide and neutralizes the endotoxicity. J Endotoxin Res 13 (2007) 150–157.

Lynn, W.A. Anti-endotoxin therapeutic options for the treatment of sepsis. J Antimicrob Chemother 41 Suppl A, (1998) 71–80.

Mayer, H., Weckesser, J. "Unusual" lipid As: structures, taxonomical relevance and potential value for endotoxin research. In Rietschel, E.T. (ed), Handbook of Endotoxin, Vol. 1: Chemistry of Endotoxin. Elsevier Science Publishers B.V., Netherlands (1984) pp. 221–247.

McCormick, J.K., Yarwoo, J.M., Schlievert, P.M. Toxic shock syndrome and bacterial superantigens: an update. Annu Rev Microbiol 55 (2001) 77–104.

Ministry of Health, S. Ministry of Health, Singapore (2005) http://www.moh.gov.sg/corp/publications/statistics/principal.do.

Minabe, M., Takeuchi, K., Kumada, H., Umemoto, T. The effect of root conditioning with minocycline HCl in removing endotoxin from the roots of periodontally-involved teeth. J Periodontol 65 (1994) 387–392.

Morita, T., Nakamura, T., Miyata, T., Iwanaga, S. Biochemical characterisation of Limulus clotting factors and inhibitors which interact with bacterial endotoxin. In Cate, J.W., Buller, H.R., Sturk, A., Levin, J. (eds). Bacterial endotoxin, structure, biomedical significance, and detection with the limulus amoebocyte lysate test. Alan R Liss Inc, NY (1985a) pp. 53–64.

Morita, T., Ohtsubo, S., Nakamura, T., Tanaka, S., Iwanaga, S., Ohashi, K., Niwa, M. Isolation and biological activities of limulus anticoagulant (anti-LPS factor) which interacts with lipopolysaccharide (LPS). J Biochem (Tokyo) 97 (1985b) 1611–1620.

Muta, T., Iwanaga, S. Clotting and immune defense in Limulidae. Progr Mol Subcell Biol 15 (1996) 154–189.

Muta, T., Miyata, T., Misumi, Y., Tokunaga, F., Nakamura, T., Toh, Y., Ikehara, Y. Iwanaga, S. Limulus Factor C: endotoxin-sensitive serine protease zymogen with a mosaic structure of complement-like, epidermal growth factor-like and lectin-like domains. J Biol Chem 266 (1991) 6554–6561.

Nakamura, T., Furunaka, H., Miyata, T., Tokunaga, F., Muta, T., Iwanaga, S., Niwa, M., Takao, T., Shimonishi, Y. Tachyplesin, a class of antimicrobial peptide from the hemocytes of the horseshoe crab (Tachypleus tridentatus). Isolation and chemical structure. J Biol Chem 263 (1988) 16709–16713.

Navas, III M.A.A., Ding, J.L., Ho, B. Inactivation of Factor C by dimethyl sulfoxide inhibits coagulation of the *Carcinoscorpius amoebocyte* lysate. Biochem Intl 21 (1990) 805–813.

Ng, P.M.L., Le Saux, A., Lee, C.M., Tan, N.S., Lu, J.H., Thiel, S., Ho, B., Ding, J.L. C-reactive protein collaborates with plasma lectins to boost immune response against bacteria. EMBO J 26 (2007) 3431–3440.

Novitsky, T.J. Discovery to commercialisation: the blood of the horseshoe crab. Oceanus 27 (1984) 13–18.

Novitsky, TJ. Limulus amoebocyte lysate (LAL) detection of endotoxin in human blood. J Endotoxin Res 1 (1994) 253–263.

Petsch, D., Anspach, F.B. Endotoxin removal from protein solutions. J Biotechnol 76 (2000) 97–119.

Pearson, F.C. III, Woodland, E. The in vitro cultivation of limulus amoebocytes. In Cohen, E., Levin, J., Bang, F.B. (eds), Progress in Clinical and Biological Research. Vol. 29 Alan R. Liss, NY. (1979), pp 93–102.

Pollack, M., Chia, J.K.S., Koles, N.L., Miller, M. Specificity and cross-reactivity of monoclonal antibodies reactive with the core and lipid A regions of bacterial lipopolysaccharide. J Infect Dis 159 (1989) 168–188.

Pui, A.W.M., Ho. B., Ding, J.L. Yeast recombinant Factor C from horseshoe crab binds endotoxin and causes bacteriostasis. J Endotoxin Res 4 (1997) 391–400.

Raetz, C.R.H. Biochemistry of endotoxins. Annu Rev Biochem 59 (1990) 129–170.

Raetz, C.R.H. Bacterial endotoxins: extraordinary lipids that activate eukaryotic signal transduction. J Bacteriol 175 (1993) 5745–5753.

Raetz, C.R.H., Brozek, K.A., Clementz, T., Coleman, J.D., Galloway, S.M., Golenbock, D.T., Hampton, R.Y. Gram-negative endotoxin: a biologically active lipid. Cold Spring Harbor Symp. Quant Biol 53 (1988) 973–982.

Raetz, C.R.H., Ulevitch, R.J., Wright, S.D., Sibley, C.H., Ding, A., Nathan, C.F. Gram-negative endotoxin: an extraordinary lipid with profound effects on eukaryotic signal transduction. FASEB J 5 (1991) 2652–2660.

Read, T., Harris, H.W., Grunfeld, C., Calhoum, M.C., Kane, J.P., Rapp, J.H. Chylomicrons enhance endotoxin excretion in bile. Infect Immun 61 (1993) 3496–3502.

Rietschel, E.T., Brade, H. Bacterial endotoxins. Sci Am 267 (1992) 54–61.

Rietschel, E.T., Brade, H., Holst, O., Brade, L, Muller-Leonnies, S., Mamat, U., Zahringer, U., Beckmann, F., Seydel, U., Brandenburg, K., Ulmer, A.J., Mattern, T., Heine, H., Schletter, S., Loppnow, H., Schonbeck, U., Flad, H.D., Hauschildt, S., Schade, U.F., Di, Padova F., Kusumoto, S., Schumann, R.R. Bacterial endotoxin: chemical constitution, biological recognition, host response and immunological detoxification. Curr Topics Microbiol Immunol 216 (1996) 39–81.

Rick, P.D. In *Escherichia coli* and *Salmonella typhimurium*, ASM Publications, Washington, DC (1987).

Roopashree, S.D., Ho, B., Ding, J.L. Expression of Carcinoscorpius rotundicauda Factor C in Pichia pastoris. Mol Mar Biol Biotechnol 5 (1996) 334–343.

Roopashree, S.D., Ho, B., Ding, J.L. Recombinant COS-1 cells express Carcinoscorpius rotundicauda Factor C. Biotech Lett 19 (1997) 357–361.

Ruiter, D.J., van der Meulen, J., Brouwer, A., Hummel, M.J., Mauw, B.J., van der Ploeg, J.C., Wisse, E. Uptake by liver cells of endotoxin following its intravenous injection. Lab Invest 45 (1981) 38–45.

Schumann, R.R., Leong, S.R., Flaggs, G.W., Gray, P.W., Wright, S.D., Mathison, J.C., Tobias, P.S., Ulevitch, R.J. Structure and function of lipopolysaccharide binding protein. Science 249 (1990) 1429–1431.

Sekiguchi, K., Nakamura, K. Ecology of the extant horseshoe crab. In Cohen, E., Levin, J., Bang, F.B. (eds), Progress in Clinical and Biological Research Vol. 29. Alan, R. Liss, NY (1979), pp. 37–45.

Sekiguchi, K., Yamamichi, Y., Costlow, J.D. Horseshoe crab developmental studies I. Normal embryonic development of *Limulus polyphemus* compared with *Tachypleus tridentatus*. In Bonaventura, J., Bonaventura, C., Tesh, S. (eds), Physiology and Biology of Horseshoe Crabs: Studies on Normal and Environmentally Stressed Animals. Alan R. Liss, NY (1982), pp. 53–73.

Sekiguchi, K., Yamamichi, Y., Seshimo, H., Sugita, H. Normal development. In Sekiguchi, K. (ed) Biology of Horseshoe Crabs. Science House, Tokyo (1988), pp. 133–181.

Sullivan, J.D. Jr., Watson, S.W. Limulus lysate of improved sensitivity and preparing the same. US Patent No. 4 107 077 (1978).

Sweet, M.J., Hume, D.A. Endotoxin signal transduction in macrophages. J Leukoc Biol 60 (1996) 8–26.

Takayama, K., Qureshi, N., Mascagni, P., Nashed, M.A., Anderson, L., Raetz, C.R.H. Fatty acyl derivatives of glucosamine 1-phosphate in *Escherichia coli* and their relation to lipid A. Complete structure of A diacyl GlcN-1-P found in a phosphatidylglycerol-deficient mutant. J Biol Chem 258 (1983) 7379–7385.

Tang, X., Marciano, D.L., Leeman, S.E., Amar, S. LPS induces the interaction of a transcription factor, LPS-induced TNF-alpha factor, and STAT6(B) with effects on multiple cytokines. Proc Natl Acad Sci USA 102 (2005) 5132–5137.

Tobias, P.S., Mathison, J.C., Ulevitch, R.J. A family of lipopolysaccharide binding proteins involved in responses to Gram-negative sepsis. J Biol Chem 263 (1988) 13479–13481.

Ulmer, A.J., Rietschel, E.T., Zahringer, U. Lipopolysaccharide: Structure, bioactivity, receptors, and signal transduction. Trends Glycosci Glycotechnol 14 (2002) 53–68.

Van Leeuwen, P.A., Boermeester, M.A., Houdijk, A.P., Ferwerda, C.C., Cuesta, M.A., Meyer, S., Wesdorp, R.I. Clinical significance of translocation. Gut 35 (1994) S28–34.

Weersink, A.J.L., van Kessel, K.P.M., van den Tol, M.E., van Strijp, J.A.G., Torensma, R., Verhoef, J., Elsbach, P., Weiss, J. Human granulocytes express a 55 kDa lipopolysaccharide-binding protein on the cell surface that is identical to the bactericidal/permeability-increasing protein. J Immunol 150 (1992) 253–263.

Westphal, O., Hann, K., Himmelspach, K. Chemistry and immunochemistry of bacterial lipopolysaccharide as cell wall antigens and endotoxin. Progr Allergy 33 (1983) 9–39.

Widener, J.W., Barlow, R.B. Decline of a horseshoe crab population on Cape Cod. Biol Bull 197 (1999) 300–302.

Wright, S.D. Multiple receptors for endotoxin. Curr Opinion Immunol 3 (1991) 83–90.

Wright, S.D., Ramos, R.A., Patel, M., Miller, D.S. Septin: a factor in plasma that opsonises lipopolysaccharide-bearing particles for recognition by CD14 on phagocytes. J Exp Med 176 (1992) 719–727.

Wright, S.D., Ramos, R.A., Tobias, P.S., Ulevitch, R.J., Mathison, J.C. CD14, a receptor for complexes of lipopolysaccharide (LPS) and LPS binding protein. Science 249 (1990) 1431–1433.

Yang, H., Young, D.W., Gusovsky, F., Chow, J.C. Cellular events mediated by lipopolysaccharide-stimulated toll-like receptor 4. MD-2 is required for activation of mitogen-activated protein kinases and Elk-1. J Biol Chem 275 (2000) 20861–20866.

Yu, B., Wright, S.D. Catalytic properties of lipopolysaccharide (LPS) binding protein. J Biol Chem 271 (1996) 4100–4105.

Zhang, J., Koh, J.Y.Y., Lu, J.H., Thiel, S., Leong, B.S.H., Sethi, S.K., He, Y.C., Ho, B., Ding, J.L. Local inflammation induces complement crosstalk which amplifies the antimicrobial response. PLoS Pathogens 5 (2009) 1–15, e1000282.

Zasloff, M. Antimicrobial peptides of multicellular organisms. Nature 415 (2002) 389–395.

Ziegler-Heitbrock, H.W., Ulevitch, R.J. CD14: cell surface receptor and differentiation marker. Immunol Today 14 (1993) 121–125.

Chapter 10
The Role of Endotoxin in Infection: *Helicobacter pylori* and *Campylobacter jejuni*

Anthony P. Moran

Abstract Both *Helicobacter pylori* and *Campylobacter jejuni* are highly prevalent Gram-negative microaerophilic bacteria which are gastrointestinal pathogens of humans; *H. pylori* colonizes the gastroduodenal compartment and *C. jejuni* the intestinal mucosa. Although *H. pylori* causes chronic gastric infection leading to gastritis, peptic ulcers and eventually gastric cancer while *C. jejuni* causes acute infection inducing diarrhoeal disease, the endotoxin molecules of both bacterial species contrastingly contribute to their pathogenesis and the autoimmune sequelae each induces. Compared with enterobacterial endotoxin, that of *H. pylori* has significantly lower endotoxic and immuno-activities, the molecular basis for which is the underphosphorylation and underacylation of the lipid A component that interacts with immune receptors. This induction of low immunological responsiveness by endotoxin may aid the prolongation of *H. pylori* infection and therefore infection chronicity. On the other hand, this contrasts with acute infection-causing *C. jejuni* where overt inflammation contributes to pathology and diarrhoea production, and whose endotoxin is immunologically and endotoxically active. Futhermore, both *H. pylori* and *C. jejuni* exhibit molecular mimicry in the saccharide components of their endotoxins which can induce autoreactive antibodies; *H. pylori* expresses mimicry of Lewis and some ABO blood group antigens, *C. jejuni* mimicry of gangliosides. The former has been implicated in influencing the development of inflammation and gastric atrophy (a precursor of gastic cancer), the latter is central to the development of the neurological disorder Guillain-Barré syndrome. Both diseases raise important questions concerning infection-induced autoimmunity awaiting to be addressed.

Keywords Bacterial pathogenesis · *Campylobacter jejuni* · *Helicobacter pylori* · Lipid A · Molecular mimicry

---

A.P. Moran (✉)
Laboratory of Molecular Biochemistry, Microbiology, School of Natural Sciences, National University of Ireland, Galway, Ireland
e-mail: anthony.moran@nuigalway.ie

X. Wang, P.J. Quinn (eds.), *Endotoxins: Structure, Function and Recognition*, Subcellular Biochemistry 53, DOI 10.1007/978-90-481-9078-2_10,

Abbreviations

AMAN	acute motor axonal neuropathy
CAMPs	cationic antimicrobial peptides
EAN	experimental ataxic neuropathy
FucT	fucosyltransferase
Gal	D-galactose
GalNAc	*N*-acetyl-D-glaactosmine
GBS	Guillain-Barré syndrome
GlcN	D-Glucosamine
GlcN3N	2,3-diamino-2,3-dideoxy-D-glucose
H^+,K^+-ATPase	H^+,K^+-adenosine triphosphatase
Ig	immunoglobulin
LacNAc	*N*-acetyl-lactosamine
Le	Lewis
LOS	lipooligosaccharide
LPS	lipopolysaccharide
MFS	Miller Fisher syndrome
Neu5Ac	*N*-acetyl-neuraminic acid
OS	oligosaccharide
*P*EtN	phosphoethanolamine
TLRs	Toll-like receptors
14:0(3-OH)	(*R*)-3-hydroxytetradecanoic acid
16:0(3-OH)	(*R*)-3-hydroxyhexadecanoic acid
18:0(3-OH)	(*R*)-3-hydroxyoctadecanoic acid
3-(14:0-*O*)-14:0	(*R*)-3-(tetradecanoyloxy)tetradecanoic acid
3-(16:0-*O*)-14:0	(*R*)-3-(hexadecanoyloxy)tetradecanoic acid
3-(12:0-*O*)-16:0	(*R*)-3-(dodecanoyloxy)hexadecanoic acid
3-(14:0-*O*)-16:0	(*R*)-3-(tetradecanoyloxy)hexadecanoic acid
3-(18:0-*O*)-18:0	(*R*)-3-(octadecanoyloxy)octadecanoic acid.

10.1 Introduction

10.1.1 Helicobacter pylori and Campylobacter jejuni Infections

Both *Helicobacter pylori* and *Campylobacter jejuni* are fastidious, microaerophilic, Gram-negative bacteria whose endotoxins play important roles in their pathogenesis. Despite their similar microaerophilic metabolisms, *C. jejuni* causes an acute infection (Blaser and Engberg, 2008), whereas *H. pylori* chronically colonizes the gastric mucosa of the stomach or gastric metaplasia in the duodenum (Ernst and Gold, 2000; Suerbaum and Michetti, 2002; Blaser and Atherton, 2004; Kusters et al., 2006).

H. pylori is a highly prevalent, bacterial gastroduodenal pathogen of humans infecting 50% of the world's population by 50 years of age (Kusters et al., 2006).

Infection once established can persist for life if left untreated and is associated with an active inflammation in the gastric mucosa, termed gastritis (Blaser and Atherton, 2004; Kusters et al., 2006). Although infection outcome is diverse, including the development and recurrence of gastritis as well as gastric and duodenal ulcers, some individuals with long-term infection develop atrophic gastritis, which can have an autoimmune contributory background, and which is a precursor state in gastric cancer development (Ernst and Gold, 2000). *H. pylori* has been proposed as a model for investigating and understanding the dynamics of bacterial persistence and parasitism in chronic infections (Blaser and Kirschner, 1999). Likewise, studies on the attributes of *H. pylori* endotoxin have produced novel insights into the structure and contributing properties of this class of molecule to chronic disease pathogenesis (see reviews, Moran, 2007, 2008).

C. jejuni is well established as the leading cause of bacterial-mediated diarrhoea worldwide (Allos, 2001); in the developed world the bacterium is the leading cause of foodborne gastroenteritis whereas in the developing world this infection is a leading cause of infant mortality due to dehydration-associated diarrhoea. Infection symptoms vary in severity and include nausea, severe or bloody diarrhoea, abdominal cramping and fever (Blaser and Engberg, 2008). *C. jejuni* colonization usually produces a self-limiting, acute infection but in some cases (1:1,000) initiates the autoimmune, polyneuropathic disorder Guillain-Barré syndrome (GBS) (Prendergast and Moran, 2000; Yuki, 2007). Importantly, *C. jejuni* is the commonest antecedent infection in these neuropathies, and expression of molecular mimicry of host gangliosides by GBS-associated isolates is considered a prerequisite for neuropathy development since this mimicry can induce pathogenic, cross-reactive antibodies central to disease development (Prendergast and Moran, 2000; Moran et al., 2002b; Ang et al., 2004; Yuki, 2005). Moreover, the characteristics of *C. jejuni* endoxtoxin contribute significantly to this phenomenon (Prendergast and Moran, 2000; Moran, 2009). Interestingly, gastric infection by *H. pylori* has also been implicated in the induction of autoimmune responses contributing to the development of gastric atrophy, and the subsequent development of gastric cancer, in which the endotoxin of the bacterium may play a contributory role (Moran, 2009). Thus, while colonization by *H. pylori* and *C. jejuni* produce chronic and acute infections, respectively, these infections can in turn induce important autoimmune sequelae.

10.1.2 Nature of H. pylori and C. jejuni Endotoxins

Like *Escherichia coli*, clinical isolates of *H. pylori* produce high-molecular-mass (smooth-form) lipopolysaccharide (LPS) composed of an outermost saccharide moiety, divided into the O-polysaccharide or O-specific chain (O-antigen) and core oligosaccharide (OS) regions, covalently linked to lipid A (Fig. 10.1). *H. pylori* smooth-form LPS possesses O-polysaccharide chains of relatively constant chain length compared with those of enterobacterial LPS (Moran et al., 1992a; Moran, 1999). Their length is determined by an enzymatic molecular ruler mechanism

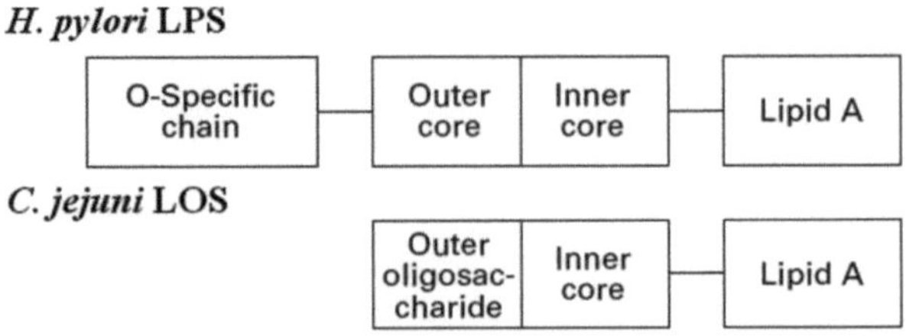

Fig. 10.1 Schematic representations of the general structures of high-molecular-mass LPS of *H. pylori* (*top*) and low-molecular-mass LOS of *C. jejuni* (*bottom*)

(Nilsson et al., 2006). Numerous subcultures on conventional solid media can induce production of low-molecular-mass, rough-form LPS in many strains but which can be reversed when strains are passaged in liquid media (Moran, 1999). The ability of such strains to revert to smooth-form LPS production and the influence of environmental factors, e.g. pH, on LPS expression and antigenic phase variation has been detailed previously (Moran, 2001b; Moran et al., 2002a; Moran and Trent, 2008).

In contrast, *C. jejuni* produces low-molecular-mass lipooligosaccharide (LOS) comprised of core oligosaccharide OS and lipid A components, but devoid of O-chain, thus resembling those of *Haemophilus* and *Neisseria* spp. (Moran, 2009) (Fig. 10.1). The core OSs of *C. jejuni* LOS are structurally diverse (Moran and Penner, 1999; Moran et al., 2000; Gilbert et al., 2008) and, in part, with serodominant capsular polysaccharides account for heat-stable antigen serospecificity (Moran et al., 2001; Karlyshev et al., 2008). This is similar to some other *Campylobacter* spp., e.g. *C. coli* and *C. lari,* where extracellular polysaccharides associated with, but independent of, LPS/LOS have been described (Moran et al., 2000). Notably, mimicry of the saccharide component of gangliosides by the outer core of *C. jejuni* LOS is well documented (Moran et al., 2000; Moran and Prendergast, 2001; Gilbert et al., 2008) and can induce cross-reactive anti-ganglioside antibodies contributing to GBS development (Prendergast and Moran, 2000; Moran et al., 2002b; Ang et al., 2004; Yuki, 2005; Kaida et al., 2009).

It is important to note that the LOS biosynthesis-encoding genes of *C. jejuni* are clustered on the bacterial genome, as occurs in many enterobacterial pathogens e.g. *E. coli and Salmonella*, whereas those of *H. pylori* are generally unclustered on the genome (Berg et al., 1997), except for those associated with GDP-fucose synthesis (Moran and Trent, 2008). Nevertheless, the genetics and characteristics of the enzymes required for biosynthesis of *H. pylori* and *C. jejuni* LPSs have been detailed in a number of reviews (Berg et al., 1997; Moran 2001a, b, 2008; Ma et al., 2006; Gilbert et al., 2008; Moran and Trent, 2008) and hence will not be the focus of this chapter.

The present review instead will discuss how the structure and properties of endotoxin from a chronically infecting bacterium, in contrast to one that causes acute infection, are adapted to fit the environmental niche of the respective microbes using *H. pylori* and *C. jejuni* as model organisms. Furthermore, the pathogenic consequences of these endotoxin structures will be discussed. Although the core OS contributes to *H. pylori* pathogenesis, which has been the subject of other reviews (Moran, 1996, 1999, 2001a,b), discussion will focus here on the lipid A moiety and O-polysacharide chain of this endotoxin, reflecting the intensive investigation of these portions of the *H. pylori* LPS molecule in recent years. Thus, in the first portion of the chapter, the structural and biological properties of the lipid A components of *H. pylori* and *C. jejuni* will be examined, and in the second part, the roles of molecular mimicry within the saccharide moieties of *H. pylori* LPS and *C. jejuni* LOS in pathogenesis, particularly autoimmune sequelae to infection, will be discussed.

10.2 Structure and Properties of *H. pylori* and *C. jejuni* Lipid A Moieties

10.2.1 Structural Analysis of H. pylori Lipid A

Using two different strains, two independent research groups examined the fine structure of lipid A from *H. pylori* rough-form LPS. One study reported the presence of a tetra-acyl lipid A (Moran et al., 1997), whereas another reported a tri-acyl lipid A with an identical phosphorylation pattern to the tetra-acyl form, but lacking acylation at position-3 of the lipid A backbone (Suda et al., 1997). A re-investigation by the latter group of the same strain found the tetra-acyl form of *H. pylori* lipid A (Suda et al., 2001). Potentially, the tri-acyl lipid A may have arisen as a degradation product of the former during chemical isolation of the lipid A (Moran, 2001b). Therefore, it may be concluded that the major molecular species in lipid A of *H. pylori* rough-form LPS is composed of a β-(1→6′)-linked D-glucosamine (GlcN) disaccharide backbone acylated by (*R*)-3-hydroxyoctadecanoic acid [18:0(3-OH)] and (*R*)-3-hydroxyhexadecanoic acid [16:0(3-OH)] at positions -2 and -3 and, (*R*)-3-(octadecanoyloxy)octadecanoic acid [3-(18:0-*O*)-18:0] at the 2′-position, and carries phosphate or phosphoethanolamine (*P*EtN) groups at position-1 (Moran et al., 1997) (Fig. 10.2).

Likewise, this mono-phosphorylated tetra-acyl lipid A predominates in *H. pylori* smooth-form LPS, but there is a second, minor constituent consisting of a bis-phosphorylated hexa-acyl lipid A which is distinguished from tetra-acyl lipid A by carrying (*R*)-3-(dodecanoyloxy)hexadecanoic acid [3-(12:0-*O*)-16:0] or (*R*)-3-(tetradecanoyloxy)hexadecanoic acid [3-(14:0-*O*)-16:0] at position-3′ and an extra phosphate group at position-4′ (Moran et al., 1997) (Fig. 10.2). Whether derived from rough- or smooth-form LPS, compared to the lipid A of *E. coli* the predominant lipid A molecular species of *H. pylori* lacks the usual 4′-phosphate group, as well as the 3′-ester-linked fatty acyl chains, thus containing only four rather than six fatty acids, which have longer chain length (16–18 carbons versus 12–14 carbons), and is derivatized with a *P*EtN residue at C-1 of the proximal glucosamine (Moran et al., 1997; Moran, 1998). Nonetheless, since *H. pylori* can synthesize a bisphosphorylated hexa-acyl lipid A, it can be deduced that this bacterium expresses enzymes

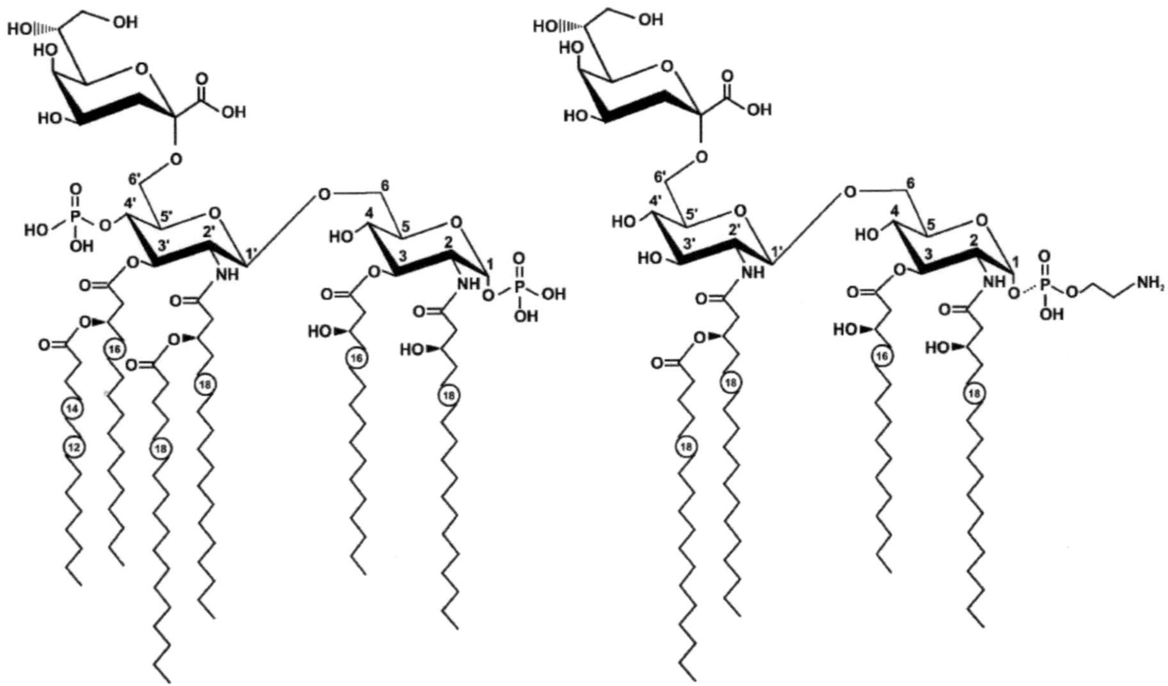

Minor lipid A species **Major lipid A species**

Fig. 10.2 Structures of the minor hexa-acyl lipid A molecular species found in *H. pylori* smooth-form LPS (*left*) and the tetra-acyl lipid A species found in *H. pylori* rough- and predominating in smooth-form LPS (*right*) (Moran et al., 1997). One 3-deoxy-D-*manno*-oct-2-ulosonic acid residue, as occurs in the *H. pylori* core OS, is shown attached to the 6′-position of lipid A. The numbers in circles refer to the number of carbon atoms in the acyl chains. Compared to the hexa-acyl minor species, the tetra-acyl molecular species lacks 4′-phosphate and is substituted at position-1 by phosphoethanolamine

(e.g. a 4′-phoshatase and 3′-acyloxyacyl deacylase) capable of re-modelling, and hence modifying to a tetra-acyl form its lipid A domain after completion of the conserved hexa-acyl lipid A biosynthetic pathway as has been reviewed elsewhere (Tran et al., 2005; Moran and Trent, 2008; Moran, 2010).

10.2.2 Structural Analysis of C. jejuni Lipid A

Although early studies reported the presence of GlcN in *C. jejuni* LPS and lipid A, more detailed investigations showed that both GlcN and 2,3-diamino-2,3-dideoxy-D-glucose (GlcN3N) occurs in LPS and lipid A of a number of *C. jejuni* strains (Moran et al., 1991a, 1992b; Moran, 1995b). Structural analysis of this so-called "mixed lipid A" of *C. jejuni* serostrain HS:2 showed that three β-(1→6′)-linked hexosamine backbones could occur; about 73% of the lipid A molecules have a disaccharide backbone of GlcN–GlcN3N, 15% have a backbone of GlcN3N–GlcN3N and 12% have a GlcN-GlcN disaccharide backbone (Moran et al., 1991b). Thus, the predomainant backbone is mixed (Fig. 10.3). However, all three disaccharide backbones are phosphorylated and acylated in the same manner (Moran et al., 1991b) producing a bisphosphorylated hexa-acyl lipid A. The backbones carry acyl chains of (*R*)-3-hydroxytetradecanoic acid [14:0(3-OH)] at positions-2 and -3, (*R*)-3-(hexadecanoyloxy)tetradecanoic acid [3-(16:0-*O*)-14:0]

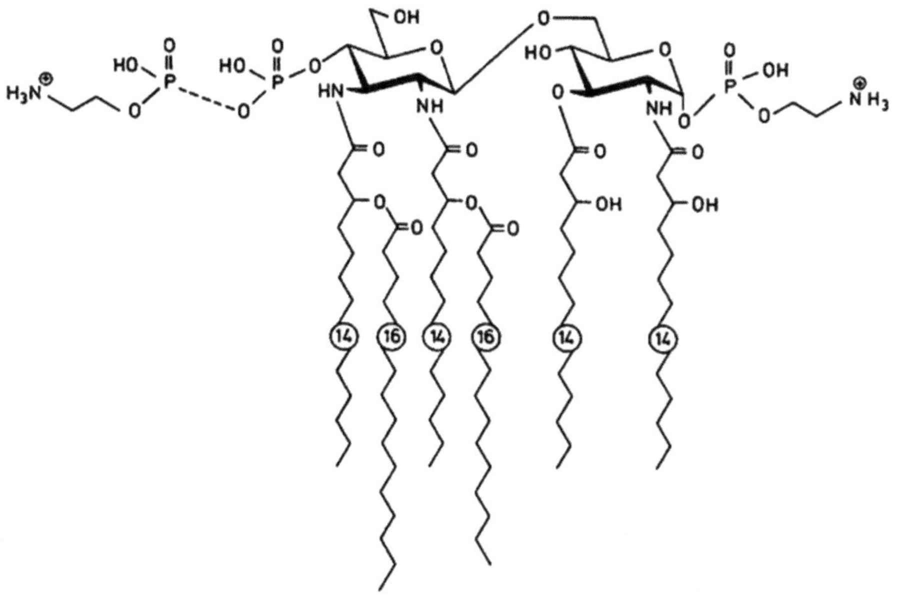

Fig. 10.3 Structure of the predominant molecular species in *C. jejuni* lipid A. The numbers in circles refer to the number of carbon atoms in the acyl chains. Dashed lines indicate partial substitution.The C-1 phosphate can also be substituted non-quantitatively by phosphethanolamine (producing a diphosphoethanolamine group) and the 16:0 on 14:0(3-OH) at C-2′ is partially replaced by 14:0. The 6′-position hydroxyl group is the attachment site for the core OS

or (*R*)-3-(tetradecanoyloxy)tetradecanoic acid [3-(14:0-*O*)-14:0] at position-2′, and 3-(16:0-*O*)-14:0 at position-3′, with a *P*EtN residue at C-1 and a phosphate group or diphosphoethanolamne pyrophosphate at C-4′ (Fig. 10.3). Moreover, analysis of lipid A from different *C. jejuni* serostrains has confirmed the occurrence of the same backbone structures with substitution patterns identical to those encountered in *C. jejuni* serostrain HS:2 (Moran et al., 2000).

Despite the presence of GlcN3N, *C. jejuni* lipid A antigenically resembles enterobacterial lipid A when tested with anti-lipid A antibodies (Moran, 1995b), reflecting that GlcN and GlcN3N are *gluco*-configured sugars and will display similar epitopes. The only structural influence exerted by the occurrence of GlcN3N in the backbone disaccharides is the presence of a higher proportion of amide-bound 14:0(3-OH) fatty acid residues (75% of the total for GlcN3N-GlcN versus 50% for a GlcN-GlcN backbone) (Moran et al., 1991b; Moran, 1997). Biosynthetically, in vitro studies have shown that *E. coli* lipid A synthase can accept both GlcN- and GlcN3N-related intermediates, and hence, the conserved hexa-acyl lipid A biosynthetic pathway is followed in producing mixed *C. jejuni* lipid A (Moran et al., 1991b; Raetz et al., 2007).

10.2.3 Molecular and Supramolecular Basis for the Contrasting Immunological Activities of C. jejuni and H. pylori Lipid A Moieties

A striking feature of *H. pylori* LPS, and which has been reviewed extensively (Moran 1999, 2001a, 2007; Moran and Trent, 2008), is the significantly lower endotoxic and immunological activities of this LPS (up to 10,000-fold lower) compared with enterobacterial LPS as the gold standard. Importantly, early studies that tested chemically modified *H. pylori* LPS-derived components in immunological assays indicated that the molecular basis for these low immuno-activities resided in the lipid A moiety, and is modulated by the saccharide component of *H. pylori* LPS, particularly the core OS (Moran, 1995a, 1996). Agreeing with the proposed hypothesis that the structure of lipid A would endow *H. pylori* LPS with these low bioactivities (Muotiala et al., 1992), the under-phosphorylation, under-acylation and substitution by long chain fatty acids in this lipid A compared with enterobacterial-derived lipid A (Moran, 1998), as descrived above, are consistent with the established structure-bioactivity relationships for lipid A with low activities (Rietschel et al., 1990, 1994). Moreover, natural and synthetic *H. pylori* lipid A preparations, with the reported under-acylation and under-phosphorylation patterns, have been shown to have low endotoxic potency and immunological activities like *H. pylori* LPS (Suda et al., 2001; Ogawa et al., 2003). Importantly, these novel structural attributes not only impact upon the established primary structure-bioactivity relationships but also upon the supramolecular conformation of *H. pylori* lipid A (Schromm et al., 2000) and other biophysical properties of lipid A that have been correlated previously with lower bioactivity (Seydel et al., 2000). These include a higher phase transition temperature and a lower inclination angle of the lipid A diglucosamine backbone to the membrane plane than encountered with enterobacterial lipid A (Moran et al., 2005b) which would influence interaction with immune receptors.

In contrast to *H. pylori*, *C. jejuni* LPS when compared with enterobacterial LPS exhibits slightly lower, but comparable, endotoxic activities in biological test systems (Moran, 1995b; Moran et al., 2000). For instance, *C. jejuni* LPS possesses 50% lower lethal toxicity in mice, 30- to 50-fold lower pyrogenicity, and 100-fold lower ability to induce tumour necrosis factor secretion than does *Salmonella* LPS (Moran, 1995b). Of note, *C. jejuni* LPS and lipid A exhibit higher phase transition temperatures than do standard *Salmonella* preparations, and therefore *C. jejuni* preparations have lower fluidity at 37°C (Moran, 1995b, 1997). Such lower fluidity of acyl chains may influence the biological and immunological activities of *C. jejuni* LPS (Moran et al., 2000). However, the states of order of the acyl chains may play a less important role in the observed biological activities than the supramolecular structure of lipid A (Moran, 1995b). The acyl chain characteristics, including the presence of a higher proportion of a longer chain fatty acid (16:0) than in enterobacterial lipid A, and the replacement of GlcN with GlcN3N may also influence the supramolecular structure of *C. jejuni* lipid A, thereby affecting biological activities (Moran, 1997). Thus, both the significantly lower bioactivities of *H. pylori* LPS, but comparable activities of *C. jejuni* LPS, to those of enterobacterial LPS can be related to the supramolecular nature of their lipid A components.

10.3 Relevance of Low Endotoxin and Lipid A Immuno-Activities to Chronic Infection: *H. pylori* as a Comparative Model

As seen with other chronic bacterial infections or colonizing commensal bacteria, the induction of low immunological responsiveness, particularly by endotoxin, may aid the prolongation of *H. pylori* infection and therefore infection chronicity (Moran, 2007). This contrasts with acute infection-causing *C. jejuni* where overt inflammation contributes to pathology and diarrhoea production, and whose endotoxin is immunologically and endotoxically active (Moran, 1995b). It has been hypothesized that *H. pylori* LPS, and its lipid A component in particular, have evolved their present structure, on the one hand, to fulfill their role in producing a functional macromolecular matrix for bacterial interaction with its environment, whereas on the other, to reduce the immune response to these essential molecules of the *H. pylori* outer membrane (Muotiala et al., 1992; Moran, 1995a, 1996). Hence, in producing a functional macromolecular matrix relevant to the microbial niche of *H. pylori,* the lipid A component which is embedded in the outer membrane, would retain essential features to maintain structural integrity of the outer membrane but undergo modification to allow long-term persistence within the specific microbial niche.

An example supporting this view, concerns resistance to cationic antimicrobial peptides (CAMPs) produced by the innate immune system. Of note, expression of human cathelicidin LL-37 (Hase et al., 2003) and human β-defensin 2 (Wada et al., 1999; Hamanaka et al., 2001) is upregulated in the gastric mucosa of *H. pylori*-infected patients. Additionally, *H. pylori* produces a ceropin-like peptide, Hp(2-20), to which it is resistant, that is derived from the amino-terminal part of its ribososmal protein L1 (Putsep et al., 1999). Thus, in order to persist in the human gastric mucosa *H. pylori* must be able to resist the action of CAMPs. In some other Gram-negative bacteria, ceropin, as an example of CAMPs, has been shown to bind to bisphosphorylated lipid A (De Lucca et al., 1995), but by masking the negative charge of their phosphate groups in lipid A with positively charged amine-containing substituents (e.g. ethanolamine and 4-amino-4-deoxy-L-arabionse) these bacteria can promote CAMP resistance. Since lipid A phosphate groups are necessary for binding of cationic peptides (Trent et al., 2006), the predominant lipid A molecular species, lacking 4′-phosphate and with *P*EtN substitution at C-1 of the *H. pylori* lipid A backbone, should promote resistance to CAMPs during host infection. Consistent with this deduction, wild-type strains of *H. pylori* are inherently resistant to polymyxin, a CAMP, but disruption of *lpxE* in *H. pylori* can result in the production of a lipid A bearing a single phosphate group at the 1-position resulting in a 25-fold increase in polymyxin sensitivity (Tran et al., 2006).

Moreover, as has been reviewed previously (Moran 2001a, 2007; Moran and Trent, 2008), the interactions of *H. pylori* LPS *via* its lipid A component with a variety of immune recognition molecules and receptors, e.g. LPS-binding protein, CD14 and Toll-like receptors (TLRs), consistently have been reported to be low. Whether *H. pylori* LPS and lipid A act as TLR4 agonists, as classically seen with the majority of LPSs of other Gram-negative bacteria, or bind TLR2 has been much debated (Smith et al., 2003, 2006; Mandell et al., 2004; Moran, 2007; Triantafilou et al., 2007; Uno et al., 2007; Moran and Trent, 2008). Nevertheless, upon initial infection of the TLR4-expressing but TLR2-deficient gastric mucosa (Ortega-Cava et al., 2003; Mandell et al., 2004), *H. pylori* is weakly or unrecognized by TLR4 and may escape detection and elimination by the immune response initially. Consistent with this concept of *H. pylori* evasion of immune detection is the very low activation by *H. pylori* flagellins FlaA and FlaB of TLR5-mediated responses (Lee et al., 2003), since functional TLR5 is expressed in the adult stomach (Schmausser et al., 2004), and TLR5 is considered to bind and respond to bacterial flagellins.

With progression of the immune response in long-term infection, a substantial inflammatory cytokine response to *H. pylori* may develop only after the infiltration of TLR2-expressing granulocytes and monocytes into the infected gastric mucosa (Moran, 2010). *H. pylori* can activate mononuclear cells by LPS-independent, as well as by TLR4-independent mechanisms, and hence non-LPS component(s) of the bacterium are the major inflammation-activating molecule(s). In general, since bacterial lipoproteins, lipopeptides and lipoteichoic acids have been considered TLR2 ligands, the numerous putative lipoproteins encoded in the *H. pylori* genome represent candidate TLR2 ligands (Moran, 2007). Additionally, although an *H. pylori* heat shock protein (Hsp60) has been implicated in activation of TLR2

(Takenaka et al., 2004), this protein has also been reported to induce interleukin-6 production by macrophages *via* a TLR2-independent mechanism (Gobert et al., 2004). Recognition of *H. pylori* peptidoglycan-derived muropeptides by the intracellular pattern recognition molecule Nod1 can occur (Viala et al., 2004) and contribute to the development of the subsequent inflammatory response.

10.4 Molecular Mimicry in *H. pylori* Endotoxin

10.4.1 H. pylori Expression of Lewis (Le) and Blood Group Antigen Mimicry

Structurally, the O-polysaccharide chains of *H. pylori* clinical isolates have a poly-*N*-acetyl-lactosamine (-LacNAc) chain decorated with multiple lateral α-L-fucose residues forming internal Le^x determinants with terminal Le^x or Le^y units (Fig. 10.4) or, in some strains with additional, D-glucose or D-galactose residues (Moran 2001a,b, 2008; Monteiro, 2001). Moreover, Le^a, Le^b, sialyl-Le^x, and H-1 antigens have been structurally described in other strains, as well as the related blood groups A and B (Fig. 10.4), but occur in association with Le^x and LacNAc chains (Monteiro et al., 2000a,b; Heneghan et al., 2000).

Expression of Le^x or Le^y antigens is a common property of *H. pylori* strains, as 80–90% of isolates from various geographical regions worldwide, that have been probed using anti-Le antibodies, express these antigens (Moran, 2008). However, as some strains that were considered non-typeable with anti-Le antibodies have been shown to express these antigens when examined in structural studies (Knirel et al., 1999), under-estimation of Le^x and Le^y expression in a population of strains can occur depending on the assay format employed (Hynes and Moran, 2000). Overall, a mosaicism of Le antigen and blood group expression can occur in the same O-chain, and thereby, along with some variability in the core OS of LPS, give rise to antigenic diversity that can be detected by antibody and lectin probing, and hence serve as a basis for strain typing (Simoons-Smit et al., 1996; Hynes et al., 1999, 2002).

Collectively, factors affecting Le antigen expression in *H. pylori*, and that can influence the biological impact of this molecular mimicry, include regulation of fucosyltransferase (FucT) genes through slipped-strand mispairing, the activity and expression levels of the functional enzymes, the preferences of the expressed enzyme for distinctive acceptor molecules, and the availability of activated sugar intermediates (see reviews, Moran, 2001b; Ma et al., 2006; Moran, 2008; Moran and Trent, 2008). With the accessibility of the crystal structure of *H. pylori* FucTs (Sun et al., 2007) further insights should be gained into the molecular recognition and functioning of these enzymes.

Significantly, *H. pylori* strains expressing predominatly Le^x and those expressing predominantly Le^y can be isolated from the same host, and extensive diversity occurs in expression of Le^x and Le^y in the O-chains of *H. pylori* strains isolated from the same human subject at different times and from different regions of the stomach (Nilsson et al., 2006). This diversity of Le expression by *H. pylori*

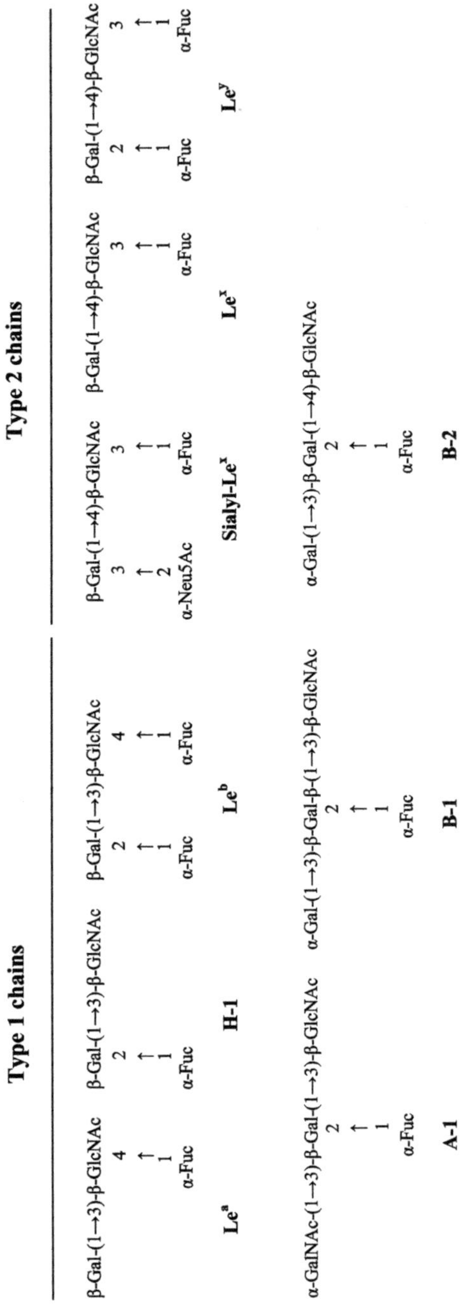

Fig. 10.4 Structures of the Lewis and blood group antigens expressed in *H. pylori* strains. See Moran (2008) for details. Abbreviations: Fuc, fucose; Gal, galactose; GalNAc, *N*-acetyl-galactosamine; GlcNAc, *N*-acetyl-glucosamine; Neu5Ac, *N*-acetyl-neuraminic (sialic) acid

subclonal isolates in one host (Nilsson et al., 2006) may reflect an ability, and potential, of the bacterium to adapt to differing micro-niches and environmental conditions within the human stomach (Moran et al., 2002a; Keenan et al., 2008). For example, pH may vary in the differing localities of the stomach, as well as varying across the microbial niches of the mucosa (i.e. pH 1-3 on the luminal surface of gastric mucus, about pH 4-5 within the mucus, and pH 7 on the epithelial cell surface) and local pH has been shown to influence relative Le^x/Le^y expression by *H. pylori* (Moran et al., 2002a; Skoglund et al., 2009). Likewise, dietary iron can vary, as can accessible iron for *H. pylori* acquisition, which in turn affects pathogenesis of *H. pylori* (Keenan et al., 2004), and can influence the quantitative expression of LPS and the nature of Le antigen mimicry occurring in *H. pylori* O-polysaccharide chains (Keenan et al., 2008). Moreover, Nilsson et al. (2008) detected genetic modifications in *H. pylori* FucT genes that could be attributable to recombination events within and between these genes that creates diversity, and which together with phase variation (Appelmelk et al., 1999), contributes to divergent LPS expression of Le antigens within a community of *H. pylori* colonizing strains.

It should not be overlooked that there are *H. pylori* strains, isolated from clinically asymptomatic individuals, that do not express Le antigens but express other glycans as their O-chains, e.g. heptans (Senchenkova et al., 2001) and polysaccharides with deoxy and branched sugars (Kocharova et al., 2000). The absence of Le antigen mimicry in these strains, and animal studies in which a genetically modified *H. pylori* strain lacking Le antigen expression failed to induce gastritis compared to the parental strain (Eaton et al., 2004), support a role for Le antigen-expressing LPS contributing to disease development.

The biological and pathogenic roles of Le antigens expressed in the O-chains of *H. pylori* LPS have been the subject of other reviews (Moran 2008, 2010; Moran and Trent 2008) and include aiding gastric adaptation and adhesion, thereby facilitating bacterial colonization, and influencing the innate and inflammatory response. Since these roles have been discussed extensively elsewhere they will not be enlarged upon here. On the other hand, despite the comparatively lower endotoxic and immunological activities of *H. pylori* endotoxin attributable to its lipid A component, and although remaining controversial, *H. pylori* has been implicated in the induction of autoreactive immune responses, including humoral responses against LPS-expressed Le antigens (Moran, 2009) that may contribute to pathology and which are worthy of further examination.

10.4.2 Anti-Le Antibodies in Autoimmune Pathogenesis

10.4.2.1 Anti-Le Antibodies and the Inflammatory Response

Anti-Le^x and anti-Le^y antibodies have been detected in the sera of *H. pylori*-infected patients, particularly those with prolonged infection, using *H. pylori* LPS (Appelmelk et al., 1996; Heneghan et al., 2001; Hynes et al., 2005). Importantly, the use of synthetic glycoconjugates in solution or on a solid phase does not optimally detect *H. pylori*-induced anti-Le antibodies (Heneghan et al., 2001), thus explaining the inability to detect these antibodies in some patient studies (Amano et al., 1997). The most appropriate assay system, allowing carbohydrate conformations to be maintained, is an inhibition enzyme-linked immunosorbent assay, in which Le-expressing LPS is immobilized and human sera are used to inhibit the reaction of anti-Le monoclonal antibodies (Heneghan et al., 2001).

Although it has become apparent that bacterial colonization density and the ensuing inflammatory response can be influenced by host expression of ABO and Le^a blood group determinants (Heneghan et al., 1998), bacterial Le^x expression is associated with peptic ulcer disease (Marshall et al., 1999), and is statistically related to neutrophil infiltration (Heneghan et al., 2000). Of relevance, neutrophils are a potential target recognized by anti-Le^x antibodies (Appelmelk et al., 1996). Neutrophils express CD15 (i.e. Le^x) on members of the adhesion-promoting glycoprotein family (CD11/CD18), and cross-linking of CD15 by *H. pylori*-induced anti-Le^x autoantibodies could potentiate polymorph adhesiveness to the endothelium (Appelmelk et al., 1996; Moran et al., 1996). Consistent with this, anti-Le^x monoclonal antibodies, including those induced by *H. pylori*, can activate and cause enhanced adherence of these cells, which may result in tissue damage and inflammation (Stöckl et al., 1993; Appelmelk et al., 1996; Moran et al., 1996). Nevertheless, further unequivocal evidence is required to fully support this pathogenic mechanism.

10.4.2.2 Anti-Le Antibodies in Gastric Atrophy

During chronic infection by *H. pylori*, induction of autoreactive antibodies (including anti-Le^y antibodies) against the gastric mucosa, particularly against the gastric proton pump, has been implicated in the pathogenesis of atrophic gastritis which is a precursor pathological state before the development of gastric cancer (Negrini et al., 1991, 1996; Appelmelk et al., 1996; Faller et al., 1997; Moran, 1999, 2001a). Anti-Le^y autoantibodies have been shown to occur in patients with atrophic gastritis and gastric cancer (Negrini et al., 1996; Heneghan et al., 2001; Hynes et al., 2005). Centrally, the β-chain of the gastric proton pump, H^+,K^+-adenosine triphosphatase (H^+,K^+-ATPase), occurring within parietal cell canaliculi, is glycosylated predominantly by Le^y, and anti-Le^y antibodies have been implicated in the pathogenic autoimmune responses in atrophic gastritis (Appelmelk et al., 1996; Moran et al., 1996; Moran, 1999). In particular, anti-Le^y *H. pylori*-induced antibodies have been shown to react with purified human and murine β-chain (Appelmelk et al., 1996), and in an adoptive transfer experiment, growth in mice of an *H. pylori*-induced anti-Le^y-secreting hybridoma resulted in gastric histopathological changes consistent with those of gastritis (Negrini et al., 1991). Also, anti-Le and anti-parietal cell-related antibodies have been induced in a transgenic mouse model of *H. pylori* infection in which gastric pathology developed (Negrini et al., 1991). Based on these data it has been inferred that Le^y mimicry plays a role in the induction of *H. pylori*-related atrophic gastritis (Moran et al., 1996; Moran, 1999, 2001a,b), but since these experiments were performed largely in mice, questions were raised as to whether a similar phenomenon occurs in humans (Appelmelk et al., 1998).

Despite this, a correlation has been established between the presence of autoantibodies in *H. pylori*-infected human subjects (Faller et al., 1997) with the degree of gastric infiltration, numbers of inflammatory cells, and glandular atrophy (Negrini et al., 1996). In the case of patients suffering from severe gastric atrophy, *H. pylori* isolates were found more likely to express Le^x/Le^y antigens, whereas isolates from individuals with near-normal mucosa were less likely to express these antigens and to induce autoantibodies in experimental animals (Negrini et al., 1996). Additionally, the anti-canalicular autoantibodies have been shown to increase significantly in patients with the duration of *H. pylori* gastritis and to correlate with gastric corpus atrophy (Vorobjova et al., 2000).

Based on the classical model of organ-specific autoimmunity in which there is a central role for increased autoanitgen (i.e. Le) presentation in the gastric mucosa because of *H. pylori*-induced damage (Appelmelk et al., 1998; Faller et al., 1998), it has been suggested that these pathogenic antibodies are the consequence, rather than the causative factor in atrophic gastritis. Such a view has been largely based on two types of experiments; the first attempting to abolish the reactivity of autoantibodies from human sera using *H. pylori* preparations, the second examining the reaction of human sera with recombinant H^+,K^+-ATPase. Confusingly, one study using *H. pylori* lysates in absorption experiments reported a decrease in anti-gastric autoreactivity (Negrini et al., 1991), whereas in another no significant reduction was observed (Faller et al., 1998), and anti-H^+,K^+-ATPase serum autoantibodies were not absorbed with *H. pylori* whole cells in another study (Ma et al., 1994). Critical examination showed that many of these experiments did not use matched isolates with the sera of the patient from whom they were isolated (Moran, 2008). Furthermore, although patient sera were shown to react with recombinant H^+,K^+-ATPase expressed in *Xenopus* oocytes, which was deduced to indicate that the reactivity observed was based on protein epitopes rather than Le glycosylation (Claeys et al., 1998), nonetheless, the nature or lack of any glycosylation occurring in the oocytes was not confirmed.

Importantly, it has been established that antigen presentation and the serological assay format that is used plays an important role in successful detection of Le antigen expression displayed in *H. pylori* O-chains (Appelmelk et al., 1996; Hynes and Moran, 2000; Heneghan et al., 2001; Moran et al., 2004). For these reasons, it has been speculated that the form of antigenic presentation of Le^y could potentially influence the ability to absorb *H. pylori*-derived anti-Le antibodies from patient sera (Moran, 1999; 2001b) and thus confound conclusions of the role of such antibodies in the development of atrophic gastritis. Bearing in mind that the outer membrane of *H. pylori* undergoes blebbing, producing outer membrane vesicles which contain LPS with associated Le^x/Le^y expression (Hynes et al., 2005; Keenan et al., 2008), presentation of Le antigens by these vesicles may play a central role in the induction of the autoreactive antibodies. In absorption experiments, using these *H. pylori*-derived vesicles and sera from gastric cancer patients, it has been possible to remove 83–100% of the anti-Le^y antibodies from the sera and cause significant reduction, though not complete ablation, of all antibodies reactive with the canaliculi of parietal cells (Hynes et al., 2005; Moran, 2009). This is consistent with carbohydrate-based but also protein-related epitopes of the H^+,K^+-ATPase being involved in the autoreactivity observed (Faller et al., 1998).

Using the available data, an interpretative model has been proposed (Moran, 2008), whereby it can be deduced that the mechanisms underlying anti-gastric autoantibody production likely include, but are not exclusive to, molecular mimicry between *H. pylori*-expressed Le^y and the gastric mucosa, as well as facilitated exposition and presentation of the bacterial epitopes to recruited immune cells within the gastric mucosa. This, in turn, will be influenced by host immune-regulation and environmental factors, and hence, complex interactions can occur. A cross-reactive antibody, such as anti-Le^y, can initiate damage to the proton pump (Appelmelk et al., 1996), with subsequent alteration in acid output. Changes in acid output would regulate LPS glycosylation (Moran et al., 2002a), thereby influencing the LPS structure of the infecting *H. pylori* strain such that phase variation occurs in the expression of Le epitopes. Thus, a secondary reduction in autoreactive antibody production could occur, thereby confounding interpretation of the role of *H. pylori*-induced anti-Le antibodies. Furthermore, the initiation of the inflammatory cascade against *H. pylori* and the activities of the virulence factors produced by the bacterium, and even anti-Le^x antibodies causing complement-mediated lysis of host cells (Appelmelk et al., 1996; Moran et al., 1996), would cause histological damage in the gastric mucosa. Subsequently, host structural epitopes would become exposed and presented to the immune system to further drive the autoreactive response, through humoral and T-cell responses to peptides (Faller et al., 1998; Appelmelk et al., 1998; D'Elios et al., 2004). Moreover, molecular mimicry by *H. pylori* peptides also may contribute to the gastric autoimmunity observed (Amedei et al., 2003). Therefore, it is unlikely that the presence of anti-Le^x/Le^y antibodies would be solely responsible for both the initiation and maintenance of the anti-gastric autoimmune response in *H. pylori*-positive patients.

10.5 Molecular Mimicry in *C. jejuni* Endotoxin

10.5.1 Expression of Gangliosiode Mimicry by C. jejuni

The discovery of sialic acid (*N*-acetyl-neuraminic acid, Neu5Ac) in the LOSs of certain *C. jejuni* serostrains (Moran et al., 1991a) induced intensive investigations of the LOS core OSs. As reviewed previously (Moran et al., 2000; Prendergast and Moran, 2000; Moran and Prendergast, 2001; Yuki, 2001; Moran et al., 2002b; Gilbert et al., 2008), these structural studies have revealed that the core OSs of certain strains mimic the carbohydrate moieties of human gangliosides. Mimicry of a variety of gangliosides has been documented, including GM_1, GD_{1a}, GD_{1b}, GD_{1c}, G_{T1a} and GM_2, GD_2, GM_2, GD_3, examples of which are shown in Fig. 10.5. The genetic determination of *C. jejuni* LOS and ganglioside expression is well established and has been reviewed in detail elsewhere (Yuki, 2007; Gilbert et al., 2008). Importantly, *C. jejuni* LOS can undergo phase variation and this impacts on the expression of ganglioside mimicry by the bacterium, both in vitro and in vivo (Guerry et al., 2002;

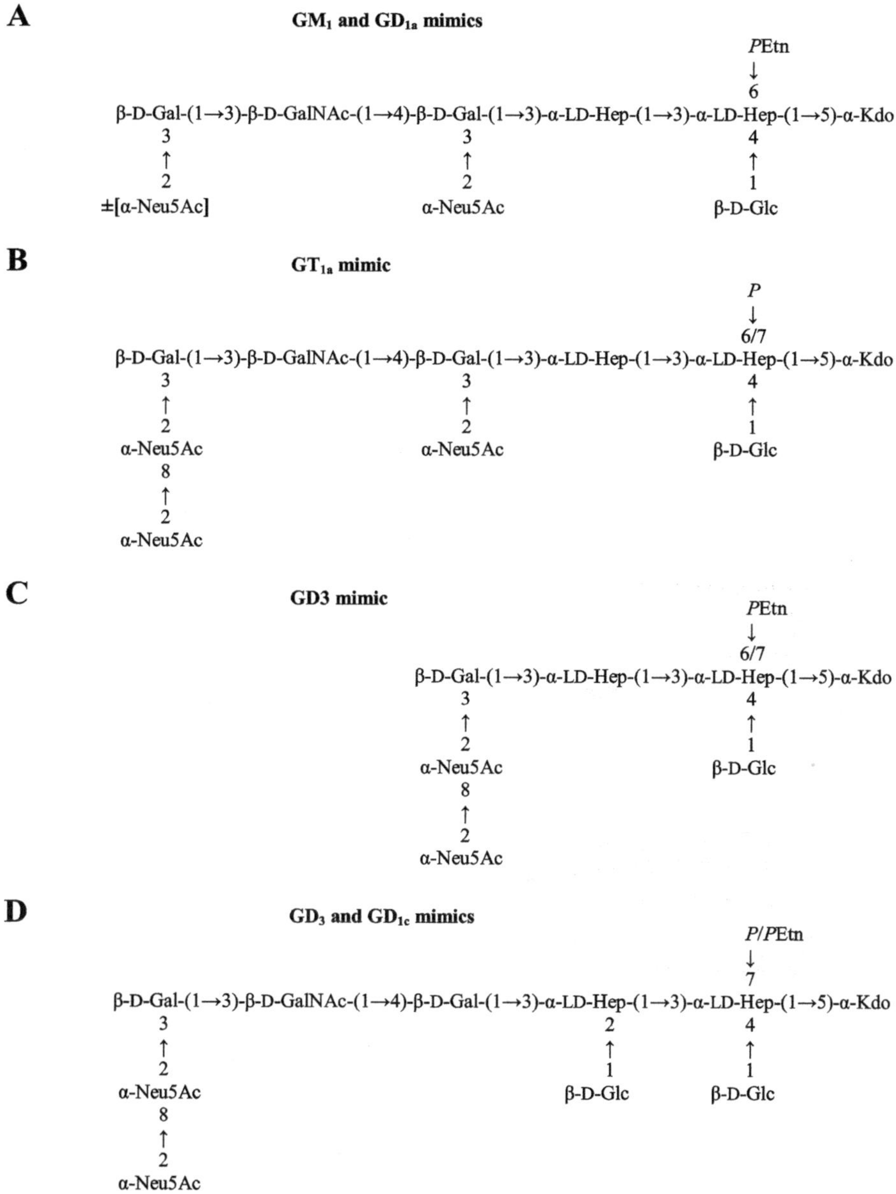

Fig. 10.5 (continued)

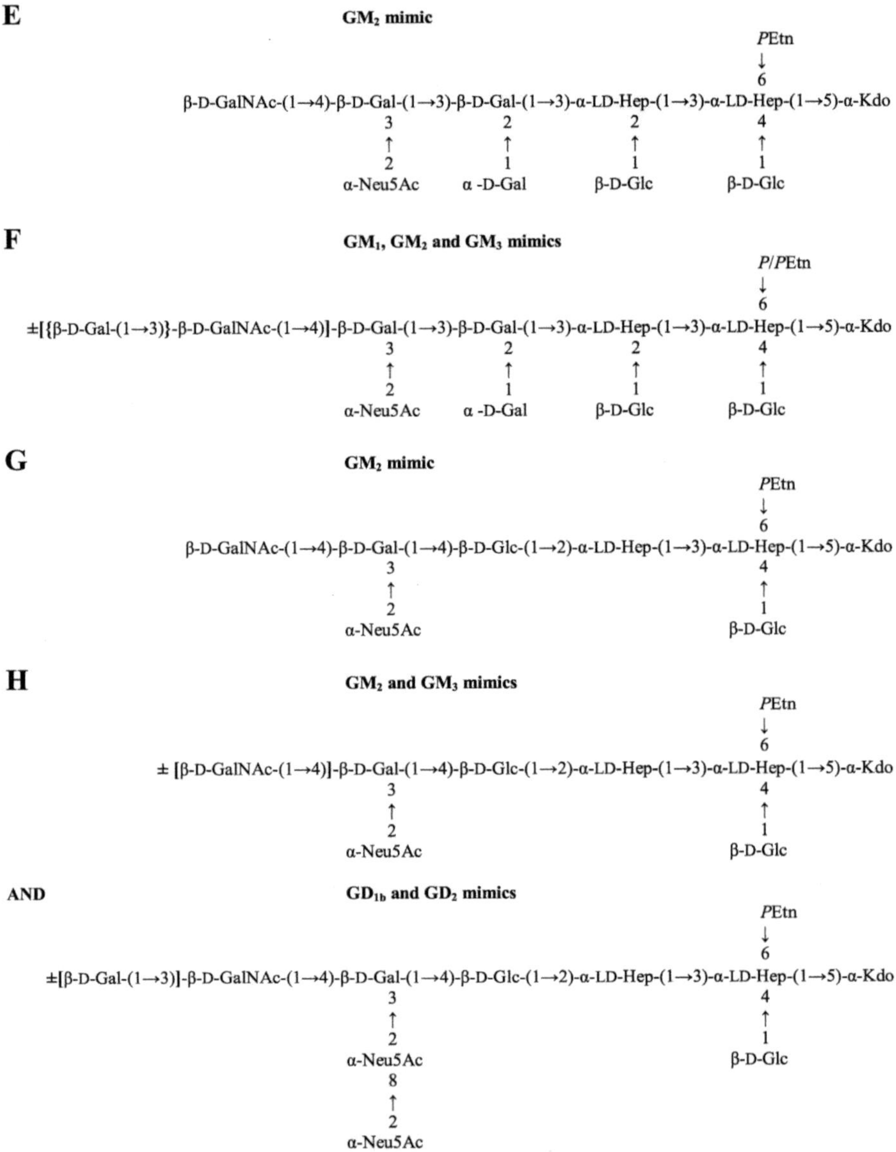

Fig. 10.5 Structures of the core OSs of *C. jejuni* LOSs from different serostrains and isolates. (**A**) Serostrains HS:4 and HS:19, GM_1 and GD_{1a} mimics; (**B**) isolate OH4384, GT_{1a} mimic; (**C**) isolate 4382, a GD3 mimic; (**D**) serostrain HS:10 and isolate PG836, GD_3 and GD_{1c} mimics; (**E**) serostrain HS:1, a GM_2 mimic; (**F**) serostrain HS:2 without the terminal disaccharide β-D-Gal-(1→3)-β-D-GalNAc, a GM_3 mimic, and NCTC 11637 (also serotype HS:2), GM_1 and GM_2 mimics; (**G**) serostrains HS:23 and HS:36, a GM_2 mimic; and (**H**) phase-variable strain 81-176, GM_2 and GM_3 mimics predominantly, but also GD_{1b} and GD_2 mimics. All sugars are in the pyranosidic form. Abbreviations: Gal, galactose; GalNAc, *N*-acetyl-galactosamine; Glc, glucose; LD-Hep, L-*glycero*-D-*manno*-heptose; Kdo, 3-deoxy-D-*manno*-oct-2-ulosonic acid; Neu5Ac, *N*-acetyl-neuraminic acid; *P*, phosphate; *P*Etn, phosphoethanolamine. For the literature from which these structures are dervived, see the reviews of Moran et al. (2000), Prendergast and Moran (2000), and Gilbert et al. (2008)

Prendergast et al., 2004). In turn, since sialyltion, and hence ganglioside expression can influence immunogenicity and serum resistance of *C. jejuni* (Guerry et al., 2000) and invasiveness (Guerry et al., 2002; Kanipes et al., 2004; Habib et al., 2009), phase variation has important implications for bacterial pathogenesis and post-infection sequelae (Moran, 2009).

10.5.2 Pathogenic Anti-Ganglioside Antibodies and C. jejuni Ganglioside Mimicry

10.5.2.1 Pathogenic Anti-Ganglioside Antibodies in GBS

Previous to the observation of ganglioside mimcry in *C. jejuni* LOS, infection with this bacterium had been associated with the development of the neurological disorder of the peripheral nervous system, GBS (see Prendergast and Moran, 2000; Yuki, 2007). This was noteworthy since gangliosides are a family of Neu5Ac-containing glycosphingolipids present in the outer leaflet of the plasma membrane and are the major surface molecules in both the peripheral and central nervous systems. Furthermore, reviewing reports over many decades it is apparent that anti-ganglioside antibodies with a variety of specificities occur in GBS patients and these are considered to play a central role in the pathogenesis of this syndrome (Prendergast and Moran, 2000; Moran and Prendergast, 2001; Yuki, 2007; Kaida et al., 2009). This view is supported by the observation that plasma exchange facilitates the rate of recovery from GBS (Guillain-Barré syndrome Study Group, 1985).

Clinically, GBS is considered an acute, progressive and symmetrical motor weakness of the extremities with loss of tendon reflexes, which has replaced polio as the commonest cause of neuromuscular paralysis in most developed countries. Although at least four distinct subtypes of the disorder are recognized (Hughes and Cornblath, 2005), of these, a severe, purely motor form known as acute motor axonal neuropathy (AMAN) and Miller Fisher syndrome (MFS) manifesting ophthalmpplegia, ataxia, and areflexia have received most attention in *C. jejuni*-associated disease. In previous literature surveys (Prendergast and Moran, 2000; Moran and Prendergast, 2001; Moran et al., 2002b; Yuki, 2007), reports have included multiple findings of anti-ganglioside antibodies to GM_1, GM_{1b}, GQ_{1b}, LM_1, GalNAc-GD_{1a}, and less frequent finding of antibodies to GM_2, GD_{1a}, GD_{1b}, GT_{1b} and asialo-GM_1. This wide range of anti-ganglioside antibody specificities reported in GBS contrasts with their limited range in chronic neuropathies, e.g. anti-GM_1 antibodies in multifocal motor neuropathy and a neurophathy induced by parenteral GM_1 therapy (Moran et al., 2002b). Nevertheless, a correlation between antecedent infection with *C. jejuni* and the presence of anti-GM_1 antibodies has been demonstrated [see Moran and Prendergast (2001) and Yuki (2007) and references therein]; generally, half of GBS patients who are sero-positive for *C. jejuni* also have anti-GM_1 antibodies. Also, AMAN is more strongly associated with *C. jejuni* infection than with other forms of GBS, and has been correlated with anti-GM_1, anti-GD_{1a} and anti-GalNAc-GD_{1a} antibodies (Prendergast and Moran, 2000; Yuki, 2007). Anti-GQ_{1b} antibodies are found in a high proportion of MFS patients, in up to 100% of cases (Willison and O'Hanlon, 2000), and importantly for pathogensis, GQ_{1b} is enriched in the cranial nerves that innervate the extraocular muscles (Yuki, 2001).

Anti-ganglioside antibodies are often present in disease-free individuals (Latov, 1990), but are not considered to be autopathogenic, since they differ from those found in GBS patients in their immunoglobulin (Ig) G subtype, and they are not derived from antigen-driven mechanisms (Paterson et al., 1995). Moreover, in contrast to the immunoglobulin IgM isotype of anti-GM_1 antibodies found in chronic neuropathies, GBS sera contain IgG or less commonly IgA isotypes (Moran and Prendergast, 2001). The IgG subclasses are restricted to IgG1 and IgG3, subclasses which had been taken to be indicative of T-cell-dependent reponses to proteins (Willison and Veitch, 1994) but which, nonetheless can be induced by immunized *C. jejuni* LOSs (Moran et al., 2001).

10.5.2.2 T-Cells in GBS Development

The long-lasting titres of IgG and IgA antibodies present in GBS (Moran and Prendergast, 2001) suggests that T-cell-dependent mechanisms are involved, and that the mechanisms of glycolipid recognition in GBS do not follow the normal rules of T-independent carbohydrate antigen recognition. Populations of T-cells bearing $\alpha\beta$ receptors are found in GBS, but lymphocytes associated with recognition of glycolipid targets, $\gamma\delta$ T-cells, are also found (Khalili-Shirazi et al., 1999). Since CD1 molecules are present on endoneurial macrophages and are upregulated in GBS, presentation of ganglioside-related molecules to T-cells could occur in association with CD1 molecules, as has been demonstrated for lipoglycans (Sieling et al., 1995; Moran and Prendergast, 1998). Thus, $\gamma\delta$ T-cells and CD1 molecules could play a role in the prominent anti-glycolipid antibody response in GBS patients. Interestingly, *C. jejuni* LOSs have been shown to increase endogenous glycosphingolipid synthesis and stimulate autoreactive, ganglioside-specific T-cells in a CD1-dependent manner (De Libero et al., 2005). Despite these observations, the extent to which T-cells contribute to the pathophysiology of GBS and the mechanism by which they do so is unexplained. They could be involved in helping B-cells to secrete autoantibodies, but generation of anti-ganglioside antibodies has been reported to be CD1-independent in mice (Matsumoto et al., 2008), although it remains possible that T-cell-derived, cytokine-dependent pathways for B-cell activation and differentiation, rather than cognate B-cell–T-cell interactions, may play a substantial role in the production of anti-ganglioside antibodies.

10.5.2.3 *C. jejuni* Ganglioside Mimicry and Anti-Ganglioside Antibodies

The observation that GBS-associated *C. jejuni* isolates exhibit ganglioside mimicry prompted a number of studies that showed the reactivity of serum antibodies in GBS patients, against not only *C. jejuni* LOS (Yuki et al., 1993; Schwerer et al., 1995; Neisser et al., 1997, 2000), but particularly against the core OS mimicking gangliosides (Prendergast et al., 1999). Cross-reactive anti-GM_1 and -GD_{1a} antibodies were observed and the sialosyl-D-galactose (Gal) structure α-Neu5Ac-(2→3)-β-Gal-(1→ proposed as the minimal motif associated with GBS development (Moran et al., 2002b). On the other hand, MFS is of great interest because of its specific correlation with anti-GQ_{1b} antibodies, although to date none of the chemically analysed LOSs from neuropathy-associated strains exhibit a complete GQ_{1b} ganglioside mimic. Strains that have been isolated from MFS patients exhibit a GD_2/GD_3 mimic or GT_{1a} and GD_{1c}-like mimicry (Moran et al., 2000; Prendergast and Moran, 2000; Yuki, 2005) and, whose LOSs show cross-reaction with MFS sera. Since none of the structurally analysed LOSs from neuropathy-associated strains have been reported to exhibit a complete GQ_{1b} ganglioside mimic, it has been suggested that the minimal required motif for MFS development is a disialosyl-Gal structure, α-Neu5Ac-(2→8)-α-Neu5Ac-(2→3)-β-Gal-(1→ (Moran et al., 2002b; Yuki, 2005).

Collectively, analyses of *C. jejuni* LPSs with anti-ganglioside antibodies has verified the presence of ganglioside-related epitopes in *C. jejuni* LPS (Schwerer et al., 1995; Neisser et al., 1997, 2000), and thus, it has been inferred that infection with ganglioside-bearing *C. jejuni* strains may elicit high titres of anti-ganglioside autoantibodies in GBS patients. Although it is possible that anti-ganglioside antibodies may be formed secondarily by auto-immunization against gangliosides released from previously damaged nerves, such a process for the generation of anti-ganglioside antibodies is unlikely in GBS, since the antibodies are present on the first day of neurological symptoms and high titres of theses antibodies are not found in patients with peripheral nerve damage due to diseases other than GBS (Prendergast and Moran, 2000). Hence, the molecular mimicry of gangliosides by *C. jejuni* LOSs has been proposed as a pathogenic mechanism in GBS whereby the immune-mediated response, particularly the humoral reponse, elicits the production of cross-reactive anti-ganglioside antibodies that target and damage the neural gangliosides.

10.5.3 Relevance of Molecular Mimicry in the Pathogenesis of GBS

10.5.3.1 Galway Postulates

Despite numerous studies describing an association of molecular mimicry by a range of viral, parasitic and bacterial pathogens with the development of a variety of autoimmune diseases (Moran, 2009), Marrack et al. (2001) concluded that in many studies, although consistent with the molecular mimicry hypothesis in autoimmune disease development, none have convincingly demonstrated that mimicry is an important mechanism in disease in humans. However, others have contended that evidence concerning *C. jejuni* ganglioside mimicry fulfils such requirements (Ang et al., 2004; Yuki, 2005). Nevertheless, in the case of infection-induced automimmunity the precepts that would need to be fulfilled include not only those of Koch's postulates for infectious disease but also those of Witebsky for autoimmunity, and thus, a combination of these has been proposed in the so-called Galway postulates

(Moran et al., 2002b). Therefore, criteria for determination of the role of molecular mimicry in infection-induced autoimmune disease are: (i) evidence for antecedent infection in the host is to be established; (ii) cross-recativity of the pathogen with a host target molecule has to be demonstrated; (iii) experimental disease reproduction in an animal by administration of the tissue target agent, a molecular mimic, or anti-target antibodies is required; and (iv) disease reproduction in an experimental host by the pathogen must be demonstrated. Despite the first three precepts having been fulfilled, the last has proven more problematic for *C. jejuni*-related GBS. This is because an animal model allowing colonization by the infectious agent and additionally the generation of a humoral response like that seen in humans with GBS, and hence inducing a relevant pathology in animals, is required.

10.5.3.2 Experimental Models

Experimental inoculation of humans with *C. jejuni* in a vaccine trial has been shown to induce anti-ganglioside antibodies, but no neuropathy was observed to develop, potentially due to the absence of a persistent anti-ganglioside antibody response, thereby reflecting the role of host factors in GBS development (Prendergast et al., 2004). This represents an important aspect that requires attention during choice of an appropraiate animal model. Experimental ataxic neuropathy (EAN) is considered the animal model for GBS because it shares a striking similarity to the classical form of GBS, acute inflammatory demyelinating polyradiculoneuropathy, in clinical symptoms and pathological findings (Notterpek and Tolwani, 1999). However, EAN does not model the AMAN form of disease with which *C. jejuni* infection is particularly associated, nor is this T-cell-mediated model an animal model for *Campylobacter*-induced GBS.

On the other hand, GD_{1b} immunization has produced sensory ataxic neuropathy in rabbits (Kusunoki et al., 1996), and the passive transfer of anti-GD_{1b} antibodies gives rise to axonal degeneration and macrophage infiltration. Intraneural injection of GBS serum or Ig into rat sciatic nerves gives variable results, but in most reports significant conduction block and demyelination at levels higher than control human serum have been observed. Of note, Illa et al. (1995) demonstrated that purified anti-GM_1 antibodies from patients who exhibited AMAN after immunization with a ganglioside preparation, recognized epitopes at the nodes of Ranvier and at the presynaptic nerve terminals of motor end-plates from human nerve biopsies. Accumulation of these antibodies at the nodes of Ranvier can cause disruption of Na^+ and K^+ channels and, thus, interfere with nerve conduction. Concerning MFS-associated anti-GQ_{1b} antisera, there is strong supportive evidence that these antibodies can mediate pathological changes. In a mouse hemi-nerve diaphragm model, administration of anti-GQ_{1b}-positive sera decreased nerve excitability and neurotransmitter release, and subsequently, induced nerve terminal paralysis (Willison and O'Hanlon, 2000), but this was not observed with anti- GQ_{1b} antibody in human limb muscle (Kuwabara et al., 2007).

Active immunization of rabbits with GM_1 ganglioside-mimicking *C. jejuni* LOSs induced the production of high titres of IgG anti-LOS antibodies that were cross-reactive with GM_1 ganglioside (Ang et al., 2000). Nevertheless, none of the animals developed overt signs of muscle weakness, but this may have been due to the short duration of the experiment. Other groups have induced anti-ganglioside antibodies with *C. jejuni* LOSs in animals (Wirguin et al., 1997; Goodyear et al., 1999). Importantly, rabbit antisera with anti-GM_1 specificity induced by a *C. jejuni* GM_1 mimic have been shown to bind at the nodes of Ranvier of human sciatic nerve in vitro (Moran et al., 2005a). This binding pattern was identical to that seen with anti-GM_1 sera from a GBS patient. Additionally, immunization of mice with GT_{1a}-containing LOS can produce a serum with anti-GQ_{1b} antibodies which bind to ganglioside-rich sites causing release of acetylcholine and complement-mediated conduction block (Goodyear et al., 1999). The latter is noteworthy since complement-activating antibodies to ganglioside complexes have high avidity and pro-inflammatory properties in GBS (Notturno et al., 2009).

An animal model in Japanese white rabbits has been established by sensitization with a bovine brain ganglioside mixture or isolated GM_1, in which rabbits developed high titre anti-GM_1 IgG antibodies and flaccid limb weakness (Yuki et al., 2001). The disease characteristics in this animal model correpond to the pathogical findings for human AMAN (Hughes and Cornblath, 2005). Building upon this model, a putative AMAN model has been reported after immunization of Japanese white rabbits with *C. jejuni* LOS bearing a GM_1-like structure (Yuki et al., 2004). After sensitization with this GM_1-like LOS, rabbits developed high titre anti-GM_1 IgG antibodies and subsequent flaccid limb weakness. Moreover, axons of these nerve fibres showed various degrees of degeneration, but demyelination and remyelination were rare. The model has been proposed as a replica of *C. jeuni* LOS-induced GBS, and even claimed as the first definitive replica of a human autoimmune disease produced by immunization with an infectious mimic (Yuki, 2005). Nonetheless, since this model requires multiple immunizations with LOS and haemocyanin as adjuvant (Yuki et al., 2004), it cannot be excluded that because of the aggressive nature of haemocyanin, auto-presentation of host gangliosides may occur in this model. Also, this is not an infective animal model of *C. jejuni*-induced GBS and thus does not fulfil the Galway postulates (Moran et al., 2002b). On the other hand, a chicken model developed by Li et al. (1996), in which outbred chickens were fed an AMAN-related *C. jejuni* isolate that exhibits GM_1/GD_{1a} mimicry, showed a neuropathy in 30% of animals and pathology similar to human AMAN. Despite this potential advance, but disappointingly, extensive experiments by others to reproduce this model have not proven successful (see Prendergast and Moran, 2000).

10.6 Conclusions and Future Perspective

Although *H. pylori* induces chronic infection whereas *C. jejuni* induces an acute infection (and to which the structural characteristics and biological properties of endotoxin, particularly their lipid A components contribute), both bacteria induce pathogenic autoimmune responses in which molecular mimicry in the saccharide component of endotoxin putatively plays an important role. Although the contributory role of ganglioside mimicry in *C. jejuni* LOS to GBS development is generally accepted, that of Le antigen mimicry by *H. pylori* in autoimmune consequences of infection is yet to be unequivocally established. By comparison of both bacterial infections and their autoimmune consequences, central issues for further consideration arise, including (i) the development of infection-based animal models for producing autoimmune sequeale, (ii) the extent and role of T-cell involvement with B-cells in the autoimmune humoral responses observed, (iii) how tolerance is broken to carbohydrate-based self-structures borne on endotoxin, and (iv) the role of other bacterial and host factors contributing with molecular mimicry in the development of autoimmunity. All represent important aspects of infection-induced autoimmunity awaiting to be addressed.

Acknowledgements The author gratefully acknowledges Science Foundation Ireland (grant no. 08/SRC/B1393) and the EU Marie Curie Programme (grant no. MTKD-CT-2005-029774) for financial support of his research.

References

Allos, B.M. *Campylobacter jejuni* infections: update on emerging issues and trends. Clin Infect Dis 32 (2001) 1201–1206.

Amano, K.-I., Hayashi, S., Kubota, T., Fujii, N., Yokota, S.-I. Reactivities of Lewis antigen monoclonal antibodies with the lipopolysaccharides of *Helicobacter pylori* strains isolated from patients with gatroduodenal diseases in Japan. Clin Diagn Lab Immnunol 4 (1997) 540–544.

Amedei, A., Bergman, M.P., Appelmelk, B.J., Azzurri, A., Benagiano, M., Tamburini, C., van der Zee, R., Telford, J.L., Vandenbroucke-Grauls, C.M., D'Elios, M.M., Del Prete, G. Molecular mimicry between *Helicobacter pylori* antigens and H^+,K^+ -adenosine triphosphatase in human gastric autoimmunity. J Exp Med 198 (2003) 1147–1156.

Ang, C.W., Endtz, H.P., Jacobs, B.C., Laman, J.D., de Klerk, M.A., van der Meché, F.G., van Doorn, P.A. *Campylobacter jejuni* lipopolysaccharides from Guillain-Barré syndrome patients induce IgG anti-GM_1 antibodies in rabbits. J Neuroimmunol 104 (2000) 133–138.

Ang, C.W., Jacobs, B.C., Laman, J.D. The Guillain-Barré syndrome: a true case of molecular mimicry. Trends Immunol 25 (2004) 61–66.

Appelmelk, B.J., Simoons-Smit, I., Negrini, R., Moran, A.P., Aspinall, G.O., Forte, J.G., De Vries, T., Quan, H., Verboom, T., Maaskant, J.J., Ghiara, P., Kuipers, E.J., Bloemena, E., Tadema, T.M., Townsend, R.R., Tyagarajan, K., Crothers, J.M., Monteiro, M.A., Savio, A., De Graaf, J. Potential role of molecular mimicry between *Helicobacter pylori* lipopolysaccharide and host Lewis blood group antigens in autoimmunity. Infect Immun 64 (1996) 2031–2040.

Appelmelk, B.J., Faller, G., Claeys, D., Kirchner, T., Vandenbroucke-Grauls, C.M.J.E. Bugs on trial: the case of *Helicobacter pylori* and autoimmunity. Immunol Today 19 (1998) 296–299.

Appelmelk, B.J., Martin, S.L., Monteiro, M.A., Clayton, C.A., McColm, A.A., Zheng, P., Verboom, T., Maaskant, J.J., van den Eijnden, D.H., Hokke, C.H., Perry, M.B., Vandenbroucke-Grauls, C.M.J.E., Kusters, J.G. Phase variation in *Helicobacter pylori* lipopolysaccharide due to changes in the lengths of poly(C) tracts in α3-fucosyltransferase genes. Infect Immun 67 (1999) 5361–5366.

Berg, D.E., Hoffman, P.S., Appelmelk, B.J., Kusters, J.G. The *Helicobacter pylori* genome sequence: genetic factors for long life in the gastric mucosa. Trends Microbiol 5 (1997) 468–474.

Blaser, M.J., Atherton, J.C. *Helicobacter pylori* persistence: biology and disease. J Clin Invest 113 (2004) 321–333.

Blaser, M.J., Engberg, J. Clinical aspects of *Campylobacter jejuni* and *Campylobacter coli* infections. In Nachamkin, I., Szymanski, C.M., Blaser, M.J. (eds), *Campylobacter*, 3rd edn. ASM Press, Washington, DC (2008), pp. 99–121.

Blaser, M.J., Kirschner, D. Dynamics of *Helicobacter pylori* colonization in relation to the host response. Proc Natl Acad Sci USA 96 (1999) 8359–8364.

Claeys, D., Faller, G., Appelmelk, B.J., Negrini, R., Kirchner, T. The gastric H^+,K^+-ATPase is a major autoantigen in chronic *Helicobacter pylori* gastritis with body mucosa atrophy. Gastroenterology 115 (1998) 340–347.

D'Elios, M.M., Appelmelk, B.J., Amedei, A., Bergman, M.P., del Prete, G. Gastric autoimmunity: the role of *Helicobacter pylori* and molecular mimicry. Trends Mol Med 10 (2004) 316–323.

De Libero, G.O., Moran, A.P, Gober, H.J., Rossy, E., Shamshiev, A., Chelnokova, O., Mazorra, Z., Vendetti, S., Sacchi, A., Prendergast, M.M., Sansano, S., Tonevitsky, A., Landmann, R., Mori, L. Bacterial infections promote T cell recognition of self-glycolipids. Immunity 22 (2005) 763–772.

De Lucca, A.J., Jacks, T.J., Brogden, K.A. Binding between lipopolysaccharide and ceropin A. Mol Cell Biochem 151 (1995) 141–148.

Eaton, K.A., Logan, S.M., Baker, P.E., Peterson, R.A., Monteiro, M.A., Altman, E. *Helicobacter pylori* with a truncated lipopolysaccharide O-chain fails to induce gastritis in SCID mice injected with splenocytes from wild-type C57BL/6 J mice. Infect Immun 72 (2004) 3925–3931.

Ernst, P.B., Gold, B.D. The disease spectrum of *H. pylori*: the immunopathogenesis of gastroduodenal ulcer and gastric cancer. Annu Rev Microbiol 54 (2000) 615–640.

Faller, G., Steininger, H., Kränzlein, J., Maul, H., Kerkau, T., Hensen, J., Hahn, E.G., Kirchner, T. Antigastric autoantibodies in *Helicobacter pylori* infection: implications of histological and clinical parameters of gastritis. Gut 41 (1997) 619–623.

Faller, G., Steininger, H., Appelmelk, B., Kirchner, T. Evidence of novel pathogenic pathways for the formation of antigastric autoantibodies in *Helicobacter pylori* gastritis. J Clin Pathol 51 (1998) 244–245.

Gilbert, M., Parker, C.T., Moran, A.P. *Campylobacter jejuni* lipooligosaccharides: structure and biosynthesis. In Nachamkin, I., Szymanski, C.M., Blaser, M.J. (eds), *Campylobacter*, 3rd edn. ASM Press, Washington, DC (2008), pp. 483–504.

Gobert, A.P., Bambou, J.-C., Werts, C., Balloy, V., Chignard, M., Moran, A.P., and Ferrero, R.L. *Helicobacter pylori* heat shock protein 60 mediates interleukin-6 production by macrophages via a Toll-like receptor (TLR)-2-, TLR-4-, and myeloid differentiation factor 88-independent mechanism. J Biol Chem 279 (2004) 245–250.

Goodyear, C.S., O'Hanlon, G.M., Plomp, J.J., Wagner, E.R., Morrison, I., Veitch, J., Cochrane, L., Bullens, R.W., Molenaar, P.C., Conner, J., Willison, H.J. Monoclonal antibodies raised against Guillain-Barré syndrome-associated *Campylobacter jejuni* lipopolysaccharides react with neuronal gangliosides and paralyze muscle-nerve preparations. J Clin Invest 104 (1999) 697–708.

Guerry, P., Ewing, C.P., Hickey, T.E., Prendergast, M.M., Moran, A.P. Sialylation of lipooligosaccharide cores affects immunogenicity and serum resistance of *Campylobacter jejuni*. Infect Immun 68 (2000) 6656–6662.

Guerry, P., Szymanski, C.M., Prendergast, M.M., Hickey, T.E., Ewing, C.P., Pattarini, D.L., Moran A.P. Phase variation of *Campylobacter jejuni* 81–176 lipooligosaccharide affects ganglioside mimicry and invasiveness in vitro. Infect Immun 70 (2002) 787–793.

Guillain-Barré syndrome Study Group. Plasmapheresis and acute Guillain-Barré syndrome. Neurology 35 (1985) 1096–1104.

Habib, I., Louwen, R., Uyttendaele, M., Houf, K., Vandenberg, O., Nieuwenhuis, E.E., Miller, W.G., van Belkum, A., De Zutter, L. Correlation between genotypic diversity, lipooligosaccharide gene locus class variation, and Caco-2 cell invasion potential of *Campylobacter jejuni* isolates from chicken meat and humans: contribution to virulotyping. Appl Environ Microbiol 75 (2009) 4277–4288.

Hamanaka, Y., Nakashima, M., Wada, A., Ito, M., Kurazono, H., Hojo, H., Nakahara, Y., Kohno, S., Hirayama, T., Sekine, I. Expression of human β-defensin 2 (hBD-2). *Helicobacter pylori* induced gastritis: antibacterial effect of hBD-2 against *Helicobacter pylori*. Gut 49 (2001) 481–487.

Hase, K., Murakami, M., Iimura, M., Cole, S.P., Horibe, Y., Ohtake, T., Obonyo, M., Gallo, R.L., Eckmann, L., Kagnoff, M.F. Expression of LL-37 by human gastric epithelial cells as a potential host defense mechanism against *Helicobacter pylori*. Gastroenterology 125 (2003) 1613–1625.

Heneghan, M.A., Moran, A.P., Feeley, K.M., Egan, E.L., Goulding, J., Connolly, C.E., McCarthy, C.F. Effect of host Lewis and ABO blood group antigen expression on *Helicobacter pylori* colonisation density and the consequent inflammatory response. FEMS Immunol Med Microbiol 20 (1998) 257–266.

Heneghan, M.A., McCarthy, C.F., Moran, A.P. Relationship of blood group determinants on *Helicobacter pylori* lipopolysaccharide with host Lewis phenotype and inflammatory response. Infect Immun 68 (2000) 937–941.

Heneghan, M.A., McCarthy, C.F., Janulaityte, D., Moran, A.P. Relationship of anti-Lewis x and anti-Lewis y antibodies in serum samples from gastric cancer and chronic gastritis patients to *Helicobacter pylori*-mediated autoimmunity. Infect Immun 69 (2001) 4774–4781.

Hughes, R.A.C., Cornblath, D.R. Guillain-Barré syndrome. Lancet 366 (2005) 1653–1666.

Hynes, S.O., Moran, A.P. Comparison of three serological methods for detection of Lewis antigens on the surface of *Helicobacter pylori*. FEMS Microbiol Lett 190 (2000) 67–72.

Hynes, S.O., Hirmo, S., Wadström, T., Moran, A.P. Differentiation of *Helicobacter pylori* isolates based on lectin binding of cell extracts in an agglutination assay. J Clin Microbiol 37 (1999) 1994–1998.

Hynes, S.O., Broutet, N., Eurohepygast Study Group, Wadström, T., Mikelsaar, M., O'Toole, P.W., Telford, J., Engstrand, L., Kamiya, S., Mentis, A.F., Moran, A.P. Phenotypic variation of *Helicobacter pylori* isolates from geographically distinct regions detected by lectin typing. J Clin Microbiol 40 (2002) 227–232.

Hynes, S.O., Keenan, J.I., Ferris, J.A., Annuk, H., Moran, A.P. Lewis epitopes on outer membrane vesicles of relevance to *Helicobacter pylori* pathogenesis. Helicobacter 10 (2005) 146–156.

Illa, I., Ortiz, N., Gallard, E., Juarez, C., Grau, J.M., Dalakas, M.C. Acute axonal Guillain-Barré syndrome with IgG antibody against motor axons following parenteral gangliosides. Ann Neurol 38 (1995) 218–224.

Kaida, K., Ariga, T., Yu, R.K. Antiganglioside antibodies and their pathophysiological effects on Guillain-Barré syndrome and related disorders – a review. Glycobiology 19 (2009) 676–692.

Kanipes, M.I., Holder, L.C., Corcoran, A.T., Moran, A.P., Guerry, P. A deep-rough mutant of *Campylobacter jejuni* 81-176 is non-invasive for intestinal epithelial cells. Infect Immun 72 (2004) 2452–2455.

Karlyshev, A.J. Wren, B.J., Moran, A.P. *Campylobacter jejuni* capsular polysaccharide. In Nachamkin, I., Szymanski, C.M., Blaser, M.J. (eds), *Campylobacter*, 3rd edn. ASM Press, Washington, DC (2008), pp. 505–521.

Keenan, J.I., Peterson, R.A., Fraser, R., Frampton, C.M., Walmsley, T.A., Allardyce, R.A., Roake, J.A. The effect of *Helicobacter pylori* infection and dietary iron deficiency on host iron homeostasis: a study in mice. Helicobacter 9 (2004) 643–650.

Keenan, J.I., Davis, K.A., Beaugie, C.R., McGovern, J.J., Moran, A.P. Alterations in *Helicobacter pylori* outer membrane and outer membrane vesicle-associated lipopolysaccharides under iron-limiting growth conditions. Innate Immun 14 (2008) 279–290.

Khalili-Shirazi, A., Gregson, N., Gray, I.A., Rees, J.H., Winer, J.B., Hughes, R.A.C. Anti-ganglioside antibodies in Guillain-Barré syndrome following recent cytomegalovirus infection. J Neurol Neurosurg Psychiatry 66 (1999) 376–379.

Knirel, Y.A., Kocharova, N.A., Hynes, S.O., Widmalm, G., Andersen, L.P., Jansson, P.-E., Moran, A.P. Structural studies on lipopolysaccharides of serologically non-typable strains of

Helicobacter pylori, AF1 and 007, expressing Lewis antigenic determinants. Eur J Biochem 266 (1999) 123–131.

Kocharova, N.A., Knirel, Y.A., Widmalm, G., Jansson, P.-E., Moran, A.P. Structure of an atypical O-antigen polysaccharide of *Helicobacter pylori* containing a novel monosaccharide 3-*C*-methyl-D-mannose. Biochemistry 39 (2000) 4755–4760.

Kusunoki, S., Iwamori, M., Chiba, A., Hitoshi, S., Arita, M., Kanazawa, I. GM_{1b} is a new member of antigen for serum antibody in Guillain-Barré syndrome. Neurology 47 (1996) 237–242.

Kusters, J.G., van Vliet, A.H.M., Kuipers, E.J. Pathogenesis of *Helicobacter pylori* infection. Clin Microbiol Rev 19 (2006) 449–490.

Kuwabara, S., Misawa, S., Takahashi, H., Sawai, S., Kanai, K., Nakata, M., Mori, M., Hattori, T., Yuki, N. Anti-GQ1b antibody does not affect neuromuscular transmission in human limb muscle. J Neuroimmunol 189 (2007) 158–162.

Latov, N. Antibodies to glycoconjugates in neurological disease. Clin Aspects Autoimmun 4 (1990) 18–29.

Lee, S.K., Stack, A., Katzowitsch, E., Aizawa, S.I., Suerbaum, S., Josenhans, C. (2003) *Helicobacter pylori* flagellins have very low intrinsic activity to stimulate human gastric epithelial cells via TLR5. Microbes Infect 5 (2003) 1345–1346.

Li, C.Y., Xue, P., Tian, W.Q., Liu, R.C., Yang, C. Experimental *Campylobacter jejuni* infection in the chicken: an animal model of axonal Guillain-Barré syndrome. J Neurol Neurosurg Psychiatr 61 (1996) 279–284.

Ma, B., Simala-Grant, J.L., Taylor, D.E. Fucosylation in prokaryotes and eukaryotes. Glycobiology 16 (2006) 158R–184R.

Ma, J.Y., Borch, K., Sjöstrand, S.E., Janzon, L., Mårdh, S. Positive correlation between H,K-adenosine triphosphatase autoantibodies and *Helicobacter pylori* antibodies in patients with pernicious anemia. Scand J Gastroenterol 29 (1994) 961–965.

Mandell, L., Moran, A.P., Cocchiarella, A., Houghton, J.M., Taylor, N., Fox, J.G., Wang, T.C., Kurt-Jones, E.A. Intact Gram-negative *Helicobacter pylori, Helicobacter felis,* and *Helicobacter hepaticus* bacteria activate innate immunity via Toll-like receptor 2 not Toll-like receptor 4. Infect Immun 72 (2004) 6446–6454.

Marrack, P., Kappler, J., Kotzin, B.L. Autoimmune disease: why and where it occurs. Nat Med 7 (2001) 899–905.

Marshall, D.G., Hynes, S.O., Coleman, D.C., O'Morain, C.A., Smyth, C.J., Moran, A.P. Lack of a relationship between Lewis antigen expression and *cag A*, CagA, *vacA* and VacA status of Irish *Helicobacter pylori* isolates. FEMS Immunol Med Microbiol 24 (1999) 79–90.

Matsumoto, Y., Yuki, N., Van Kaer, L., Furukawa, K., Hirata, K., Sugita, M. Cutting edge: Guillain-Barré syndrome-associated IgG responses to gangliosides are generated independently of CD1 function in mice. J Immunol 180 (2008) 39–43.

Monteiro, M.A. *Helicobacter pylori*: a wolf in sheep's clothing. The glyconjugate families of *Helicobacter pylori* lipopolysaccharides expressing histo-blood groups: structure, biosynthesis, and role in pathogenesis. Adv Carbohydr Chem Biochem 57 (2001) 99–158.

Monteiro, M.A., Appelmelk, B.J., Rasko, D.A., Moran, A.P., Hynes, S.O., MacLean, L.L., Chan, K.H., St. Michael, F., Logan, S.M., O'Rourke, J., Lee, A., Taylor, D.E., Perry, M.B. Lipopolysaccharide structures of *Helicobacter pylori* genomic strains 26695 and J99, mouse model *H. pylori* Sydney strain, *H. pylori* P466 carrying sialyl Lewis X, and *H. pylori* UA915 expressing Lewis B. Classification of *H. pylori* lipopolysacccharides into glycotype families. Eur J Biochem 267 (2000a) 305–320.

Monteiro, M.A., Zheng, P.-y., Ho, B., Yokota, S.-i., Amano, K.-i., Pan, Z.-j., Berg, D.E.,, Chan, K.H., MacLean, L.L., Perry, M.B. Expression of histo-blood group antigens by lipopolysaccharides of *Helicobacter pylori* strains from Asian hosts: the propensity to express type 1 blood-group antigens. Glycobiology 10 (2000b) 701–713.

Moran, A.P. Cell surface characteristics of *Helicobacter pylori*. FEMS Immunol Med Microbiol 10 (1995a) 271–280.

Moran, A.P. Biological and serological characterization of *Campylobacter jejuni* lipopolysaccharides with deviating core and lipid A structures. FEMS Immunol Med Microbiol 11 (1995b) 121–130.

Moran, A.P. The role of lipopolysaccharide in *Helicobacter pylori* pathogenesis. Aliment Pharmacol Ther 10 Suppl 1 (1996) 39–50.

Moran, A.P. Structure and conserved characteristics of *Campylobacter jejuni* lipopolysaccharides. J Infect Dis 176 Suppl 2 (1997) S115–S121.

Moran, A.P. The products of *Helicobacter pylori* that induce inflammation. Eur J Gastroenterol Hepatol 10 Suppl 1 (1998) S3–S8.

Moran, A.P. *Helicobacter pylori* lipopolysaccharide-mediated gastric and extragastric pathology. J Physiol Pharmacol 50 (1999) 787–805.

Moran, A.P. *Helicobacter pylori* lipopolysaccharides. In Achtman, M., Suerbaum, S. (eds), *Helicobacter pylori:* Molecular and Cellular Biology. Horizon Scientific Press, Wymondham (2001a), pp. 207–226.

Moran, A.P. Molecular structure, biosynthesis and pathogenic roles of lipopolysaccharides. In Mobley, H.L.T., Mendz, G.L., Hazell, S.L. (eds), *Helicobacter pylori:* Physiology and Genetics. ASM Press, Washington, DC (2001b), pp. 81–95.

Moran, A.P. Lipopolysaccharide in bacterial chronic infection: insights from *Helicobacter pylori* lipopolysaccharide and lipid A. Int J Med Microbiol 297 (2007) 307–319.

Moran, A.P. Relevance of fucosylation and Lewis antigen expression in the bacterial gastroduodenal pathogen *Helicobacter pylori*. Carbohydr Res 343 (2008) 1952–1965.

Moran, A.P. Molecular mimicry of host glycosylated structures by bacteria. In Moran, A.P., Holst, O., Brennan, P.J., von Itzstein, M. (eds), Microbial Glycobiology: Structures, Relevance and Applications. Academic Press, London (2009), pp. 847–870.

Moran, A.P. Lipopolysaccharides of *Helicobacter pylori*: importance in gastric adaptation and pathogenesis. In Sutton, P., Mitchell, H.M. (eds), *Helicobacter pylori* in the 21st Century. CABI Publishers, Wallingford (2010), in press.

Moran, A.P., Penner, J.L. Serotyping of *Campylobacter jejuni* based on heat-stable antigens: relevance, molecular basis and implications in pathogenesis. J Appl Microbiol 86 (1999) 361–377.

Moran, A.P., Prendergast, M.M. Molecular mimicry in *Campylobacter jejuni* lipopolysaccharides and the development of Guillain-Barré syndrome. J Infect Dis 178 (1998) 1549–1550.

Moran, A.P., Prendergast, M.M. Molecular mimicry in *Campylobacter jejuni* and *Helicobacter pylori* lipopolysaccharides: contribution of gastrointestinal infections to autoimmunity. J Autoimmun 16 (2001) 241–256.

Moran, A.P., Trent, M.S. *Helicobacter pylori* lipopolysaccharides and Lewis antigens. In Yamaoka, Y. (ed), *Helicobacter pylori*: Molecular Genetics and Cellular Biology. Caister Academic Press, Norfolk (2008), pp. 7–36.

Moran, A.P., Rietschel, E.T., Kosunen, T.U., Zähringer U. Chemical characterization of *Campylobacter jejuni* lipopolysaccharides containing *N*-acetylneuraminic acid and 2,3-diamino-2,3-dideoxy-D-glucose. J Bacteriol 173 (1991a) 618–626.

Moran, A.P., Zähringer, U., Seydel, U., Scholz, D., Stütz, P., Rietschel, E.T. Structural analysis of the lipid A component of *Campylobacter jejuni* CCUG 10936 (serotype O:2) lipopolysaccharide. Description of a lipid A containing a hybrid backbone of 2-amino-2-deoxy-D-glucose and 2,3-diamino-2,3-dideoxy-D-glucose. Eur J Biochem 198 (1991b) 459–469.

Moran, A.P., Helander, I.M., Kosunen, T.U. Compositional analysis of *Helicobacter pylori* rough-form lipopolysaccharides. J Bacteriol 174 (1992a) 1370–1377.

Moran, A.P., Scholz, D., Zähringer, U. Application of methylation analysis in the determination of the structure of disaccharides containing 2,3-diamino-2,3-dideoxy-D-glucose (GlcN3N) associated with the backbone of lipid A. Carbohydr Res 231 (1992b) 309–316.

Moran, A.P., Appelmelk, B.J., Aspinall, G.O. Molecular mimicry of host structures by lipopolysaccharides of *Campylobacter* and *Helicobacter* spp.: implications in pathogenesis. J Endotoxin Res 3 (1996) 521–531.

Moran, A.P., Lindner, B., Walsh, E.J. Structural characterization of the lipid A component of *Helicobacter pylori* rough- and smooth-form lipopolysaccharides. J Bacteriol 179 (1997) 6453–6463.

Moran, A.P, Penner, J.L., Aspinall, G.O. *Campylobacter* lipopolysaccharides. In Nachamkin, I., Blaser, M.J. (eds), *Campylobacter*, 2nd edn. ASM Press, Washington DC (2000), pp. 241–257.

Moran, A.P., Kosunen, T.U., Prendergast, M.M. Biochemical and serological analysis of the heat-stable antigens involved in serotyping *Campylobacter jejuni/coli*. Int J Med Microbiol 291 Suppl 31 (2001), 72.

Moran, A.P., Knirel, Y.A., Senchenkova, S.N., Widmalm, G., Hynes, S.O., Jansson, P.-E. Phenotypic variation in molecular mimicry between *Helicobacter pylori* lipopolysaccharides and human gastric epithelial cell surface glycoforms. Acid-induced phase variation in Lewis^x and Lewis^y expression by *H. pylori* lipopolysaccharides. J Biol Chem 277 (2002a) 5785–5795.

Moran, A.P., Prendergast, M.M., Hogan, E.L. Sialosyl-galactose: a common denominator of Guillain-Barré and related disorders? J Neurol Sci 196 (2002b) 1–7.

Moran, A.P., Perepelov, A.V., Knirel, Y.A., Amano, K.-i., McGovern, J.J., Jansson, P.-E. Lewis expression on LPS of Japanese *Helicobacter pylori*-associated cancer strains and importance of serological test format for detection. J Endotoxin Res 10 (2004) 315.

Moran, A.P., Annuk, H., Prendergast, M.M. Antibodies induced by ganglioside-mimicking *Campylobacter jejuni* lipooligosaccharides recognise epitopes at the nodes of Ranvier. J Neuroimmunol 165 (2005a) 179–185.

Moran, A.P., Schromm, A.B., Andrä, J., Koch, M.H.J., Seydel, U., Brandenburg, K. Influence of endotoxic activity of *Helicobacter pylori* lipopolysaccharide on physicochemical parameters. Helicobacter 10 (2005b) 489.

Muotiala, A., Helander, I.M., Pyhälä, L., Kosunen, T.U., Moran, A.P. Low biological activity of *Helicobacter pylori* lipopolysaccharide. Infect Immun 60 (1992) 1714–1716.

Negrini, R., Lisato, L., Zanella, I., Cavazzini, L., Gullini, S., Villanacci, V., Poiesi, C., Albertini, A., Ghielmi, S. *Helicobacter pylori* infection induces antibodies cross-reacting with human gastric mucosa. Gastroenterology 101 (1991) 437–445.

Negrini, R., Savio, A., Poiesi, C., Appelmelk, B.J., Buffoli, F., Paterlini, A., Cesari, P., Graffeo, M., Vaira, D., Franzin, G. Antigenic mimicry between *Helicobacter pylori* and gastric mucosa in the pathogenesis of body atrophic gastritis. Gastroenterology 111 (1996) 655–665.

Neisser, A., Bernheimer, H., Berger, T., Moran, A.P., Schwerer, B. Serum antibodies against gangliosides and *Campylobacter jejuni* lipopolysaccharides in Miller Fisher syndrome. Infect Immun 65 (1997) 4038–4042.

Neisser, A., Schwerer, B., Bernheimer, H., Moran, A.P. Ganglioside-induced antiganglioside antibodies from a neuropathy patient cross-react with lipopolysaccharides of *Campylobacter jejuni* associated with Guillain-Barré syndrome. J Neuroimmunol 102 (2000) 85–88.

Nilsson, C., Skoglund, A., Moran, A.P., Annuk, A., Engstrand, L., Normark, S. An enzymatic ruler modulates Lewis antigen glycosylation of *Helicobacter pylori* LPS during persistent infection. Proc Natl Acad Sci USA 103 (2006) 2863–2868.

Nilsson, C., Skoglund, A., Moran, A.P., Annuk, H., Engstrand, L., Normark, S. Lipopolysaccharide diversity evolving in *Helicobacter pylori* communities through genetic modifications in fucosyltransferases. PLoS One 3 (2008), e3811.

Notterpek, L., Tolwani, R.J. Experimental models of peripheral neuropathies. Lab Animal Sci 49 (1999) 588–599.

Notturno, F., Luciani, M., Caporale, C.M., Ciarelli, A., Uncini, A. Antibodies to ganglioside complexes in Guillain-Barré syndrome: clinical correlates, fine specificity and complement activation. Int J Immunopathol Pharmacol 22 (2009) 437–445.

Ogawa, T., Asai, Y., Sakai, Y., Oikawa, M., Fukase, K., Suda, Y., Kusumoto, S., Tamura T. Endotoxic and immunobiological activities of a chemically synthesized lipid A of *Helicobacter pylori* strain 206-1. FEMS Immunol Med Microbiol 36 (2003) 1–7.

Ortega-Cava, C.F., Ishihara, S., Rumi, M.A., Kawashima, K., Ishimura, N., Kazumori, H., Udagawa, J., Kadowaki, Y., Kinoshita, Y. Strategic compartmentalization of Toll-like receptor 4 in the mouse gut. J Immunol 170 (2003) 3977–3985.

Paterson, G., Wilson, G., Kennedy, P.G.E., Willison, H.J. Analysis of anti-GM_1 antibodies cloned from motor neuropathy patients demonstrates diverse variable region gene usage with extensive somatic mutation. J Immunol 155 (1995) 3049–3059.

Prendergast, M.M., Moran, A.P. Lipopolysaccharides in the development of the Guillain-Barré syndrome and Miller Fisher syndome forms of acute inflammatory peripheral neuropathies. J Endotoxin Res 6 (2000) 341–359.

Prendergast, M.M., Willison, H.J., Moran, A.P. Human monoclonal IgM antibodies to GM_1 ganglioside show diverse cross-reactivity with lipopolysaccharides of *Campylobacter jejuni* strains associated with Guillain-Barré syndrome. Infect Immun 67 (1999) 3698–3701.

Prendergast, M.M., Tribble, D.R., Baqar, S., Scott, D.A., Ferris, J.A., Walker, R.I., Moran, A.P. In vivo phase variation and serological response to lipooligosaccharide of *Campylobacter jejuni* in experimental human infection. Infect Immun 72 (2004) 916–922.

Putsep, K., Branden, C.I., Boman, H.G., Normark, S. Antibacterial peptide from *H. pylori*. Nature 398 (1999) 671–672.

Raetz, C.R., Reynolds, C.M., Trent, M.S., Bishop, R.E. Lipid A modification systems in Gram-negative bacteria. Annu Rev Biochem 76 (2007) 295–329.

Rietschel, E.T., Brade, L., Holst, O., Kulschin, V.A., Lindner, B., Moran, A.P., Schade, U.F., Zähringer, U., Brade, H. Molecular structure of bacterial endotoxin in relation to bioactivity. In Nowotny, A., Spitzer, J.J., Ziegler, E.J. (eds), Cellular and Molecular Aspects of Endotoxin Reactions. Vol. I. Elsevier Science Publishers B.V., Amsterdam (1990), pp. 15–32.

Rietschel, E.T., Kirikae, T., Schade, F.U., Mamat, U., Schmidt, G., Lopponow, H., Ulmer, A.J., Zähringer, U., Seydel, U., Di Padova, F. Bacterial endotoxin: molecular relationships of structure to activity and function. FASEB J 8 (1994) 217–225.

Schmausser, B., Andrulis, M., Endrich, S., Lee, S.K., Josenhans, C., Müller-Hermelink, H.K., Eck, M. Expression and subcellular distribution of toll-like receptors TLR4, TLR5 and TLR9 on the gastric epithelium in *Helicobacter pylori* infection. Clin Exp Immunol 136 (2004) 521–526.

Schromm, A.B., Brandenburg, K., Loppnow, H., Moran, A.P., Koch, M.H.J., Rietschel, E.T., Seydel, U. Biological activities of lipopolysaccharides are determined by the shape of their lipid A portion. Eur J Biochem 267 (2000) 2008–2013.

Schwerer, B., Neisser, A., Polt, R.J., Bernheimer, H., Moran, A.P. Antibody crossreactivities between gangliosides and lipopolysaccharides of *Campylobacter jejuni* serotypes associated with Guillain-Barré syndrome. J Endotoxin Res 2 (1995) 395–403.

Senchenkova, S.N., Zatonsky, G.V., Hynes, S.O., Widmalm, G., Andersen, L.P., Knirel, Y.A., Jansson, P.-E., Moran, A.P. Structure of a D-*glycero*-D-*manno*-heptan from the lipopolysaccharide of *Helicobacter pylori*. Carbohydr Res 331 (2001) 219–224.

Seydel, U., Oikawa, M., Fukase, K., Kusumoto, S., Brandenburg, K. Intrinsic conformation of lipid A is responsiblke for agonistic and antagonistic activity. Eur J Biochem 267 (2000) 3032–3039.

Sieling, P.A., Chatterjee, D., Porcelli, S.A., Prigozy, T.I., Mazzaccaro, R.J., Soriano, T., Bloom, B.R., Brenner, M.B., Kronenberg, M., Brennan, P.J., Modlin, R.L. CD1-restricted T-cell recognition of microbial lipoglycans. Science 269 (1995) 227–230.

Simoons-Smit, I.M., Appelmelk, B.J., Verboom, T., Negrini, R., Penner, J.L., Aspinall, G.O., Moran, A.P., Fei Fei S., Bi-shan, S., Rudnica, W., Savio, A., De Graaff, J. Typing of *Helicobacter pylori* with monoclonal antibodies against Lewis antigens in lipopolysaccharide. J Clin Microbiol 34 (1996) 2196–2200.

Skoglund, A., Bäckhed, H.K., Nilsson, C., Björkholm, B., Normark, S., Engstrand, L. A changing gastric environment leads to adaptation of lipopolysaccharide variants in *Helicobacter pylori* populations during colonization. PLoS One 4 (2009) e5885.

Smith, M.F., Mitchell, A., Li, G., Ding, S., Fitzmaurice, A.M., Ryan, K., Crowe, S., Goldberg, J.B. Toll-like receptor (TLR) 2 and TLR5, but not TLR4, are required for *Helicobacter pylori*-induced NF-κB activation and chemokine expression by epithelial cells. J Biol Chem 278 (2003) 32552–32560.

Smith, S., Moran, A.P., Kelleher, D.P. Lipopolysaccharides from *Helicobacter pylori* NCTC 11637 upregulate IL-8 and VEGF via Toll-like receptor 2 in an epithelial cell system. Gastroenterology 130 Suppl 2 (2006), A526.

Stöckl, J., Majdic, O., Rosenkranz, A., Fiebiger, E., Kniep, B., Stockinger, H., Knapp, W. Monoclonal antibodies to the carbohydrate structure Lewis^x stimulate the adhesive activity of leukocyte integrin CD11b/CD18 (CR3, Mac-1, $\alpha_m\beta_2$) on human granulocytes. J Leukocyte Biol 53 (1993) 541–549.

Suda, Y., Ogawa, T., Kashihara, W., Oikawa, M., Shimoyama, T., Hayashi, T., Tamura, T., Kusumoto, S. Chemical structure of lipid A from *Helicobacter pylori* strain 206-1 lipolysaccharide. J Biochem (Tokyo) 121 (1997) 1129–1133.

Suda, Y., Kim, Y.M., Ogawa, T., Yasui, N., Hasegawa, Y., Kashihara, W., Shimoyama, T., Aoyama, K., Nagat, K., Tamura, T., Kusomoto, S. Chemical structure and biological activity of a lipid A component from *Helicobacter pylori* strain 206. J Endotoxin Res 7 (2001) 95–104.

Suerbaum, S., Michetti, M. *Helicobacter pylori* infection. N Engl J Med 347 (2002) 1175–1186.

Sun, H.Y., Lin, S.W., Ko, T.P., Pan, J.F., Liu, C.L., Lin, C.N., Wang, A.H., Lin, C.H. Structure and mechanism of *H. pylori* fucosyltransferase. A basis for lipopolysaccharide variation and inhibitor design. J Biol Chem 282 (2007) 9973–9982.

Takenaka, R., Yokota, K., Ayada, K., Mizuno, M., Zhao, Y., Fujinami, Y., Lin, S.N., Toyokawa, T., Okada, H., Shiratori, Y., Oguma, K. *Helicobacter pylori* heat shock protein 60 induces inflammatory responses through the Toll-like receptor-triggered pathway in cultured human gastric epithelial cells. Microbiology 150 (2004) 3913–3922.

Tran, A.X, Stead, C.M., Trent, M.S. Remodeling of *Helicobacter pylori* lipopolysaccharide. J Endotoxin Res 11 (2005) 161–166.

Tran, A.X., Whittimore, J.D., Wyrick, P.B., McGrath, S.C., Cotter, R.J., Trent, M.S. The lipid A 1-phosphatase of *Helicobacter pylori* is required for resistance to the antimicrobial peptide polymyxin. J Bacteriol 188 (2006) 4531–4541.

Trent, M.S., Stead, C.M., Tran, A.X., Hankins, J.V. Diversity of endotoxin and its impact on pathogenesis. J Endotoxin Res 12 (2006) 205–223.

Triantafilou, M., Gamper, F.G., Lepper, P.M., Mouratis, M.A., Schumann, C., Harokopakis, E., Schifferle, R.E., Hajishengallis, G., Triantafilou, K. Lipopolysaccharides from atherosclerosis-associated bacteria antagonize TLR4, induce formation of TLR2/1/CD36 complexes in lipid rafts and trigger TLR2-induced inflammatory responses in human vascular endothelial cells. Cell Microbiol 9 (2007) 2030–2039.

Uno, K., Kato, K., Atsumi, T., Suzuki, T., Yoshitake, J., Morita, H., Ohara, S., Kotake, Y., Shimosegawa, T., Yoshimura T. Toll-like receptor (TLR) 2 induced through TLR4 signaling initiated by *Helicobacter pylori* cooperatively amplifies iNOS induction in gastric epithelial cells. Am J Physiol Gastrointest Liver Physiol 293 (2007) G1004–G1012.

Viala, J., Chaput, C., Boneca, I.G., Cardona, A., Girardin, S.E., Moran, A.P., Athman, R., Mémet, S., Huerre, M.R., Coyle, A.J., DiStefano, P.S., Sansonetti, P.J., Labigne, A., Bertin, J., Philpott, D.J., Ferrero, R.L. Nod1 responds to peptidoglycan delivered by the *Helicobacter pylori cag* pathogenicity island. Nat Immunol 5 (2004) 1166–1174.

Vorobjova, T., Faller, G., Maaroos, H.-I., Sipponen, P., Villako, K., Uibo, R., Kirchner, T. Significant increase in antigastric autoantibodies in a long-term follow-up study of *H. pylori* gastritis. Virchows Arch 437 (2000) 37–45.

Wada, A., Mori, N., Oishi, K., Hojo, H., Nakahara, Y., Hamanaka, Y., Nagashima, M., Sekine, I., Ogushi, K., Niidome, T., Nagatake, T., Moss, J., Hirayama, T. Induction of human β-defensin-2 mRNA expression by *Helicobacter pylori* in human gastric cell line MKN45 cells on *cag* pathogenicity island. Biochem Biophys Res Commun 263 (1999) 770–774.

Willison, H.J., O'Hanlon, G.M. Anti-glycospingolipid antibodies and Guillain-Barré syndrome. In Nachamkin, I., Blaser, M.J. (eds), *Campylobacter*. 2nd edn. ASM Press, Washington, DC (2000), pp. 259–285.

Willison, H.J., Veitch, J. Immunoglobulin subclass distribution and binding characteristics of anti-GQ_{1b} antibodies in Miller Fisher syndrome. J Neuroimmunol 50 (1994) 159–165.
Wirguin, I., Briani, C., Suturkova-Milosevic, L., Fisher, T., Della-Latta, P., Chalif, P., Latov, N. Induction of anti-GM_1 antibodies by *Campylobacter jejuni* lipopolysaccharides. J Neuroimmunol 78 (1997) 138–142.
Yuki, N. Infectious origins of, and molecular mimicry in, Guillain-Barré and Fisher syndromes. Lancet Infect Dis 1 (2001) 29–37.
Yuki, N. Carbohydrate mimicry: a new paradigm of autoimmune diseases. Curr Opin Immunol 17 (2005) 577–582.
Yuki, N. Ganglioside mimicry and peripheral nerve disease. Muscle Nerve 35 (2007) 691–711.
Yuki, N., Taki, T., Inagaki, F., Kasama, T., Takahashi, M., Saito, K., Handa, S., Miyatake, T. A bacterium lipopolysaccharide that elicits Guillain-Barré syndrome has a GM_1 ganglioside-like structure. J Exp Med 178 (1993) 1771–1775.
Yuki, N., Yamada, M., Koga, M., Odaka, M., Susuki, K., Tagawa, Y., Ueda, S., Kasama, T., Ohnishi, A., Hayashi, S., Takahashi, H., Kamijo, M., Hirata, K. Aninal model of axonal Guillain-Barré syndrome induced by sensitization with GM1 ganglioside. Ann Neurol 49 (2001) 712–720.
Yuki, N., Susuki, K.,Koga, M., Nishimoto, Y., Odaka, M., Hirata, K., Taguchi, K., Miyatake, T., Furukawa, K., Kobata, T., Yamada, M. Carbohydrate mimicry between human ganglioside GM1 and *Campylobacter jejuni* lipooligosaccharide causes Guillain-Barré syndrome. Proc Natl Acad Sci USA 101 (2004) 11404–11409.

Chapter 11
The Role of *Pseudomonas* Lipopolysaccharide in Cystic Fibrosis Airway Infection

Samuel M. Moskowitz and Robert K. Ernst

Abstract *Pseudomonas aeruginosa* (PA) is a ubiquitous environmental Gram-negative bacterium found in soil and water. This opportunistic pathogen can cause infections in individuals with impaired phagocytic function, such as those with burns, exposure to chemotherapy, or cystic fibrosis (CF). PA infects the lungs of most individuals with CF, and is associated with severe progressive pulmonary disease that is the major cause of premature death in this disorder. The specific adaptations of PA to the CF airway responsible for bacterial persistence and antibiotic tolerance are not completely understood but may include increased alginate production (i.e., mucoid phenotype), biofilm formation, and specific lipid A modifications. During adaptation to the CF airway, PA synthesizes a variety of lipid A structures that alter host innate immune responses and promote bacterial persistence and chronic infection. The synthesis of specific lipid A structures is attributable to bacterial enzymes that: (1) remove the 3OH-C10:0 acyl chain from the 3-position (PagL); (2) add a C16:0 acyl chain to the 3OH-C10:0 chain at the 3'-position (PagP); (3) add C12:0 and 2OH-C12:0 acyl chains to the 3OH-C12:0 chains at the 2- and 2'-positions (HtrB and LpxO); and (4) add aminoarabinose to phosphate groups at the 1- and 4'-positions (PmrH, PmrF, PmrI, PmrJ, PmrK, and PmrE). These lipid A modifications represent an essential aspect of PA adaptation to the CF airway.

Keywords *Pseudomonas aeruginosa* · Cystic fibrosis · Lipopolysaccharide · Lipid A · Chronic infection · Antibiotic resistance · Innate immunity · inflammation

Abbreviations

PA	*Pseudomonas aeruginosa*
CF	cystic fibrosis
LPS	lipopolysaccharide

R.K. Ernst (✉)
Department of Microbial Pathogenesis, School of Dentistry, University of Maryland – Baltimore, 650 W. Baltimore Street - 8 South, Baltimore, MD, 21201, USA
e-mail: rkernst@umaryland.edu

X. Wang, P.J. Quinn (eds.), *Endotoxins: Structure, Function and Recognition*, Subcellular Biochemistry 53, DOI 10.1007/978-90-481-9078-2_11,

11.1 Introduction

In human infection, *Pseudomonas aeruginosa* (PA) is acquired from environmental reservoirs, and can cause both acute and chronic infections depending on the clinical context (Lyczak et al., 2000). In human hosts with damaged epithelial barriers (e.g., burns) or immune system defects (e.g., post-chemotherapy), acute PA infections of the skin, lungs, urinary tract, or blood can be rapidly fatal. In contrast, in specific clinical conditions such as cystic fibrosis (CF), diffuse panbronchiolitis, HIV infection, and idiopathic bronchiectasis, PA can cause chronic airway infections that persist for months or years. In the setting of long-standing infection, PA isolates from CF patients are phenotypically similar to isolates from other chronic lung infections but distinct from acute infection isolates.

One such distinguishing characteristic of CF isolates is the frequent occurrence of rough or deep rough variants. Such variants represent changes in the structure of lipopolysaccharide (LPS), the major constituent of the outer leaflet of the Gram-negative outer membrane. LPS is an important pathogenic factor of Gram-negative organisms and consists of three distinct regions: O-antigen, core, and lipid A. Both O-antigen and core consist of polysaccharide chains, whereas lipid A consists of fatty acid and phosphate moieties bonded to a central glucosamine dimer. Rough variants reflect loss of O-antigen, whereas deep rough variants reflect loss of core oligosaccharide. In PA, these changes are associated with a switch from serum resistance to serum sensitivity (Hancock et al., 1983). This switch indicates that the key structural component of PA LPS in chronic infection is its lipid A moiety, the focus of this review.

11.2 Cystic Fibrosis is an Important Disease of Children and Young Adults

CF is most common lethal autosomal recessive disorder of Caucasians, affecting about 1 in 3,200 persons, and is caused by mutation of the gene encoding the cystic fibrosis transmembrane conductance regulator (CFTR) (Moskowitz et al., 2008). The CFTR gene encodes a protein kinase A-activated epithelial cell ion channel that is selective for chloride and bicarbonate and also regulates sodium (and potentially other) channels in the plasma membrane. Thus, deficiency of CFTR results in imbalances in epithelial chloride and bicarbonate, as well as additional epithelium- and channel-specific electrolyte abnormalities. While CF affects multiple organ systems, including the gastrointestinal and reproductive tracts and the sweat glands, morbidity and mortality are primarily related to disease of the respiratory tract. In the respiratory tract, CFTR deficiency results in lack of chloride and bicarbonate secretion, and also excessive sodium absorption owing to hyperactivity of the epithelial sodium channel (ENaC) that CFTR regulates. Because the respiratory epithelium is quite water permeable, these ion imbalances result in dehydration (decreased depth) of the airway surface liquid layer that bathes cilia responsible for mucus transport across the epithelial surface (Knowles and Boucher, 2002). These same electrolyte abnormalities also adversely affect mucus glands within the respiratory tract, leading to secretion of respiratory mucus with a higher concentration of mucins and thus greater viscosity than normal (Ballard et al., 1999; Jayaraman et al., 2001). Owing to this combination of thicker airway secretions and impaired mucociliary transport, mucus plaques tend to adhere to the surface of the CF airway and become a nidus of infection.

According to the CF Foundation National Patient Registry, the median age of survival for a person with CF in 2007 was ~37 years (Cystic Fibrosis Foundation 2009). As more advances have been made in the treatment of CF airway disease, the number of adults with CF has steadily grown. As of 2007, more than 45% of the CF population in the US was age 18 and older. This increase in median survival is attributable to advances in nutritional and respiratory management. Nonetheless, pulmonary disease remains the principal cause of morbidity and mortality. Some individuals with CF still die from lung disease during their teenage years, and for those who survive into adulthood, the severity of lung disease and intensity of daily treatments needed to manage it are major determinants of functional status. These treatments include a variety of inhaled and oral medications such as hypertonic saline, mucolytics, and antibiotics. Antibiotic resistance of CF pathogens is a major problem, as is the generally unfavorable pharmacokinetics of most antibiotics with respect to the airway compartment. Clearly, improved therapies are needed that specifically target key steps in the development of CF airway infection and inflammation.

11.2.1 Persistent Inflammation and Chronic Infection are the Hallmarks of CF Pulmonary Disease

CF lung disease begins in infancy or early childhood, and is characterized by chronic bacterial infection and severe inflammation that leads to progressive destruction of the lungs. The airways, rather than the lung parenchyma, are the primary sites of inflammation and infection (Davis et al., 1996). CF lung disease is characterized by secretion of copious amounts of thick mucus and migration of a predominantly neutrophilic infiltrate into the airways. This chronic neutrophilic infiltrate is accompanied by chronic lymphocytic and mononuclear cell infiltrates within adjacent submucosal tissues (van Heeckeren et al., 1997). Bronchoalveolar lavage fluid obtained from teenagers and adults with CF show increased levels of the pro-inflammatory cytokines TNF-α, IL-1, and IL-8 and decreased levels of the anti-inflammatory cytokine IL-10 as compared to normal individuals (Bonfield et al., 1995). Nevertheless, these cytokine and neutrophil responses are largely ineffective in eliminating the bacterial airway infection. Most distinctive about CF lung disease is its microbiology. Although *Staphylococcus aureus* is the most common pathogen in the first five years of life, the opportunistic pathogen PA is by far the most important pathogen in severe progressive CF lung disease, infecting the majority of CF patients in later years (Rajan and Saiman, 2002).

11.2.2 *P. aeruginosa* Is an Important Pathogen in the CF Airway

Epidemiological evidence of the importance of PA in CF lung disease comes from the National Cystic Fibrosis Patient Registry (Cystic Fibrosis Foundation, 2009). Analysis of this database revealed that in 2007, PA was cultured from the airways of about 25% of toddlers and 70–80% of adults with CF. Furthermore, a combined serologic and microbiologic clinical study suggested that PA infection occurs at least intermittently in nearly all patients with CF by age 3 years (Burns et al., 2001). In addition, the bacterial load in PA infection tends to be higher than for other pathogens, up to 10^9 cfu/gm of sputum (Tummler et al., 1997). These data suggest that both in terms of numbers of patients affected and severity of the effect, control of PA infection should be a major goal of efforts to combat CF lung disease.

11.2.3 Adaptation of *P. aeruginosa* to the Airway Is Important for CF Lung Disease

Previous clinical studies have shown that following initial infection with environmental isolates, specific bacterial phenotypes are selected within the CF airway environment. These characteristics are manifested in both children and adults (mean age >9 years) and include loss of flagellar-dependent motility (Mahenthiralingam et al., 1994), loss of O-antigen and other LPS changes (Hancock et al., 1983), increased auxotrophy (Thomas et al., 2000), decreased secretion of virulence factors (Tummler et al., 1997), inability to produce pyocyanin and phage (Romling et al., 1994), antibiotic resistance (Burns et al., 1999), mucoidy (Henry et al., 1992; Boucher et al., 1997), and formation of biofilms (Singh et al., 2000). Infection with mucoid PA has been associated with worse respiratory prognosis (Li et al., 2005). Mucoidy is also observed in several other conditions including bronchiectasis (Pujana et al., 1999) and chronic urinary catheterization (Reid et al., 1992; Goto et al., 1999). Many of these adaptations are observed in PA isolates from the airways of older patients with CF; in contrast, an early adaptation is the synthesis of specific structures of lipid A, the bioactive component of LPS (Ernst et al., 1999).

11.2.4 *P. aeruginosa* Lipid A Structures in CF Clinical Isolates are Distinct from Those Seen in Acute Clinical Infections and Isolates from the Environment

The structure of PA lipid A has been elucidated using both mass spectrometry (MS) and NMR techniques (Goldman et al., 1988; Kulshin et al., 1991; Karunaratne et al., 1992). As with lipid A of other Gram-negative organisms, PA lipid A consists of a ß-(1',6)-linked diglucosamine backbone with phosphates at the 1 and 4' positions, amide-linked fatty acids at the 2 and 2' positions, and ester-linked fatty acids at the 3 and 3' positions (Fig. 11.1). The chain lengths of the fatty acids attached to the PA

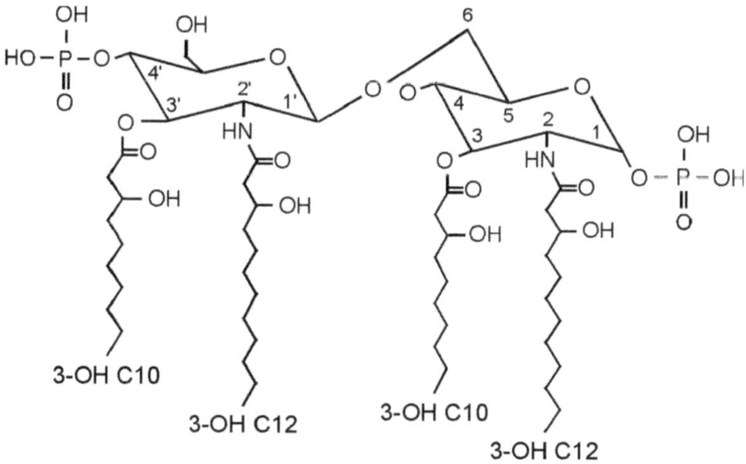

Fig. 11.1 Chemical structure of *P. aeruginosa* lipid IV_A

lipid A (C12/C10) are generally two carbons shorter than those of *S. typhimurium* and *E. coli* lipid A (C14/C12).

MS analysis of lipid A from laboratory-adapted PA strains grown in rich medium reveals a dominant ion species (*m/z*) of mass 1447 (Fig. 11.2a), corresponding to a penta-acylated molecule. This is the major structural form of lipid A isolated from laboratory-adapted PA strains (PAO1, PAK, PA14) grown in rich medium (Bhat et al., 1990; Ernst et al., 1999; Moskowitz et al., 2004). Additional lipid A species observed by MS are dependent on growth conditions (concentration of magnesium or iron, low pH, and reduced oxygen levels), and the strain from which the lipid A was isolated (Goldman et al., 1988). In PA strains, penta-acylated lipid A species are usually most abundant; however, specific lipid A structures often differ between CF (*m/z* 1447, Fig. 11.2a) and non-CF isolates (*m/z* 1419, Fig. 11.2d).

PA adaptation to CF airways often leads to constitutive expression of lipid A modifications that are regulated in isolates from acute infection (blood, ear, eye, and urinary tract) or environmental isolates. Previously, our laboratory has shown that minimally passaged PA isolated from children with CF as young as 3 month of age have unique lipid A structures as compared to laboratory-adapted strains, isolates from patients with acute clinical infections (blood, ear, eye, urinary tract), or bronchiectasis, a chronic non-CF airway infection (Ernst et al., 1999). MS analysis of lipid A isolated from greater than 110 minimally-passaged clinical PA isolates, grown under conditions in which laboratory-adapted PA strains do not modify their lipid A (magnesium replete growth medium – 1 mM), demonstrated the addition of palmitate (C16:0) (*m/z* 1685, Fig. 11.2b), resulting in hexa-acylated lipid A. Interestingly, growth of PA acute infection, bronchiectasis, or laboratory-adapted isolates in magnesium-limited medium (low magnesium – Fig. 11.2b) resulted in the synthesis of lipid A species with palmitate, indicating that the enzymatic pathways necessary for their synthesis are intact and inducible in these non-CF clinical isolates.

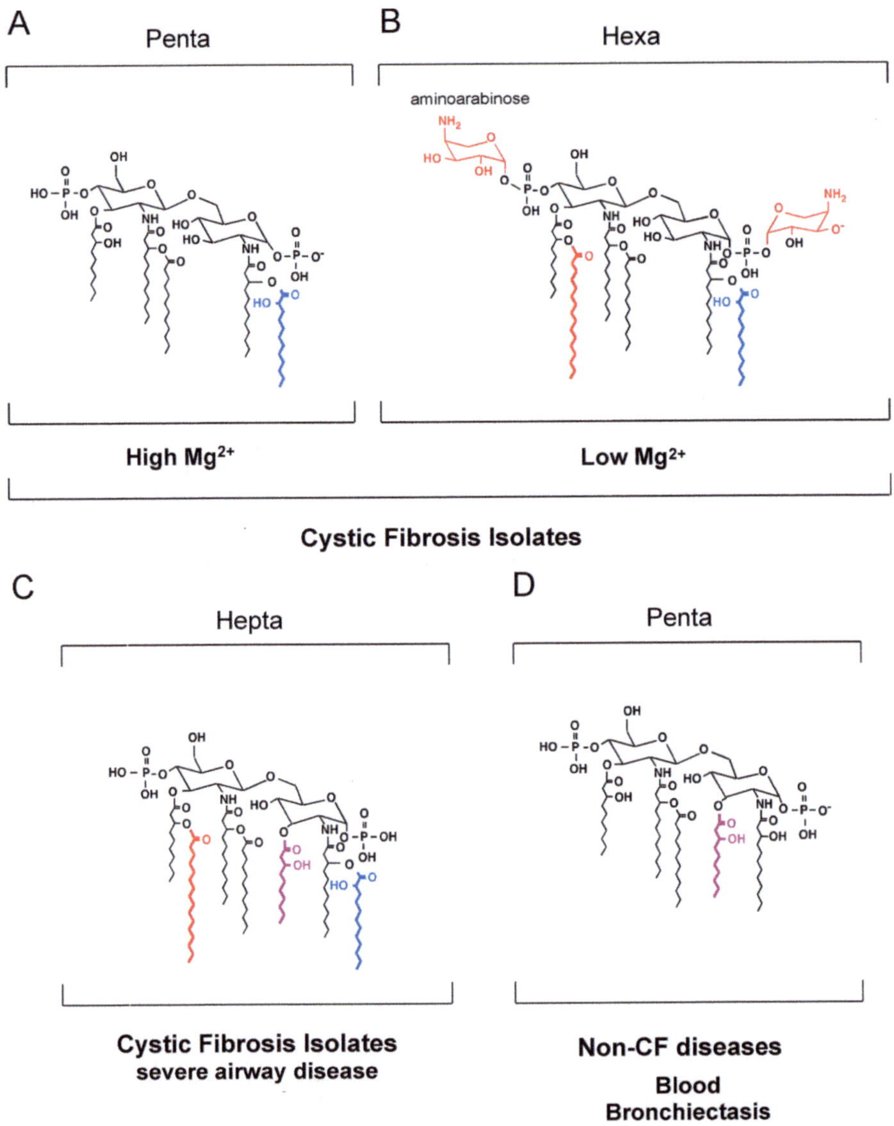

Fig. 11.2 Chemical structure of PA penta-, hexa, and hepta-acylated lipid A from (**a** and **b**) CF isolates, (**c**) a subset of isolates from patients with severe CF pulmonary disease, and (**d**) acute clinical infection or non-CF bronchiectasis isolates

More recent studies have shown that a novel hepta-acylated lipid A (*m/z* 1855, Fig. 11.2c) is present in a subset of clinical isolates from patients with severe CF pulmonary disease. Formation of hepta-acylated lipid A results from loss of an enzymatic activity (PogL) that ordinarily deacylates the 3-position of the diglucosamine backbone. Retention of this fatty acid at this position is associated with enhanced resistance to β-lactam antibiotics but not to aminoglycosides.

These results indicated that clinical PA isolates from various infections display marked heterogeneity with respect to the acylation state of their lipid A, reflecting differences in selective pressure that these infections impose on PA lipid A synthesis and structural modifications. Further study of the synthesis and regulation of lipid A modifications that are associated with CF lung disease is needed to understand their role in antibiotic resistance and other adaptations that are relevant to this specialized niche.

11.2.5 Synthesis of Cystic Fibrosis-Specific Lipid A Modifications in P. aeruginosa

Based on analyses of lipid A structures synthesized by various PA strains, the following biosynthetic scheme for the generation of CF-associated lipid A molecules from lipid IV_A is proposed (Fig. 11.3) (Ernst et al., 2003). Starting from PA lipid IV_A (Fig. 11.3a), a lauryltransferase, HtrB2 catalyzes the acyl-oxy-acyl addition of a laurate group (C12:0) to the amide-linked 3OH-C12:0 at the 2'-position (Fig. 11.3b). A second lauryltransferase, HtrB1, combined with the lipid A-specific hydroxylase, LpxO, catalyzes the addition of a 2-hydroxylaurate group (2OH-C12:0) to the 3OH-C12:0 at the 2-position (Fig. 11.3c) to generate the hexa-acylated lipid A species, *m/z* 1617 (Fig. 11.3d). Both of these chromosomally-encoded lauryltransferases are required for the synthesis of the hexa-acylated lipid A species shown in Fig. 11.3d. Hydroxylation of the laurate group (C12:0) that is attached to the amide-linked 3OH-C12:0 at the 2'-position, resulting in lipid A species with two 2-hydroxylaurate groups, may also be observed.

A deacylase, PagL, the gene for which has recently been identified (Trent et al., 2001; Geurtsen et al., 2005), catalyzes the removal of 3-hydroxydecanoate (3OH-C10:0) at the 3-position of the hexa-acylated structure to generate the penta-acylated lipid A species, *m/z* 1447 (Fig. 11.3e) that is often the predominant lipid A structure in CF isolates. A palmitoyltransferase, PagP, can modify this penta-acylated species through the acyl-oxy-acyl addition of palmitate (C16:0) at the 3' position to generate a second hexa-acylated species, *m/z* 1685 (Fig. 11.3f). Finally, a hepta-acylated lipid A species, m/z 1855 (Fig. 11.3 g) is observed in a subset of PA isolates from patients with severe CF pulmonary disease, presumably the result of loss of deacylase activity (PagL) in these isolates. The key steps in the synthesis of CF-specific lipid A include the addition of 2-hydroxylaurate (2OH-C12) and palmitate (C16:0), as well as deacylation of the 3-position fatty acid, leading to the synthesis of hexa-acylated lipid A species shown in Fig. 11.3d, f, and the penta-acylated lipid A species shown in Fig. 11.3e, respectively.

These enzymatic activities include three acyltransferases (HtrB1, HtrB2, PagP) and a deacylase (PagL). Depending on the specific PA isolate background, these activities can be classified as inducible (lipid A structures only observed under specific growth conditions that induce modification in these isolates), constitutive (lipid A structures always observed under any growth condition in these

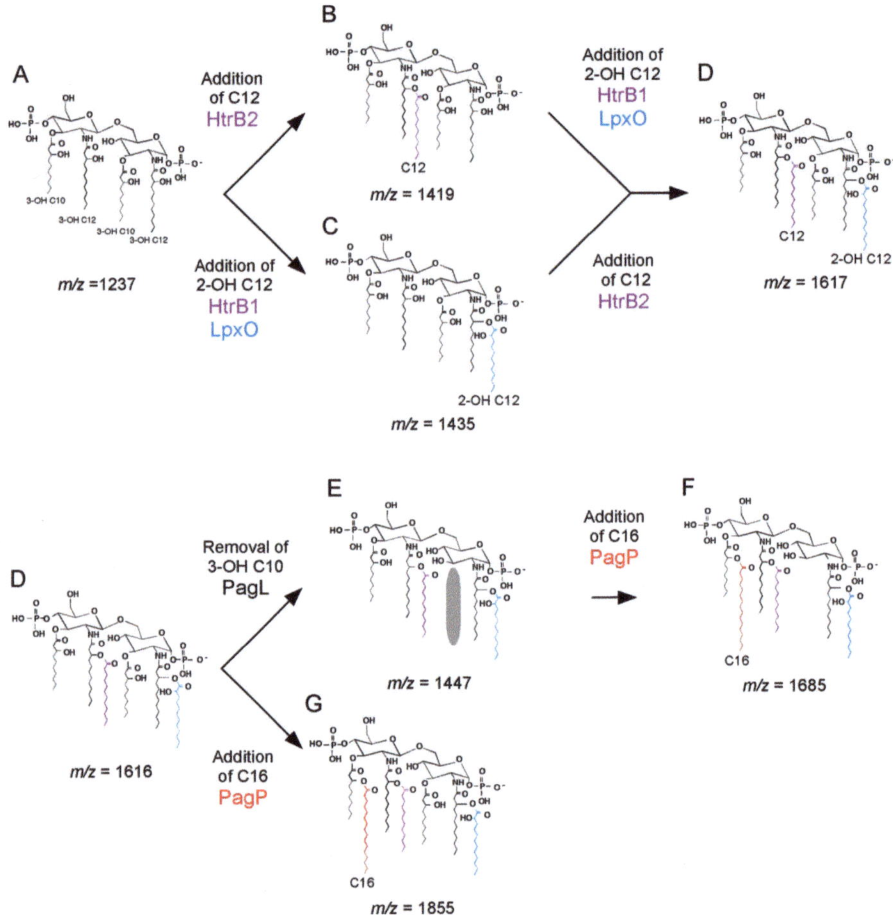

Fig. 11.3 Biosynthesis of *P. aeruginosa* lipid A in cystic fibrosis patient isolates. For all structures, the mass/charge ratio of the singly charged lipid A species is indicated

isolates), or deficient (lipid A structures never observed under any growth condition in these isolates). In CF isolates, lipid A modifications that are constitutively expressed are typically stable after repeated passage under non-inducing growth conditions.

11.2.6 P. aeruginosa Lipid A Modifications Promote CAP Resistance

PA has the genetic capacity to modify the structure of its lipid A as a mechanism for resisting the effects of elements of host innate immunity, such as cationic antimicrobial peptides (CAPs). PA accomplishes this in part through the addition of aminoarabinose to the terminal phosphates of its lipid A (Bhat et al., 1990; Ernst et al., 1999; Moskowitz et al., 2004). PA possesses the two-component regulatory system, PhoPQ (sensor-kinase, PhoQ and phosphorylated transcriptional activator, PhoP). PhoP regulates structural lipid A alterations (addition of palmitate and aminoarabinose) essential for CAP resistance (Ernst et al., 1999; Macfarlane et al., 1999; Macfarlane et al., 2000). The addition of aminoarabinose to PA lipid A is also regulated by a second two-component regulatory system, PmrAB, that contributes to resistance to polymyxins, a family of acylated cyclic CAMPs (McPhee et al., 2003; Moskowitz et al., 2004; Pamp et al., 2008).

Both the PhoPQ and PmrAB systems regulate induction of the *pmrH* operon (McPhee et al., 2003), which encodes enzymes necessary for the synthesis of aminoarabinose as well as its transfer to the 1- and 4'-phosphate groups of lipid A (Fig. 11.4). In PA laboratory-adapted strains, addition of aminoarabinose ($\Delta m/z$ +131) to the penta- and hexa-acylated species (m/z 1447 or m/z 1685) can be observed under appropriate inducing conditions, resulting in lipid A species of m/z 1578 or m/z 1816, respectively.

In polymyxin-resistant clinical isolates, in addition to constitutive aminoarabinosylation, loss of the lipid A modifying enzyme LpxO, an oxygenase required for the synthesis of 2-OH fatty acids (Gibbons et al., 2000) appears to contribute to CAP resistance, as many such isolates exhibit relative or absolute deficiency

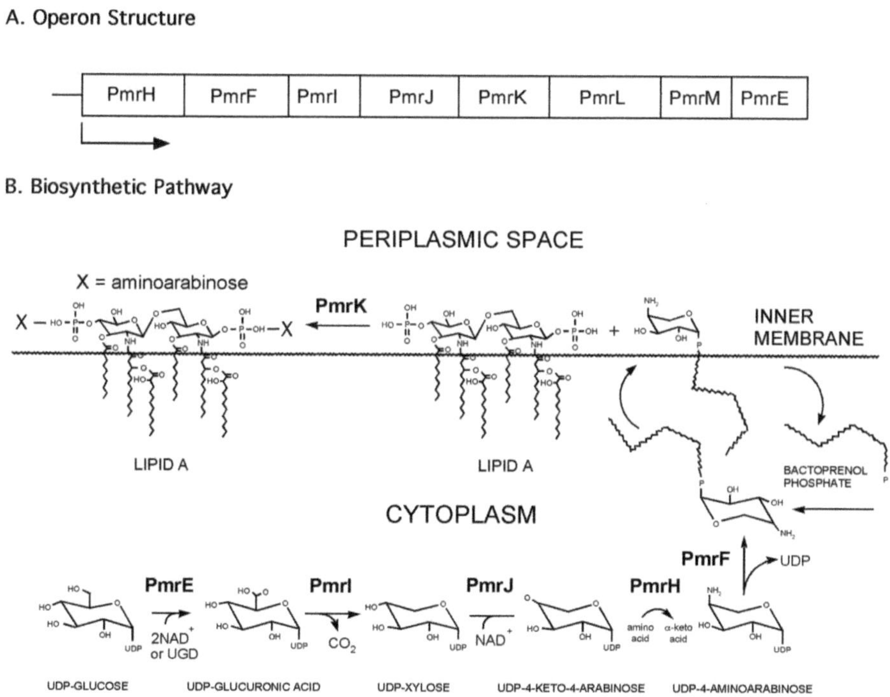

Fig. 11.4 Biosynthesis of aminoarabinose-modified lipid A in *P. aeruginosa*

of 2-hydroxylaurate (Moskowitz et al., 2000). Thus, addition of aminoarabinose to a hexa-acylated species, *m/z* 1601, a penta-acylated species, *m/z* 1431, and a hexa-acylated species, *m/z* 1669 (corresponding to loss of a hydroxyl group from the structures shown in Fig. 11.3d–f, respectively), is commonly seen in polymyxin-resistant clinical isolates.

Occasional polymyxin-resistant clinical isolates appear to have lost the second lauryltransferase, HtrB1. Lipid A from these isolates does not contain either of the corresponding hexa-acylated lipid A species, *m/z* 1601 or 1617 (Fig. 11.3d), indicating the presence of only a single laurate (C12:0), presumably in acyl-oxy-acyl linkage to the 2'-position 3OH-C12:0. Rarely, polymyxin-resistant clinical isolates may lose the deacylase, PagL. However, as compared to constitutive aminoarabinosylation and deficient 2-hydroxylation of laurate, these additional changes do not seem to be strongly associated with CAP resistance, as they may also be observed in polymyxin-susceptible CF isolates. Interestingly, in polymyxin-resistant clinical isolates that have lost HtrB1, deacylation of the penta-acylated lipid A species shown in Fig. 11.3b (*m/z* 1419) can still occur, as well as palmitoylation of both the resulting tetra-acylated species (*m/z* 1249) and the preceding penta-acylated species, resulting in MS peaks at *m/z* 1487 and *m/z* 1657 respectively. This indicates that the later enzymatic steps in CF-specific lipid A synthesis (Fig. 11.3) are not strictly dependent on the earlier steps.

11.2.7 *P. aeruginosa* Lipid A Modifications Modulate Host Inflammatory Responses

Persistent lung infections in CF cause the airways to produce a heightened proinflammatory response. Over time, this repetitive cycle progressively damages the lungs. Lipid A is a major stimulator of inflammatory responses through a receptor complex that includes TLR4, the GPI anchored protein CD14, and the secreted protein component MD-2 (Palsson-McDermott and O'Neill, 2004; Miller et al., 2005). PA can synthesize a variety of lipid A structures and these structures have varying inflammatory stimulatory properties in human and mouse cells through a specific domain of TLR4 (Hajjar et al., 2002; Ernst et al., 2003). PA synthesizes a more highly-acylated lipid A structure containing palmitate during adaptation to the cystic fibrosis airway that stimulates increased NF-κB-mediated responses suggesting that the acylation state may affect LPS-mediated responses (Ernst et al., 1999; Hajjar et al., 2002; Ernst et al., 2003). The fact that this adaptation significantly increases the inflammatory properties of the PA lipid A for humans leads to the hypothesis that an initial lack of recognition of non-palmitoylated lipid A may promote colonization in humans. Therefore, understanding the synthesis of lipid A that occurs in CF clinical isolates should provide important information about the complex host-bacterial interactions that take place within the lungs of CF patients infected with PA.

11.3 Conclusions

PA isolates from CF patients constitutively synthesize lipid A with diverse structural modifications. The synthesis of these structures may make a critical contribution to the pathogenesis of CF lung disease. PA with unique lipid A could contribute to CF lung disease in two ways: by increasing host inflammatory responses, and by increasing bacterial resistance to antibiotics or to elements of host innate immunity such as CAPs. Future investigation of the relevance and regulation of these lipid A structural modifications to CF pulmonary disease will require the construction and testing of PA mutant strains unable to synthesize specific lipid A structures. The components of the bacterial machinery responsible for lipid A structural modifications represent potential drug targets that may enable the development of specific therapies to reduce inflammation and render PA more susceptible to host cell killing.

Acknowledgments The review was supported by grants from the U.S. National Institutes of Health (NIH) to SMM (R01AI067653) and RKE (R01AI047938)

References

Ballard, S.T., Trout, L., Bebok, Z., Sorscher, E.J., Crews, A. CFTR involvement in chloride, bicarbonate, and liquid secretion by airway submucosal glands. Am J Physiol 277 (1999) L694–699.

Bhat, R., Marx, A., Galanos, C., Conrad, R.S. Structural studies of lipid A from *Pseudomonas aeruginosa* PAO1: occurrence of 4-amino-4-deoxyarabinose. J Bacteriol 172 (1990) 6631–6636.

Bonfield, T.L., Konstan, M.W., Burfeind, P., Panuska, J.R., Hilliard, J.B. Berger, M., Normal bronchial epithelial cells constitutively produce the anti-inflammatory cytokine interleukin-10, which is down-regulated in cystic fibrosis. Am J Respir Cell Mol Biol 13 (1995) 257–261.

Boucher, J.C., Yu, H., Mudd, M.H., Deretic, V. Mucoid *Pseudomonas aeruginosa* in cystic fibrosis: characterization of muc mutations in clinical isolates and analysis of clearance in a mouse model of respiratory infection. Infect Immun 65 (1997) 3838–3846.

Burns, J.L., Gibson, R.L., McNamara, S., Yim, D., Emerson, J., Rosenfeld, M., Hiatt, P., McCoy, K., Castile, R., Smith, A.L., Ramsey, B.W. Longitudinal assessment of *Pseudomonas aeruginosa* in young children with cystic fibrosis. J Infect Dis 183 (2001) 444–452.

Burns, J.L., Van Dalfsen, J.M., Shawar, R.M., Otto, K.L., Garber, R.L., Quan, J.M., Montgomery, A.B., Albers, G.M., Ramsey, B.W., Smith, A.L. Effect of chronic intermittent administration of inhaled tobramycin on respiratory microbial flora in patients with cystic fibrosis. J Infect Dis 179 (1999) 1190–1196.

Cystic Fibrosis Foundation. CFF Patient Registry. Patient Registry Annual Data Report 2007. Bethesda, Maryland (2009).

Davis, P.B., Drumm, M., Konstan, M.W. Cystic fibrosis. Am J Respir Crit Care Med 154 (1996) 1229–1256.

Ernst, R.K., Hajjar, A.M., Tsai, J.H., Moskowitz, S.M., Wilson, C.B., Miller, S.I. *Pseudomonas aeruginosa* lipid A diversity and its recognition by Toll-like receptor 4. J Endotoxin Res 9 (2003) 395–400.

Ernst, R.K., Yi, E.C., Guo, L., Lim, K.B., Burns, J.L., Hackett, M., Miller, S.I. Specific lipopolysaccharide found in cystic fibrosis airway *Pseudomonas aeruginosa*. Science 286 (1999) 1561–1565.

Geurtsen, J., Steeghs, L., Hove, J.T., van der Ley, P., Tommassen, J. Dissemination of lipid A deacylases (pagL) among gram-negative bacteria: identification of active-site histidine and serine residues. J Biol Chem 280 (2005) 8248–8259.

Gibbons, H.S., Lin, S., Cotter, R.J., Raetz, C.R. Oxygen requirement for the biosynthesis of the S-2-hydroxymyristate moiety in *Salmonella typhimurium* lipid A. Function of LpxO, A new Fe2+/alpha-ketoglutarate-dependent dioxygenase homologue. J Biol Chem 275 (2000) 32940–32949.

Goldman, R.C., Doran, C.C., Kadam, S.K., Capobianco, J.O. Lipid A precursor from *Pseudomonas aeruginosa* is completely acylated prior to addition of 3-deoxy-D-manno-octulosonate. J Biol Chem 263 (1988) 5217–5223.

Goto, T., Nakame, Y., Nishida, M., Ohi, Y., Bacterial biofilms and catheters in experimental urinary tract infection. Int J Antimicrob Agents 11 (1999) 227–231.

Hajjar, A.M., Ernst, R.K., Tsai, J.H., Wilson, C.B., Miller, S.I. Human Toll-like receptor 4 recognizes host-specific LPS modifications. Nat Immunol 3 (2002) 354–359.

Hancock, R.E.W., Mutharia, L.M., Chan, L., Darveau, R.P., Speert, D.P., Pier, G.B. *Pseudomonas aeruginosa* isolates from patients with cystic fibrosis: a class of serum sensitive, nontypable strains deficient in lipopolysaccharide side chains. Infect Immun 42 (1983) 170–177.

Henry, R.L., Mellis, C.M., Petrovic, L. Mucoid *Pseudomonas aeruginosa* is a marker of poor survival in cystic fibrosis. Pediatr Pulmonol 12 (1992) 158–161.

Jayaraman, S., Joo, N.S., Reitz, B., Wine, J.J. Verkman, A.S. Submucosal gland secretions in airways from cystic fibrosis patients have normal [Na(+)] and pH but elevated viscosity. Proc Natl Acad Sci USA 98 (2001) 8119–8123.

Karunaratne, D.N., Richards, J.C., Hancock, R.E.W. Characterization of lipid A from *Pseudomonas aeruginosa* O-antigenic B band lipopolysaccharide by 1D and 2D NMR and mass spectral analysis. Arch Biochem Biophys 299 (1992) 368–376.

Knowles, M.R., Boucher, R.C. Mucus clearance as a primary innate defense mechanism for mammalian airways. J Clin Invest 109 (2002) 571–577.

Kulshin, V.A., Zahringer, U., Lindner, B., Jager, K.E., Dmitriev, B.A., Rietschel, E.T. Structural characterization of the lipid A component of *Pseudomonas aeruginosa* wild-type and rough mutant lipopolysaccharides. Eur J Biochem 198 (1991) 697–704.

Li, Z., Kosorok, M.R., Farrell, P.M., Laxova, A., West, S.E., Green, C.G., Collins, J., Rock, M.J., Splaingard, M.L. Longitudinal development of mucoid *Pseudomonas aeruginosa* infection and lung disease progression in children with cystic fibrosis. Jama 293 (2005) 581–588.

Lyczak, J.B., Cannon, C.L., Pier, G.B. Establishment of *Pseudomonas aeruginosa* infection: lessons from a versatile opportunist. Microbes Infect 2 (2000) 1051–1060.

Macfarlane, E.L., Kwasnicka, A., Hancock, R.E. Role of *Pseudomonas aeruginosa* PhoP-phoQ in resistance to antimicrobial cationic peptides and aminoglycosides. Microbiology 146(Pt 10) (2000) 2543–2554.

Macfarlane, E.L., Kwasnicka, A., Ochs, M.M., Hancock, R.E. PhoP-PhoQ homologues in *Pseudomonas aeruginosa* regulate expression of the outer-membrane protein OprH and polymyxin B resistance. Mol Microbiol 34 (1999) 305–316.

Mahenthiralingam, E., Campbell, M.E., Speert, D.P. Nonmotility and phagocytic resistance of *Pseudomonas aeruginosa* isolates from chronically colonized patients with cystic fibrosis. Infect Immun 62 (1994) 596–605.

McPhee, J.B., Lewenza, S., Hancock, R.E. Cationic antimicrobial peptides activate a two-component regulatory system, PmrA-PmrB, that regulates resistance to polymyxin B and cationic antimicrobial peptides in *Pseudomonas aeruginosa.* Mol Microbiol 50 (2003) 205–217.

Miller, S.I., Ernst, R.K., Bader, M.W. LPS, TLR4 and infectious disease diversity. Nat Rev Microbiol 3 (2005) 36–46.

Moskowitz, S.M., Burns, J.L., Nguyen, C.D., Høiby, N., Ernst, R.K., Miller, S.I. Polymyxin resistance and lipid A structure of *Pseudomonas aeruginosa* isolated from colistin-treated and colistin-naîve cystic fibrosis patients. Pediatr Pulmonol Suppl 20 (2000) 272.

Moskowitz, S.M., Chmiel, J.F., Sternen, D.L., Cheng, E., Gibson, R.L., Marshall, S.G., Cutting, G.R. Clinical practice and genetic counseling for cystic fibrosis and CFTR-related disorders. Genet Med 10 (2008) 851–868.

Moskowitz, S.M., Ernst, R.K., Miller, S.I. PmrAB, a two-component regulatory system of *Pseudomonas aeruginosa* that modulates resistance to cationic antimicrobial peptides and addition of aminoarabinose to lipid A. J Bacteriol 186 (2004) 575–579.

Palsson-McDermott, E.M., O'Neill, L.A. Signal transduction by the lipopolysaccharide receptor, Toll-like receptor-4. Immunology 113 (2004) 153–162.

Pamp, S.J., Gjermansen, M., Johansen, H.K., Talker-Nielsen, T. Tolerance to the antimicrobial peptide colistin in *Pseudomonas aeruginosa* biofilms is linked to metabolically active cells, and depends on the pmr and mexAB-oprM genes. Mol Microbiol 68 (2008) 223–240.

Pujana, I., Gallego, L., Martin, G., Lopez, F., Candela, J., Cisterna, R. Epidemiological analysis of sequential *Pseudomonas aeruginosa* isolates from chronic bronchiectasis patients without cystic fibrosis. J Clin Microbiol 37 (1999) 2071–2073.

Rajan, S., Saiman, L. Pulmonary infections in patients with cystic fibrosis. Semin Respir Infect 17 (2002) 47–56.

Reid, G., Charbonneau-Smith, R., Lam, D., Kang, Y.S., Lacerte, M., Hayes, K.C. Bacterial biofilm formation in the urinary bladder of spinal cord injured patients. Paraplegia 30 (1992) 711–717.

Romling, U., Fiedler, B., Bosshammer, J., Grothues, D., Greipel, J., von der Hardt, H., Tummler, B. Epidemiology of chronic *Pseudomonas aeruginosa* infections in cystic fibrosis. J Infect Dis 170 (1994) 1616–1621.

Singh, P.K., Schaefer, A.L., Parsek, M.R., Moninger, T.O., Welsh, M.J., Greenberg, E.P. Quorum-sensing signals indicate that cystic fibrosis lungs are infected with bacterial biofilms. Nature 407 (2000) 762–764.

Thomas, S.R., Ray, A., Hodson, M.E., Pitt, T.L. Increased sputum amino acid concentrations and auxotrophy of *Pseudomonas aeruginosa* in severe cystic fibrosis lung disease. Thorax 55 (2000) 795–797.

Trent, M.S., Pabich, W., Raetz, C.R., Miller, S.I. A PhoP/PhoQ-induced Lipase (PagL) that catalyzes 3-O-deacylation of lipid A precursors in membranes of *Salmonella typhimurium*. J Biol Chem 276 (2001) 9083–9092.

Tummler, B., Bosshammer, J., Breitenstein, S., Brockhausen, I., Gudowius, P., Herrmann, C., Herrmann, S., Heuer, T., Kubesch, P., Menus, F., Romling, U., Schmidt, K.D., Spangenberg, C., Walter, S. Infections with *Pseudomonas aeruginosa* in patients with cystic fibrosis. Behring Inst Mitt (1997) 249–255.

van Heeckeren, A., Walenga, R., Konstan, M.W., Bonfield, T., Davis, P.B., Ferkol, T. Excessive inflammatory response of cystic fibrosis mice to bronchopulmonary infection with *Pseudomonas aeruginosa*. J Clin Invest 100 (1997) 2810–2815.

Chapter 12
Development of Small-Molecule Endotoxin Sequestering Agents

Sunil A. David and Diptesh Sil

Abstract Sepsis, otherwise referred to as "blood poisoning" is a serious clinical problem, the incidence of which continues to rise in the US and worldwide despite advances in antimicrobial chemotherapy. The primary trigger in Gram-negative sepsis is endotoxin, a lipopolysaccharide (LPS) constituent of the outer membrane of all Gram-negative bacteria. The structurally highly conserved glycolipid called lipid A is the active moiety of LPS. Lipid A is composed of a hydrophilic, *bis*-phosphorylated di-glucosamine backbone, and a hydrophobic polyacyl domain. The *bis*-anionic, amphiphilic nature of lipid A enables it to interact with a variety of cationic hydrophobic ligands, including polymyxin B, a toxic peptide antibiotic which binds to lipid A and neutralizes endotoxicity. Having determined the structural basis of the interaction of polymyxin B with lipid A, our long-term goal has been to rationally design non-peptidic, nontoxic, small-molecule LPS-sequestrants. Our efforts began with defining the central pharmacophore that determined LPS-recognition and -neutralization properties in small molecules, which led to the discovery of a novel lipopolyamine lead, DS-96. DS-96 is an effective LPS-neutralizer, rivaling polymyxin B in a panel of vitro assays, as well as in protecting animals against endotoxicosis. Structure-activity relationships in our effort to rationally design endotoxin sequestering agents, preclinical assessment of hits and leads, and approaches to overcoming issues with toxicity are described in this chapter.

Keywords Sepsis · Septic shock · Endotoxin · Lipopolysaccharide · Lipopolyamine · Alkylpolyamine · Polymyxin B · Pharmacokinetics · Pharmacodynamics · Cytokine · p38MAPK · Prodrug

---

S.A. David (✉)
Department of Medicinal Chemistry, University of Kansas, Multidisciplinary Research Building, Room 320D, 2030 Becker Drive, Lawrence, KS 66047, USA
e-mail: sdavid@ku.edu

X. Wang, P.J. Quinn (eds.), *Endotoxins: Structure, Function and Recognition*, Subcellular Biochemistry 53, DOI 10.1007/978-90-481-9078-2_12,

Abbreviations

AUC	area under curve
hERG	human Ether-à-go-go related Gene
IL-1β	interleukin-1β
IL-6	interleukin-6
LBP	LPS-binding acute-phase plasma protein
LPS	lipopolysaccharide
Mϕ	macrophages
NOE	nuclear Overhauser effect
NF-κB	nuclear factor κB
P38 MAPK	mitogen-activated kinase p38
PK/PD	pharmacokinetics/pharmacodynamics
PMB	polymyxin B
TNF-α	tumor necrosis factor α

12.1 Introduction

Sepsis, or "blood poisoning" in lay terminology, is a common and serious clinical problem. While fewer than 100 cases were reported prior to 1920 (Felty and Keefer, 1924), it is now the thirteenth leading cause of overall mortality (Gelfand and Shapiro, 1993) and the number one cause of deaths in the intensive care unit (Gasche et al., 1995) accounting for some 200,000 fatalities in the US annually (1990). While the incidence continues to rise in the US (Martin et al., 2003) (Fig. 12.1) and worldwide (Moss and Martin, 2004) due to increased invasive procedures, immunosuppression and cytotoxic chemotherapy, mortality has essentially remained unchanged at about 45% (Cross and Opal, 1994) despite tremendous strides in antimicrobial chemotherapy, due to the lack of specific therapy aimed at the pathophysiology of sepsis.

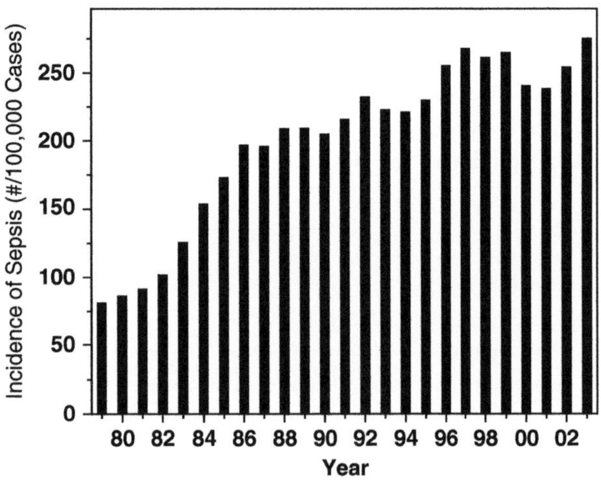

Fig. 12.1 Incidence of Sepsis in the U.S. Data provided by Greg Martin (Martin et al., 2003)

12.2 Endotoxin, the Trigger in Gram-Negative Sepsis

The primary trigger in the Gram-negative septic shock syndrome is endotoxin, a constituent of the outer membrane of Gram-negative bacteria. Endotoxins consist of a polysaccharide portion and a lipid called lipid A, and are therefore also called lipopolysaccharides (LPS) (Fig. 12.2). The polysaccharide portion consists of an O-antigen-specific polymer of repeating oligosaccharide units, the composition of which is highly varied among Gram-negative bacteria. A relatively well-conserved core hetero-oligosaccharide covalently bridges the *O*-antigen-specific chain with lipid A (Rietschel et al., 1994). Total synthesis of the structurally highly conserved lipid A has been shown to be the active moiety of LPS (Rietschel et al., 1987).

12.2.1 Host Responses to Endotoxin

Whereas LPS itself is chemically inert, the presence of LPS in blood (endotoxemia), often a consequence of antibiotic therapy of preexisting bacterial infections, sets off a cascade of exaggerated host responses, which, under normal, homeostatic conditions, serve to orchestrate innate immune defenses. It is the uncontrolled, overwhelming, and precipitous systemic inflammatory response that ultimately manifests clinically in the frequently fatal shock syndrome characterized by

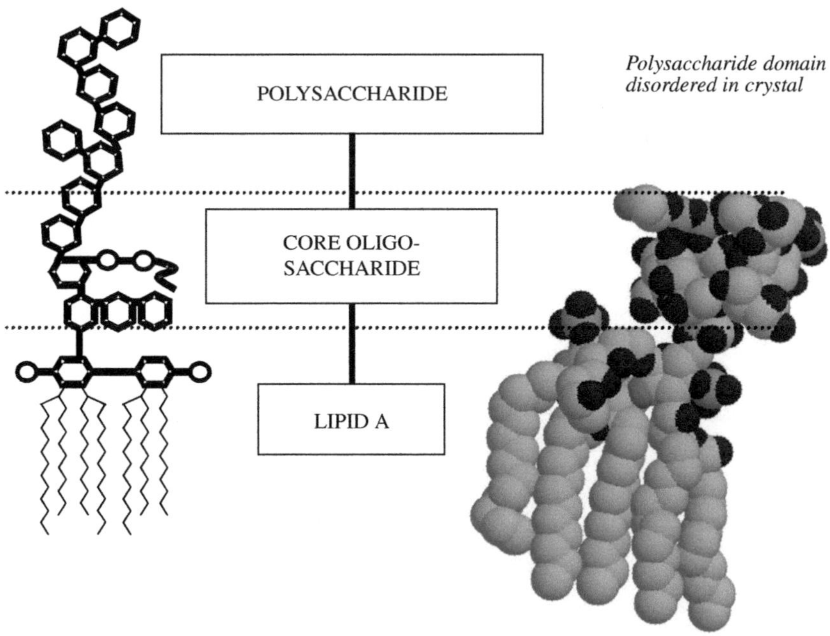

Fig. 12.2 Schematic (*left*), and crystal structure (*right*) of lipopolysaccharide (LPS). Atoms are shown colored in standard CPK scheme

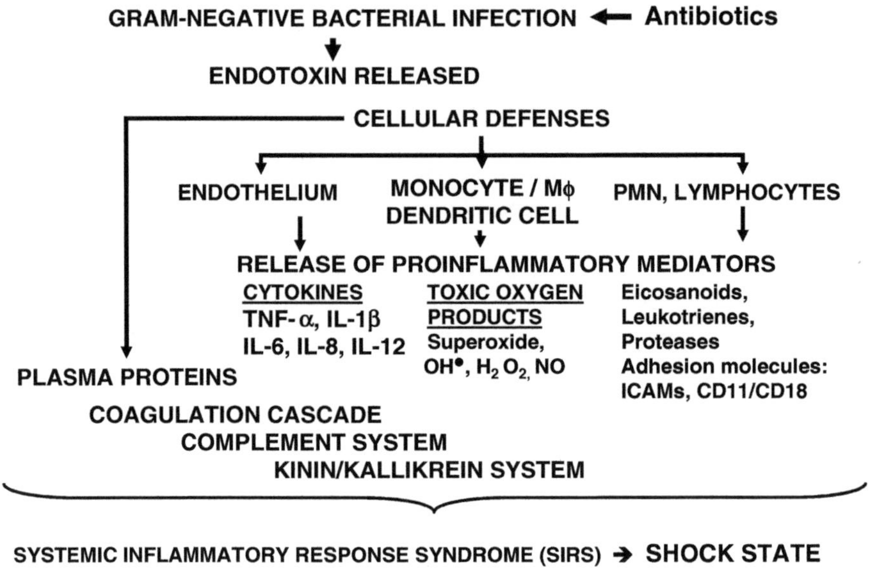

Fig. 12.3 Schematic representation of immune activation by LPS

endothelial damage, coagulopathy, loss of vascular tone, myocardial dysfunction, tissue hypoperfusion, and multiple-system organ failure (Balk and Bone, 1989; Bone et al., 1992; Bone, 1993). LPS activates almost every component of the cellular and humoral (plasma protein) limbs of the immune system (Fig. 12.3), resulting in the production of a plethora of proinflammatory mediators, important among which are the cytokines tumor necrosis factor(TNF-α), interleukin 1β (IL-1), and IL-6, secreted mainly by monocytes and macrophages (Mϕ) (Dinarello, 1991, 1996). These cytokines and other mediators act in concert, amplifying the resultant generalized inflammatory processes. Endotoxemia and its sequelae may arise even in the absence of Gram-negative bacterial infections, conditions such as trauma (Saadia et al., 1990), burns (Jones II et al., 1991), and splanchnic ischemia during cardiac surgery (Rocke et al., 1987) increased intestinal permeability, resulting in the spillover into the portal circulation of LPS from the colon which is abundantly colonized by Gram-negative bacteria.

Our understanding of basic mechanisms underlying the cellular response to LPS has increased vastly in recent years. Important research contributions include LPS delivery to the cell surface via an LPS-binding acute-phase plasma protein (LBP) (Betz Corradin et al., 1992; Fenton and Golenbock, 1998; Gallay et al., 1994; Poltorak et al., 1998; Wahl et al., 1979), subsequent recognition by CD14 (Arditi et al., 1993; Astiz et al., 1996; Wright et al., 1990; Wright, 1995), initiation of signal transduction by toll-like receptor-4, (Baker et al., 1992; Kaisho and Akira, 2006; Kawai and Akira, 2007; Uematsu and Akira, 2006), and downstream cellular activation events mediated by mitogen-activated kinase p38 and c-Jun N-terminal kinase, leading to nuclear translocation of NF-κB (Bohuslav et al., 1998;

Geppert et al., 1994; Han et al., 1994; Haruhiko et al., 1988; Kawasaki et al., 1994; Unger et al., 1984), resulting in cytokine mRNA transcription. These advances will likely offer novel therapeutic possibilities in the future. However, after more than two decades of intensive effort at evaluating more than 30 investigational compounds, specific therapeutic options for sepsis have remained elusive. In November 2001, the FDA approved drotrecogin alfa (Xigris^{TM}, recombinant human activated protein C), an anticoagulant that ameliorates disseminated intravascular coagulation. In a landmark clinical trial, drotrecogin alfa was found to lower the absolute mortality rate to 24.7% from 30.8% in the placebo group (Bernard et al., 2001). A recent, large, multi-center clinical trial concluded that "the absence of a beneficial treatment effect, coupled with an increased incidence of serious bleeding complications, indicates that drotrecogin alfa-activated should not be used in patients with severe sepsis..." (Abraham et al., 2005). The economic impact and cost-effectiveness of Xigris^{TM} is also of concern, with the cost of treating patients with an APACHE II score of 24 or less (severe sepsis) being $575,054 per life-year gained (Manns et al., 2002). Clinical trials of recent years aimed at blocking various proinflammatory mediators including TNF-α, IL-1β, platelet-activating factor, and prostaglandins produced by the activated cellular components (Fig. 12.2) have all been disappointing (Zeni et al., 1997) (Fig. 12.4), suggesting that targeting downstream cellular inflammatory processes once immune activation has already progressed is unlikely to be of benefit.

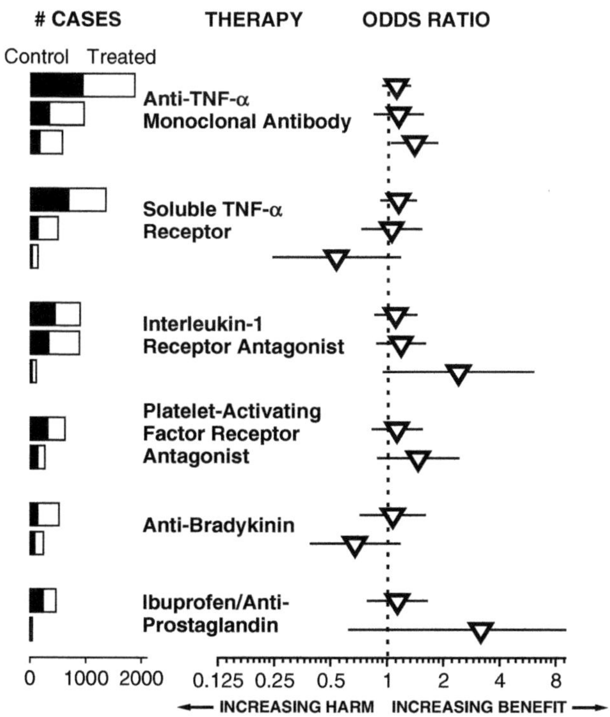

Fig. 12.4 Outcomes of clinical trials of anti-inflammatory agents in sepsis

12.2.2 Complexation of LPS by Macromolecules: A Failed Therapeutic Strategy?

As mentioned earlier, the polysaccharide portion of LPS is highly variable and serologically distinct for each strain of the same species of Gram-negative organisms. Although anti-O-polysaccharide antibodies afford protection in experimental models where animals are challenged with homologous bacteria (Kim et al., 1988; Siegel, 1995), these are not likely to be of significant clinical value since sepsis runs an acute course before the pathogen is identified and appropriate specific immunotherapy is instituted. The biologically active part of LPS, lipid A, as well as the core oligosaccharide portion are structurally highly conserved across Gram-negative genera, and thus are attractive targets for sequestration, and elimination of circulating LPS would, in principle, prevent the activation of inflammatory cascades (Wortel et al., 1991; Ziegler et al., 1982; Ziegler, 1988; Ziegler and Smith, 1992). Experimental studies as early as 1968 suggested that antibodies directed toward epitopes in the core region of LPS may be broadly cross-protective against a range of Gram-negative organisms (Chedid et al., 1968). However, neither human (HA-1A) (Ziegler et al., 1991) nor murine (E5) (Bone et al., 1995) anti-lipid A monoclonal antibodies afforded significant protection in large, multiple, placebo-controlled clinical trials (Cross and Opal, 1994). In the wake of these failures, the subject of several editorials, it became apparent that these monoclonal antibodies had been expedited through clinical trials without rigorous preclinical evaluation. Both HA-1A and E5 exhibited low intrinsic binding affinities to LPS (Warren et al., 1993) ($<10^4 M^{-1}$), neutralized LPS poorly (Baumgartner et al., 1990), bound promiscuously to a wide range of hydrophobic ligands such as lipoproteins and cardiolipin (Helmerhorst et al., 1998), as well as to a variety of human B cell and erythrocyte proteins (Bhart et al., 1993; Helmerhorst et al., 1998), and proved to be toxic in a canine model of septic shock (Quezado et al., 1993). Although efforts at developing core region-directed antibodies continued (Di Padova et al., 1993; Le Roy et al., 1999), disappointing results obtained subsequently could point to intrinsic problems with lipid antigens: poor immunogenicity, inaccessibility of neutralizing epitopes, the generation of nonspecific cross-reactive antibodies against irrelevant hydrophobic epitopes (Vaarala et al., 1988), and potential problems with the antibody molecule itself: predominant intravascular compartmentalization, and possible tissue damage induced by activation of complement.

12.2.3 The Paradigm of Non-Immunologic Sequestration of LPS by Small Molecules: A More Accessible Strategy?

The structurally invariant and biologically active center of LPS, lipid A, is a logical therapeutic target for neutralization. Lipid A is composed of a hydrophilic, negatively charged *bis*-phosphorylated di-glucosamine backbone, and a hydrophobic domain of 6 (*E. coli*) or 7 (*Salmonella*) acyl chains (Galanos and Lüderitz,

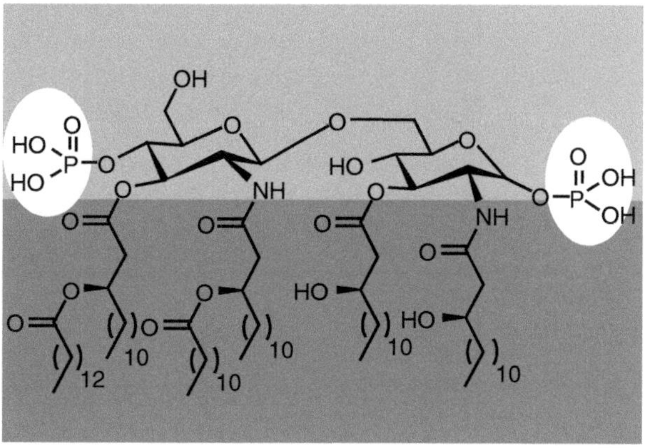

Fig. 12.5 Structure of lipid A. The presence of anionic (*white*), hydrophilic (*light gray*), and hydrophobic (*dark gray*) domains enable the binding of cationic amphipaths to lipid A

1984) (Fig. 12.5). The anionic amphiphilic nature of lipid A enables it to interact with a variety of cationic hydrophobic ligands (Peterson et al., 1985; Rocque et al., 1988; Vaara and Vaara, 1983). Polymyxin B (PMB), a cationic amphiphilic cyclic decapeptide antibiotic isolated from *Bacillus polymyxa* (Storm and Rosenthal, 1977), has long been recognized to bind lipid A (Morrison and Jacobs, 1976), and neutralize its toxicity in animal models of endotoxemia (Durando et al., 1994; Stokes et al., 1989; Yao et al., 1995). Although PMB is a commonly-used topical antibiotic, it is potently nephro- and oto-toxic, which, while precluding its utility as an LPS-neutralizer in patients with sepsis, has stimulated the search for nontoxic PMB analogs (Porro et al., 1998; Rustici et al., 1993), PMB derivatives (Vaara, 1983; Viljanen et al., 1991), as well as other structurally diverse cationic amphiphilic peptides (Iwagaki et al., 2000; Jerala and Porro, 2004; Porro, 1994; Rustici et al., 1993; Scott et al., 2000) as candidate LPS-binding agents. Notably, a hemoperfusion cartridge based on polymyxin B covalently immobilized via one of its NH_2 groups to a polystyrene based fiber became available in Japan in late 2000 for clinical use ("Toraymyxin", Toray Industries Inc., Tokyo) (Nakamura et al., 1999, 2002, 2003). Whilst the utility of Toraymyxin provides a clinically validated proof-of-principle for the value of sequestering circulating LPS, opportunities for extracorporeal hemoperfusion are rare. The typical patient is often profoundly hypotensive with circulatory failure, refractory even to maximal vasopressor regimens, underlining the need to develop alternate strategies for LPS sequestration.

Our long-term goal has been to utilize the structural information in the LPS-PMB complex to rationally design non-peptide, small-molecule LPS-sequestrants. We have elected to focus on targeting the lipid A moiety of LPS rather than on downstream inflammatory processes which have all met with failure (Quezado et al., 1995; Zeni et al., 1997). We first elucidated the solution structure of PMB, both in its free, aqueous, as well as LPS-bound states (Fig. 12.6) (Bhattacharjya et al., 1997). We initially evaluated peptides, both naturally occurring (Bhattacharjya et al., 1997; David et al., 1992, 1993), and de novo synthesized (Bhattacharjya et al., 1997; David, 2001), testing specific hypotheses pertaining to structural correlates

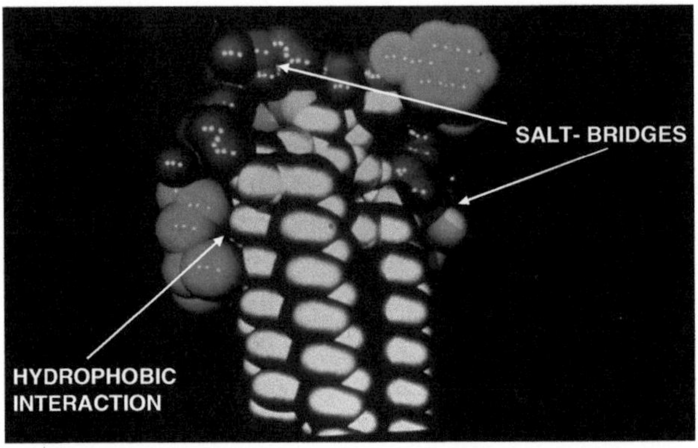

Fig. 12.6 A transfer-NOE derived space-fill model of the polymyxin B (PMB) -lipid A complex. Bidendate ionic H-bonds between pairs of the γ-amino groups of Dab residues of PMB and the phosphates of lipid A, as well as hydrophobic interactions between the methyloctanoate group of PMB and the polyacyl domain of lipid A stabilize the interactions

of lipid A binding, and later extended those design principles to small molecules (David et al., 1994, 1995b). This approach firstly led to the definition of a crucial pharmacophore that determined LPS-recognition and –neutralization properties in small molecules (David et al., 1994, 1995b), and, subsequently, to the discovery of a novel lipopolyamine lead which was shown to be an effective LPS-neutralizer (David et al., 1999).

12.3 Lipopolyamines as Endotoxin Sequestrants

Our desire in evaluating lipopolyamines as potential endotoxin sequestering molecules was based on two simple heuristics which had been experimentally tested and validated:

Heuristic 1: An optimal distance of ∼14 Å is necessary between protonatable functions in *bis*-cationic molecules for simultaneous ionic interactions with the glycosidic phosphates on lipid A, it is the *bis*-cationic scaffold that is the principal determinant of binding affinity (Fig. 12.7) (David et al., 1994, 1995b).

Heuristic 2: Binding is necessary, but not sufficient for activity, and an additional, appropriately positioned hydrophobic group is obligatory for the interaction to manifest in neutralization of endotoxicity (Blagbrough et al., 2000; David et al., 2002).

For the sake of brevity, salient aspects of key structure-activity relationships in one homologous series of twelve mono- and *bis*-acyl homologated spermine analogues (Fig. 12.8) will be first discussed (Miller et al., 2005). We addressed two questions in this study: (i) what is the optimal hydrophobic chain length for effective anti-endotoxic activity, (ii) are symmetrical *bis*-acyl spermines more effective than mono-acyl compounds? We found that a carbon number of 14–16 is optimal in mono-acyl spermines (Fig. 12.9) which are, in general, as potent as

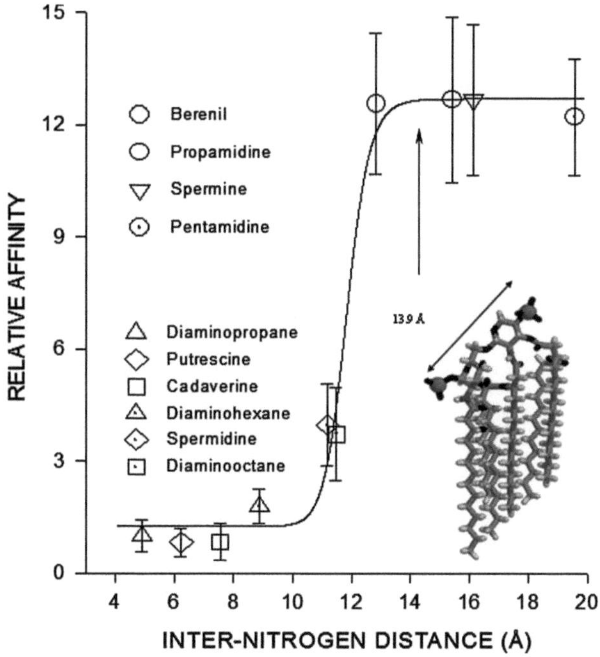

Fig. 12.7 Relationship between inter-nitrogen distance of α,ω–diaminoalkanes, polyamines, and bisamidines, and binding affinity to lipid A. The inflection point of the sigmoidal curve coincides with the inter-phosphate distance of lipid A (inset)

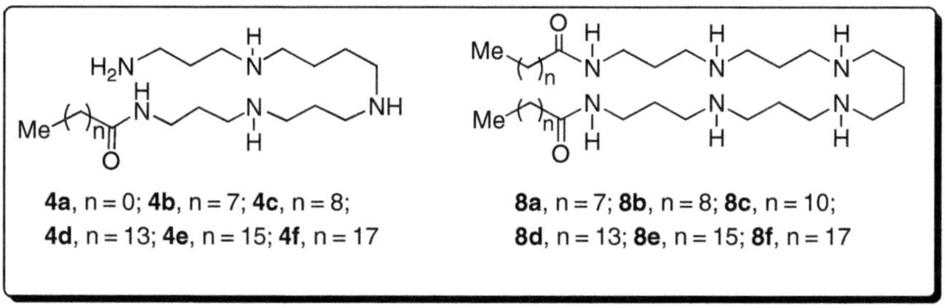

Fig. 12.8 Structures of mono- and *bis*-acyl polyamine analogs

their *bis*-homologs and, in addition, show less surface activity (lower nonspecific cytotoxicity) and possess physical properties that are better suited for parenteral administration. In comparing the NO and TNF-α inhibition profiles with the LPS-binding affinities for this congeneric series, with the exception of **4a**, all of the other mono-acyl compounds bind LPS with ED_{50} values between ~1–2 μM, while only the longer acyl chain compounds (**4d–f**) are biologically active (Fig. 12.10). This result emphasizes the necessity of employing a biological primary screen in tandem

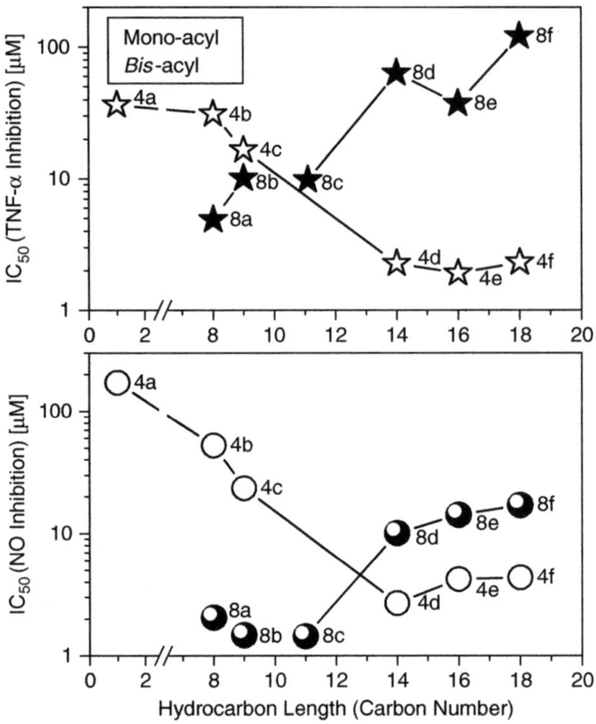

Fig. 12.9 Correlation between carbon number of the hydrocarbon group in mono- and *bis*-acyl and TNF-α inhibition in human blood (*top*) and NO inhibition in murine J774A.1 cells (*bottom*)

with the displacement assay in order to derive reliable structure-activity relationships in LPS-sequestering compounds. The apparent inverse correlation between ED_{50} and neutralization potency for the *bis*-acyl **8** series was simply a consequence of the poor solubility of the **8** homologs. Free aqueous concentrations of **8** are progressively retarded with increasing chain length, diminishing binding and, consequently, neutralization (Fig. 12.10). We characterized the protective effects of **4e** (the most potent compound in the human TNF-α inhibition assay), **8a**, (most active in inhibiting NO release in J774A.1 cells), and **8c**, which was of lower potency than either **4e** or **8a** in both assays (for comparison). A supralethal dose (twice the dose causing 100% lethality) of 200 ng/mouse was administered intraperitoneally (i.p.) and separate i.p. injections of graded doses of compound, and lethality was observed at 24 h. The highly soluble **4e** was dissolved in saline. The poor solubility of **8a** and **8c** necessitated the administration in 50% DMSO. As is evident from Fig. 12.11, a clear dose-response is observed, with **4e** affording complete (statistically significant) protection at the 100 or 200 μg/mouse dose, and partial protection at the 50 μg dose. Both the *bis*-compounds are inferior to **4e**, underlining the importance of favorable physical properties enabling adequate plasma concentrations of free drug to effectively sequester LPS.

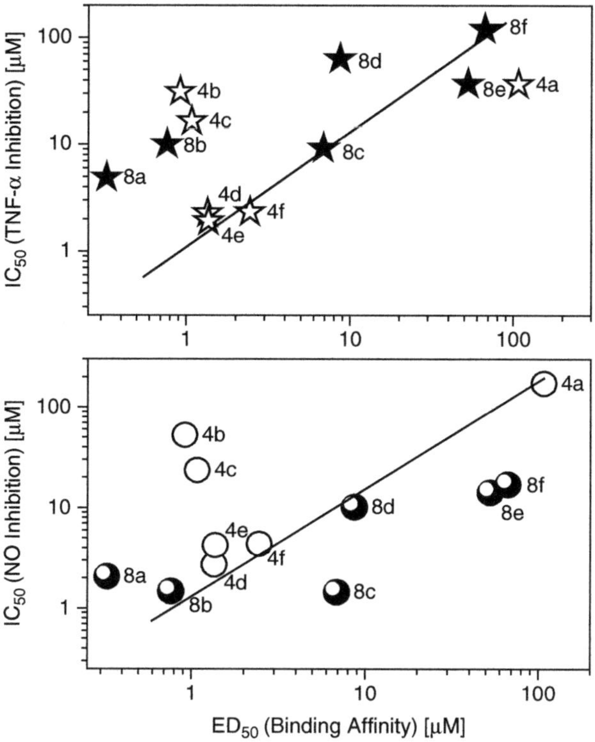

Fig. 12.10 Correlation of binding affinity of the acylhomospermines determined by BC fluorescent probe displacement with NO inhibition in murine J774A.1 cells (*Bottom*) and TNF-α inhibition in human blood (*Top*)

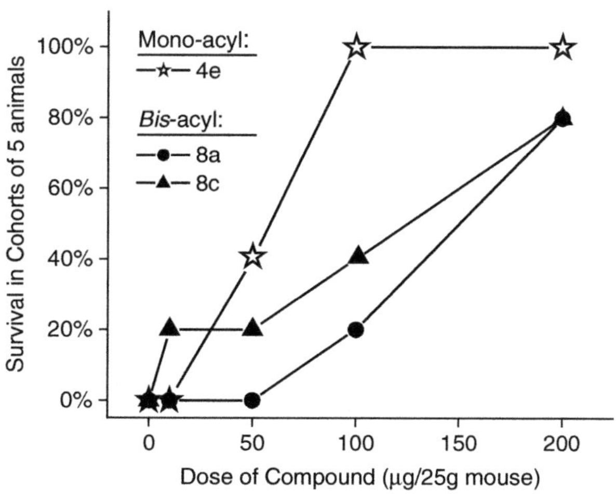

Fig. 12.11 Protective effects of representative mono- and *bis*-acyl compounds in a murine model of septic shock

Compounds such as **4e** bind HSA with a K_D (determined by isothermal titration calorimetry) of 1.98 μM, and a ligand:protein stoichiometry of 5:1 (Fig. 12.12). The binding sites on HSA were determined to be the multiple fatty acid binding pockets (Curry et al., 1998) because of specific displacement of dansylsarcosine and *cis*-parinaric acid, but not warfarin by the lipopolyamines (Nguyen et al., 2008). The consequence of the interaction is a complete abrogation of the surface-activity (and hemolytic activity) of these cationic amphipathic compounds (Balakrishna et al., 2006; Miller et al., 2005) (Fig. 12.12). Gratifyingly, the complexation of the lipopolyamines with HSA results in no change in potency in a variety of in vitro assays (Fig. 12.12) as well as in the murine model of endotoxic shock, implying that either the K_{off} rate of the lipopolyamine:HSA complex is very rapid, or that compounds such as **4e** bind to and sequester LPS as a ternary complex with albumin, as has been observed in the case of polymyxin B (David et al., 1995a, David, 1999). These fortuitous observations have led to a formulation compatible with systemic administration of lipopolyamines. We now routinely solubilize newer analogs in isotonic saline containing physiological concentrations of albumin. Dose-response profiles are identical to that obtained with DMSO solutions with the advantage that repeated intravenous or intraperitoneal injections result in no observable thrombophlebitis or sterile peritonitis in mice.

In addition, significant intrinsic antibacterial activity against representative Gram-positive and -negative bacteria (*E. coli* ATCC 9637 and *S. aureus* ATCC 13709) were observed for the lipopolyamines (Balakrishna et al., 2006). Particularly noteworthy was the observation that some of these compounds sensitized Gram-negative bacteria to otherwise impermeable antimicrobials such as rifampin by almost 4000-fold, by perturbing the structural integrity of both outer and inner membranes (Balakrishna et al., 2006). Our primary interest in examining these compounds lay not so much in evaluating their intrinsic antimicrobial properties, but rather in understanding the mechanisms and structure-activity relationships underlying their putative membrane permeabilizing action, and in exploring the possibility of employing such compounds as adjuncts to conventional antimicrobial chemotherapy for purposes of sequestering endotoxin released as a consequence of Gram-negative bacterial lysis. The necessity of having to use potent, rapid-acting, bactericidal antibiotics in the therapy of serious systemic Gram-negative infections on the one hand, and the problem of antibiotic-induced endotoxin release (see Fig. 12.3) being a major contribution to the resultant mortality has long been recognized as a therapeutic paradox (Bucklin et al., 1994; Crosby et al., 1994; Hurley, 1992; Hurley, 1995). A compound that synergizes with conventional antibiotics, and sequesters LPS released into systemic circulation as a consequence of microbicidal action would therefore be of potential value (Jackson et al., 1994; Prins et al., 1994, 1995) in the prophylaxis of Gram-negative sepsis.

Fig. 12.12 *Top:* Interaction of **4e** with human serum albumin as probed by isothermal titration calorimetry, a single-site model yielded a stoichiometry of 5:1 of **4e**:HSA with a k_D of ~2 μM. *Middle*: Inhibition of hemolysis of **4e** by HSA. *Bottom:* Identical potency of NO inhibition of **4e** solubilized in DMSO or in HSA

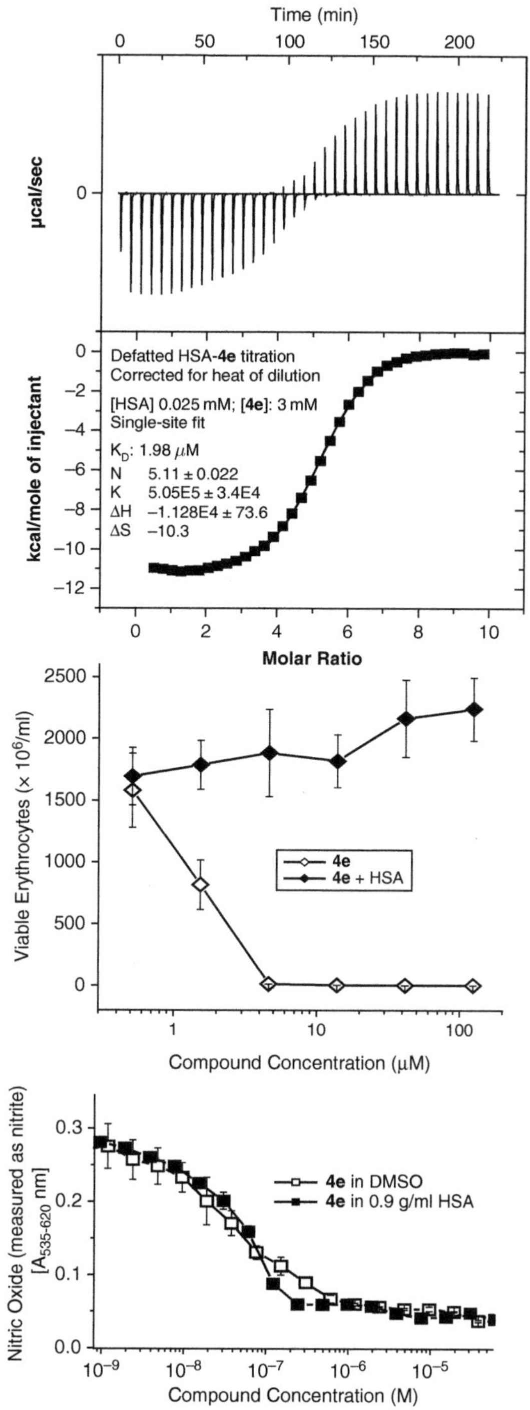

12.3.1 Further SAR Lessons Learned En Route to DS-96, an N-Alkylhomospermine Lipopolyamine

Having determined that a non-symmetric, mono-substituted analogue with a hydrophobic appendage of carbon length 16 (**4e**) was optimal, we evaluated analogues of **4e** with the C_{16} moiety linked to the homospermine backbone via ureido and carbamate functionalities, hypothesizing that these compounds would provide for a more favorable pharmacodynamic profile on account of the attenuated lability of these linkages to hydrolytic cleavage in vivo. This proved not to be fruitful for two reasons: (i) these compounds exhibited lower solubility, and significantly higher in vitro cytotoxic index as assessed by the XTT assay, (ii) furthermore, an improved $t_{1/2}$ seemed superfluous since time-course experiments in mice suggested an excellent pharmacodynamic profile for **4e**, with significant protection occurring even when the compound was administered up to 6 h prior to lethal LPS challenge (Miller et al., 2005).

We next tested the role of the H-bond donor atoms (-NH-) on the polyamine scaffold. Analogues with polypropylene glycol-type scaffolds [NH_2-$(CH_2)_3$-O-$(CH_2)_3$-O-$(CH_2)_3$-O-$(CH_2)_3$-NH-Alkyl group] were evaluated. These were amongst the worst compounds we have ever encountered, being scarcely soluble, and exhibiting low binding affinities and negligible LPS-neutralizing properties (unpublished data). Although initially disappointing, these results unambiguously pointed to the importance of preserving H-bond donor groups on the scaffold, a heuristic that was reinforced when we evaluated DS-96 (discussed below).

Several iterative design-and-test cycles of SAR studies ultimately helped converge and focus our efforts on evaluating compounds with C_{16}-*N*-alkyl polyamine motifs. In our initial studies, in a two-step conventional reductive amination sequence, the amine 1 was condensed with hexadecanal (1.1 eq.), followed by subsequent treatment of the intermediate imine with sodium borohydride. Surprisingly, the major product isolated from the above reaction was found to be the *N*-alkenyl substituted amine 2 (Fig. 12.13), the formation of which probably involves isomerization of the initially formed imine to the corresponding enamine, followed by condensation of the enamine with a second molecule of the aldehyde and subsequent dehydration. This could be prevented by performing the reductive alkylation in the presence of sodium cyanoborohydride in an acidic medium, leading to the desired *N*-alkyl lipopolyamine **5** in good yield. Preliminary evaluation of **AK-1a** proved superior to that of **4e**. This led to a detailed exploration of compounds shown in Fig. 12.14. Dose-response profiles indicated a clear segregation of activity based on structural class. As expected, the inhibitory potencies of *bis*-substituted compounds were worse than mono-substituted analogues. Importantly, this SAR led to the identification of DS-96 (**AK-5a** in Fig. 12.14), a C_{16}-*N*-alkyl-homospermine compound as an exciting lead with an IC_{50} value of 37 nM, which is indistinguishable from that of polymyxin B, the "gold standard" for LPS sequestrants, within experimental error (Fig. 12.14).

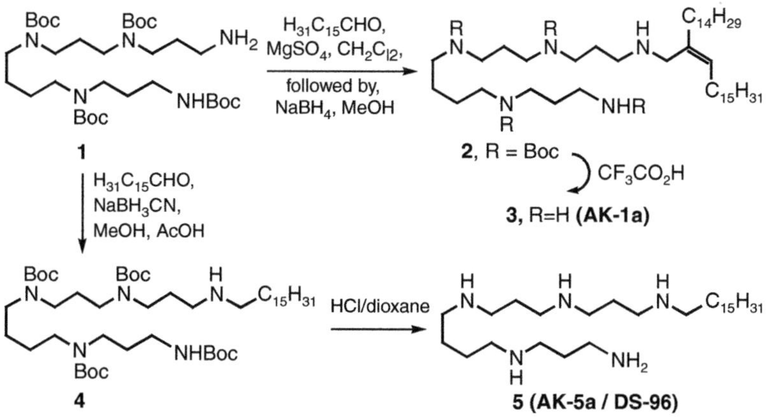

Fig. 12.13 Reductive amination conditions yielding an unexpected *N*-alkenyl and the desired *N*-alkyl homospermine analogues

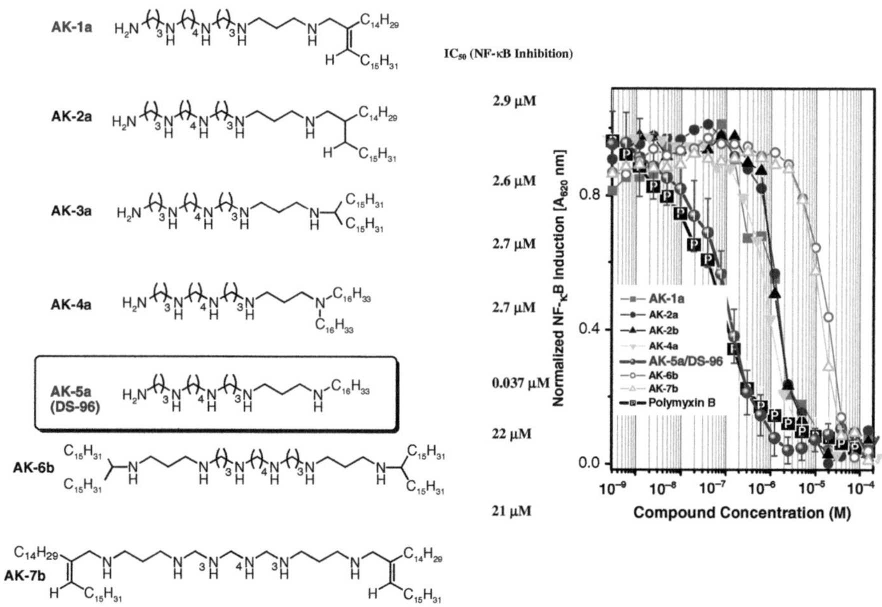

Fig. 12.14 *Left*: Structures of mono-substituted alkylpolyamines (**AK1a-5a**), two *bis*-substituted analogues (**AK6b, 7b**) were also synthesized and examined for SAR confirmation. *Right:* Inhibition of LPS (100 ng/mL)-induced NF-κB induction in HEK-293 cells stably transfected with *Tlr4*, *MD-2*, CD14, and an NF-κB-secreted alkaline phosphatase reporter gene construct. Polymyxin B was used as the positive control

Physical properties such as solubility and bioavailability alone could not adequately explain why mono-alkyl compounds such as DS-96 proved to be vastly better than, for instance, AK-6b, for both compounds could be adequately solubilized as complexes with albumin. In an effort to understand the structural basis for the observed SAR results, computational modeling and docking studies were undertaken, using the crystal structure of LPS (Ferguson et al., 1998). These results have been most instructive in that they not only predict both how a terminal free amine as well as the -NH- (H-bond donor, but not -acceptor, see SAR discussion) functionalities on the scaffold dictate the binding geometry of DS-96 to LPS, but also provide a rational design strategy for developing second-generation leads. Modeling studies indicate that the binding enthalpy is driven by strong salt-bridges between the free terminal primary amine, and additional ionic H-bonds between the intervening secondary amines and the carboxylates of the 3-deoxy-D-manno-octulosonic acid (KDO) serve to anchor the linear scaffold in the cleft between the lipid A glycosidic backbone and the inner core region of LPS (Fig. 12.15).

DS-96 is the most active compound identified by us to date. The compound behaves in every in vitro and in vivo assay, at least equivalent to, if not better than, polymyxin B (Sil et al., 2007). Both DS-96 and PMB bind LPS with equal affinity (Sil et al., 2007), and inhibit LPS-induced NF-κB transactivation with an IC_{50} of ~30 nM (Fig. 12.16a). DS-96 was also found to be active against LPS isolated from a broad range of Gram-negative bacteria (Fig. 12.16c). That the mechanism of action of DS-96 is that of a true LPS-sequestrant was confirmed not only by its specificity toward LPS in that non-LPS stimuli such as TNF-α, PMA and Tlr-2 agonists such

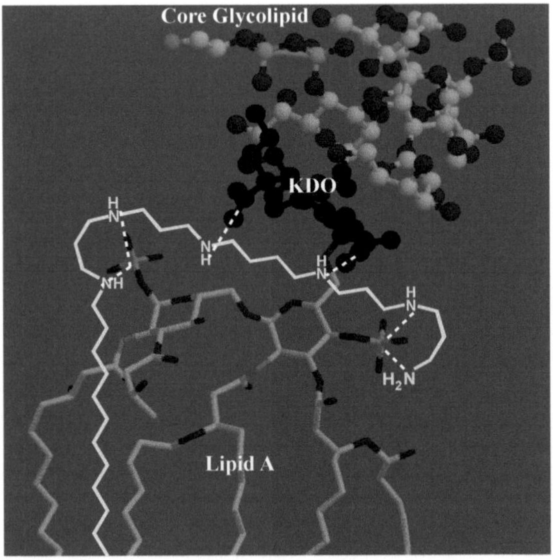

Fig. 12.15 Model of the interaction of DS-96 (in green) with LPS. Note bidentate ionic H-bonds (salt-bridges, interrupted lines) between the amines and the phosphate groups on lipid A (sticks), and additional salt-bridges between the internal secondary amines and the carboxylates of the KDO residues of LPS

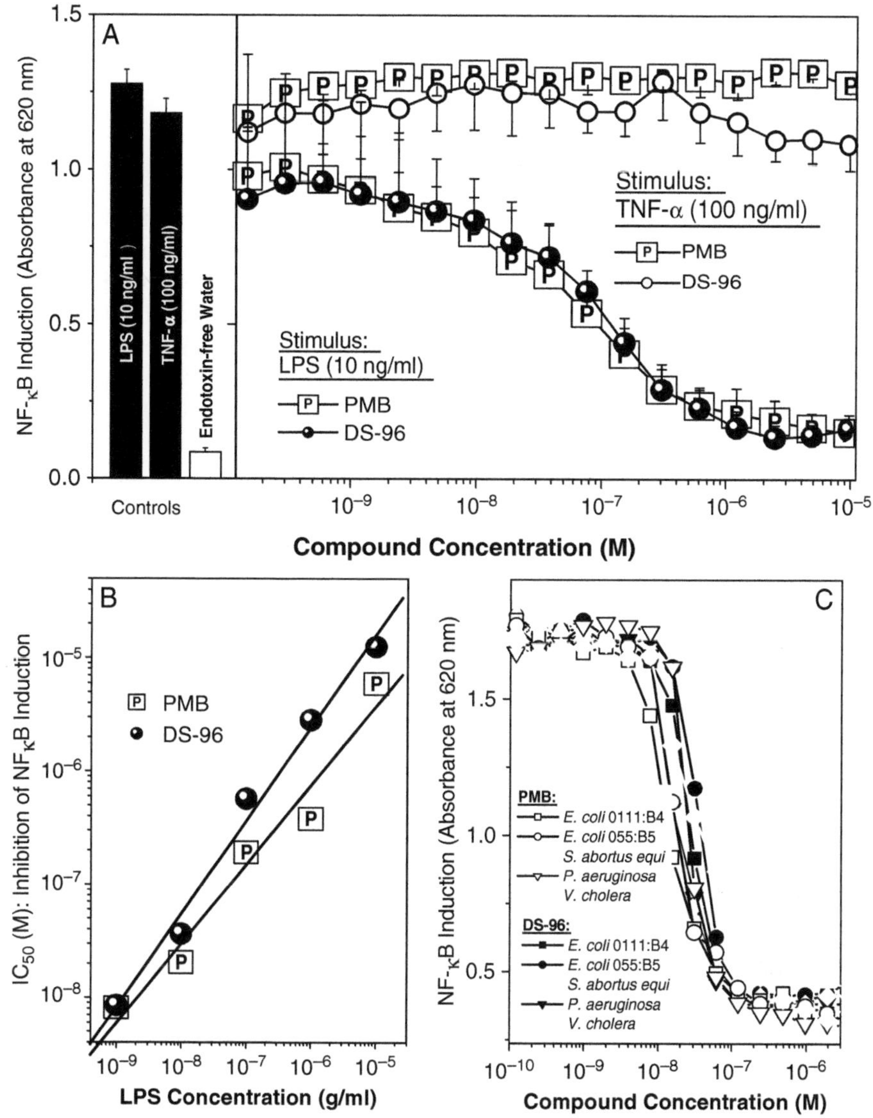

Fig. 12.16 (**a**) Inhibition of LPS-induced (but not TNF-α–induced) NF-κB induction by DS-96 and PMB with identical IC_{50} values of 31 nM, showing specificity for LPS and equipotency with PMB. (**b**) Schild-plot analyses showing a linear relationship between IC_{50} values in the NF-κB assay and the concentration of LPS used as stimulus (law of mass action effects confirm mechanism of action of DS-96 as an LPS sequestrant). (**c**) DS-96 neutralizes LPS isolated from a broad range of Gram-negative organisms

as PAM_2CSK_4 are unaffected, but also by Schild-plot analyses which show that the compound conforms to expected law of mass action behavior (Fig. 12.16b).

12.3.2 Activity of DS-96 in Human Blood

It was important to verify that the compound was active in ex vivo models of innate immune activation and inflammation using whole human blood. The inhibition of LPS-induced p38 MAPK phosphorylation (Han et al., 1994) in neutrophils, an early event in cellular signaling initiated upon recognition of LPS by its receptor, Tlr-4, is clearly evident, the IC_{50} values for PMB and DS-96 being very similar (~4 nM). The profiles for inhibiting CD11b upregulation in neutrophils are also very similar. Both compounds inhibit proinflammatory cytokine production in human blood, with DS-96 being more potent than PMB (Fig. 12.17).

12.3.3 In Vivo Potency, Pharmacodynamics and Pharmacokinetics of DS-96

In the murine model of septic shock, DS-96 afforded significantly better protection than PMB at 4 mg/kg ($p<0.01$, Ch-square test (see manuscript)) (Fig. 12.18a). We confirmed that the protection afforded by DS-96 was attributable to attenuated LPS-induced cytokine production. DS-96 was without any effect on lethality induced by 100 ng/animal of recombinant murine TNF-α. Furthermore, an in vivo Schild-type profile was observed with multiples of LD_{100} doses of LPS requiring escalating doses of DS-96 for protection which clearly show the specificity of the compound as an LPS-sequestrant (Fig. 12.18b). In time-course experiments, DS-96 was maximally effective when administered concurrent to, or up to 4 h prior to LPS administration, partial protection persisted even up to 8 h prior to LPS challenge implying a favorable half-life (Fig. 12.19). This has been confirmed by pharmacokinetic (PK) experiments (see below).

12.3.4 Toxicity of DS-96

It is to be noted that the pharmacodynamics (PD) data (Fig. 12.19) clearly indicate that if DS-96 is to ever to find utility in the clinic as a LPS-sequestrant, it will have to be administered prior to the onset of sepsis, as a prophylactic, before the onslaught of the innate immune response. It is mandatory for a prophylactic regimen to have a very wide margin of safety and, consequently, we have paid particular attention to eliciting any potential toxicity ascribable to DS-96. Daily i.p. administration to rats at doses of 0, 5, and 10 mg/kg (free base equivalent) of DS-96 (either as TFA or HCl salt; the 10 mg/kg free base dose represents 10X 100% protective dose in mice on a body weight basis) for five days resulted in a dose-dependent

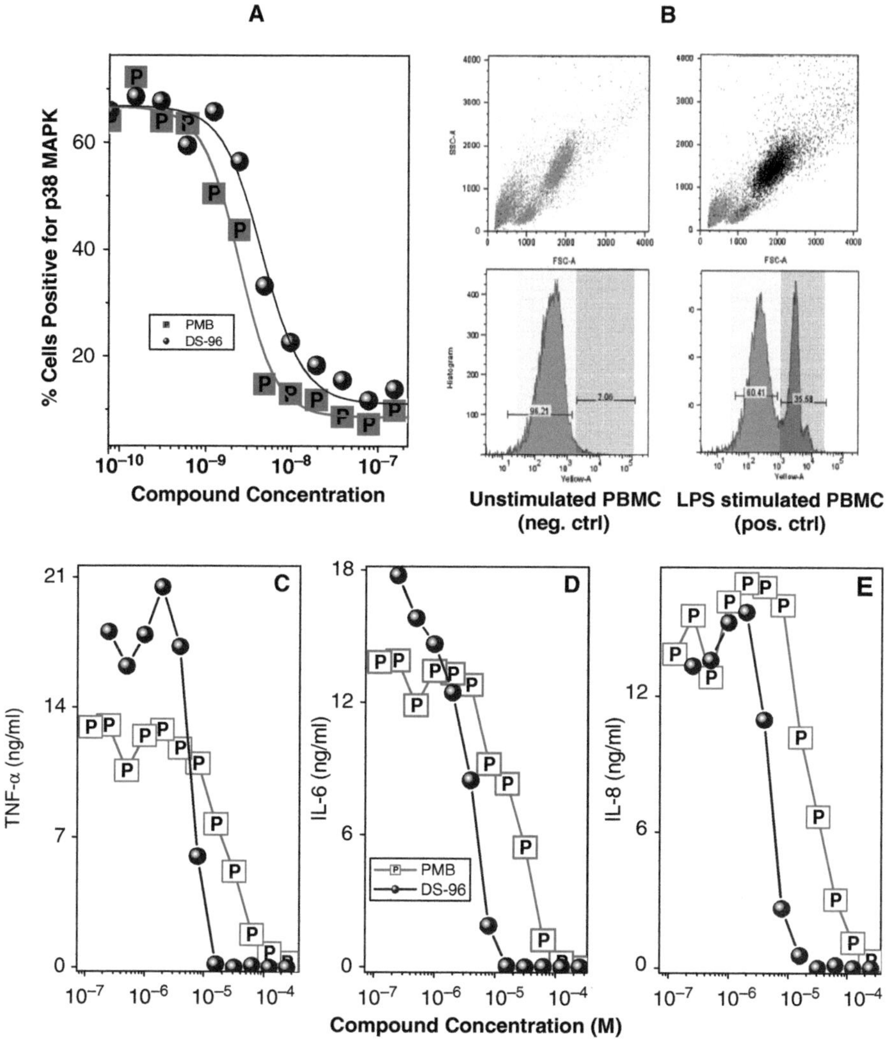

Fig. 12.17 Inhibition of phosphorylation of p38 MAP kinase in neutrophils (**a**, flow cytometry) and cytokine production (**c–e**, flow-cytometric bead array) in whole human blood stimulated with 100 ng/mL LPS, with either DS-96 or PMB (**b**) Forward scatter/side scatter profile and the gating for p38 MAPK-negative and positive gates obtained on unstimulated (negative control) and LPS-stimulated (positive control) cells, respectively

decrease in hematocrit, RBC counts, hemoglobin content, accompanied by a marked degree of compensatory reticulocytosis. These changes were especially marked at 10 mg/kg dose (Fig. 12.20), and are indicative of significant intravascular hemolysis. Although we do not observe measurable hemolysis in human blood ex vivo at up to 50 μM concentrations of DS-96, we did observe pronounced changes in erythrocytic morphology with the appearance of crenellations and membrane blebbing at 10 μM (Fig. 12.21).

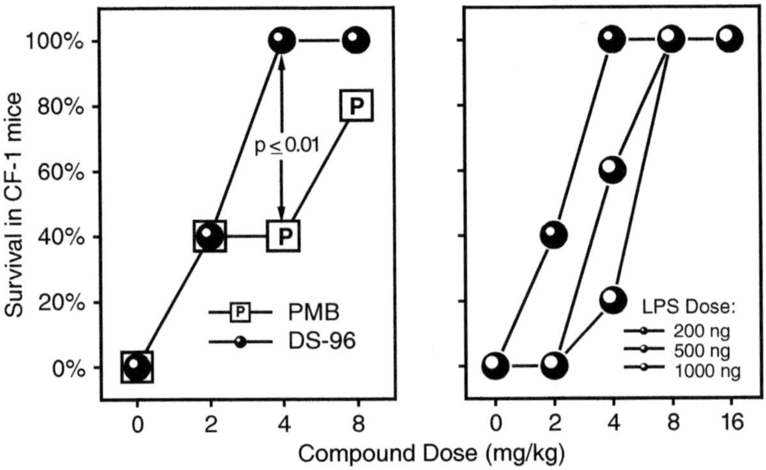

Fig. 12.18 (**a**) Comparison of in vivo potency of PMB and DS-96: dose-dependent increase in survival in mice challenged with a 2X LD_{100} of LPS (200 ng/animal). (**b**) Schild-type response in vivo: dose-dependence of survival in mice challenged with escalating supralethal doses of LPS (200, 500, 1000 ng/animal). The LD_{100} dose of LPS was determined to be 100 ng/mouse

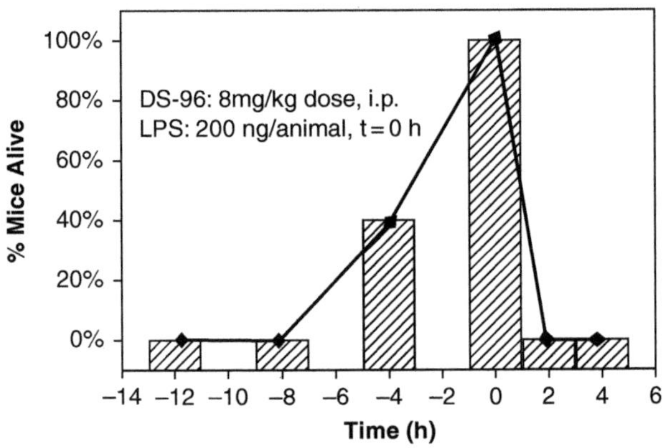

Fig. 12.19 Time-course (pharmacodynamics) of protection conferred by 4 mg/kg of DS-96 administered to cohorts of 5 mice each at various times prior to (12, -8, -4 h), concurrent with (0 h), and following (+2, +4 h) supralethal (200 ng/mouse; 2 X LD_{100} dose) LPS challenge[110]

Gratifyingly, no changes in absolute or differential WBC counts or thrombocytopenia (indices of myelosupression), alterations in bleeding and clotting times, or derangements in bilirubin, hepatic transaminases, blood urea nitrogen/creatinine, or serum electrolytes were noted. Some polyamine derivatives have been shown to cause derangement of polyamine metabolism as a consequence of polyamine oxidase inhibition (Wallace and Fraser, 2004), and we were therefore keen to examine the effects of subacute high-dose treatment with DS-96. No differences in

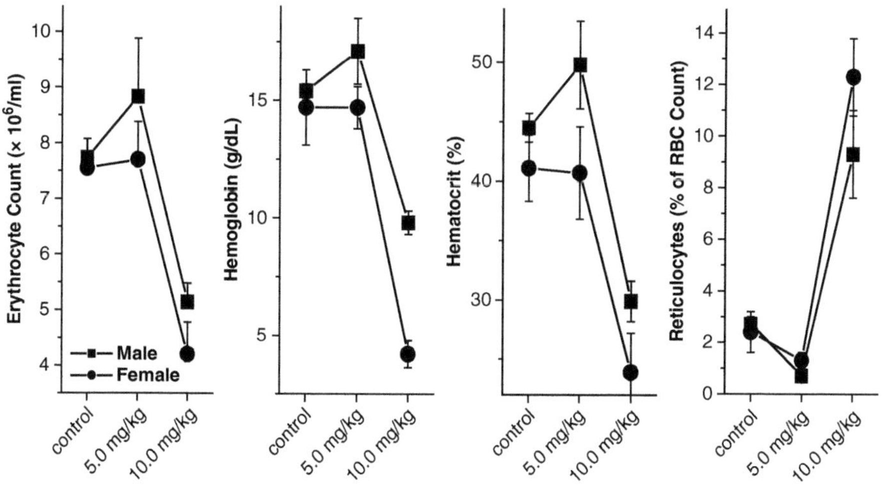

Fig. 12.20 Consequences of intravascular hemolysis in rats receiving graded doses of DS-96 i.p. daily for five days

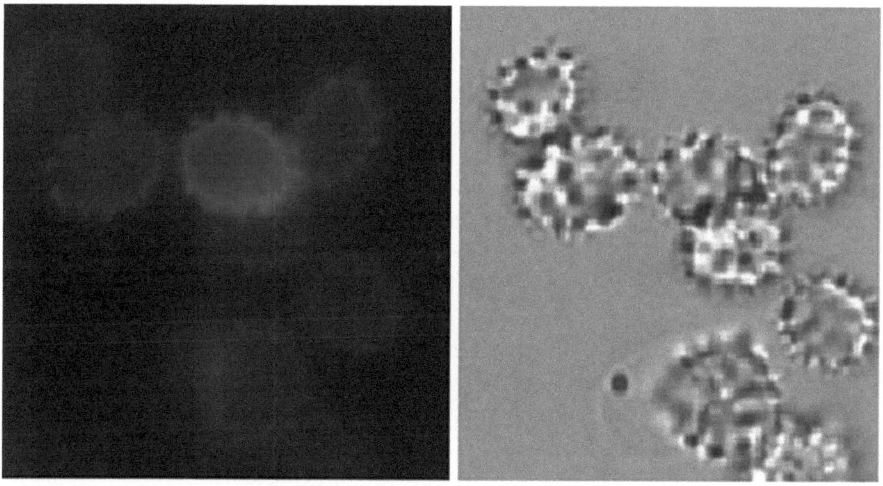

Fig. 12.21 Effect of DS-96-rhodamine conjugate on human erythrocytes at 10 μM: Fluorescence (*left*) and phase-contrast (*right*) micrographs

polyamine levels (putrescine, spermidine, spermine, N^1-acetylspermidine, HPLC detection after derivatization with fluorescamine (Udenfriend et al., 1972) between control mice and those that received 10 mg/kg per day DS-96 for five days were noted.

A more detailed pharmacoprofiling of DS-96, which included a CEREP^{TM} profile for off-target receptor binding (radioligand displacement assays), abbreviated

Ames screen (5, 10, 50, and 100 μM) for mutagenicity, (Maron and Ames, 1983) and hERG channel inhibition at a single concentration (1 M) for torsadogenic/Q-T prolongation activity. (Zhou et al., 1998a, b) In summary, the CEREP screen indicated binding to several amine transporters which is not surprising in light of the polyamine nature of DS-96. The compound was non-mutagenic in *His*-reverted *S. typhimurium* strain T98 with or without S9 rat microsomes. A 24% tail-current inhibition (classified as low to moderate inhibition) was observed in the hERG whole cell patch-clamp experiments at 1 M of DS-96, a value that is comparable to that of propranolol and carvedilol (Kawakami et al., 2006), commonly-used nonselective β-adrenergic blockers. Mention may be made that polyamines, ubiquitous constituents of all mammalian tissues, are well known to modulate the inward rectifying K^+ channels (Kurata et al., 2006; Loussouarn et al., 2005; Williams, 1997a, b). It is possible, even likely, therefore, that spermine itself, present in plasma in millimolar concentrations (Antrup and Seiler, 1980; Seiler and Knodgen, 1978), may exert effects on the hERG channel. These data collectively indicate that the toxicity of the cationic, amphipathic DS-96 is primarily due to its surface-active properties.

12.3.5 Pharmacokinetics of DS-96 and Further Developments

We had sought to address two questions: (a) what is the elimination half-life of DS-96 in the murine model, and does the $t\frac{1}{2}$ correspond to pharmacodynamic data shown in Fig. 12.19; (b) what is the therapeutic plasma concentration of DS-96 that corresponds to full protection against endotoxemic challenge in mice. DS-96 at a dose of 200 μg/mouse (8 mg/kg) was administered to CF1 mice via i.p. and i.v. routes. Plasma concentrations of DS-96 were determined by LC-MS/MS using a deuterated DS-96 internal standard (Nguyen et al., 2008; Shrestha et al., 2008). The elimination $t\frac{1}{2}$ in mice is about 400 min (Fig. 12.22), which is consistent with the observed pharmacodynamic (in vivo efficacy) data shown in Fig. 12.19. The observed concentration-versus-time profile of DS-96 in the mouse i.p. model suggests that a plasma concentration of 0.5–1.5 μg/mL corresponds to complete protection by a dose of 200 ng/animal of LPS in the D-galactosamine-primed model of endotoxin-induced lethality.

We observed that DS-96 was absorbed unexpectedly rapidly from the intraperitoneal cavity, giving rise to surprisingly high plasma levels of the compound. Extrapolation of the expected C_{max} values at 10X the therapeutic dose (i.e., 40 mg/kg) yields a peak plasma concentration of ~10 μg/mL. Surmising that it was the high peak plasma DS-96 levels that was causing intravascular hemolysis, and not the cumulative exposure ($AUC_{0\text{-}\infty}$), we tested this hypothesis by repeated (every hour, for ten hours) i.p. administration of 4 mg/kg of DS-96, and found no evidence of toxicity. Next, we administered a pertrifluoroacetamide derivative (prodrug) of DS-96 (Fig. 12.23), also at 40 mg/kg, with the rationale being that the trifluoroacetyl group would be cleaved by plasma amidases gradually, thereby blunting the C_{max} values of DS-96, but leaving the overall exposure to the drug unaltered.

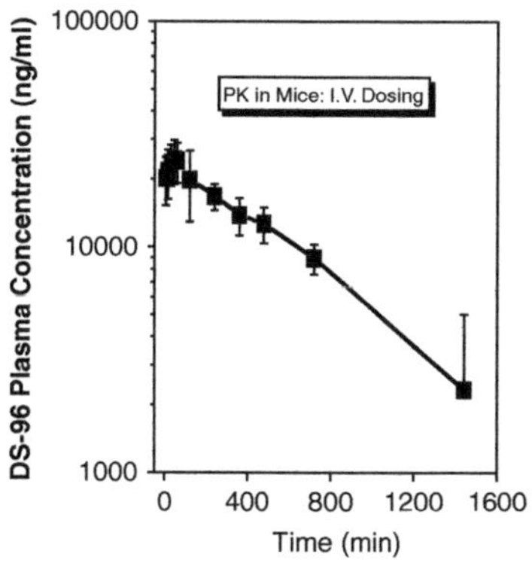

PK Parameter	Mouse, I.P.	Mouse, I.V.
Apparent Volume of Distribution V_d (mL)	190.2	9.3
Clearance/kg Cl (ml /min/kg)	6.856	0.613
Elimination half life $t_{1/2}$ (min)	428	387
C_{max} ng/mL	1366 ± 216	24320 ± 5348
$AUC_{(0-\infty)}$ ng•min•ml	910,389	16,323,904

n = 5 animals per time point

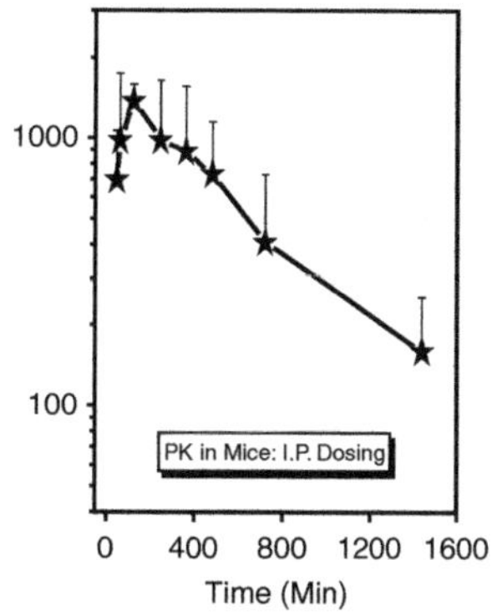

Fig. 12.22 Plasma concentrations in mice with a single i.v. (*left*) and i.p. dosing (*right*) of DS-96 at 4 mg/kg (100% protective dose against lethal endotoxemia). PK parameters are shown tabulated. Data represent mean and SD from 5 animals per time point for each experiment

Fig. 12.23 Nontoxic pertrifluoroacetamide prodrug of DS-96

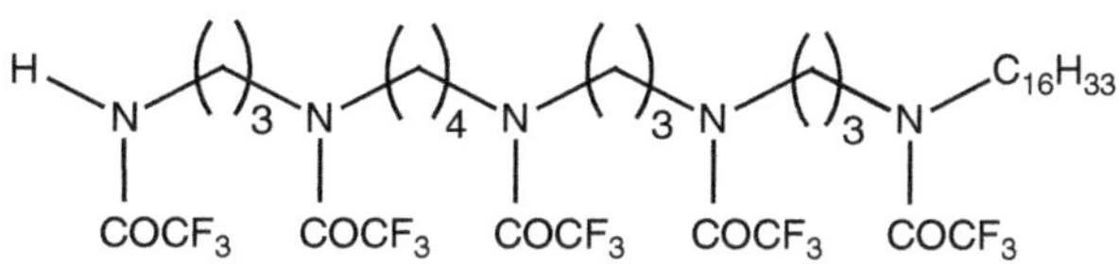

Again, we found no evidence of acute toxicity. These data collectively indicate that the toxicity of DS-96 is a manifestation of its peak plasma concentrations, and suggest that a suitable prodrug of DS-96 may greatly widen the therapeutic index and margin of safety.

12.4 Conclusions and Future Prospects

Although sepsis does not have the impact upon the consciousness of the lay public that diseases like cancer and AIDS do, it is a very common and serious clinical problem of global public health importance, as morbidity and mortality statistics clearly indicate. Unfortunately, the failure of the clinical trials on anti-LPS monoclonal antibodies in the early 90's and its consequences on the morale and initiative of the then-developing biotechnology industry has done little to sustain and fortify efforts to seek viable alternatives. As has become evident from very recent meta-analyses of Xigris, efficacious and affordable anti-sepsis strategies are in dire need. Over the course of the last several years, we have characterized the structural and mechanistic basis of the molecular recognition of endotoxin and shown that small-molecule mimics of polymyxin B may offer a feasible alternative as exemplified by the activity profile of DS-96. However, toxicity issues remain to be addressed. Our studies thus far point to peak plasma concentrations as playing a key role in acute toxicity observed at multiples of therapeutic doses. This hypothesis is being tested by controlling the rate of systemic exposure to the lipopolyamine using a prodrug approach.

References

Abraham, E., Laterre, P.F., Garg, R., Levy, H., Talwar, D., Trzaskoma, B.L., Francois, B., Guy, J.S., Bruckmann, M., Rea-Neto, A., Rossaint, R., Perrotin, D., Sablotzki, A., Arkins, N., Utterback, B.G., Macias, W.L. Drotrecogin alfa (activated) for adults with severe sepsis and a low risk of death. N Engl J Med 353 (2005) 1332–1341.

Antrup, H., Seiler, N. On the turnover of polyamines spermidine and spermine in mouse brain and other organs. Neurochem Res 5 (1980) 123–143.

Arditi, M., Zhou, J., Dorio, R., Rong, G.W., Goyert, S.M., Kim, K.S. Endotoxin-mediated endothelial cell injury and activation: Role of soluble CD14. Infect Immun 61 (1993) 3149–3156.

Astiz, M., Saha, D., Lustbader, D., Lin, R., Rackow, E. Monocyte response to bacterial toxins, expression of cell surface receptors, and release of anti-inflammatory cytokines during sepsis. J Lab Clin Med 128 (1996) 594–600.

Baker, P.J., Hraba, T., Taylor, C.E., Myers, K.R., Takayama, K., Qureshi, N., Stuetz, P., Kusumoto, S. Hasegawa, A., Structural features that influence the ability of lipid A and its analogs to abolish expression of suppressor T cell activity. Infect Immun 60 (1992) 2694–2701.

Balakrishna, R., Wood, S.J., Nguyen, T.B., Miller, K.A., Suresh Kumar, E.V., Datta, A., David, S.A. Structural correlates of antibacterial and membrane-permeabilizing activities in acylpolyamines, Antimicrob. Agents Chemother 50 (2006) 852–861.

Balk, R.A., Bone, R.C. The septic syndrome. Definition and clinical implications. Crit Care Clin 5 (1989) 1–8.

Baumgartner, J.D., Heumann, D., Gerain, J., Weinbreck, P., Grau, G.E., Glauser, M.P. Association between protective efficacy of anti-lipopolysaccharide (LPS) antibodies and suppression of LPS-induced Tumor necrosis factor à and interleukin 6. Comparison of O side chain-specific antibodies with core LPS antibodies. J Exp Med 171 (1990) 889–896.

Bernard, G.R., Vincent, J.L., Laterre, P.F., LaRosa, S.P., Dhainaut, J.F., Lopez-Rodriguez, A., Steingrub, J.S., Garber, G.E., Helterbrand, J.D., Ely, E.W., Fisher, C.J., Jr. Efficacy and safety of recombinant human activated protein C for severe sepsis. N Engl J Med 344 (2001) 699–709.

Betz Corradin, S., Mauel J., Gallay, P., Heumann, D., Ulevitch, R.J., Tobias, P.S. Enhancement of murine macrophage binding of and response to bacterial lipopolysaccharide (LPS) by LPS-binding protein. J Leukocyte Biol 52 (1992) 363–368.

Bhart, N.M., Bieber, M.M., Chapman, C.J. Human anti-lipid A monoclonal antibodies bind to human B cells and i-antigen on core red blood cells. J Immunol 151 (1993) 5011–5021.

Bhattacharjya, S., David, S.A., Mathan, V.I., Balaram, P. Polymyxin B nonapeptide: Conform -ations in water and in the lipopolysaccharide-bound state determined by two-dimensional NMR and molecular dynamics. Biopolymers 41 (1997) 251–265.

Blagbrough, I.S., Geall, A.J., David, S.A. Lipopolyamines incorporating the teraamine spermine bound to an alkyl chain, sequester bacterial lipopolysaccharide. Bioorg Med Chem Lett 10 (2000) 1959–1962.

Bohuslav, J., Kravchenko, V.V., Parry, G.C., Erlich, J.H., Gerondakis, S., Mackman, N., Ulevitch, R.J. Regulation of an essential innate immune response by the p50 subunit of NF-kappaB. J Clin Invest 102 (1998) 1645–1652.

Bone, R.C., Gram-negative sepsis: a dilemma of modern medicine. Clin Microbiol Rev 6 (1993) 57–68.

Bone, R.C., Balk, R.A., Fein, A.M., Perl, T.M., Wenzel, R.P., Reines, H.D., Quenzer, R.W., Iberti, T.J., Macintyre, N., Schein, R.M. A second large controlled clinical study of E5, a monoclonal antibody to endotoxin: results of a prospective, multicenter, randomized, controlled trial. The E5 Sepsis Study Group [see comments]. Crit Care Med 23 (1995) 994–1006.

Bone, R.C., Sprung, C.L., Sibbald, W.J. Definitions for sepsis and organ failure. Crit Care Med 20 (1992) 724–726.

Bucklin, S.E., Fujihara, Y., Leeson, M.C., Morrison, D.C., Differential antibiotic-induced release of endotoxin from gram- negative bacteria. Eur J Clin Microbiol Infect Dis 13 Suppl. 1 (1994) 43–51.

Centers for Diseases Control. Increases in national hospital discharge survey rates for septicemia – United States, 1979–1987. MMWR 39 (1990) 31–34.

Chedid, L., Parant, M., Parant, F. A proposed mechanism for natural immunity to enterobacterial pathogens. J Immunol 100 (1968) 292–306.

Crosby, H.A., Bion, J.F., Penn, C.W., Elliott, T.S.J. Antibiotic-induced release of endotoxin from bacteria in vitro. J Med Microbiol 40 (1994) 23–30.

Cross, A., Opal, S.M. Therapeutic intervention in sepsis with antibody to endotoxin: is there a future?. J Endotoxin Res 1 (1994) 57–59.

Curry, S., Mandelkow, H., Brick, P., Franks, N., Crystal structure of human serum albumin complexed with fatty acid reveals an asymmetric distribution of binding sites. Nat Struct Biol 5 (1998) 827–835.

David, S.A. The interaction of lipid A and lipopolysaccharide with human serum albumin. In Brade, H., Opal, S.M., Vogel, S.N., Morrison, D.C. (eds), Endotoxins in Health and Disease. Marcel Dekker, New York, NY (1999) 413–22.

David, S.A. Towards a rational development of anti-endotoxin agents: novel approaches to sequestration of bacterial endotoxins with small molecules. J Molec Recognit 14 (2001) 370–387.

David, S.A., Balaram, P., Mathan, V.I. Interaction of basic amphiphilic polypeptide antimicrobials, gramicidin S, tyrocidin and efrapeptin, with endotoxic lipid A. Med Microbiol Lett 2 (1993) 42–47.

David, S.A., Balaram, P., Mathan, V.I. Characterization of the interaction of lipid A and lipopolysaccharide with human serum albumin: implications for an endotoxin-carrier function for albumin. J Endotoxin Res 2 (1995a) 99–106.

David, S.A., Bechtel, B., Annaiah, C., Mathan, V.I., Balaram, P. Interaction of cationic amphiphilic drugs with lipid A: Implications for development of endotoxin antagonists. Biochim Biophys Acta 1212 (1994) 167–175.

David, S.A., Mathan, V.I., Balaram, P. Interaction of melittin with endotoxic lipid A. Biochim Biophys Acta 1123 (1992) 269–274.

David, S.A., Mathan, V.I., Balaram, P. Interactions of linear dicationic molecules with lipid A: Structural requisites for optimal binding affinity. J Endotoxin Res 2 (1995b) 325–336.

David, S.A., Perez, L., Infante, M.R. Sequestration of bacterial lipopolysaccharide by bis(args) gemini compounds. Bioorg Med Chem Lett 12 (2002) 357–360.

David, S.A., Silverstein, R., Amura, C.R., Kielian, T., Morrison, D.C. Lipopolyamines: novel antiendotoxin compounds that reduce mortality in experimental sepsis caused by gram-negative bacteria. Antimicrob Agents Chemother 43 (1999) 912–919.

Di Padova, F., Brade, H., Barclay, G.R., Poxton, I.R., Liehl, E., Schuetze, E., Kocher, H.P., Ramsay, G., Schreier, M.H., McClelland, D.B.L., Rietschel, E.T. A broadly cross-protective monoclonal antibody binding to *Escherichia coli* and *Salmonella* lipopolysaccharide. Infect Immun 61 (1993) 3863–3872.

Dinarello, C.A. The proinflammatory cytokines interleukin-1 and tumor necrosis factor and treatment of the septic shock syndrome. J Infect Dis 163 (1991) 1177–1184.

Dinarello, C.A. Cytokines as mediators in the pathogenesis of septic shock. Curr Top Microbiol Immunol 216 (1996) 133–165.

Durando, M.M., MacKay, R.J., Linda, S., Skelley, L.A. Effects of polymyxin B and *Salmonella typhimurium* antiserum on horses given endotoxin intravenously. Am J Vet Res 55 (1994) 921–927.

Felty, A.R., Keefer, C.S. Bacillus coli sepsis: A clinical study of twenty-eight cases of bloodstream infection by the colon bacillus. JAMA 82 (1924) 1430–1433.

Fenton, M.J., Golenbock, D.T. LPS-binding proteins and receptors. J Leukoc Biol 64 (1998) 25–32.
Ferguson, A.D., Hofmann, E., Coulton, J., Diedrichs, K., Welte, W. Siderophore-mediated iron transport: Crystal structure of FhuA with bound lipopolysaccharide. Science (1998) 2215–2220.
Galanos,C., Lüderitz, O. 1984. Lipopolysaccharide: Properties of an amphipathic molecule. In Handbook of endotoxin Proctor,R.A. Ed. E.T.Rietschel, editor. Elsevier Science Publisher, Amsterdam. 46–55
Gallay, P., Barras, C., Tobias, P.S., Calandra, T., Glauser, M.P., Heumann, D. Lipopolysaccharide (LPS)-binding protein in human serum determines the tumor necrosis factor response of monocytes to LPS. J Infect Dis 170 (1994) 1319–1322.
Gasche, Y., Pittet, D., Sutter, P. Outcome and prognostic factors in bacteremic sepsis. In Sibbald, W.J., Vincent J.L. (eds), Clinical Trials for Treatment of Sepsis. Springer, Berlin (1995) 35–51.
Gelfand, J.A., Shapiro, L. Cytokines and sepsis: pathophysiology and therapy. New Horizons 1 (1993) 13–22.
Geppert, T.D., Whitehurst, C.E., Thompson, P., Beutler, B. Lipopolysaccharide signals activation of tumor necrosis factor biosynthesis through the ras/raf-1/MEK/MAPK pathway. Mol Med 1 (1994) 93–103.
Han, J., Lee, J.D., Bibbs, L., Ulevitch, R.J. A MAP kinase targeted by endotoxin and hyperosmolarity in mammalian cells. Science 265 (1994) 808–811.
Haruhiko, T., Kotani, S., Tanaka, S., Ogawa, T., Takahashi, I., Tsujimoto, M., Komuro, T., Shiba, T., Kusumoto, S., Kusunose, N., Hasegawa, A., Kiso, M. Structural requirements of lipid A species in activation of clotting enzymes from the horseshoe crab, and the human complement cascade. Eur J Biochem 175 (1988) 573–580.
Helmerhorst, E.J., Maaskant, J.J., Appelmelk, B.J. Anti-lipid A monoclonal antibody centoxin (HA-1A) binds to a wide variety of hydrophobic ligands. Infect Immun 66 (1998) 870–873.
Hurley, J.C. Antibiotic-induced release of endotoxin: A reappraisal. Clin Infect Dis 15 (1992) 840–854.
Hurley, J.C. Antibiotic-induced release of endotoxin. A therapeutic paradox. Drug Saf 12 (1995) 183–195.
Iwagaki, A., Porro, M., Pollack, M. Influence of synthetic antiendotoxin peptides on lipopolysaccharide (LPS) recognition and LPS-induced proinflammatory cytokine responses by cells expressing membrane-bound CD14. Infect Immun 68 (2000) 1655–1663.
Jackson, J.J., Kropp, H., Hurley, J.C. Influence of antibiotic class and concentration on the percentage of release of lipopolysaccharide from *Escherichia coli.* J Infect Dis 169 (1994) 471–473.
Jerala, R., Porro, M. Endotoxin neutralizing peptides. Curr Top Med Chem 4 (2004) 1173–1184.
Jones II W., Minei, J., Barber, A., fahey T., Shires III G.T. Splanchnic vasoconstriction and bacterial translocation after thermal injury. Am J Physiol 261 (1991) H1190–H1196.
Kaisho, T., Akira, S. Toll-like receptor function and signaling. J Allergy Clin Immunol 117 (2006) 979–987.
Kawai, T., Akira, S. TLR signaling. Semin Immunol 19 (2007) 24–32.
Kawakami, K., Nagatomo, T., Abe, H., Kikuchi, K., Takemasa, H., Anson, B.D., Delisle, B.P., January, C.T., Nakashima, Y. Comparison of HERG channel blocking effects of various beta-blockers – implication for clinical strategy. Br J Pharmacol 147 (2006) 642–652.
Kawasaki, S., Moriguchi, R., Sekiya, K., Nakai, T., Ono, E., Kume, K., Kawahara, K. The cell envelope structure of the lipopolysaccharide-lacking gram-negative bacterium Sphingomonas paucimobilis. J Bacteriol 176 (1994) 284–290.
Kim, K.S., Kang, J.H., Cross, A.S., Kaufman, B., Zollinger, W., Sadoff, J. Functional activities of monoclonal antibodies to the O side chain of *Escherichia coli* lipopolysaccharides in vitro and in vivo. J Infect Dis 157 (1988) 47–53.
Kurata, H.T., Marton, L.J., Nichols, C.G. The polyamine binding site in inward rectifier K+ channels. J Gen Physiol 127 (2006) 467–480.
Le Roy, D., Di Padova, F., Tees, R., Lengacher, S., Landmann, R., Glauser, M.P., Calandra, T., Heumann, D. Monoclonal antibodies to murine lipopolysaccharide (LPS)-binding protein

(LBP) protect mice from lethal endotoxemia by blocking either the binding of LPS to LBP or the presentation of LPS/LBP complexes to CD14. J Immunol 162 (1999) 7454–7460.

Loussouarn, G., Marton, L.J., Nichols, C.G. Molecular basis of inward rectification: structural features of the blocker defined by extended polyamine analogs. Mol Pharmacol 68 (2005) 298–304.

Manns, B.J., Lee, H., Doig, C.J., Johnson, D., Donaldson, C. An economic evaluation of activated Protein C treatment for severe sepsis. N Engl J Med 347 (2002) 993–1000.

Maron, D.M., Ames, B.N. Revised methods for the *Salmonella* mutagenicity test. Mutat Res 113 (1983) 173–215.

Martin, G.S., Mannino, D.M., Eaton, S., Moss, M. The epidemiology of sepsis in the United States from 1979 through 2000. N Engl J Med 348 (2003) 1546–1554.

Miller, K.A., Suresh Kumar, E.V.K., Wood, S.J., Cromer, J.R., Datta, A., David, S.A. Lipopolysaccharide sequestrants: structural correlates of activity and toxicity in novel acylhomospermines. J Med Chem 48 (2005) 2589–2599.

Morrison, D.C., Jacobs, D.M. Binding of polymyxin B to the lipid A portion of bacterial lipopolysaccharides. Immunochemistry 13 (1976) 813–818.

Moss, M., Martin, G.S. A global perspective on the epidemiology of sepsis. Intensive Care Med 30 (2004) 527–529.

Nakamura, T., Ebihara, I., Shoji, H., Ushiyama, C., Suzuki, S., Koide, H. Treatment with polymyxin B-immobilized fiber reduces platelet activation in septic shock patients: decrease in plasma levels of soluble P-selectin, platelet factor 4 and beta-thromboglobulin. Inflamm Res 48 (1999) 171–175.

Nakamura, T., Ushiyama, C., Suzuki, Y., Inoue, T., Shoji, H., Shimada, N., Koide, H. Combination therapy with polymyxin B-immobilized fibre haemoperfusion and teicoplanin for sepsis due to methicillin-resistant Staphylococcus aureus. J Hosp Infect 53 (2003) 58–63.

Nakamura, T., Ushiyama, C., Suzuki, Y., Shoji, H., Shimada, N., Koide, H. Hemoperfusion with polymyxin B immobilized fibers for urinary albumin excretion in septic patients with trauma. ASAIO J 48 (2002) 244–248.

Nguyen, T.B., Suresh Kumar, E.V., Sil, D., Wood, S.J., Miller, K.A., Warshakoon, H.J., Datta, A., David, S.A. Controlling plasma protein binding: structural correlates of interactions of hydrophobic polyamine endotoxin sequestrants with human serum albumin. Mol Pharm 5 (2008) 1131–1137.

Peterson, A., Hancock, R.E.W., McGroarty, E.J., Binding of polycationic antibiotics and polyamines to lipopolysaccharides of *Pseudomonas aeruginosa.* J Bacteriol 164 (1985) 1256–1261.

Poltorak, A., He, X., Smirnova, I., Liu, M.Y., Huffel, C.V., Du, X., Birdwell, D., Alejos, E., Silva, M., Galanos, C., Freudenberg, M., Ricciardi, C.P., Layton, B., Beutler, B. Defective LPS signaling in C3H/HeJ and C57BL/10ScCr mice: mutations in Tlr4 gene. Science 282 (1998) 2085–2088.

Porro, M. Structural basis of endotoxin recognition by natural polypeptides. Trends Microbiol 2 (1994) 65–66.

Porro, M., Rustici, A., Velucchi, M., Agnello, D., Villa, P., Ghezzi, P. Natural and synthetic polypeptides that recognize the conserved lipid a binding site of lipopolysaccharides. Prog Clin Biol Res 397 (1998) 315–325.

Prins, J.M., van Agtmael, M.A., Kuijper, E.J., van Deventer, S.J., Speelman, P. Antibiotic-induced endotoxin release in patients with gram- negative urosepsis: a double-blind study comparing imipenem and ceftazidime. J Infect Dis 172 (1995) 886–891.

Prins, J.M., Van Deventer, S.J.H., Kuijper, E.J., Speelman, P. Clinical relevance of antibiotic-induced endotoxin release. Antimicrob Agents Chemother 38 (1994) 1211–1218.

Quezado, Z.M.M., Natanson, C., Alling, D.W. A controlled trial of HA-1A in a canine model of gram-negative septic shock. JAMA 269 (1993) 2221–2227.

Quezado, Z.M.N., Banks, S.M., Natanson, C. New strategies for combatting sepsis: the magic bullets missed the mark ... but the search continues. Trends Biotech 13 (1995) 56–63.

Rietschel, E.T., Kirikae, T., Schade, F.U., Mamat, U., Schmidt, G., Loppnow, H., Ulmer, A.J., Zähringer, U., Seydel, U., Di Padova, F., et al. Bacterial endotoxin: molecular relationships of structure to activity and function. FASEB J 8 (1994) 217–225.

Rietschel, E.T., Brade, H., Brade, L., Brandenburg, K., Schade, U.F., Seydel, U., Z„hringer U., Galanos, C., Lüderitz, O., Westphal, O., Labischinski, H., Kusumoto, S., Shiba, T. Lipid A, the endotoxic center of bacterial lipopolysaccharides: Relation of chemical structure to biological activity. Prog Clin Biol Res 231 (1987) 25–53.

Rocke, D., Gaffin, S., Wells, M., Koen, Y., Brocke-Utine, J. Endotoxemia associated with cardiopulmonary bypass. J Thorac Cardiovasc Surg 93 (1987) 832–837.

Rocque, W.J., Fesik, S.W., Haug, A., McGroarty, E.J. Polycation binding to isolated lipopolysaccharide from antibiotic-hypersuceptible mutant strains of *Escherichia coli.* Antimicrob Agents Chemother 32 (1988) 308–313.

Rustici, A., Velucchi, M., Faggioni, R., Sironi, M., Ghezzi, P., Quataert, S., Green, B., Porro, M. Molecular mapping and detoxification of the lipid A binding site by synthetic peptides. Science 259 (1993) 361–365.

Saadia, R., Schein, M., MacFarlane, C., Boffard, K. Gut barrier function and the surgeon. Br J Surg 77 (1990) 487–492.

Scott, M.G., Vreugdenhil, A.C., Buurman, W.A., Hancock, R.E., Gold, M.R., Cationic antimicrobial peptides block the binding of lipopolysaccharide (LPS) to LPS binding protein. J Immunol 164 (2000) 549–553.

Seiler, N., Knodgen, B. Determination of di-and polyamines by high-performance liquid chromatographic separation of their 5-dimethylaminonaphthalene-1-sulfonyl derivatives. J Chromatogr 145 (1978) 29–39.

Shrestha, A., Li, R., Sil, D., Pardeshi, N.N., Schwarting, N., Schorno, K.S., Rajewski, R.A., Datta, A., David, S.A. Pharmacokinetics of DS-96, an alkylpolyamine lipopolysaccharide sequestrant, in rodents. J Pharm Sci 97 (2008) 5376–5385.

Siegel, J.P. Antiendotoxin antibodies. Ann Intern Med 122 (1995) 315–316.

Sil, D., Shrestha, A., Kimbrell, M.R., Nguyen, T.B., Adisechan, A.K., Balakrishna, R., Abbo, B.G., Malladi, S., Miller, K.A., Short, S., Cromer, J.R., Arora, S., Datta, A., David, S.A. Bound to shock: protection from lethal endotoxemic shock by a novel, nontoxic, alkylpolyamine lipopolysaccharide sequestrant. Antimicrob Agents Chemother 51 (2007) 2811–2819.

Stokes, D.C., Shenep, J.L., Fishman, M.L., Hidner, W.K., Bysani, G.K., Rufus, K. Polymyxin B prevents lipopolysaccharide-induced release of tumor necrosis factor-a from alveolar macrophages. J Infect Dis 160 (1989) 52–57.

Storm, D.R., Rosenthal, K. Polymyxin and related peptide antibiotics. Ann Rev Biochem 46 (1977) 723–763.

Udenfriend, S., Stein, S., Bohlen, P., Dairman, W., Leimgruber, W., Weigle, M. Fluorescamine: a reagent for assay of amino acids, peptides, proteins, and primary amines in the picomole range. Science 178 (1972) 871–872.

Uematsu, S., Akira, S. Toll-like receptors and innate immunity. J Mol Med 84 (2006) 712–725.

Unger, F.M., Christian, R., Schulz, G., Waldstten F., Brade, H., Zhringer, U., Rietschel, E.T. Structure of the 3-Deoxyoctulosonate (Kdo-) Region of Lipopolysaccharide from *Salmonella minnesota* Re 595, XIIth International Carbohydrate Symposium, July 1–7, 1984, Utrecht, The Netherlands Abstract Book (1984) 348.

Vaara, M. Polymyxin B nonapeptide complexes with lipopolysaccharide. FEMS Microbiol Lett 18 (1983) 117–121.

Vaara, M., Vaara, T. Polycations as outer membrane disorganizing agents. Antimicrob Agents Chemother 24 (1983) 114–122.

Vaarala, O., Vaara, M., Palosuo, T. Effective inhibition of cardiolipin-binding antibodies in gram-negative infections by bacterial lipopolysaccharide. Scand J Immunol 28 (1988) 607–612.

Viljanen, P., Matsunaga, H., Kimura, Y., Vaara, M. The outer membrane permeability-increasing action of deacylpolymyxins. J Antibiot Tokyo 44 (1991) 517–523.

Wahl, L.M., Rosenstreich, D.L., Glode, L.M., Sandberg, A.L., Mergenhagen, S.E. Defective prostaglandin synthesis by C3H/HeJ mouse macrophages stimulated with endotoxin preparation. Infect Immun 23 (1979) 8–13.
Wallace, H.M., Fraser, A.V. Inhibitors of polyamine metabolism: review article. Amino Acids 26 (2004) 353–365.
Warren, H.S., Amato, S.F., Fitting, C., Black, K.M., Loiselle, P.M., Pasternack, M.S., Cavaillon, J.M. Assessment of ability of murine and human anti-lipid A monoclonal antibodies to bind and neutralize lipopolysaccharide. J Exp Med 177 (1993) 89–97.
Williams, K. Interactions of polyamines with ion channels. Biochem J 325 (Pt 2) (1997a) 289–297.
Williams, K. Modulation and block of ion channels: a new biology of polyamines. Cell Signal 9 (1997b) 1–13.
Wortel, C.H., Ziegler, E.J., Van Deventer, S.J.H. Therapy of Gram-negative sepsis in man with anti-endotoxin antibodies: A review. In Sturk, A. (ed), Bacterial Endotoxins: Cytokine Mediators and New Therapies for Sepsis. Wiley-Liss, Inc., (1991) 161–78.
Wright, S.D. CD14 and innate recognition of bacteria. J Immunol 155 (1995) 6–8.
Wright, S.D., Ramos, R.A., Tobias, P.S., Ulevitch, R.J., Mathison, J.C. CD14, a receptor for complexes of lipopolysaccharide (LPS) and LPS binding protein. Science 249 (1990) 1431–1433.
Yao, Y.M., Tian, H.M., Sheng, Z.Y., Wang, Y.P., Yu, Y., Sun, S.R., Xu, S.H. Inhibitory effects of low-dose polymyxin B on hemorrhage-induced endotoxin/bacterial translocation and cytokine formation in rats. J Trauma 38 (1995) 924–930.
Zeni, F., Freeman, B., Natanson, C. Anti-inflammatory therapies to treat sepsis and septic shock: A reassessment. Crit Care Med 25 (1997) 1097–1100.
Zhou, Z., Gong, Q., Epstein, M.L., January, C.T. HERG channel dysfunction in human long QT syndrome. Intracellular transport and functional defects. J Biol Chem 273 (1998a) 21061–21066.
Zhou, Z., Gong, Q., Ye, B., Fan, Z., Makielski, J.C., Robertson, G.A., January, C.T. Properties of HERG channels stably expressed in HEK 293 cells studied at physiological temperature. Biophys J 74 (1998b) 230–241.
Ziegler, E.J. Protective antibody to endotoxin core: The emperor's new clothes?. J Infect Dis 158 (1988) 286–290.
Ziegler, E.J., Fisher, C.J., Jr., Sprung, C.L., Straube, R.C., Sadoff, J.C., Foulke, G.E., Wortel, C.H., Fink, M.P., Dellinger, R.P., Teng, N.N.H., Allen, I.E., Berger, H.J., Knatterud, G.L., LoBuglio, A.F., Smith, C.R., Study Group, A.S. Treatment of gram-negative bacteremia and septic shock with HA- 1A human monoclonal antibody against endotoxin–A randomized, double-blind, placebo-controlled trial. N Engl J Med 324 (1991) 429–436.
Ziegler, E.J., McCutchan, J.A., Fierer, J., Glauser, M.P., Sadoff, J.C., Douglas, H., Braude, A.I. Treatment of gram-negative bacteremia and shock with human antiserum to a mutant *Escherichia coli*. N Engl J Med 307 (1982) 1225–1230.
Ziegler, E.J., Smith, C.R. Anti-endotoxin monoclonal antibodies. N Engl J Med 326 (1992) 1165.

Chapter 13
Development of an Anti-Endotoxin Vaccine for Sepsis

Alan S. Cross

Abstract Gram-negative bacterial lipopolysaccharide (LPS, endotoxin) is an important initiator of sepsis, a clinical syndrome that is a leading cause of death in intensive care units. Vaccines directed against core LPS structures that are widely conserved among Gram-negative bacteria (GNB) have been developed for the treatment and/or prevention of sepsis. Killed whole bacterial vaccines (*E. coli* O111:B4, J5 [Rc chemotype] mutant and *S. minnesota*, Re chemotype) protected mice against experimental sepsis. Human J5 immune antisera reduced the mortality from GNB sepsis in a large controlled clinical trial; however, subsequent clinical studies with antiendotoxin antibodies did not demonstrate protective efficacy in sepsis. Multiple clinical studies have since demonstrated a correlation between the level of circulating antibodies to LPS core and morbidity and mortality in different clinical settings. We therefore developed a subunit vaccine by combining detoxified J5 LPS (J5 dLPS) with the outer membrane protein (OMP) from group B *N. meningitidis*. This vaccine was highly efficacious in experimental models of sepsis and progressed to phase 1 clinical trial. While well-tolerated, this vaccine induced only 3–4-fold increases in anti-J5 dLPS antibody. Addition of the TLR9 agonist, oligodeoxynucleotide with a CpG motif, as adjuvant to the vaccine increased antibody levels in mice and the vaccine/CpG combination will progress to phase 1 human study. Additional vaccines in which the core glycolipid was either conjugated to carrier protein or incorporated into liposomes have been developed, but have not progressed to clinical trial. Should an antiendotoxin vaccine become available, a new immunization strategy directed towards distinct populations at risk will be required.

Keywords Endotoxin · Gram-negative bacterial lipopolysaccharide · Sepsis · Vaccine

A.S. Cross (✉)
Center for Vaccine Development, University of Maryland School of Medicine, 685 W. Baltimore Street, HSF 1, Suite 480, Baltimore, MD 21201, USA
e-mail: across@medicine.umaryland.edu

X. Wang, P.J. Quinn (eds.), *Endotoxins: Structure, Function and Recognition*, Subcellular Biochemistry 53, DOI 10.1007/978-90-481-9078-2_13,

Abbreviations

BPI	bactericidal permeability-increasing protein
ENP	endotoxin neutralizing protein
FCA	Freund's Complete Adjuvant
GNB	gram-negative bacteria
ICU	intensive care unit
IVIG	immunoglobulin for intravenous use
J5 dLPS	detoxified J5 lipopolysaccharide
KDO	keto-deoxy octulosonic acid
LPS	lipopolysaccharide
MAb	monoclonal antibody
OMP	outer membrane protein
ODN CPG	synthetic oligodeoxynucleotide with unmethylated CpG motif
OS	oligosaccharide

13.1 Introduction

Lipopolysaccharide (LPS, or endotoxin), a constituent of the outer membrane of Gram-negative bacteria, is a principal initiator of sepsis. Sepsis, a leading cause of death in non-cardiac intensive care units (ICU), is a clinical syndrome characterized by physical findings of systemic inflammation (e.g. fever, hypotension, rapid heart rate, elevated white blood cell count) associated with a suspected or proven infection. If untreated, sepsis may progress to shock, multi-organ failure and death. Since the mortality rate of sepsis has been essentially unchanged for decades despite progress in supportive care and the development of potent new antimicrobials, there has been a concerted effort to devise additional measures to supplement the current standard of care. Generally, these strategies have targeted either the initiator of the septic response in order to limit immune stimulation, or the host immune response itself (e.g. anti-cytokines). Historically, since a working knowledge of bacterial pathogens preceded our understanding of the immune system, early investigators developed therapeutic approaches that targeted bacteria or their various component molecules.

13.2 Endotoxin as a Therapeutic Target

Endotoxin was initially described in the 19th century, but became the subject of intensive investigation since the 1920s. The clinical syndrome of Gram-negative bacterial sepsis was not described until the early 1950s (Waisbren, 1951). The focus on endotoxin as a critical initiator of sepsis occurred in large part because of a serendipitous confluence of events: an attempt to understand the mechanism of action of endotoxin as a component of Coley's toxin used to treat cancer and the use of typhoid vaccine as fever therapy (Bahador and Cross, 2007); the development of a chemical procedure to purify endotoxin (Luderitz et al., 1971); advances in bacterial genetics that detailed the sequence of enzymatic activities by which LPS was synthesized (Osborn, 1966; Subbaiah and Stocker, 1964); and the serendipitous observation that a spontaneously occurring mutation of C3H mice resulted in a lack of responsiveness to LPS fascinated a generation of investigators long before the basis for this phenotype (i.e. mutation in Toll-like receptor 4) was understood (Sultzer, 1976).

Before 1960, Gram negative bacteria were an unusual cause of clinical infections. In this latter regard, there were relatively few antimicrobials available to treat the increasingly common Gram-negative bacteria, particularly among patients immunocompromised by either their underlying disease (e.g. renal failure, diabetes) or therapies (e.g. steroids, chemotherapy). By 1960 a major shift to Gram-negative bacteria as pathogens causing serious infections, usually in hospitalized patients, was well-documented (Rogers, 1959; Finland et al., 1959). Thus it is not surprising that initial attempts to treat sepsis focused on the LPS molecule itself. Indeed, as the concepts of innate and adaptive immunity developed, LPS was often used as the prototypic agonist to further study the immune response.

13.3 Endotoxin Structure and Immunity

The structure/function activities of LPS have been well-determined. There are three major components of the LPS: lipid A, responsible for the biologic activity of endotoxin, is the most highly conserved structure among species of *Enterobacteriaceae*; a conserved but more variable core region characterized by unusual sugars, heptoses and keto-deoxy-octulosonic acid (KDO); and a chain of repeating distinct sugar molecules that are attached to the LPS core. This chain of sugar molecules (the O antigen) provides bacteria their unique serologic specificity (i.e. it is not widely shared).

The biosynthetic pathways of lipid A formation have now been described in detail as well as the structure-function relationships (Raetz, 1984; Takada and Kotuni, 1992), and biologically active lipid A has been synthesized in vitro. Extensive work on the different core structures has been published (Table 13.1) (see Bhattacharjee and Cross, 1999). Early studies with the purified LPS established that when injected into animals, the immunodominant response was against the O antigen (i.e. it was specific against the serotype of the homologous bacterial strain). Relatively little antibody was made in response to the lipid A or core when delivered as part of a complete LPS molecule. Our laboratory confirmed this observation in patients with naturally-acquired Gram-negative bacteremia (Cross et al., 1989). In contrast, if one administered an LPS from a strain of bacteria that was deficient in one of the enzymes used to add either the O sugar repeat units or portions of the core region (i.e. not shielded by O antigen), then the antibody response would be against the core regions of the LPS. Chedid et al. observed that immunization with vaccines in which the core LPS regions were exposed could protect against lethal infection

Table 13.1 Structures of inner and outer cores of lipopolysaccharides from *Salmonella* RC, *Escherichia coil* R3, and *E. coli* J5 rough mutants[a]

Core type	Outer core	Inner core
Salmonella minnesota Rc	D-Glcα1-	→3 LD-Hepα1→3 LD-Hepα1→5Kdoα-[Lipid A] 7 ↑ 1 LD-Hepα (on first LD-Hepα); 4 ↑ 2 Kdoα2→4Kdoα (on Kdoα)
E. coli R3	D-Glcα1→2D-Glcα1→2D-Galα1→3D-Glcα1 3 ↑ 1 D-GlcNAcα (on D-Glcα1)	→3LD-Hepα1→3LD-Hepα1→5Kdo-[Lipid A] 7 ↑ 1 LD-Hepα 7 ↑ 1 D-GlcNAcα (on first LD-Hepα); 4 ↑ 2 Kdo 4 ↑ 2 Kdo (on Kdo)
E. coli J5	D-GlcNAcα1-	→7LD-Hepα1→7LD-Hepα1→3LD-Hepα1→5Kdoα-[Lipid A] 3 ↑ 1 D-Glcα (on second LD-Hepα); 4 ↑ 2 Kdo (on Kdoα)

[a] Data from Lugowski (1996)

This table was reproduced with permission from Infectious Disease Clinics of North America, Volume 13, by Apurba K. Bhattacharjee and Alan S. Cross, "Vaccines and antibodies in the prevention and treatment of sepsis", pages 355–369, Copyright Elsevier, 1999.

with heterologous bacteria (Chedid et al., 1968). Extensive work by Braude and colleagues with a mutant strain of *E. coli* O111:B4 and by McCabe and colleagues with *Salmonella minnesota* (Re mutant), each of whose core regions were exposed to the immune system, established the foundation for subsequent clinical studies of anti-endotoxin antibodies (Braude et al., 1960, 1981; Bruins et al. 1977; Johns et al., 1983; McCabe, 1972; McCabe et al., 1988). These observations served as the basis for the development of broadly cross-reactive endotoxin vaccines.

13.4 Initial Studies of Anti-LPS Core Antibodies as Therapy for Gram-Negative Bacterial Infections

Despite both the directness and simplicity of the hypothesis, namely that antibody directed toward a common, toxic moiety of endotoxin has therapeutic potential, the concept of anti-endotoxin antibody has been difficult to prove and mired in controversy.

In a seminal study, Braude and colleagues observed in rabbits that a localized infection without bacteremia was accompanied by fever indicative of a systemic response to the infection. The systemic signs of the localized infection were similar to those seen following administration of purified endotoxin (Braude et al., 1960).The fever abated with the development of a peripheral circulating antibody response against the LPS of that same infecting (i.e. homologous) strain. Thus, the component of the infecting bacteria, probably the LPS, entered the circulation, caused a systemic inflammatory response which receded upon the appearance of antibodies directed against the LPS of that strain. This was clear evidence that anti-endotoxin antibodies could ameliorate the systemic response to infection, even in the absence of bacteria in the blood. Tate et al. (1966) subsequently stimulated such anti-endotoxin antibodies by immunization with an O-deficient mutant of *E. coli* O112 (for review of further preclinical experiments see Cross and Opal, 1994).

The Braude group then obtained a mutant of *E. coli* O111:B4 that lacked an enzyme, gal-epimerase, which linked the O side chain to the core region. This mutant of the Rc chemotype, termed the J5 mutant, lacked the O polysaccharide thereby exposing the immune system to the LPS core. Immunization with heat-killed J5 bacteria generated antibodies that recognized the core LPS region of not only *E. coli* but to other Gram-negative bacteria (Braude et al., 1981). Active immunization with this heat-killed whole bacterial vaccine induced protection against lethal experimental sepsis caused by *Klebsiella* and *Pseudomonas*, and passive administration of antisera raised with this vaccine also protected animals against lethal sepsis caused by heterologous bacteria. Based on these strong preliminary data, this San Diego research group then conducted a clinical trial to test the efficacy of J5 antisera in reducing mortality in patients with Gram-negative bacterial sepsis.

Healthy subjects were immunized with the heat-killed J5 vaccine and their sera harvested. In a multi-center, randomized controlled clinical trial, patients with suspected Gram-negative bacterial sepsis were given either pre-immune or post-immunization sera (Ziegler et al., 1982). All patients received the standard antimicrobial and supportive care for sepsis. Those patients with Gram-negative bacterial sepsis who received the post-immunization sera had a significantly lower mortality rate than those who received pre-immune sera (22% [23/103] vs. 39% [42/109]. P=0.011). This beneficial effect was observed with progressively more severe cases of sepsis, such that those patients who required pressor agents who had the highest mortality (77% [30/39]), had reduced mortality if they received the immune sera (44%, [18/41], P=0.003).

Although this study demonstrated a convincing clinical effect of the post-immunization sera, these investigators could not correlate the clinical response with anti-J5 LPS antibody titer (based on hemagglutinating antibody levels). In addition, the antigen in the vaccine responsible for inducing the protection was not clearly identified. Consequently, the protective moiety in the sera as well as the mechanism of action was, and remains, in doubt. Nevertheless, this study spawned a series of clinical studies with antisera elicited with the J5 killed bacterial vaccine. In one study (Baumgartner et al., 1985), patients admitted to an ICU were given multiple doses of J5 antisera or non-immune sera and followed for development of sepsis. Further, at the onset of sepsis, patients received another dose of sera. While receipt of J5 antisera did not prevent the onset of subsequent infection, a beneficial effect of the J5 antisera was demonstrated in patients who developed sepsis. Subsequent studies with passive administration of J5 immune sera, however, were not as positive. A Swiss-Dutch consortium administered the heat-killed vaccine to healthy subjects, harvested the plasma and prepared an immunoglobulin for intravenous use (IVIG). Administration of a single infusion of this material to patients with sepsis failed to show any benefit; however, the post-immune sera that was processed into IVIG had only a two-fold increase in anti-J5 LPS antibody before fractionation compared to plasma from non-immunized controls (Calandra et al., 1988). Thus, while antibody levels were not measured in the patients, it is unlikely that patients received adequate levels of anti-J5 IgG. In another study, investigators in France gave children hospitalized with purpura fulminans, a condition associated with meningococcal sepsis, J5 plasma or non-immune plasma at the onset of infection. Here, too, no benefit was demonstrated; however when measured at 6 hr after infusion, there was no evidence of any increase in anti-J5 LPS antibody over baseline, perhaps due to antibody consumption (J5 Study Group, 1991). In another study, blood donors were screened for high levels of naturally-acquired anti-core LPS (S. *minnesota* Re 595) antibody. The high-titered material was pooled and made into a hyperimmune IVIG for use in a clinical trial and compared to standard IVIG. When given prophylactically to patients undergoing surgery, it was unable to prevent infection, sepsis or death (Cometta et al., 1992). In the absence of documented infection, the levels of anti-core LPS antibody at 2 days was <50% that of levels obtained at 2 hr post infusion. Thus in all these studies, it was unlikely that adequate levels of anti-endotoxin antibodies were present (Cross et al., 1999). Consequently, in the absence of documentation that patients achieved adequate levels of anti-core antibody, it was premature to reject the hypothesis that anti-endotoxin antibodies might be effective adjunctive therapy for sepsis.

13.4.1 Anti-Core LPS Antibody Levels

A number of studies, however, clearly established a relationship between the level of anti-core LPS antibody at onset of sepsis and outcome (Fomsgaard et al., 1989; Goldie et al., 1995; Nys et al., 1993), and a decrease in these anti-core LPS antibodies during a septic episode clearly predicted a poor outcome (Pollack et al., 1983). Two small clinical studies did demonstrate that maintenance of "adequate levels" of anti-endotoxin antibodies led to a decrease in circulating endotoxin levels and increased survival. Schedel and colleagues administered a polyclonal, non-hyperimmune immunoglobulin that contained IgA, IgG and IgM isotypes. The decrease in septic mortality (1/27 vs. 9/28, $p<0.01$) was correlated with a decrease in circulating anti-endotoxin activity (Schedel et al., 1991). A series of studies by Bennett-Guerrero and colleagues suggested that intraoperative splanchnic ischemia during surgery, even cardiac surgery not involving the abdomen, leads to high intraoperative plasma concentration of endotoxin which is associated with post-operative organ dysfunction, and even multi-organ failure (Bennett-Guerrero et al., 1997, 2000). Such observations from different centers over extended periods of time served as an impetus to the continued study of anti-LPS antibodies for their potential usefulness in the prevention or treatment of sepsis. Unlike the case with inhibitors of host mediator responses (anti-TNF and anti-IL-1 preparations), anti-endotoxin antibody therapy has the theoretic advantage of inhibiting an initiator of sepsis without compromising host defenses.

13.4.2 Anti-Lipid A Antibodies

Given the lack of efficacy in these followup studies of J5 immunization, other strategies were sought. With the development of monoclonal antibody (MAb) technology, investigators at both Xoma and Centocor made antibodies against lipid A, the most highly conserved region of LPS molecule and the portion responsible for its biologic activity, including the induction of inflammatory mediators associated with sepsis. Clinical trials in septic patients were conducted with each MAb. In neither trial was there a significant improvement in survival (Ziegler et al., 1991; Greenman et al., 1991). In retrospect, this finding should not have been surprising since multiple pre-clinical studies with polyclonal anti-lipid A antibodies were unable to document protection against sepsis (Cross and Opal, 1994). This may be attributed to the fact that lipid A portion of LPS is buried within the outer membrane of Gram-negative bacteria, and therefore not accessible to the antibodies. There was an anecdotal report that the anti-lipid A MAb may have been effective in a case of endotoxemia in a human subject.

13.5 Anti Core Monoclonal Antibodies

In addition to monoclonal antibodies directed towards the lipid A portion of LPS (e.g. HA-1A and E5), MAbs have been developed against core regions of Gram negative bacterial LPS. Nnalue et al. described a broadly reactive MAb directed against the inner core heptose disaccharide of *Salmonella* LPS which bound to 123 of 126 clinical isolates of *Salmonella* and 11 of 73 *E. coli* strains (Nnaule et al., 1992). The most extensively studied of these MAbs, WN1 222-5, bound to all clinical isolates of *E. coli* tested, to some *Citrobacter, Enterobacter* and *Klebsiella* isolates, but not to *P. aeruginosa* (DiPadova et al., 1993). This MAb was generated by immunization with a mixture of LPS preparations (R1-4, Ra and a rough strain of *E. coli* O18). This MAb bound to *E. coli* J5 but not to lipid A and was protective in murine sepsis models when challenge was administered intravenously, but not intraperitoneally. Since this antibody bound the LPS core but not lipid A, yet neutralized the biologic activity of LPS, the investigators speculated that the binding of the MAb to the core induced conformational changes in LPS structure such that access of immune cells to the lipid A was blocked. This antibody has not entered clinical development. More recently the minimal structure of the epitope to which the MAb WN1 222-5 bound was identified. This information may provide the structural basis for the rational development of a potential vaccine against *E. coli* LPS (Muller-Loennies et al., 2003; Muller-Loennies et al., 2007).

Non-vaccine generated anti-endotoxin reagents that bind and/or neutralize LPS, such as bactericidal permeability increasing protein (BPI), endotoxin neutralizing protein (ENP), and cationic peptide 18 conjugated to immunoglobulin G (CAP18-IgG) have been developed for the adjunctive therapy of sepsis, but only BPI has progressed to clinical trials (Levin et al., 2000).

13.6 Development of J5 Subunit Vaccine

Following the series of studies with the J5 whole bacterial vaccine, most of which failed to observe the protective effect observed in the Ziegler study, as well as the failure of the anti-lipid A MAbs, investigators and biotech companies largely abandoned the development of anti-endotoxin strategies. Some suggested that this area of investigation was the “Bermuda triangle” of biotechnology. Given the notion that sepsis represents a dysregulated innate immune system, it is not surprising that many of the adjunctive therapies proposed and/or tested have targeted various components of the immune system. Efforts largely turned to modulating the activity of cytokines and other inflammatory mediators that were becoming recognized as associated with sepsis. Reagents were developed against tumor necrosis factor-α and the IL-1 receptor and tested as therapy for sepsis. In addition, inhibitors of other mediators such as platelet activating factor, bradykinin, nitric oxide and many others underwent clinical trials (see Opal and Cross, 1999). None of these trials demonstrated a significant improvement from septic mortality for any of these therapies, with the lone exception of recombinant activated protein C (dotrecogin alpha) (Bernard et al., 2001). Indeed, some studies were stopped because of an increase in mortality due to fungal or Gram-positive pathogens. Thus, modulation of the immune response by these therapies may have left the patients susceptible to other pathogens.

13.6.1 Initial Studies with Unadjuvanted Vaccine

With the disappointing results from these interventions aimed at modifying the host response, our group at WRAIR decided to re-examine the work of Braude, Ziegler and colleagues. Review of many studies with the J5 bacterial vaccine suggested that the inability of J5 bacterial vaccine to protect may have been due to insufficient levels of antibody achieved and/or the consumption of the antibody during fulminant sepsis (Cross et al., 1999). Another problem was the lack of a convenient animal model. Although Braude and colleagues found anti-J5 antibody provided protection in rabbit models of sepsis, these animals were large, unwieldy and difficult to use in numbers sufficient for statistical analysis.

We therefore developed a neutropenic rat model of sepsis which reproduced many of the characteristics of sepsis: untreated, the rats were resistant to lethal infection when administered *Pseudomonas* by gavage, and induction of neutropenia alone did not alter the susceptibility to sepsis; however, if rats were pretreated with antibiotics to overcome the colonization resistance from the normal flora, then the neutropenic rats developed typhilitis, fever and bacteremia by approximately day 5 after bacterial challenge. Untreated, the animals died; however, administration of antiserum raised against the whole, killed *E. coli* J5 mutant bacterium (obtained from Ziegler et al.) improved the survival in these rats compared to non-immune serum (Collins et al., 1989). This model differs from many animal models of sepsis in that rats, unlike mice, develop fever, and therefore can be treated at the onset of infection. Since the onset of sepsis cannot be determined in murine models of sepsis, treatments are often administered as prophylaxis, before bacterial challenge or inciting event (e.g. cecal ligation/puncture).

To address the issue raised in the original clinical trial, i.e. was the protection due to antibody, we fractionated the immune sera raised against the heat-killed bacterial vaccine and found that the IgG and IgM fractions of the immune sera could mediate the protection. Nine of 16 rats treated with purified IgM and 13/20 rats treated with IgG fractions survived compared to 0/25 rats treated with pre-immune sera from the same rabbits ($p<0.001$). Further, when the IgG fraction was affinity purified over an agarose column to which J5 LPS was bound, the subsequently eluted anti-J5 LPS antibody was highly protective in the neutropenic rats (6/8 rats survived). Thus, we concluded that IgG antibody directed against the LPs of the J5 *E. coli* was able to mediate protection. Interestingly, the J5 immune sera did not have increased levels of antibody against either lipid A or the challenge *P. aeruginosa* strain. Importantly, we noted that the protection was related to the amount of antisera administered (Bhattacharjee et al., 1994).

Given that anti-endotoxin antibodies may provide benefit under various clinical conditions, we prepared a subunit LPS vaccine from the J5 mutant. Previous experience with whole, killed Gram negative bacterial vaccines (e.g. typhoid, cholera and *Salmonella* Re vaccines) demonstrated a high degree of reactogenicity when given to human subjects (Schwartzer et al., 1988). We therefore decided to purify the J5 LPS away from the other components of the bacteria that may have contributed to the reactogenicity and to detoxify the purified LPS. This de-O-acylation was achieved by alkaline treatment of the J5 LPS which removed the ester-linked fatty acids (but leaving the amide-linked fatty acids). By this time the contribution of phosphate groups and fatty acids to the biologic activity of the lipid A moiety of LPS was clearly defined. The resulting detoxified LPS (dLPS) was less pyrogenic in rabbit pyrogenicity tests (Bhattacharjee et al., 1996a). This preparation was poorly immunogenic in mice, however, with only a 2–4-fold increase in antibody above baseline (Bhattacharjee et al., 1996b). In an attempt to enhance the immunogenicity, we conjugated the LPS to diphtheria toxoid. This preparation increased the response 7–16-fold in the absence of any antibody response to lipid A. A colleague, Dr. Wendell Zollinger, had pioneered the use of group B meningococcal outer membrane protein (OMP) as a component of bacterial vaccines, and suggested that we administer the J5 dLPS along with the OMP as a non-covalent complex. The J5 dLPS moiety complexed with the OMP via a hydrophobic interaction, with the hydrophilic portion of the complex on the outer face which facilitated its aqueous solubility. Immunization with this formulation elicited a 37–142-fold increase in antibody titer at 4 weeks after the first dose. (Bhattacharjee et al., 1996b). We therefore used this formulation for both active and passive studies of sepsis.

The IgG fraction prepared from J5 dLPS/OMP immunized rabbits protected against lethal *Pseudomonas* sepsis in the neutropenic rat model (16/26 survived vs. 0/20 controls, $p<0.001$). The elicited anti-J5 dLPS IgG bound to several heterologous clinical isolates of Gram-negative bacteria. Of interest, this vaccine also raised anti-OMP antibodies that were highly bactericidal against the homologous group B *N. meningitidis* strain.

Given the success of passively administered anti-J5 dLPS/OMP sera, we actively immunized rats before rendering them neutropenic. Following oral challenge with either *Klebsiella* or *P. aeruginosa,* two heterologous strains of Gram negative bacteria, we observed 13/28 (48%) survival in immunized rats when challenged with *P. aeruginosa* vs. 2/29 (7%) survival in non-immunized controls, and 64% (9/14) survival in immunized vs. 13% (2/15) in control *Klebsiella*-infected rats. Immunized rats had a decreased organ bacterial burden and lower levels of circulating endotoxin at the onset of fever. (Cross et al., 2001).

Based on these data, we prepared a lot of J5 dLPS/OMP vaccine under cGMP conditions, repeated the immunogenicity and protection assays, and then progressed to phase I clinical trials in healthy subjects (Cross et al., 2003). Twenty-four subjects were given 5, 10 and 25 mcg of J5 dLPS/OMP complex vaccine intramuscularly at time 0, 28 and 56 days. We did not observe any fever, arthralgias or other systemic signs following any of the immunizations. Local reactions were those typically seen following licensed immunizations, such as the influenza vaccine. While the vaccine previously induced >20-fold increase in antibody levels over baseline in mice, rats and rabbits in previous studies, in human subjects we detected only a 3–4-fold increase in IgG, IgM and IgA antibody levels over baseline. A booster dose of vaccine at 12 months to a few of the subjects did not result in any further increase in antibody levels. Plasma obtained from high and low responders, however, did lower the amount of TNF-α elicited in human whole blood in response to LPS (Cross et al., 2004). Since we previously demonstrated that the protective effect of passively administered antisera was dose-dependent (Bhattacharjee et al., 1994), we concluded that a greater antibody response would be required. We therefore initiated studies of the vaccine in conjunction with vaccine adjuvants.

13.6.2 J5 dLPS/OMP Vaccine and Adjuvants

Initial preclinical studies with the J5 dLPS/OMP vaccine revealed that alum, the saponin-based adjuvant, QS-21, and oil-in-water adjuvant MF-59 were not effective in increasing the antibody response (Bhattacharjee et al., 1996a). Studies in human subjects with a hepatitis vaccine, which like our J5 dLPS/OMP vaccine, is given as a 3 dose regimen, demonstrated that the synthetic oligodeoxynucleotide containing an unmethylated CpG motif (ODN CPG), a potent TLR9 agonist, both accelerated and enhanced the immune response to the hepatitis B vaccine (Cooper et al., 2004). Unlike subjects receiving vaccine alone, recipients of the CpG/vaccine combination achieved "protective" levels of anti-hepatitis B antibody after the second of three scheduled vaccine doses. The vaccine/adjuvant regimen was also well-tolerated, with no serious adverse events reported and no long-term effects at a 5 year followup. Consequently, we studied the adjuvant effect of CPG with our vaccine.

Administration of the J5 dLPS/OMP complex vaccine with CPG increased the antibody response in mice 5-fold more than in mice immunized with vaccine alone (Opal et al., 2005). Interestingly, unlike the case with the hepatitis vaccine, addition of alum to the vaccine/CPG combination blocked the increased antibody response. This suggests that the alum may cover up an important vaccine epitope. Active immunization of mice with the vaccine and CPG combination led to significantly improved survival in cecal ligation/puncture models of polymicrobial sepsis in mice. This was accompanied by a decreased organ bacterial burden as seen earlier. Of particular interest, there was specific consumption of anti-J5 dLPS antibody during sepsis, which suggests that there was specific epitope binding and clearance in vivo. These experimental data support the hypothesis that supplemental administration of anti-J5 dLPS antibody may improve the likelihood of survival during sepsis, even after earlier active immunization. Based on these studies, we are planning to perform a phase 1 study with J5 dLPS/OMP vaccine and CPG ODN.

13.7 Other Anti-Endotoxin Vaccines

Whole cell, killed bacterial vaccines previously given to humans were generally of limited immunogenicity. While many of the adverse reactions with these preparations were described as acceptable, the frequency of systemic reactions (25%) in healthy individuals might give one pause before using them in larger patient populations. Several studies have administered whole bacterial, killed vaccines to human subjects with the purpose of inducing anti-core LPS antibody (Dale et al., 1992; Calandra et al. 1988; Schwartzer et al., 1988; Baumgartner et al., 1991), but in only one study was the individual antibody response analyzed (Schwartzer et al., 1988): a $\geq$ 4-fold increase in IgG and IgM antibody was observed in 8/16 and 9/16 vaccinees, respectively, but this response was short-lived and not increased by revaccination 30 days later. This also was the only study that reported the individual acceptability of the vaccine: all subjects had transient local reactions and 7/16 had $\geq$1 systemic reaction.

Covalent conjugate vaccines have been prepared with oligosaccharide cores of *E. coli* R1, R2, R3 and J5 (Rc) and with *Salmonella* Ra linked to tetanus toxoid (Lugowski et al., 1996a, b). Antisera generated with these vaccines with Freund's Complete Adjuvant (FCA) showed the antibodies reacted with conserved core oligosaccharide epitopes of smooth LPS of identical or related core types in ELISA and immunoblot assays (Lugowski et al., 1996a). Anti-OS R1 antiserum reacted with a free form of smooth LPS and inhibited the TNF stimulatory activity in vitro and in vivo. Active immunization protected 4–100% of mice against *P. aeruginosa* challenge (Stanislavsky et al., 1997). This Polish group also showed that a covalent conjugate vaccine of *E. coli* R4 core OS induced antibody that reacted with LPS present on live, intact smooth bacteria and mediated the uptake and killing of bacteria by macrophages in vitro (Lugowski et al., 2000). Serologic studies with antisera raised with these vaccines suggest heterogeneity in the antibody response to the *E. coli* core. These vaccines have not advanced to human studies in the published literature.

In addition to the development of covalent conjugate vaccines with LPS core structures, multiple Ra core LPS structures have been formulated into multilamellar liposomal preparations. This vaccine elicited antibodies that recognized heterologous GNB and protected mice from lethal challenge with *E. coli* LPS (Erridge et al., 2002). While these formulations elicited antibody to multiple different core LPS structures, they have not progressed to clinical development. This formulation does induce anti-LPS core antibodies in chickens and protects them against lethal challenge with *E. coli* (Dissanayake et al., 2009).

It is well-established in experimental Gram negative bacterial sepsis that treatment with type-specific (i.e. homologous) O antibodies is superior to treatment with antibodies directed against the LPS core (Bailat et al., 1997; McCabe, 1972) Given the large number of different serotypes among the genera of Gram negative bacteria, the use of polyclonal O type-specific antibodies has not been considered feasible. For example, there are >100 serotypes of *E. coli* alone based on the structure of O-polysaccharide. However, seroepidemiologic studies of isolates retrieved from cases of Gram negative bacteremia revealed that only a relatively small number of serotypes are associated with bacteremia (Donta et al., 1996). These observations led to the development of polyvalent vaccines against the most prevalent bacteremic serotypes. A 12-valent *E. coli* and 8-valent *Ps*eudomonas vaccine have progressed through phase 1 clinical trials. While a 23-valent *Klebsiella* capsular polysaccharide vaccine was similarly developed, subsequent data suggested that an O-polysaccharide based vaccine could be developed from only 3 *Klebsiella* O types (Trautmann et al., 2004). Thus a broadly based type-specific vaccine directed against Gram negative bacterial isolates frequently cultured from clinical cases of bacteremia could be formulated on the basis of complete O-polysaccharide LPS preparations.

13.8 Immunization Strategies with Anti-Endotoxin Vaccines

While public health vaccines are used in the general population for prevention of infections in the community, endotoxin vaccines, if successful, would be targeted for specific populations, such as occupations subjected to a high likelihood of trauma (e.g. soldiers, firemen, police, fishermen, lumber industry), or perhaps individuals who will have complicated abdominal or genitourinary tract surgery (Bennett-Guerrero et al., 1997; Goldie et al., 1995). In an earlier study with experimental *Pseudomonas* and *Klebsiella* vaccines, we found that patients who suffered acute trauma were as responsive to these vaccines as healthy volunteers (Campbell et al., 1996). This suggests that acutely injured or burned patients, with their TH2 immune response bias (e.g. IL-4, IL-10 which are associated with antibody response), may be another target population. Since sepsis is often a later complication of wounds, there may be sufficient time to develop a protective antibody response, particularly if an adjuvant can accelerate that vaccine response. Unlike the case with passive immunization, increased levels of antiendotoxin antibodies were present after active immunization for >3 months.

Some patients may present to the hospital with sepsis and not have time to develop an anti-endotoxin antibody response after immunization. Such patients may benefit from the passive administration of high-titered antibody to core LPS, as was the case in the Ziegler study (Ziegler et al., 1982). In addition, as was evident in earlier studies of sepsis (Opal et al., 2005; J5 study group, 1991; Goldie et al., 1995; Fomsgaard et al., 1989; Nys et al., 1993; Schedel et al., 1991), and confirmed experimentally, there may be consumption of antibody during fulminant sepsis. Thus, an anti-endotoxin vaccine may also be used to prepare antibodies in healthy subjects for passive administration, either in the absence of previous patient immunization, or as a supplement at the time of sepsis for individuals previously immunized. Baumgartner et al. added an additional infusion of anti-J5 serum to surgical patients who developed sepsis despite prophylactic infusions, and reported an improved survival (Baumgartner et al., 1985).

13.9 Other Potential Applications of Anti-Endotoxin Vaccine

Anti-core LPS vaccines have long been used in the veterinary industry, where bovine mastitis has had a major impact milk production on dairy farms, and Gram-negative sepsis among pigs and foals carries a high morbidity and mortality. Several whole core-LPS vaccines, whole bacterial (both J5 [J-5 bacterin] and *Salmonella* Re [IMMVAC]) and detoxified conjugate vaccines have been administered in multi-dose regimens with adjuvants.

Decreased levels of anti-endotoxin antibodies have been associated with a deleterious outcome in a number of conditions in which sepsis was not present, such as cardiovascular surgery (Goldie et al., 1995) and allogeneic bone marrow transplantation (Cohen et al., 1987). Preliminary observations also suggest that anti-endotoxin antibodies may affect the outcome from experimental heat stroke (Gathiram et al., 1987), acute radiation injury syndrome (Wells et al., 1990) and even exercise tolerance in marathon runners and race horses (Brock-Utne et al., 1988; Bosenberg et al., 1988). The common element may be endotoxemia from a compromised, now permeable bowel. For example, with exercise race horses developed endotoxemia, perhaps on the basis of relative gut ischemia as the blood flow was directed to the muscles (Baker et al., 1988). Horses that were well-trained had lower levels of endotoxemia, perhaps indicating that the training may have generated anti-core endotoxin antibodies via transient episodes (i.e. "immunizations") of endotoxemia. This same rationale for attributing a role for endotoxin has been applied to heat stroke in humans (Gathiram et al., 1987). Heat stroke is an important problem in military training camps where soldiers may undergo extreme physical activity during periods of high ambient temperature. Morbidity during this training could be due to an increase in endotoxemia originating from a leaky gut (Brock-Utne et al., 1988). Indeed, core body temperatures in excess of 105°F have been recorded in marathon runners. Of interest, the sensation of "hitting the wall" well into a race is similar to the symptoms observed during infusion of endotoxin into human subjects. It is not clear whether the excessive mortality to elderly patients who suffer heat stroke during a heat wave may be attributed in part to endotoxemia. Given the impact of radiation on gut integrity, it is not surprising that this is accompanied by endotoxemia (Wells et al., 1990).

Although the clinical relevance of these observations remains to be better demonstrated, antiendotoxin strategies could have some additional therapeutic value. These preliminary observations lend credence to Ivan Bennett's insightful comment over 30 years ago that endotoxin may play a larger role in human health and disease. "It has been said in jest that endotoxins will probably turn out to be the cause of most human diseases now classified as idiopathic and that they may also be the cause of human health. . .these possibilities have yet to be excluded" (Bennett, 1964).

Acknowledgements The work in this review was supported in part by the National Institutes of Health (2RO1 AI42181-04 AI). The author is grateful for the excellent assistance of Karen Webster in the preparation of this manuscript.

References

Bahador, M., Cross, A.S. From therapy to experimental model: A hundred years of endotoxin administration to human subjects. J Endotox Res (now Innate Immunity) 13 (2007) 251–279.

Baker, B., Gaffin, S.L., Wells, M., Wessels, B.C., Brock-Utne, J.G. Endotoxemia in racehorses following exercise. J S Afr Vet Assoc 59 (1988) 64–66.

Bailat, S., Heumann, D., LeRoy, D. Baumgartner, J-D., Rietschel, E.T., Glauser, M.P., DiPadova, F., Similarities and disparities between core-specific and O-side chain specific antilipopolysaccharide monoclonal antibodies in models of endotoxemia and bacteremia in mice. Infect Immun 65 (1997) 811.

Baumgartner, J-D., Glauser, M.P., McCutchan, J.A, Ziegler, E.J., van Melle, G., Klauber, M.R., Vogt, M., Muehlen, E., Luethy, R., Chiolero, R., Preventiion of gram-negative shock and death in surgical patients by antibody to endotoxin core glycolipid. Lancet 2 (1985) 59–63.

Baumgartner, J.D., Heumann, D., Calandra, T., Glauser, M.P., Antibodies to lipopolysaccharides after immunization of humans with the rough mutant *Escherichia coli* J5. J Infect Dis 163 (1991) 769.

Bennett, I.L., Jr. Approaches to the mechanisms of endotoxin action. In Braun, W., Landy, M. (eds), Bacterial Endotoxins: Proceedings of a Symposium held at the Institute of Microbiology of Rutgers, the State University. New Brunswick, Institute of Microbiology, Rutgers, the State University (1964), pp. xiii–xvi.

Bennett-Guerrero, E., Ayuso, L., Hamilton-Davies, C., White, W.D., Barclay, G.R., Smith, P.K., King, S.A., Muhlbaier, L.H. Newman, M.F., Mythen, M.G., Relationship of preoperative antiendotoxin core antibodies and adverse outcomes following cardiac surgery. JAMA 277 (1997) 646.

Bennett-Guerrero, E., McIntosh, T.J., Barclay, G.R., Snyder, D.S., Gibbs, R.J., Mythen, M.G., Poxton, I.R. Preparation and preclinical evaluation of a novel liposomal comlete-core lipopolysaccharide vaccine. Infect Immun 68 (2000) 6202–6208.

Bernard, G.R., Vincent, J.L., Laterre, P.F., LaRosa, S.P., Dhainaut, J.F., Lopez-Rodriguez, A., Steingrub, J.S., Garber, G.E., Helterbrand, J.D., Ely, E.W., Fisher, C.J., Jr. Efficacy and safety of recombinant human activated protein C for severe sepsis. N Engl J Med 344 (2001) 699–709.

Bhattacharjee, A.K., Cross, A.S., Opal, S.M., Warren, H.S., Sadoff, J.C., New trends in *Escherichia coli* O111:B4 J5 mutant (Rc Chemotype) vaccine development for use in gram-negative bacillary sepsis. In Morrison, D.C., Ryan, J.L. (eds), Novel Therapeutic Strategies in the Treatment of Sepsis. Marcel Dekker, New York (1996a), p. 31.

Bhattacharjee, A.K., Opal, S.M., Palardy, J.E., Drabick, J.J., Collins, H., Taylor, R., Cotton, A., Cross, A. Affinity-purified *Escherichia coli* J5 lipopolysaccharide-specific IgG protects neutropenic rats against gram negative bacterial sepsis. J Infect Dis 170 (1994) 622–629.

Bhattacharjee, A.K., Opal, S.M., Taylor, R., Naso, R., Semenuk, M., Zollinger, W.D. Moran, E.E., Young, L., Hammack, C., Sadoff, J.C., Cross, A.S. A noncovalent complex vaccine prepared with detoxified *Escherichia coli* J5 (Rc Chemotype) lipopolysaccharide and Neisseria meningitidis Group B outer membrane protein produces protective antibodies against gram-negative bacteremia, J Infect Dis 173 (1996b) 1157–1163.

Bhattacharjee, A.K., Cross, A.S. Vaccines and antibodies in the prevention and treatment of sepsis. Infect Dis Clin N Am 13 (1999) 355–369.

Bosenberg, A.T., Brock-Utne, J.G., Gaffin, S.L., Wells, M.T., Blake, G.T. Strenuous exercise causes systemic endotoxemia. J Appl Physiol 65 (1988) 106–108.

Braude, A.E., Jones, J.L., Douglas, H. The behavior *Escherichia coli* endotoxin (somatic antigen) during infectious arthritis, J Immunol 90 (1960) 297–312.

Braude, A..I., Ziegler, E.J., McCutchan, J.A., Douglas, H. Immunization against nosocomial infection. Am J Med 70 (1981) 463–466.

Bruins, S.C., Stumacher, R., Johns, M.A., McCabe, W.R. Immunization with R mutants of *Salmonella minnesota*. III. Comparison of the protective effect of immunization with lipid A and the Re mutant. Infect Immun 17 (1977) 16–20.

Brock-Utne, J.G., Gaffin, S.L., Wells, M.T., Gathiram, P., Sohar, E., James, M.F., Morrell, D.F., Norman, R.J. Endotoxemia in exhausted runners after a long-distance race. S Afr Med J 73 (1988) 533.

Calandra, T., Glauser, M.P., Schellekens, J., Verhoef, J. Swiss Dutch, J5 Immunoglobulin Study Group, Treatment of Gram-negative septic shock with human IgG antibody to *Escherichia coli* J5: A prospective, double-blind, randomized trial, J Infect Dis 158 (1988) 312–319.

Campbell, W.N., Hendrix, E., Cryz, S.J., Cross, A.S. Immunogenicity of a 24-valent *Klebsiella* capsular polysaccharide vaccine and an 8-valent *Pseudomnas* O-polysaccharide conjugate vaccine administered to acute trauma victims, Clin Infect Dis 23 (1996) 179–181.

Chedid, L., Parant, M., Parant, F., Boyer, F. A proposed mechanism for natural immunity to enterobacterial pathogens. J Immunol 100 (1968) 292–301.

Cohen J, Moore, R.H., Al Hashimi, S., Jones, L., Apperley, J.F., Aber, V.R. Antibody titers to a rough-mutant strain of *Escherichia coli* in patients undergoing allogeneic bone marrow transplantation. Evidence of a protective effect against graft-versus-host disease. Lancet 1 (1987) 8–11.

Collins, H.H., Cross, A.S., Dobeck, A., Opal, S.M., McClain, J.B., Sadoff, J.C. Oral ciprofloxacin and monoclonal antibody to lipopolysaccharide protect leukopenic rats from lethal infection with *Pseudomonas aeruginosa*, J Infect Dis 159 (1989) 1073–1082.

Cometta, A., Baumgartner, J-D., Lee, M.L., Hanique, G., Glauser, M-P. IVIG Collaborative Study Group, Prophylactic intravenous administration of standard immune globulin as compared with core-lipopolysaccharide immune globulin in patients at high risk of postsurgical infection, N Engl J Med 327 (1992) 234–240.

Cooper, C.L., Davis, H.L., Morris, M.L., Efler, S.M., Adhami, M.A., Krieg, A.M., Cameron, D.W., Heathcote, J. CPG 7909, an immunostimulatory TLR9 agonist oligodeoxynucleotide, as adjuvant to Engerix-B HBV vaccine in healthy adults: A double blind phase I/II study. J Clin Immunol 24 (2004) 693–701.

Cross, A., Sidberry, H., Sadoff, J.C. The human antibody response during natural bacteremic infection with gram-negative bacilli against lipopolysaccharide core determinants. J Infect Dis 160 (1989) 225–236.

Cross, A.S., Opal, S. Therapeutic intervention in sepsis with antibody to endotoxin: Is ther a future? J Endotox Res 1 (1994) 57.

Cross, A.S., Opal, S.M,, Bhattacharjee, A.K., Donta, S.T., Peduzzi, P.N., Furer E, Que, J.U., Cryz, S.J., Jr. Immunotherapy of sepsis: Flawed concept or faulty implementation? Vaccine 17 (1999) S13–S21.

Cross, A.S., Opal, S.M., Warren, H.S., Palardy, J.E., Glaser, K., Parejo, N.A., Bhattacharjee, A.K. Active immunization with a detoxified *Escherichia coli* J5 lipopolysaccharide group B meningococcal outer membrane protein complex vaccine protects animals from experimental sepsis. J Infect Dis 183 (2001) 1079–1086.

Cross, A.S., Opal, S.M., Palardy, J.E., Drabick, J.J., Warren, H.S., Huber, C., Cook, P., Bhattacharjee, A.K. Phase I study of detoxified *Escherichia coli* J5 lipopolysaccharide (J5dLPS)/group B meningococcal outer membrane protein (OMP) complex vaccine in human subjects. Vaccine 21 (2003) 4576–4588.

Cross, A.S., Opal, S., Cook, P., Drabick, J., Bhattacharjee, A. Development of an anti-core lipopolysaccharide vaccine for the prevention and treatment of sepsis. Vaccine 22 (2004) 812–817.

Dale, P.A. McQuillen, D.P., Gulati, S., Rice, P.A. Human vaccination with *Escherichia coli* J5 mutant induces cross-reactive bactericidal antibody against *Neisseria gonorrhoeae* lipooligosaccharide. J Infect Dis 166 (1992) 316.

DiPadova, F.E., Brade, H., Barclay, G.R., Poxton, I.R., Liehl, E., Schuetze, E., Kocher, H.P., Ramsay, G., Schrier, M.H., McClelland, D.B. A broadly cross-protective monoclonal antibody binding to *Escherichia coli* and *Salmonella* lipopolysaccharides. Infect Immun 61 (1993) 3863–3872.

Dissanayake, D.R., Wijewardana, T.G., Gunawardena, G.A., Poxton, I.R. Potential use of liposome-encapsulated mixture of lipopolysaccharide core types (R1, R2, R3 and R4) of *Escherichia coli* in controlling colisepticemia in chickens. J Med Microbiol (2009) (epublished).

Donta, S.T., Peduzzi, P., Cross, A.S., Sadoff, J., Haakenson, C., Cryz, S.J., Jr., Kauffman, C., Bradley, S., Gafford, G., Elliston, D., Beam, T.R., John, J.F., Jr., Ribner, B., Cantey, R., Welsh, C.H., Ellison, R. III, Young, E.J., Hamill, R.J., Leaf, H., Schein, R.M., Mulligan, M.,

Johnson, C., Abrutyn, E., Griffiss, J.M., Slagle, D. Immunoprophylaxis against *Klebsiella* and *Pseudomonas aeruginosa* infections. J Infect Dis 174 (1996) 537–543.

Erridge, C., Stewart, J., Bennett-Guerrero E, McIntosh TJ, Poxton, I.R. The biological activity of a liposomal complete core LPS vaccine. J Endotoxin Res 8 (2002) 39–46.

Finland, M., Jones, W.F., Jr., Barnes, M.W. Occurrence of serious bacterial infections since introduction of antibacterial agents. JAMA 170 (1959) 2188–2197.

Fomsgaard, A., Baek, L., Fomsgaard, J.S., Engquist, A. Preliminary study on treatment of septic shock patients with antilipopolysaccharide IgG from blood donors. Scand J Infect Dis 21 (1989) 697–708.

Gathiram P, Wells, M.T., Brock-Utne, J.G., Gaffin, S.L. Antilipopolysaccharide improves survival in primates subjected to heat stroke. Circulatory Shock 23 (1987) 157.

Goldie, A.S., Fearon, K.C.H., Ross, J.A., Barclay, R., Jackson, R.E., Grant, I.S., Ramsay, G., Blyth, A.S., Howie, J.C. Natural cytokine antagonists and endogenous antiendotoxin core antibodies in sepsis syndrome, JAMA 274 (1995) 172–177.

Greenman, R.L., Schein, R.M.H., Martin, M.A., Wenzel, R.P., MacIntyre, N.R., Emmanuel, G., Chmel, H., Kohler, R.B., McCarthy, M., Plouffe, J. A controlled clinical trial of E5 murine monoclonal IgM antibody to endotoxin in the treatment of gram-negative sepsis. The XOMA Sepsis Study Group. JAMA 266 (1991) 1097–1102.

J5 Study Group, Treatment of severe infectious purpura in children with human plasma from donors immunized with *Escherichia coli* J5: A prospective, double-blind study. J Infect Dis 165 (1991) 695–701.

Johns, M., Skehill, A., McCabe, W.R. Immunization with rough mutants of *Salmonella minnesota*. IV. Protection by antisera to O and rough antigens against endotoxin. J Infect Dis 147 (1983) 57–62.

Levin, M., Quint, P.A., Goldstein, B., Barton, P., Bradley, J.S., Shemie, S.D, Yeh, T., Kim, S.S., Cafaro, D.P., Scannon, P.J., Giroir, B.P. Recombinant bactericidal/permeability increasing protein (rBPI21) as adjunctive treatment for children with severe meningococcal sepsis: A randomised trial. rBPI21 Meningococcal Sepsis Study Group. Lancet 356 (2000) 961–967.

Luderitz, O., Westphal, O. Staub, A.M., Kikaido, H. Isolation and chemical and immunological characterization of bacterial lipopolysacchairdes. In. Weinbaum, G., Kadis, S., Ajl, S.J. (eds), Microbial Toxins. Vol. 4. Academic Press, New York (1971), pp. 145–233.

Lugowski, C., Jachymek, W., Niedziela, T., Rowinski, S. Serological characterization of anti-endotoxin sera directed against the conjugates of oligosaccharide core of *Escherichia coli* type R1, R2, R3, J5 and *Salmonella* Ra with tetanus toxoid. FEMS Immunol Med Microbiol 16 (1996a) 21–30.

Lugowski, C., Niedziela, T., Jachymek, W. Anti-endotoxin antibodies directed against *Escherichia coli* R1 oligosaccharide core-tetanus toxoid conjugate bind to smooth live bacteria and smooth lipopolysaccharides and attenuate their tumor necrosis factor stimulating activity, FEMS Immunol Med Microbiol 16 (1996b) 31–38.

Lugowski, C., Czaja, J., Jachymek, W., Niedziela, T. Lakomska, J. Serological characterization of anti-endotoxin serum directed against the conjugate of oligosaccharide core of *Escherichia coli* type R4 with tetanus toxoid. J Endotoxin Res 6 (2000) 166 (Abstr).

McCabe, W.R. Immunization with R mutants of, S. minnesota. I. Protection against challenge with heterologous Gram-negative bacilli. J Immunol 108 (1972) 601–610.

McCabe, W.R., DeMaria, A., Jr., Berberich, H., Johns, M.A. Immunization with rough mutants of *Salmonella minnesota*: Protective activity of IgM and IgG antibody to the R595 (Re chemotype) mutant. J Infect Dis 158 (1988) 291–300.

Muller-Loennies, S., Brade, L., MacKenzie, C.R., DiPadova, F.E., Brade, H. Identification of a cross-reactive epitope widely present in lipopolysaccharide from enterobacteria and recognized by the cross-protective monoclonal antibody WN1 222–5. J Biol Chem 278 (2003) 25618–25627.

Muller-Loennies, S., Brade, L., Brade, H. Neutralizing and cross-reactive antibodies against enterobacterial lipopolysaccharide. Int J Med Microbiol 297 (2007) 321–340.

Nnaule, N.A., Lind, S.M., Lindberg, A.A. The disaccharide L-alpha-D-heptosel 7-L-alpha-D-heptosel of the inner core domain of *Salmonella* lipopolysaccharide is accessible to antibody and is the epitope of a broadly reactive monoclonal antibody, J Immunol 149 (1992) 2722–2728.

Nys, M., Damas, P., Joassin, L., Lamy, M. Sequential anti-core glycolipid immunoglobulin antibody activities in patients with and without septic shock and their relation to outcome, Ann Surg 217 (1993) 300–306.

Opal, S.M., Cross, A.S. Clinical trials for severe sepsis. Past failures and future hopes. Infect Dis Clin North Am 13 (1999) 285–297.

Opal, S.M. Palardy, J.E. Chen, W.H., Parejo, N.A., Bhattacharjee, A.K., Cross, A.S, Active immunization with a detoxified endotoxin vaccine protects against lethal polymicrobial sepsis: Its use with CpG adjuvant and potential mechanisms. J Infect Dis 192 (2005) 2074–2080.

Osborn, M.J. Biosynthesis and structure of the core region of lipopolysaccharide in *Salmonella typhimurium*. Ann N Y Acad Sci 133 (1966) 375–383.

Pollack, M., Huang, A.I., Prescott, R.K., Young, L.S., Hunter, K.W., Cruess, D.F., Tsai, C.M. Enhanced survival in *Pseudomonas aeruginosa* septicemia associated with high levels of circulating antibody to *Escherichia coli* endotoxin core. J Clin Invest 72 (1983) 1874–1881.

Raetz, C.R.H. The enzymatic synthesis of lipid A: Molecular structure and biologic function of monosaccharide precursors. Reviews of Infect Dis 6 (1984) 463.

Rogers, D.E. The changing pattern of life-threatening microbial disease. N Engl J Med 261 (1959) 677–683.

Schedel, I., Dreikhausen, U., Nentwig, B., Hockenschneider, M., Rauthmann, D., Balikcioglu, S., Coldeway, R., Deicher, H. Treatment of gram-negative septic shock with an immunoglobulin preparation: A prospective, randomized clinical trial. Crit Care Med 9 (1991) 1104–1113.

Schwartzer, T.A., Alcid, D.V., Numsuwan, V., Gocke, D.J. Characterization of the human anti-body response to an *Escherichia coli* O111:B4 (J5) vaccine. J Infect Dis 158 (1988) 1135–1136.

Stanislavsky, E.S., Makarenko, T.A., Kholodkova, E.V., Lugowski, C. R-form lipopolysaccharidess of gram-negative bacteria as possible vaccine antigens. FEMS Immunol Med Microbiol 18 (1997) 139–145.

Subbaiah, T.V., Stocker, B.A.D. Rough mutants of *Salmonella typhimurium*. 1. Genetics. Nature 201 (1964) 1298–1299.

Sultzer, B.M. Genetic analysis of lymphocyte activation by lipopolysaccharide endotoxin. Infect Immun 13 (1976) 1579–1584.

Takada, H., Kotuni, S. Structure function relationships of lipid, A. In Morrison, D.C., Ryan, J.L. (eds), Bacterial Endotoxic Lipopolysaccharides. CRC Press, Boca Raton, FL (1992), p. 107.

Tate, W.J., Douglas, H., Braude, A.I. Protection against lethality of, *E. coli* endotoxin with 'O' antiserum. Ann N Y Acad Sci 133 (1966) 746–762.

Trautmann, M., Held, T.K., Cross, A.S. O antigen seroepidemiology of *Klebsiella* clinical isolates and implications for immunoprophylaxis of *Klebsiella* infections. Vaccine 22 (2004) 818–821.

Waisbren, B.A. Bacteremia due to gram-negative bacilli other than the *Salmonella*. Arch Intern Med 88 (1951) 467–488.

Wells, M.T., Gaffin, S.L., Wessels, B.C., Brock-Utne, J.G., Jordaan, J.P., van den Ende, J. Anti-LPS antibodies reduce endotoxemia in whole body ^{60}Co irradiated primates: A preliminary report. Aviat Space Environ Med 61 (1990) 802.

Ziegler, E.J., Fisher, C.J., Jr., Sprung, C.L., Straube, R.C., Sadoff, J.C., Foulke, G.E., Wortel, C.H., Fink, M.P., Dellinger, R.P., Teng, N.P. Treatment of gram-negative bacteremia and septic shock with HA-1A human monoclonal antibody against endotoxin. A randomized, double-blind, placebo-controlled trial, The HA-1A Sepsis Study Group. New Engl J Med 324 (1991) 429–436.

Ziegler, E.J., McCutchan, J.A., Fierer, J., Glauser, M.P., Sadoff, J.C., Douglas, H., Braude, A.I. Treatment of gram negative bacteremia and shock with human antiserum to a mutant *Escherichia coli*. New Engl J Med 307 (1982) 1225–1230.

Chapter 14
Synthetic and Natural TLR4 Agonists as Safe and Effective Vaccine Adjuvants

Christopher B. Fox, Martin Friede, Steven G. Reed, and Gregory C. Ireton

Abstract Natural derivatives and synthetic analogues of lipopolysaccharide are potent stimulators of the mammalian immune system. Retained adjuvant activity with reduced toxicity was obtained by the development of monophosphoryl lipid A (MPL®), which is approved for use in several vaccine products. Ongoing research and development of synthetic TLR4 agonists may offer increased purity and biological activity with reduced cost. Extensive research has elucidated the mechanism of action of TLR4 agonists and structure-function relationships. Moreover, the formulation of TLR4 agonists has been shown to significantly affect the type and magnitude of elicited immune response. TLR4 agonists comprise a promising class of adjuvants for safe and effective vaccines.

Keywords Lipopolysaccharide · TLR4 agonist · Monophosphoryl lipid A · Glucopyranosyl lipid adjuvant · adjuvant formulation

Abbreviations

AGP	aminoalkyl glusosaminide 4-phosphate
CTL	cytotoxic T lymphocyte
DLPC	1,2-dilauroyl-sn-glycero-3-phosphocholine
DMPC	1,2-dimyristoyl-sn-glycero-3-phosphocholine
DPPC	1,2-dipalmitoyl-sn-glycero-3-phosphocholine
DPPG	1,2-dipalmitoyl-sn-glycero-3-phospho-(1′-rac-glycerol)
DPTAP	1,2-dipalmitoyl-3-trimethylammonium-propane
DSC	differential scanning calorimetry
FTIR	Fourier-transform infrared spectroscopy
GLA	glucopyranosyl lipid A
GLA-SE	stable emulsion containing glucopyranosyl lipid adjuvant
HPV	human papilloma virus

G.C. Ireton (✉)
Infectious Disease Research Institute, 1124 Columbia St, Ste 400, Seattle, WA 98104, USA
e-mail: gireton@idri.org

X. Wang, P.J. Quinn (eds.), *Endotoxins: Structure, Function and Recognition*, Subcellular Biochemistry 53, DOI 10.1007/978-90-481-9078-2_14,

ITC	isothermal titration calorimetry
LBP	lipopolysaccharide-binding protein
LPS	lipopolysaccharide
LRR	leucine-rich repeat
Mal	MyD88-adapter-like
MAPK	mitogen-activated protein kinase
MPL	monophosphoryl lipid A
MPL-AF	aqueous formulation of monophosphoryl lipid A
MPL-SE	stable emulsion containing monophosphoryl lipid A
MyD88	myeloid differentiation primary response gene (88)
NF-κβ	nuclear factor-κβ
PAMP	pathogen-associated molecular pattern
PRR	pattern-recognition receptor
TLR	Toll-like receptor
TNF-α	tumor necrosis factor-α
TRIF	TIR-domain-containing adapter-inducing interferon-β
TRAM	TRIF-related adapter molecule

14.1 Introduction

Several molecules derived from pathogenic and non-pathogenic microbes interact with receptors on cells of the mammalian immune system. Endotoxin (lipopolysaccharide, LPS) derived from Gram-negative bacteria was long ago described as a potent stimulus of antibody responses, and extensive biochemical, biophysical, and immunological studies were undertaken to define the effects of LPS on the immune system and to attempt to disassociate, at the molecular level, the toxic (inflammatory) properties of LPS from its adjuvant effects. Among the early studies demonstrating that LPS could be a potent adjuvant for protein antigens was a report in 1956 by Johnson et al. (1956), who also demonstrated that LPS could exert its biological activity if administered to a different site than the antigen (Johnson, 1964). This observation led to the conclusion that the adjuvant activity of LPS was systemic, rather than local, unlike aluminium or oil-based adjuvants which only worked if co-administered with the antigen. LPS, and its active component lipid-A, were however too toxic to be used as an adjuvant for vaccines, so this early discovery was followed by many years of research to separate the adjuvant activity from the pyrogenicity and toxicity of the parent LPS compounds.

The first preparation of an LPS derivative which retained immunostimulatory activity yet had significantly reduced toxicity was made in 1979 by Ribi et al. (1979) who found that mild acid hydrolysis of LPS reduced the pyrogenicity without affecting tumor inhibition activity. This reduced-toxicity product was shown to be the 4′-monophosphoryl derivative of lipid-A (MPL®) (Qureshi et al., 1982). Ribi′s group later demonstrated that further alkaline hydrolysis of MPL resulted in de-acylation at the 3′ position, producing a molecule with even lower toxicity yet retaining full immunstimulatory and adjuvant activity (Myers et al., 1990). This preparation, referred to as MPL® or 3D-MPL to indicate the absence of the 3′acyl chain has been extensively reviewed (Ulrich and Myers, 1995). The early clinical development of MPL® focused to a large extent on its potential use in cancer therapy: MPL® mixed with the cell wall skeleton (CWS) of *Mycobacterium phlei*, formulated in an oil-in-water emulsion was referred to as DETOX™ and was tested in combination with melanoma cell lysates in several phase II clinical trials for immunotherapy of melanoma (Elliott et al., 1993; Mitchell et al., 1990). Although extended survival times were observed, and despite receiving regulatory approval in Canada, development of this product (Melacine) was not continued.

In contrast, prophylactic vaccines containing MPL® have been developed and have received regulatory approval, and numerous others are in active development. The first such product was GlaxoSmithKline's (GSK) Fendrix®, a hepatitis B vaccine, and the second is GSK's Cervarix® vaccine against human papilloma virus (HPV). This vaccine, which is approved in many countries including the USA, contains the proprietary adjuvant system AS04 which is a combination of MPL® and aluminium hydroxide (Garcon, 2005). AS04 is also used in Fendrix®. The manufacturer reports that the addition of MPL® to the vaccine enhances the intensity and duration of the antibody response, and may also be responsible for the broad cross-strain protection afforded by this vaccine (Paavonen et al., 2009). GSK has also included MPL® in combination with other immunostimulants in several other adjuvant systems which are in late clinical development for various indications: AS01, a combination of MPL® with the saponin QS21 in a liposomal formulation is a component of a malaria vaccine currently in phase 3 clinical trials (Lell et al., 2009) and in AS02, a combination of MPL® with QS21 and an oil-in-water emulsion is under clinical evaluation for cancer vaccines (Atanackovic et al., 2008).

In addition to its use in the GSK vaccine pipeline, MPL® is also used by Allergy Therapeutics Ltd. (ATL) in a therapeutic pollen-allergy vaccine, Polinex Quatro®, which is approved in several European countries (Gawchik and Saccar, 2009). In this vaccine it is thought that the MPL®, formulated with an alkyl tyrosine co-adjuvant, assists in driving the immune response towards a Th1 response and away from the allergy-mediating Th2 response. Although MPL® has achieved significant success, its use is not without challenges. The material is derived from bacteria and exhibits heterogeneity in terms of degree of acylation, containing tetra-, penta-, and hexa-acyl forms of the 3′-de-acalyted MPL®. Although all of these are recognized by murine cells, the non-hexa forms appear to be less active on human cells. Hence careful verification of the composition of the MPL® is important, as variation in content could affect biological activity. The use of a pure synthetic hexa-acyl form of MPL® could overcome these challenges.

In parallel to the approach of detoxifying LPS through hydrolysis, other groups attempted to separate the toxicity and adjuvanticity by chemically synthesizing a range of lipid-A analogues and testing these for immune modulating and inflammatory activities (Johnson, 1994). These studies showed that manipulation of the fatty acid content and phosphorylation affected both toxicity and adjuvant action and provided insight into the structure-function relationship between adjuvant activity and toxicity. None of these synthetic lipid-A derivatives (referred to variously as LA-14-PP or compound 406, LA-14-HP or compound 405, LA-14-PH or compound 504, etc) has ever been included in approved vaccines.

Interestingly, several synthetic monosaccharide derivatives (derivatives of lipid Y, the reducing monosaccharide unit of lipid A from S. minnesota) also displayed adjuvant activity in rodents, while being essentially devoid of toxicity (Galanos et al., 1986; Kotani et al., 1986; Kumazawa et al., 1985). These tri- and tetra-acylated molecules were not developed for clinical application, however the biological activity of these monosaccharide forms did lead to the development of a new class of synthetic lipid A analogues referred to as aminoalkyl glusosaminide 4-phosphates (AGPs). This class of molecules was developed at Ribi Immunochem and Corixa Corp. (USA) by replacing the reducing sugar in MPL with an aminoalkyl aglycone unit, thus permitting six fatty acyl chains to be presented on a monosaccharide backbone. Numerous analogues with varying acyl chain lengths were made, some being identified as agonists and others as antagonists of lipid A activity. One of these, RC-529 (Table 14.1) was shown to have adjuvant activity comparable to MPL® (Evans et al., 2003) and has been developed as an adjuvant for use in vaccines. This molecule, in combination with aluminium hydroxide was used as the adjuvant in the approved recombinant hepatitis B vaccine Supervax™ produced by Crucell.

Another approach to using lipid A as a safe adjuvant for vaccines was developed by the group of Carl Alving which observed that when lipid A was formulated in liposomes, the toxicity was significantly reduced (Alving and Rao, 2008). This formulation was tested successfully in the clinic as an adjuvant for a malaria vaccine (Fries et al., 1992b), however no further clinical development was undertaken and the non-toxic derivatives or analogues of lipid-A are now preferred.

Several other non-toxic lipid A analogues have been developed and tested for adjuvant activity. The Infectious Disease Research Institute and Immune Design Corp. are developing vaccines containing a synthetic hexa-acylated TLR4 agonist, glucopyranosyl lipid A (GLA) (Anderson et al., 2010; Baldwin et al., 2009a, b; Bertholet et al., 2009; Reed et al., 2009). The company Biomira (USA) has developed a range of synthetic MPL derivatives (Jiang et al., 2007) including a novel tri-lipid acyl group (Jiang et al., 2002). These molecules have been shown in preclinical studies to have adjuvant activity but have not yet undergone clinical development. OM Pharma in Switzerland developed a tri-acyl derivative of E. Coli lipid-A (OM-174) (Brandenburg et al., 2000) and also a synthetic form based on a tri-acylated di-phosphorylated pseudopeptide backbone (OM-294-DP) (Martin et al., 2006). These molecules have not undergone clinical development. The Japanese company Eisai has developed a range of lipid-A analogues, of which the synthetic compound E6020, containing a hexa-acylated acyclic backbone (Ishizaka and Hawkins, 2007; Morefield et al., 2007), is being developed as an adjuvant for use in vaccines. Another Japanese company ONO is developing the compound referred to as ONO-4007 (2-Deoxy-3-O-(9-phenylnonanoyl)-2-[3(S)-(9-phenylnonanoyloxy)tetradecanamido]-4-O-sulfo-D-glucopyranose) which is a tri-acylated acyclic sulphonated backbone. This compound has been used in phase 1 clinical studies for anti-tumor activity. Table 14.1 provides an overview of the chemical structures, formulation, and development status of several TLR4 agonists.

Table 14.1 TLR4 agonists currently in approved vaccines or clinic evaluation

Molecule	Structure	Formulation	Company	Product Name	Indication	Development Stage
MPL	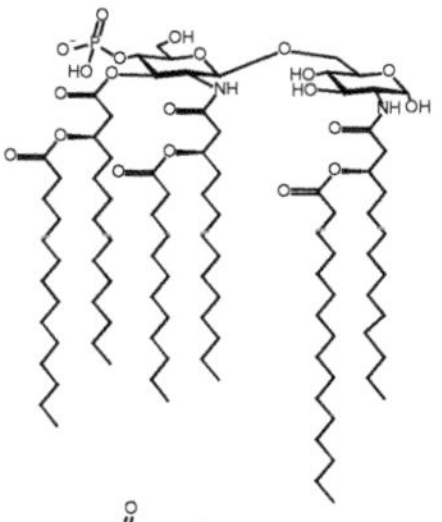	alum-adsorbed alum-adorbed tyrosine-adsorbed liposomal with QS21 liposomal	GSK GSK ALT GSK Oncothyreon	 Cervarix® Fendrix® Pollinex®	HPV HBV Allergy Malaria Cancer	Approved Approved Approved Phase III Phase III
RC-529	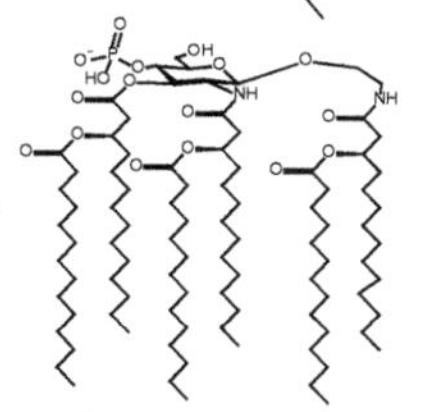	alum-adsorbed	Dynavax	SuperVax®	HBV	Approved
Lipid A	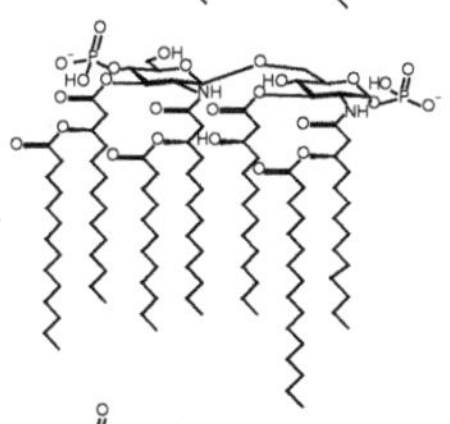	liposomal	WRAIR		Malaria	Phase I
GLA	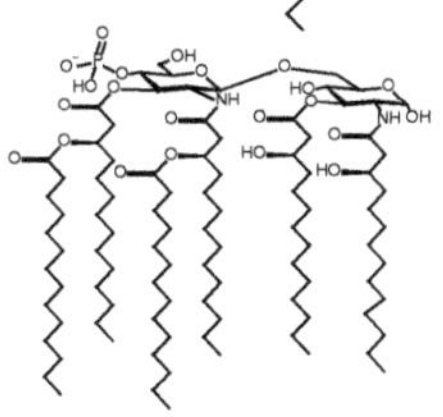	emulsion	Immune Design		Influenza	Phase I
ONO-4007	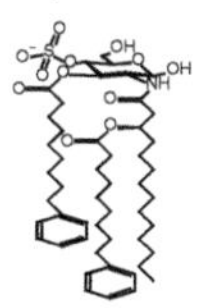	ethanol co-solvent	Ono Pharmaceutical			Phase I
OM-174	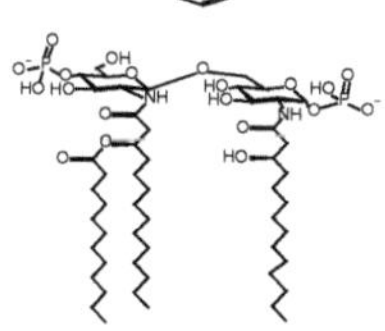	aqueous	OM Pharma			Preclinical
E6020	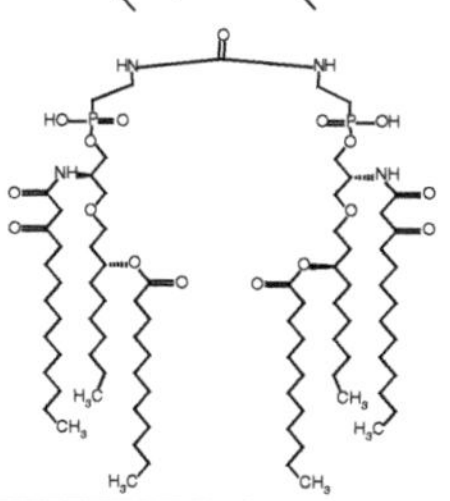	alum-adsorbed or emulsion	Eisai			Preclinical

Note: Table is meant to be representative and not exhaustive.

14.2 Immune Recognition of Lipopolysaccharide (LPS) and Related Molecules

Cells of the vertebrate innate immune system are able to rapidly respond to invading pathogens by recognizing conserved pathogen-associated molecular patterns (PAMPs) through a complex network of membrane bound and cellular pattern-recognition receptors (PRRs) (Janeway and Medzhitov, 2002; Kumar, 1991; Medzhitov, 2007). These PAMPs are usually essential components necessary for the survival of the pathogen. Recognition of PAMPs by PRRs in various cellular compartments triggers host defense mechanisms including the release of inflammatory cytokines and type I interferons to limit further infection (Beutler, 2009; Janeway and Medzhitov, 2002). The responses of the innate immune system are important not only to eliminate pathogens but also to develop pathogen-specific adaptive immunity, which is mediated by B and T cells (Beutler, 2009; Pasare and Medzhitov, 2004).

LPS released from the outer membrane of invading Gram-negative bacteria is recognized by the innate immune system and plays a critical role in triggering host responses to Gram-negative infections by stimulating the release of pro-inflammatory cytokines from various target cells. Because of its potent activity, LPS has been implicated in a variety of diseases such as septic shock (Beutler and Rietschel, 2003; Cohen and Druilhe, 2002). The pattern-recognition receptor for LPS belongs to a class of type I trans-membrane glycoproteins known as Toll-like receptors (TLRs) (Medzhitov et al., 1997; Poltorak et al., 1998). Thus far, 13 members of the TLR family have been identified recognizing a wide variety of microbial products with conserved structural patterns present in bacteria, fungi and viruses (Matsushima et al., 2007). The extracellular domains of TLRs consist of leucine-rich repeats (LRRs) with a horseshoe like shape (Choe et al., 2005; Kim et al., 2007; Park et al., 2009; Zhao et al., 2007). Binding of agonist ligands induces Toll-like receptor dimerization and is believed to trigger the recruitment of specific adaptor proteins to the intracellular domains, initiating a signaling cascade (Liu et al., 2008; Park et al., 2009; Zhao et al., 2007).

The structure of LPS is well defined and consists of lipid A, core oligosaccharide and an O side chain (Raetz and Whitfield, 2002). The lipid A portion of LPS represents the conserved molecular pattern of LPS and is the main inducer of potent inflammatory responses and septic shock (Galanos et al., 1986; Tsujimoto et al., 1989). Lipid A consists of a ß(1-6) linked di-glucosamine that varies among gram negative bacteria with respect to its number and length of fatty acid side chains, and in number and secondary modification of terminal phosphates. Immune recognition of lipid A is initiated by extraction of monomers from LPS aggregates by LPS-binding protein (LBP) in the serum (Miyake, 2006). Monomeric LPS is transferred from LBP to another accessory protein CD14, which in turn transfers LPS to MD2, a secreted glycoprotein that associates with the extracellular domain of Toll-like receptor 4 (TLR4) to form the heterodimeric receptor that is responsible for the physiological recognition of LPS (Kobayashi et al., 2006; Miyake, 2006; Nagai et al., 2002).

Lipid A bound to the TLR4/MD2 complex activates two distinct intracellular signaling pathways which have come to be known by the names of the TLR4-proximal adaptor proteins, MyD88 and TRIF (Akira and Takeda, 2004; Akira et al., 2006). The MyD88 pathway functions to activate mitogen-activated protein kinase (MAPK) and nuclear factor-κβ (NF-κβ) dependent pro-inflammatory responses, while the TRIF pathway activates kinases responsible for type I interferon responses (Barton and Kagan, 2009). Requirements for these adaptors in TLR4-mediated signaling have been defined primarily in knock-out and induced mutant studies. Of the TLRs identified to date, only TLR4 has been demonstrated to activate both signaling pathways, which may account for the potent inflammatory effects of LPS.

14.3 Lipid A Structure and Activity

The TLR4–MD2 hetero-dimer has complex ligand specificity. It can be activated by structurally diverse LPS molecules, and minor changes in synthetic derivatives of LPS can abolish their endotoxic potency (Raetz and Whitfield, 2002; Rietschel et al., 1994). The diversity in potency of LPS is derived from variance within lipid A, as observed in both the number and the length of fatty acid side chains and the presence of terminal phosphate residues with a variety of modifications. Optimal lipid A potency is achieved with bi-phosphorylated, hexa-acylated, lipid A species (Raetz and Whitfield, 2002). Lipid A moieties that deviate from this pattern often demonstrate a significant decrease in endotoxic activity (Alexander and Rietschel, 2001).

Structure function analysis of LPS isolates from different bacterial species and of synthetic lipid A derivatives indicates that the length and number of acyl chains are critical for TLR4 activation (Hajjar et al., 2002; Kusumoto et al., 2003). Studies on the activity of lipid A indicate that hexa-acylated lipid A from *Escherichia coli* with side chains maximally stimulates TLR4 and altering the number or length of the attached fatty acids or altering the charge of lipid A can reduce the magnitude of activation. Lipid A molecules with five or seven acyl chains are ~100 fold less active, while lipid A with four acyl chains lacks agonistic activity altogether, and instead is antagonistic(Hajjar et al., 2002; Kusumoto et al., 2003). Even among hexa-acylated lipid A molecules, acyl chain length is critical to TLR4 activation. It has been demonstrated that acyl chains 12–14 carbons in length have full potency, with rapid decline in activities with chain lengths less than 10 or greater than 16 (Johnson et al., 1999; Stover et al., 2004).

Two phosphate groups in lipid A also greatly affect the endotoxic activity of LPS (Rietschel et al., 1994). Deletion of either phosphate reduces endotoxic activity ~100-fold, with the resulting monophosphoryl lipid A (MPL) being a weak activator of the human innate immune response (Caglar et al., 2009; Tsujimoto et al., 1989). It appears that the requirement for negative charge is a general requirement as substitutions of the phosphates with other negatively charged groups has only minor effects (Seydel et al., 2005; Ulmer et al., 1992). MPL has been reported to preferentially activate the TLR4-TRAM-TRIF-based signaling pathway but not the TLR4-Mal-MyD88 pathway (Mata-Haro et al., 2007). This has recently been confirmed with a synthetic hexa-acylated *E. coli* MPL and highly purified diphosphate lipid A to confirm that removal of a single phosphate was sufficient to shift TLR4 activation from balanced MyD88 and TRIF contributions, to a biased TRIF pathway.

Examination of recent crystal structures of TLR4/MD2 bound to LPS and the antagonistic tetra-acylated compound Eritoran shed light on the structural requirements of lipid A binding and TLR4 activation (Kim et al., 2007; Ohto et al., 2007; Park et al., 2009). In the structures containing antagonistic tetra-acylated ligands, the four lipid chains of Eritoran or lipid IVa are bound in a hydrophobic pocket on MD2 (see Fig. 14.1). MD2 has a β-cup fold structure consisting of two antiparallel β-sheets, which forms the hydrophobic pocket for ligand binding (Kim et al., 2007; Ohto et al., 2007). In the Eritoran/TLR4/MD2 structure, no direct contacts are made between the ligand and TLR4, and receptor dimerization is not observed. In contrast, Park et al. demonstrated that in the absence of ligand TLR4 and MD2

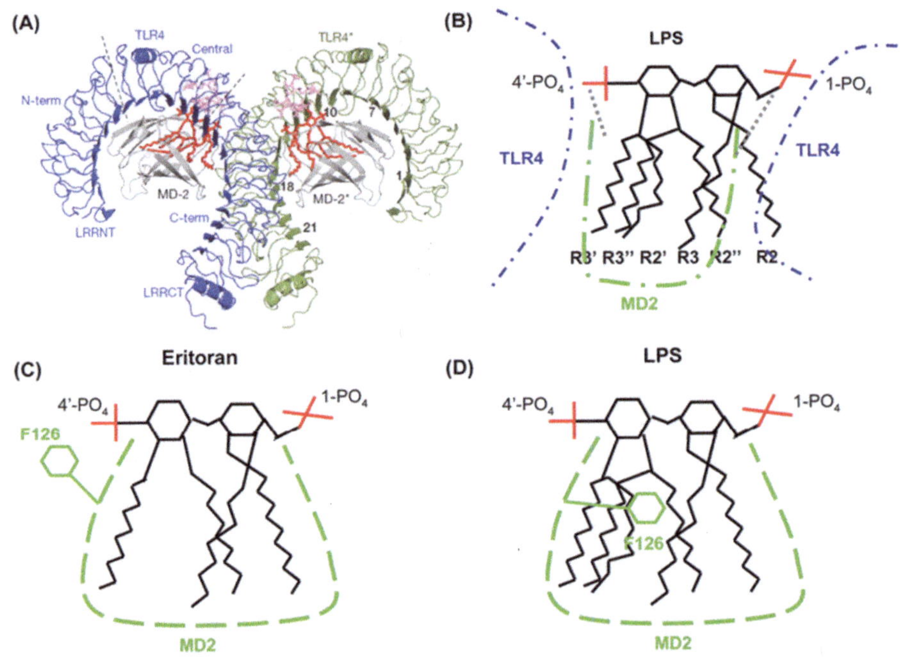

Fig. 14.1 (**a**) The crystal structure of TLR4–MD2 complex binding to LPS reported by Park et al. LRRCT, C-terminal Leucine rich repeat; LRRNT, N-terminal leucine rich repeat. (**b**) Five of the acyl chains of LPS bind to MD2 while the remaining chain (R2) is exposed to the surface and it binds TLR4. The1- and 4-phosphates contribute to TLR4 dimer formation through binding interactions with lysines and an arginine residue on both TLR4 and MD2. (**c**) When Eritoran is bound to TLR4–MD2, the MD2 Phe126 loop is exposed to the solvent area. (**d**) Upon TLR4–MD2 binding LPS, Phe126 forms hydrophobic interactions with lipid chains R2 and R3 and with TLR4 resulting in a structural shift in the Phe126 loop which positions the R2 lipid chain to interact with TLR4 allowing dimerization with TLR4 to occur. Reproduced with permission from S. Carpenter and L. A. J. O'Neill (2009) Biochem J 422, 1–10

associate, but dimerization of the TLR4/MD2 complex with another TLR4/MD2 complex only occurs following binding of LPS. The structure shows the lipid A portion of LPS binding to two copies of TLR4 and MD2 arranged symmetrically. TLR4 dimerization is facilitated through hydrophilic and hydrophobic interactions with LPS. Within this structure, five of the lipid chains of LPS are buried within the hydrophobic pocket of MD2 in a manner similar to Eritoran. The remaining lipid chain is on the surface of MD2 where it forms a hydrophobic interaction with phenylalanine residues on TLR4. Binding of LPS causes structural changes within a loop region on MD2 (Phe126) leading to hydrophilic interactions between TLR4 and MD2 that further stabilize the complex. The structural change in the Phe126 loop region of MD2 following LPS stimulation is essential for the formation of the dimers and subsequently the initiation of downstream signaling (Park et al., 2009). This finding is supported by mutational analysis of Phe126 in MD2 showing that it prevents dimerization and abolishes downstream signaling (Kobayashi et al., 2006). The 1- and 4'phosphate groups on the diglucosamine form ionic interactions with positively charged residues on both TLR4 and MD2, and the adjacent TLR4 within the dimer complex (Park et al., 2009). The extra lipid chain inserted into the binding pocket of MD2 in LPS compared with Eritoran results in a displacement of the phosphorylated glucosamines and allows the phosphate groups to associate with TLR4 and MD2.

14.4 Formulation Effects on TLR4 Agonist Activity

The formulation of synthetic and natural TLR4 agonists is a critical parameter in their adjuvant activity. As discussed above, these molecules stimulate an immune signaling cascade through binding to the membrane protein TLR4. However, it has been found that differences in preparation methods and delivery vehicles can have a profound effect on TLR4 agonist biological activity. For example, a monomeric TLR4 agonist showed significantly decreased immunostimulating capacity compared to aggregates of the same compound (Mueller et al., 2004). In addition, mixed preparations of a hexa-acylated lipid A (agonistic) with a tetra-acylated lipid A (antagonistic) were found to be more effective in stimulating an enhanced immune response in human mononuclear cells compared to a homogenous preparation of either lipid A compound alone (Mueller et al., 2004). The tetra-acylated lipid A could be replaced in the mixture by a phospholipid molecule with similar results. Moreover, to maintain their biological activity, the two compounds had to be mixed together in organic phase and then transferred to an aqueous phase, illustrating the importance of the formulation preparation method. Similarly, it is interesting to note that bacterial phosphatidylethanolamine is a common contaminant of lipid A derivations and may therefore influence biological activity (Alving and Rao, 2008). Thus, interactions between components of lipid A aggregates and preparation methods can have important effects on biological activity and should be taken into account when evaluating TLR4 agonists.

Given the importance of molecular aggregates in TLR4 agonist activity, it is not surprising that the structural geometry and phase behavior of TLR4 agonists are also important factors for adjuvanticity (Mueller et al., 2004; Seydel et al., 2000). Phase structure of lipids such as the TLR4 agonists is strongly influenced by environmental conditions, such as temperature or the presence of salts. TLR4 agonists with inverted conical or hexagonal structures have been shown to have increased potency over cylindrical or conical molecules (Seydel et al., 2000). Thus, the presence of divalent cations in solutions of LPS induce ordered, lamellar structures instead of inverted hexagonal structures, resulting in less acyl chain mobility and cytokine response (Garidel et al., 2005). Likewise, binding of hemoglobin to LPS was found to increase the hydrophobic cross-section of the molecule, causing a greater inverted conical geometric profile and inducing more TNF-α production (Jurgens et al., 2001).

Because synthetic and natural TLR4 agonists are generally hydrophobic, it is difficult to solubilize these molecules in aqueous solution. Therefore, it is common to formulate the adjuvants as aqueous nanosuspensions of molecular aggregates or micelles. For instance, an aqueous nanosuspension of MPL® or GLA can be mixed in organic solvent with a 0.25% molar ratio of phospholipid, such as dipalmitoyl phosphatidylcholine (Anderson et al., 2010; Childers et al., 2000; Crane, 2002). After evaporating the solvent, the lipids are hydrated, heated, and sonicated or microfluidized to form small particles. A similar method involves hydrating lyophilized MPL® with an aqueous solution containing triethanolamine, followed by heating and sonication (Baldridge and Crane, 1999). Aqueous formulations of TLR4 agonists have demonstrated efficacy as vaccine adjuvants with many different antigens and have progressed to clinical trials (Childers et al., 2000; Reed et al., 2009). Because these aqueous preparations include a minimum of excipients and are relatively straightforward to prepare, they are a logical first choice for a safe and effective TLR4 agonist formulation (Baldridge and Crane, 1999).

In addition to aqueous formulations, TLR4 agonists can be adsorbed to aluminium hydroxide. TLR4 agonists such as MPL® and GLA tend to bind strongly to aluminium hydroxide particles, which are micron-sized crystalline aggregates (Zhu et al., 2009). This is due, at least in part, to electrostatics: the aluminum gel is positively charged at neutral pH whereas the TLR4 agonist is anionic (Alving and Matyas, 2005). Other phenomena may also play an important role in the binding mechanism, such as ligand exchange and hydrophobic interactions (Alving and Matyas, 2005). AS04, an MPL®-alum adjuvant, has been recently approved by the FDA for the HPV vaccine Cervarix®, and has been extensively used in Europe for years.

Liposomal formulations may also be used to deliver TLR4 agonists. Consisting of spherical vesicles formed by the self-assembly of phospholipid bilayers, liposomes are a versatile, biocompatible vaccine adjuvant formulation. The wide variety of available phospholipid molecules that are employed to make liposomes may have significant effects on the structure and biological activity of the adjuvant.

As an amphiphilic molecule similar to a phospholipid, the TLR4 agonist is presumably incorporated into the vesicular lipid wall. Interestingly, the presence of hydrophilic saccharide groups and charge on TLR4 agonists may even improve liposome stability (Davidsen et al., 2006). Another benefit of liposomal formulation is demonstrated by the finding that endotoxic activity of lipid A was largely reduced by liposome incorporation while adjuvant activity was maintained (Matyas et al., 2000). Several studies have utilized liposome-associated TLR4 agonists to induce higher antibody and CTL immune responses (Fries et al., 1992a; Richards et al., 1998). As mentioned above, GSK's AS01 is a liposomal formulation containing MPL®. The influence of liposome chain saturation, charge, and headgroup structure on adjuvant structure and localization should be taken into account. For example, an in vitro cytokine assay employing GLA liposomal formulations composed of various lipids showed that lipid chain length and charge altered the in vitro immunostimulatory behavior of the TLR4 agonist (Fig. 14.2). Thus, increasing lipid chain length from 12 carbons (DLPC) to 16 carbons (DPPC) in GLA liposomes corresponded with a decreased immunostimulatory cytokine response. Also, an anionic liposomal formulation (DPPC/DPPG) of GLA induced greater cytokine response than a cationic liposomal formulation (DPPC/DPTAP) of GLA.

Oil-in-water (o/w) emulsions are a very effective category of formulations for delivery of TLR4 agonists. O/w emulsions consist of biocompatible oil droplets emulsified with one or more surfactants in an aqueous solution. Due to their amphipathic structure, it is supposed that TLR4 agonists associate at the interface of the oil droplets. Indeed, HPLC analysis on aqueous extractions of an emulsion containing a synthetic TLR4 agonist found that the adjuvant was associated with the oil phase (Fig. 14.3a) (Anderson et al., 2010). This was confirmed by zeta potential measurements, which indicated the presence of the anionic TLR4 agonist at the oil/water interface since it caused a more negative particle charge (Fig. 14.3b) (Anderson et al., 2010). MPL® has been formulated in various oil-in-water emulsions, including AS02 and MPL®-SE, which are in clinical trials (Reed et al., 2009). In our studies using a stable emulsion of GLA (GLA-SE) we have observed a consistent pattern, using antigens from influenza virus, *M. tuberculosis*, *Plasmodium*, and *Leishmania* of o/w emulsions generating, in mouse and non-human primates, strong Th2 responses, while the addition of GLA to these emulsions, such as GLA-SE, dramatically down regulates Th2 responses, and strongly up-regulates Th1. Another synthetic TLR4 agonist emulsion in clinical trials is CoVaccineHT™ (Adams, 2008; Bodewes et al., 2009).

Comparing immune responses elicited by different formulations illustrates important effects. For instance, a preclinical study evaluating liposome, emulsion, and alum-liposome formulations of MPL® with QS21 (a saponin adjuvant) in the context of a malaria antigen revealed some interesting results (Pichyangkul et al., 2004). While all formulations were safe and immunogenic, the liposome and emulsion formulations elicited the highest amount of malaria antigen-specific antibody. An adjuvanted malaria vaccine trial in children demonstrated that the liposomal formulation of MPL® and QS21 was less reactogenic and induced higher anti-parasite antibodies than the emulsion formulation (Owusu-Agyei et al., 2009).

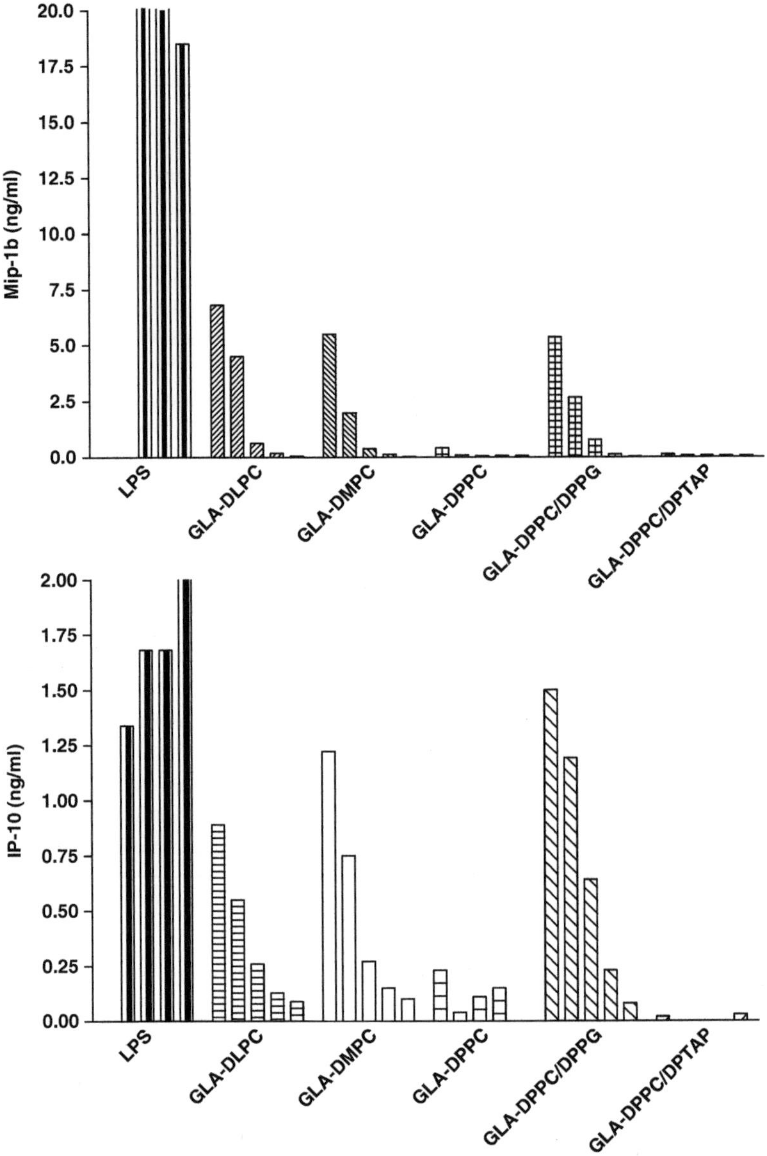

Fig. 14.2 Cytokine responses to a human macrophage cell line (MonoMac 6) exposed to liposomal formulations of GLA. Liposomes contained the lipids indicated on the x-axis, cholesterol, and GLA. Adjuvant concentration was serially diluted by factors of 5 for each group, from left to right. LPS is the positive control. Variations in adjuvant activity are apparent depending on the phospholipid composition of the liposomes

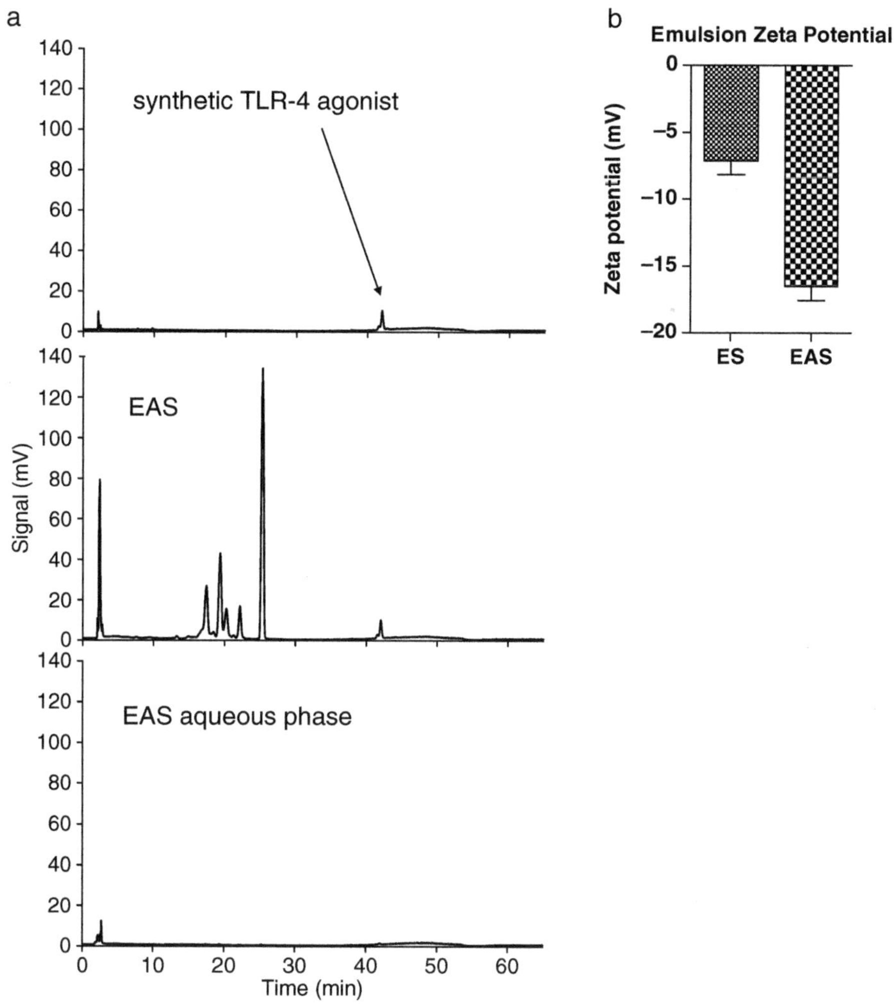

Fig. 14.3 Analytical characterization of an emulsion formulation of GLA (EAS) indicates the localization of the TLR4 agonist. (**a**) Aqueous phase extraction followed by HPLC indicates that GLA is found in the oil phase of the emulsion. (**b**) A decrease in zeta potential due to incorporation of GLA in the emulsion indicates that the TLR4 agonist associates with the oil droplet interface. ES represents the emulsion alone. Reproduced with permission from Anderson et al. (2010) Coll Surf B 75, 123–132

In another example, MPL®-SE was generally found to elicit higher antibody titers and cytokine responses than MPL®-AF, and qualitative differences in immune responses between the two formulations were also apparent (Hui and Hashimoto, 2008).

In vitro comparisons of the adjuvant activity of different formulation platforms is difficult to interpret. For example, the aqueous GLA suspension is usually the most potent formulation in vitro compared to particulate formulations such as emulsions or liposomes (Anderson et al., 2010). However, in vivo immune response of vaccine adjuvants is generally more potent with particulate formulations (Reed et al., 2009). Of course, this is not always the case, and each antigen must be tested to find the ideal adjuvant formulation. Differences between in vitro and in vivo results are not surprising given the complex factors involved in the in vivo immune response, which are impossible to replicate in an in vitro system (Anderson et al., 2010; Simberg et al., 2004). Nevertheless, in vitro comparison of formulations may provide useful data, especially when comparing formulations with similar overall structure but minor excipient variations (such as liposomes composed of different lipids, see Fig. 14.2).

It is important that thorough analytical characterization is performed in order to rationally design second generation formulations that exploit the most biologically active form of the adjuvant. Moreover, particulate formulations are often quite heterogeneous and should not be assumed to constitute a uniform, idealized structure. For example, o/w emulsions may contain various structures besides emulsified oil droplets, such as liposomes, micelles, and monomers (Floyd, 1999; Liu and Liu, 1995; McClements, 2007; McClements and Decker, 2000; Norden et al., 2001). Thus, the formulated TLR4 agonist could be interacting with multiple types of particles in a single formulation. Several analytical techniques are useful for TLR4 agonist formulation characterization. Vibrational spectroscopy techniques such as FTIR and Raman can indicate compound localization, structural conformation, and phase structure of both TLR4 agonists and formulation excipients. Microcalorimetry methods such as DSC and ITC are ideal for detecting phase transitions and ligand binding thermodynamics. These methods can be complemented by highly sensitive fluorescence spectroscopy, although chemical labeling may be necessary in this case. Dynamic light scattering and HPLC provide essential data regarding particle size and chemical stability or concentration (see Fig. 14.3a). As mentioned above, zeta potential measurements indicate particle charge (see Fig. 14.3b). Electron microscopy is useful for formulation morphological characterization. It is best to use a suite of complementary techniques in order to better understand adjuvant-formulation interactions and structure.

14.5 Conclusion

In summary, there has been significant progress in the elucidation of structure/function relationships of bacterial endotoxin-derived molecules with regards to interactions with cells of the mammalian immune system. Breakthrough studies by Ribi Immunochem, and later GSK, led to the development of the first approved vaccines to contain TLR agonists (MPL®). The demonstration that MPL® could be used to make safe and effective adjuvants has led to the development of synthetic molecules that may offer several advantages over natural products in terms of biological activity and cost of goods. Regardless of the source or structure of the TLR4 agonist, the ways in which they are formulated dramatically influence biological activity. TLR4 agonists, particularly MPL®, represent a safe and effective class of molecules, clearly demonstrated to have enormous potential as vaccine adjuvants.

References

Adams, J.U. Hypertension's shot in the arm. Nature Biotech 26 (2008) 1327–1329.

Akira, S., Takeda, K. Toll-like receptor signaling. Nat Rev Immunol 4 (2004) 499–511.

Akira, S., Uematsu, S., Takeuchi, O. Pathogen recognition and innate immunity. Cell 124 (2006) 783–801.

Alexander, C., Rietschel, E.T. Bacterial lipopolysaccharides and innate immunity. J Endotoxin Res 7 (2001) 167–202.

Alving, C.R., Matyas, G.R. Design and selection of vaccine adjuvants: principles and practice. In Kaufmann, S.H.E., Lambert, P.H. (eds), The Grand Challenge for the Future: Vaccines for Poverty-Related Diseases from Bench to Field. Birkhauser Verlag, Basel, Switzerland (2005), pp. 99–118.

Alving, C.R., Rao, M. Lipid A and liposomes containing lipid A as antigens and adjuvants. Vaccine 26 (2008) 3036–3045.

Anderson, R.C., Fox, C.B., Dutill, T.S., Shaverdian, N., Evers, T.L., Poshusta, G.R., Chesko, J., Coler, R.N., Friede, M., Reed, S.G., Vedvick, T.S. Physicochemical characterization and biological activity of synthetic TLR4 agonist formulations. Coll Surf B: Biointerfaces 75 (2010) 123–132.

Atanackovic, D., Altorki, N.K., Cao, Y., Ritter, E., Ferrara, C.A., Ritter, G., Hoffman, E.W., Bokemeyer, C., Old, L.J., Gnjatic, S. Booster vaccination of cancer patients with MAGE-A3 protein reveals long-term immunological memory or tolerance depending on priming. Proc Nat Acad Sci USA 105 (2008) 1650–1655.

Baldridge, J.R., Crane, R.T. Monophosphoryl lipid A (MPL) formulations for the next generation of vaccines. Methods 19 (1999) 103–107.

Baldwin, S.L., Bertholet, S., Kahn, M., Zharkikh, I., Ireton, G.C., Vedvick, T.S., Reed, S.G., Coler, R.N. Intradermal immunization improves protective efficacy of a novel TB vaccine candidate. Vaccine 27 (2009a) 3063–3071.

Baldwin, S.L., Shaverdian, N., Goto, Y., Duthie, M.S., Raman, V.S., Evers, T., Mompoint, F., Vedvick, T.S., Bertholet, S., Coler, R.N., Reed, S.G. Enhanced humoral and Type 1 cellular immune responses with Fluzone adjuvanted with a synthetic TLR4 agonist formulated in an emulsion. Vaccine 27 (2009b) 5956–5963.

Barton, G.M., Kagan, J.C. A cell biological view of Toll-like receptor function: regulation through compartmentalization. Nat Rev Immunol 9 (2009) 535–542.

Bertholet, S., Goto, Y., Carter, L., Bhatia, A., Howard, R.F., Carter, D., Coler, R.N., Vedvick, T.S., Reed, S.G. Optimized subunit vaccine protects against experimental leishmaniasis. Vaccine 27 (2009) 7036–7045.

Beutler, B., Rietschel, E.T. Innate immune sensing and its roots: the story of endotoxin. Nat Rev Immunol 3 (2003) 169–176.

Beutler, B.A. TLRs and innate immunity. Blood 113 (2009) 1399–1407.

Bodewes, R., Geelhoed-Mieras, M.M., Heldens, J.G.M., Glover, J., Lambrecht, B.N., Fouchier, R.A.M., Osterhaus, A.D.M.E., Rimmelzwaan, G.F. The novel adjuvant CoVaccineHT™ increases the immunogenicity of cell-culture derived influenza A/H5N1 vaccine and induces the maturation of murine and human dendritic cells in vitro. Vaccine 27 (2009) 6833–6839.

Brandenburg, K., Lindner, B., Schromm, A., Koch, M.H., Bauer, J., Merkli, A., Zbaeren, C., Davies, J.G., Seydel, U. Physicochemical characteristics of triacyl lipid A partial structure OM-174 in relation to biological activity. Eur J Biochem 267 (2000) 3370–3377.

Caglar, C., Casella, R.C., Eaves, C.A., Matsuzawa, A., Ichijo, H., Mitchell, T.C. Selective activation of the p38 MAPK pathway by synthetic monophosphoryl lipid A. J Biol Chem 284 (2009) 31982–31991.

Childers, N.K., Miller, K.L., Tong, G., Llarena, J.C., Greenway, T., Ulrich, J.T., Michalek, S.M. Adjuvant activity of monophosphoryl lipid A for nasal and oral immunization with soluble or liposome-associated antigen. Infect Immun 68 (2000) 5509–5516.

Choe, J., Kelker, M.S., Wilson, I.A. Crystal structure of human toll-like receptor 3 (TLR3) ectodomain. Science 309 (2005) 581–585.

Cohen, J., Druilhe, P. Immunogenic compositions comprising liver stage malarial antigens. WIPO WO/2002/038176 (2002).

Crane, R.T. Aqueous immunologic adjuvant compositions of monophosphoryl lipid A. U.S. Patent 6,491,919 B2 (2002).

Davidsen, J., Andersen, P., Rosenkrands, I. Compositions and methods for stabilizing lipid based adjuvant formulations using glycolipids. U.S. Patent US 2006/0008519 A1 (2006).

Elliott, G.T., McLeod, R.A., Perez, J., von Eschen, K.B. Interim results of a phase II multicenter clinical trial evaluating the activity of a therapeutic allogeneic melanoma vaccine (theraccine) in the treatment of disseminated malignant melanoma. Semin Surg Oncol 9 (1993) 264–272.

Evans, J.T., Cluff, C.W., Johnson, D.A., Lacy, M.J., Persing, D.H., Baldridge, J.R. Enhancement of antigen-specific immunity via the TLR4 ligands MPL^{TM} adjuvant and Ribi.529. Expert Rev Vaccines 2 (2003) 219–229.

Floyd, A.G. Top ten considerations in the development of parenteral emulsions. Pharm Sci Tech Today 2 (1999) 134–143.

Fries, L.F., Gordon, D.M., Richards, R.L., Egan, J.E., Hollingdale, M.R., Gross, M., Silverman, C., Alving, C.R. Liposomal malaria vaccine in humans: a safe and potent adjuvant strategy. Proc Natl Acad Sci USA 89 (1992a) 358–362.

Fries, L.F., Gordon, D.M., Richards, R.L., Egan, J.E., Hollingdale, M.R., Gross, M., Silverman, C., Alving, C.R. Liposomal malaria vaccine in humans: a safe and potent adjuvant strategy. Proc Nat Acad Sci USA 89 (1992b) 358–362.

Galanos, C., Luderitz, O., Freudenberg, M., Brade, L., Schade, U., Rietschel, E.T., Kusumoto, S., Shiba, T. Biological activity of synthetic heptaacyl lipid A representing a component of *Salmonella minnesota* R595 lipid A. Eur J Biochem 160 (1986) 55–59.

Garcon, N. Development and evaluation of AS04, a novel and improved adjuvant system containing MPL and aluminum salt. In Schijns, V., O'Hagan, D. (eds), Immunopotentiators in Modern Vaccines. Academic Press, London (2005), pp. 161–178.

Garidel, P., Rappolt, M., Schromm, A.B., Howe, J., Lhner, K., Andra, J., Koch, M.H., Brandenburg, K. Divalent cations affect chain mobility and aggregate structure of lipopolysaccharide from *Salmonella minnesota* reflected in a decrease of its biological activity. Biochim Biophys Acta 1715 (2005) 122–131.

Gawchik, S.M., Saccar, C.L. Pollinex Quattro Tree: Allergy vaccine. Expert Opin Biol Ther 9 (2009) 377–382.

Hajjar, A.M., Ernst, R.K., Tsai, J.H., Wilson, C.B., Miller, S.I. Human Toll-like receptor 4 recognizes host-specific LPS modifications. Nat Immunol 3 (2002) 354–359.

Hui, G.S., Hashimoto, C.N. Adjuvant formulations possess differing efficacy in the potentiation of antibody and cell mediated responses to a human malaria vaccine under selective immune genes knockout environment. Int Immunopharm 8 (2008) 1012–1022.

Ishizaka, S.T., Hawkins, L.D. E6020: a synthetic Toll-like receptor 4 agonist as a vaccine adjuvant. Exp Rev Vaccines 6 (2007) 773–784.

Janeway, C.A., Jr., Medzhitov, R. Innate immune recognition. Annu Rev Immunol 20 (2002) 197–216.

Jiang, Z.H., Bach, M.V., Budzynski, W.A., Krantz, M.J., Koganty, R.R., Longenecker, B.M. Lipid A structures containing novel lipid moieties: Synthesis and adjuvant properties. Bioorg Med Chem Lett 12 (2002) 2193–2196.

Jiang, Z.H., Budzynski, W.A., Qiu, D., Yalamati, D., Koganty, R.R. Monophosphoryl lipid A analogues with varying 3-O-substitution: Synthesis and potent adjuvant activity. Carbohydr Res 342 (2007) 784–796.

Johnson, A. Molecular adjuvants and immunomodulators: New approaches to immunization. Clin Microbiol Rev 7 (1994) 277–289.

Johnson, A.G. Adjuvant action of bacterial endotoxins on the primary antibody response. In Landy, M., Braun, W. (eds), Bacterial Endotoxins. Rutgers University Press, New Brunswick, CT (1964), pp. 252–262.

Johnson, A.G., Gains, S., Landy, M. Studies on the O antigen of *Salmonella ryphosa* V. Enhancement of antibody response to protein antigens by purified lipopolysaccharide. J Exp Med 103 (1956) 225–246.

Johnson, D.A., Keegan, D.S., Sowell, C.G., Livesay, M.T., Johnson, C.L., Taubner, L.M., Harris, A., Myers, K.R., Thompson, J.D., Gustafson, G.L., Rhodes, M.J., Ulrich, J.T., Ward, J.R., Yorgensen, Y.M., Cantrell, J.L., Brookshire, V.G. 3-O-Desacyl monophosphoryl lipid A derivatives: Synthesis and immunostimulant activities. J Med Chem 42 (1999) 4640–4649.

Jurgens, G., Muller, M., Koch, M.H., Brandenburg, K. Interaction of hemoglobin with enterobacterial lipopolysaccharide and lipid A. Physicochemical characterization and biological activity. Eur J Biochem 268 (2001) 4233–4242.

Kim, H.M., Park, B.S., Kim, J.I., Kim, S.E., Lee, J., Oh, S.C., Enkhbayar, P., Matsushima, N., Lee, H., Yoo, O.J., Lee, J.O. Crystal structure of the TLR4-MD-2 complex with bound endotoxin antagonist Eritoran. Cell 130 (2007) 906–917.

Kobayashi, M., Saitoh, S., Tanimura, N., Takahashi, K., Kawasaki, K., Nishijima, M., Fujimoto, Y., Fukase, K., Akashi-Takamura, S., Miyake, K. Regulatory roles for MD-2 and TLR4 in ligand-induced receptor clustering. J Immunol 176 (2006) 6211–6518.

Kotani, S., Takada, H., Takahashi, I., Ogawa, T., Tsujimoto, M., Shimauchi, H., Ikeda, T., Okamura, H., Tamura, T., Harada, K. Immunobiological activities of synthetic lipid A analogs with low endotoxicity. Infect Immun 54 (1986) 673–682.

Kumar, V.V. Complementary molecular shapes and additivity of the packing parameter of lipids. Proc Natl Acad Sci USA 88 (1991) 444–448.

Kumazawa, Y., Matsuura, M., Homma, J.Y., Nakatsuru, Y., Kiso, M., Hasegawa, A. B cell activation and adjuvant activities of chemically synthesized analogues of the nonreducing sugar moiety of lipid A. Eur J Immunol 15 (1985) 199–201.

Kusumoto, S., Fukase, K., Fukase, Y., Kataoka, M., Yoshizaki, H., Sato, K., Oikawa, M., Suda, Y. Structural basis for endotoxic and antagonistic activities: investigation with novel synthetic lipid A analogs. J Endotoxin Res 9 (2003) 361–366.

Lell, B., Agnandji, S., von Glasenapp, I., Haertle, S., Oyakhiromen, S., Issifou, S., Vekemans, J., Leach, A., Lievens, M., Dubois, M.C., Demoitie, M.A., Carter, T., Villafana, T., Ballou, W.R., Cohen, J., Kremsner, P.G. A randomized trial assessing the safety and immunogenicity of AS01 and AS02 adjuvanted RTS,S malaria vaccine candidates in children in Gabon. PLoS ONE 4 (2009) e7611.

Liu, F., Liu, D. Long-circulating emulsions (oil-in-water) as carriers for lipophilic drugs. Pharm Res 12 (1995) 1060–1064.

Liu, L., Botos, I., Wang, Y., Leonard, J.N., Shiloach, J., Segal, D.M., Davies, D.R. Structural basis of toll-like receptor 3 signaling with double-stranded RNA. Science 320 (2008) 379–381.

Martin, O.R., Zhou, W., Wu, X., Front-Deschamps, S., Moutel, S., Schindl, K., Jeandet, P., Zbaeren, C., Bauer, J.A. Synthesis and immunobiological activity of an original series of acyclic lipid A mimics based on a pseudodipeptide backbone. J Med Chem 49 (2006) 6000–6014.

Mata-Haro, V., Cekic, C., Martin, M., Chilton, P.M., Casella, C.R., Mitchell, T.C. The vaccine adjuvant monophosphoryl lipid A as a TRIF-biased agonist of TLR4. Science 316 (2007) 1628–1632.

Matsushima, N., Tanaka, T., Enkhbayar, P., Mikami, T., Taga, M., Yamada, K., Kuroki, Y. Comparative sequence analysis of leucine-rich repeats (LRRs) within vertebrate toll-like receptors. BMC Genomics 8 (2007) 124.

Matyas, G.R., Wassef, N.M., Rao, M., Alving, C.R. Induction and detection of antibodies to squalene. J Immunol Methods 245 (2000) 1–14.

McClements, D.J. Critical review of techniques and methodologies for characterization of emulsion stability. Crit Rev Food Sci Nutri 47 (2007) 611–649.

McClements, D.J., Decker, E.A. Lipid oxidation in oil-in-water emulsions: impact of molecular environment on chemical reactions in heterogeneous food systems. J Food Sci 65 (2000) 1270–1282.

Medzhitov, R. Recognition of microorganisms and activation of the immune response. Nature 449 (2007) 819–826.

Medzhitov, R., Preston-Hurlburt, P., Janeway, C.A., Jr. A human homologue of the Drosophila Toll protein signals activation of adaptive immunity. Nature 388 (1997) 394–397.

Mitchell, M.S., Harel, W., Kempf, R.A., Hu, E., Kan-Mitchell, J., Boswell, W.D., Dean, G., Stevenson, L. Active-specific immunotherapy for melanoma. J Clin Oncol 8 (1990) 856–869.

Miyake, K. Roles for accessory molecules in microbial recognition by Toll-like receptors. J Endotoxin Res 12 (2006) 195–204.

Morefield, G.L., Hawkins, L.D., Ishizaka, S.T., Kissner, T.L., Ulrich, R.G. Synthetic Toll-like receptor 4 agonist enhances vaccine efficacy in an experimental model of toxic shock syndrome. Clin Vaccine Immunol 14 (2007) 1499–1504.

Mueller, M., Lindner, B., Kusumoto, S., Fukase, K., Schromm, A.B., Seydel, U. Aggregates are the biologically active units of endotoxin. J Biol Chem 279 (2004) 26307–26313.

Myers, K., Truchot, A., Ward, J., Hudson, Y., Ulrich, J. A critical determinant of lipid A endotoxic activity. In Nowotny, A., Spitzler, J., Ziegler, E. (eds), Cellular and Molecular Aspects of Endotoxin Reactions. Elsevier, Amsterdam (1990), pp. 145–156.

Nagai, Y., Akashi, S., Nagafuku, M., Ogata, M., Iwakura, Y., Akira, S., Kitamura, T., Kosugi, A., Kimoto, M., Miyake, K. Essential role of MD-2 in LPS responsiveness and TLR4 distribution. Nat Immunol 3 (2002) 667–672.

Norden, T.P., Siekmann, B., Lundquist, S., Malmsten, M. Physicochemical characterisation of a drug-containing phospholipid-stabilised o/w emulsion for intravenous administration. Eur J Pharm Sci 13 (2001) 393–401.

Ohto, U., Fukase, K., Miyake, K., Satow, Y. Crystal structures of human MD-2 and its complex with antiendotoxic lipid IVa. Science 316 (2007) 1632–1634.

Owusu-Agyei, S., Ansong, D., Asante, K., Owusu, S.K., Owusu, R., Brobby, N.A.W., Dosoo, D., Akoto, A.O., Osei-Kwakye, K., Adjei, E.A., Boahen, K.O., Sylverken, J., Adjei, G., Sambian, D., Apanga, S., Kayan, K., Vekemans, J., Ofori-Anyinam, O., Leach, A., Lievens, M., Demoitie, M.A., Dubois, M.C., Cohen, J., Ballou, W.R., Savarese, B., Chandramohan, D., Gyapong, J.O., Milligan, P., Antwi, S., Agbenyega, T., Greenwood, B., Evans, J. Randomized controlled trial of RTS,S/AS02D and RTS,S/AS01E malaria candidate vaccines given according to different schedules in Ghanaian children. PLoS One 4 (2009) e7302.

Paavonen, J., Naud, P., Salmeron, J., Wheeler, C.M., Chow, S.N., Apter, D., Kitchener, H., Castellsague, X., Teixeira, J.C., Skinner, S.R., Hedrick, J., Jaisamram, U., Limson, G., Garland, S., Szarewski, A., Romanowski, B., Aoki, F.Y., Schwarz, T.F., Poppe, W.A., Bosch, F.X., Jenkins, D., Hardt, K., Zahaf, T., Descamps, D., Struyf, F., Lehtinen, M., Dubin, G., Group, H.P.S., Greenacre, M. Efficacy of human papillomavirus (HPV)-16/18 AS04-adjuvanted vaccine against cervical infection and precancer caused by oncogenic HPV types (PATRICIA): final analysis of a double-blind, randomised study in young women. Lancet 374 (2009) 301–314.

Park, B.S., Song, D.H., Kim, H.M., Choi, B.S., Lee, H., Lee, J.O. The structural basis of lipopolysaccharide recognition by the TLR4-MD-2 complex. Nature 458 (2009) 1191–1195.

Pasare, C., Medzhitov, R. Toll-like receptors: Linking innate and adaptive immunity. Microbes Infect 6 (2004) 1382–1387.

Pichyangkul, S., Gettayacamin, M., Miller, R.S., Lyon, J.A., Angov, E., Tongtawe, P., Ruble, D.L., Heppner, D.G., Jr., Kester, K.E., Ballou, W.R., Diggs, C.L., Voss, G., Cohen, J.D., Walsh, D.S. Pre-clinical evaluation of the malaria vaccine candidate *P. falciparum* $MSP1_{42}$ formulated with novel adjuvants or with alum. Vaccine 22 (2004) 3831–3840.

Poltorak, A., He, X., Smirnova, I., Liu, M.Y., Van Huffel, C., Du, X., Birdwell, D., Alejos, E., Silva, M., Galanos, C., Freudenberg, M., Ricciardi-Castagnoli, P., Layton, B., Beutler, B. Defective LPS signaling in C3H/HeJ and C57BL/10ScCr mice: mutations in Tlr4 gene. Science 282 (1998) 2085–2088.

Qureshi, N., Takayama, K., Ribi, E. Purification and structural determination of non-toxic lipid-A from lipopolysaccharide of *Salmonella typhimurium*. J Biol Chem 257 (1982) 11808–11815.

Raetz, C.R., Whitfield, C. Lipopolysaccharide endotoxins. Annu Rev Biochem 71 (2002) 635–700.

Reed, S.G., Bertholet, S., Coler, R.N., Friede, M. New horizons in adjuvants for vaccine development. Trends Immunol 30 (2009) 23–32.

Ribi, E., Parker, R., Strain, S., Mizuno, Y., Nowomy, A., Von Eschen, K., Cantrell, J., McLaughlin, C., Hwang, K., Goren, M. Peptides as requirement for immunotherapy of the guinea-pig line 10 tumor with endotoxins. Cancer Immunol Immunother 12 (1979) 91–96.

Richards, R. L., Rao, M., Wassef, N.M., Glenn, G.M., Rothwell, S.W., Alving, C.R. Liposomes containing Lipid A serve as an adjuvant for induction of antibody and cytotoxic T-cell responses against RTS,S malaria antigen. Infect Immun 66 (1998) 2859–2865.

Rietschel, E.T., Kirikae, T., Schade, F.U., Mamat, U., Schmidt, G., Loppnow, H., Ulmer, A.J., Zahringer, U., Seydel, U., Di Padova, F., et al. Bacterial endotoxin: molecular relationships of structure to activity and function. FASEB J 8 (1994) 217–225.

Seydel, U., Oikawa, M., Fukase, K., Kusumoto, S., Brandenburg, K., Intrinsic conformation of lipid A is responsible for agonistic and antagonistic activity. Eur J Biochem 267 (2000) 3032–3039.

Seydel, U., Schromm, A.B., Brade, L., Gronow, S., Andra, J., Muller, M., Koch, M.H., Fukase, K., Kataoka, M., Hashimoto, M., Kusumoto, S., Brandenburg, K. Physicochemical characterization of carboxymethyl lipid A derivatives in relation to biological activity. FEBS J 272 (2005) 327–340.

Simberg, D., Weisman, S., Talmon, Y., Barenholz, Y. DOTAP (and other cationic lipids): chemistry, biophysics, and transfection. Crit Rev Ther Drug Carr Syst 21 (2004) 257–317.

Stover, A.G., Da Silva Correia, J., Evans, J.T., Cluff, C.W., Elliott, M.W., Jeffery, E.W., Johnson, D.A., Lacy, M.J., Baldridge, J.R., Probst, P., Ulevitch, R.J., Persing, D.H., Hershberg, R.M. Structure-activity relationship of synthetic toll-like receptor 4 agonists. J Biol Chem 279 (2004) 4440–4449.

Tsujimoto, M., Kotani, S., Okunaga, T., Kubo, T., Takada, H., Kubo, T., Shiba, T., Kusumoto, S., Takahashi, T., Goto, Y. Enhancement of humoral immune responses against viral vaccines by a non-pyrogenic 6-O-acylmuramyldipeptide and synthetic low toxicity analogues of lipid A. Vaccine 7 (1989) 39–48.

Ulmer, A.J., Heine, H., Feist, W., Kusumoto, S., Kusama, T., Brade, H., Schade, U., Rietschel, E.T., Flad, H.D. Biological activity of synthetic phosphonooxyethyl analogs of lipid A and lipid A partial structures. Infect Immun 60 (1992) 3309–3314.

Ulrich, J., Myers, K. Monophosphoryl lipid A as an adjuvant. In Powell, M., Newman, M. (eds), Vaccine Design: The subunit and adjuvant approach. Plenum Press, New York (1995), pp. 495–524.

Zhao, B., Rong, Y.-Z., Huang, X.-H., Shen, J.-S. Experimental and theoretical study on the structure and electronic spectra of imiquimod and its synthetic intermediates. Bioorg Med Chem Lett 17 (2007) 4942–4946.

Zhu, D., Huang, S., Gebregeorgis, E., McClellan, H., Dai, W., Miller, L., Saul, A. Development of a direct alhydrogel formulation immunoassay (DAFIA). J Immunol Methods 344 (2009) 73–78.

Chapter 15
Targeting Endotoxin in the Treatment of Sepsis

Jean-Sebastien Rachoin, Christa A. Schorr, and R. Phillip Dellinger

Abstract The role of endotoxin in the genesis of sepsis has long been recognized and multiple treatments aimed at neutralizing it have been studied. Endotoxin can be bound by antibodies (whose role as a therapeutic agent is unlikely), binding proteins such as BPI or human lactoferrin (effectiveness debated and promising respectively) and phospholipid emulsion (which has not improved outcomes in a recent study). Alternatively, the action of endotoxin could be blocked by lipid A analogs (initial study showed no overall benefit and another large trial is near completion targeting a subpopulation of that study). Finally, endotoxin can be bound by polymyxin B embedded in hemoperfusion cartridges. The later treatment has been used for more than a decade in Japan. Since both pre-clinical rationale and studies support the targeting of endotoxin to ameliorate the pro-inflammatory and pro-coagulation response of severe sepsis, this therapeutic intervention is being pursued.

Keywords Sepsis · Endotoxin · Lipopolysaccharide · Hemoperfusion · Gram-negative · Treatment

Abbreviations

PAMPs	pathogen-associated molecular patterns
PRRs	pattern recognition receptors
IL	interleukin
TLR	toll-like receptor
BP	binding protein
LPS	lipopolysaccharide
BPI	bactericidal/permeability increasing protein
HL	human lactoferrin

R.P. Dellinger (✉)
Critical Care Medicine, Cooper University Hospital, Camden, NJ 08103, USA
e-mail: dellinger-phil@cooperhealth.edu

X. Wang, P.J. Quinn (eds.), *Endotoxins: Structure, Function and Recognition*, Subcellular Biochemistry 53, DOI 10.1007/978-90-481-9078-2_15,

PMX B	polymyxin B
HP	hemoperfusion
CLP	cecal ligation and perforation
MAP	mean arterial pressure

15.1 Introduction

Our understanding of the pathophysiology of sepsis has continued to grow over the years and although mortality in severe sepsis has declined it still remains unacceptably high (Townsend et al., 2008). The pre-eminent role for endotoxin, also called lipopolysaccharide (LPS), in the genesis of sepsis has long been recognized and continues to generate interest for targeting neutralization of endotoxin in the treatment of sepsis (Braude et al., 1963).

In the present manuscript, we will discuss the mechanism by which endotoxin initiates the sepsis cascade, the rationale for targeting LPS, and the most significant treatments that have been studied: antibodies, vaccines, binding peptides, lipid A analog, phospholipids, and polymyixin B hemoperfusion.

15.2 Endotoxin and the Initiation of the Sepsis Cascade

In humans, the innate immunity, that is our ability to generate a non-specific immune reaction in response to any antigen, starts by the recognition of specific antigenic structures on microorganisms: pathogen-associated molecular patterns (PAMPs) by pattern recognition receptors (PRRs). After binding to the PAMPs, the PRRs trigger a series of intracellular signaling cascades that result in the release of pro-inflammatory cytokines (such as tumor necrosis factor α, interleukin [IL]-1, IL-6, IL-8, high-mobility group box protein 1) (Cinel and Dellinger, 2007; Fukui et al., 2003; Lin and Yeh, 2005; Taveira da Silva et al., 1993), cellular apoptotic mediators (Fukui et al., 2003; Lin and Yeh, 2005), anti-inflammatory mediators (IL-10,transforming growth factor β) (Volk et al., 1996), and mediators of procoagulation. The systemic repercussions of an imbalance between pro-inflammatory/pro-coagulant (severe sepsis and septic shock) and anti-inflammatory (immune suppression) agents can be dire. Recently the ability of infection induced mediator release in response to trigger apoptosis is also recognized (Cantaluppi et al., 2008; Fukui et al., 2003; Ito et al., 2009).

Among one of the most potent PAMP's is endotoxin. Endotoxin is a prominent anchor on the cell membrane of Gram negative bacteria and is composed of three regions, lipid A (a significantly variable construct among the different types of Gram negative bacteria), core oligosaccharides and distal polysaccharides (Raetz et al., 2007; Raetz and Whitfield, 2002). Endotoxin produces cellular stimulation through interaction with the toll-like receptor 4 (TLR4) and TLR4's association with the CD14 and MD2 receptors, present on macrophages, monocytes and endothelial cells (Bosshart and Heinzelmann, 2007; Lin and Yeh, 2005).

15.3 Rationale for Targeting Endotoxin in Sepsis

With one exception sepsis therapy focusing primarily on the pro-inflammatory mediators has not produced positive results (Zeni et al., 1997). Hence, targeting the sepsis cascade at earlier steps may yield better outcomes.

Endotoxin is present in both Gram-negative and Gram-positive infections (Marshall et al., 2004). The presence of endotoxemia in infections other than Gram negative likely reflects translocation of endotoxin across the gastrointestinal

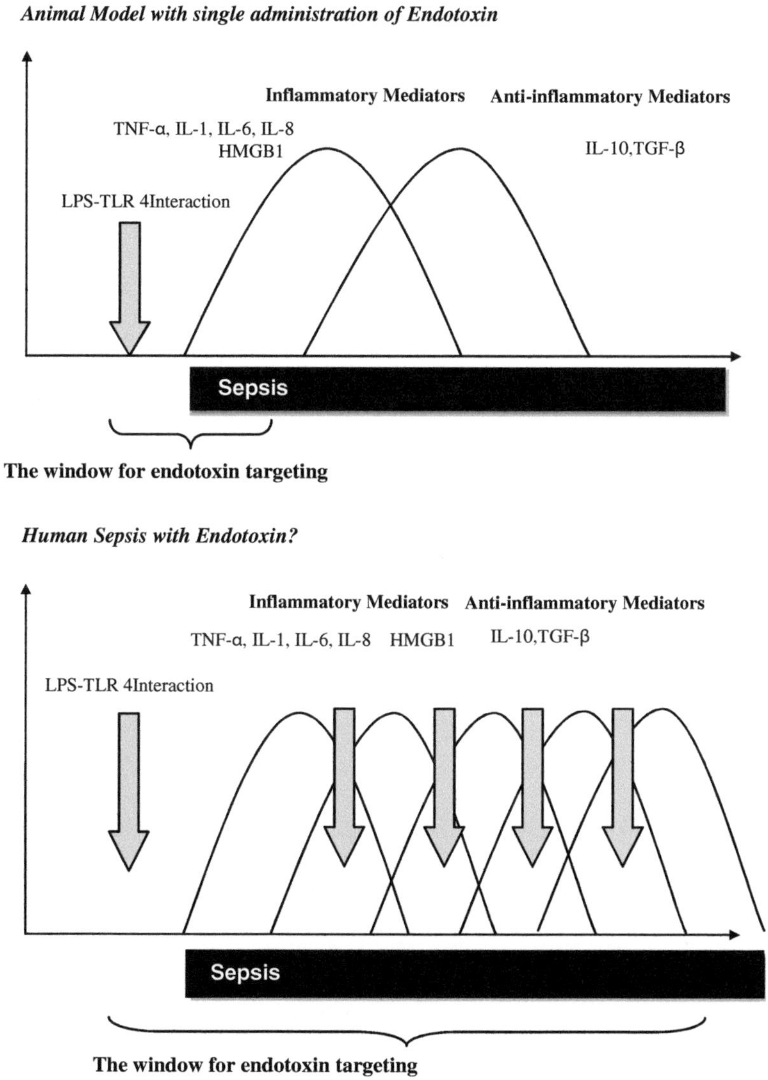

Fig. 15.1 Time-line of events in sepsis

mucosa. In the MEDIC study, endotoxin levels were shown to be associated with likelihood of Gram negative infection, presence of severe sepsis and risk of mortality where a cutoff value of 0.4 endotoxin activity assay units or greater was considered to be an intermediate level and 0.6 a high level (Marshall et al., 2004). Moreover, presence of circulating endotoxin has been linked with increased severity of illness and acute respiratory distress syndrome (ARDS) (Hurley, 1995; Parsons et al., 1989).

The time-line of events in sepsis in both animal models and human sepsis as well as the window for treatments targeting endotoxin are shown in Fig. 15.1.

Treatments aimed to counteract the action of endotoxin likely remain a viable investigative pathway.

15.4 Potential for Targeting Endotoxin in Sepsis

After endotoxin attaches to a binding protein (BP) it connects to TLR4 in order to generate an intracellular immune response. The presence of both a cellular wall CD14 receptor and a cell wall linked MD2 is necessary for adequate stimulation. The attenuation of the body's response to endotoxin has proceeded along multiple investigative pathways: (1) antibodies/vaccines that will facilitate neutralization of the endotoxin by antibodies, (2) endotoxin binding peptides (BPI or human lactoferrin), (3) Lipid A analogs which will bind to the TLR 4 and impede its activation, (4) phospholipid emulsion as an endotoxin scavenger (5) hemoperfusion through a cartridge containing Polymyxin B (Fig. 15.2).

15.5 Anti-Endotoxin Antibodies

The lipid A component of endotoxin varies between different Gram-negative organisms but the core part is a relatively constant structure (Raetz and Whitfield, 2002). Studies have demonstrated a correlation between the presence of antibodies against the core part of the endotoxin and survival in patients with Gram negative sepsis (Pollack et al., 1983; Zinner and McCabe, 1976). Hence, an antibody that could be effective against a wide variety of Gram negative organisms would target this core part of endotoxin.

Several randomized controlled trials performed in patients with Gram negative sepsis studied the effect of monoclonal antibody against endotoxin with the presumption that the antibody would target the core portion of endotoxin. The E5 compound is a murine (mouse) monoclonal IgM antibody against the core part of the endotoxin of the J5 mutant *E. Coli*. Although developed using this specific strain, animal studies showed that it was effective against a wider range of gram-negative microorganisms (Inada et al., 1993; Romulo et al., 1993). Early human trials supported survival benefit but subsequent larger studies failed to confirm these results (Angus et al., 2000; Bone et al., 1995; Greenman et al., 1991). HA-1A, a human IgM monoclonal antibody that binds the lipid A domain, showed

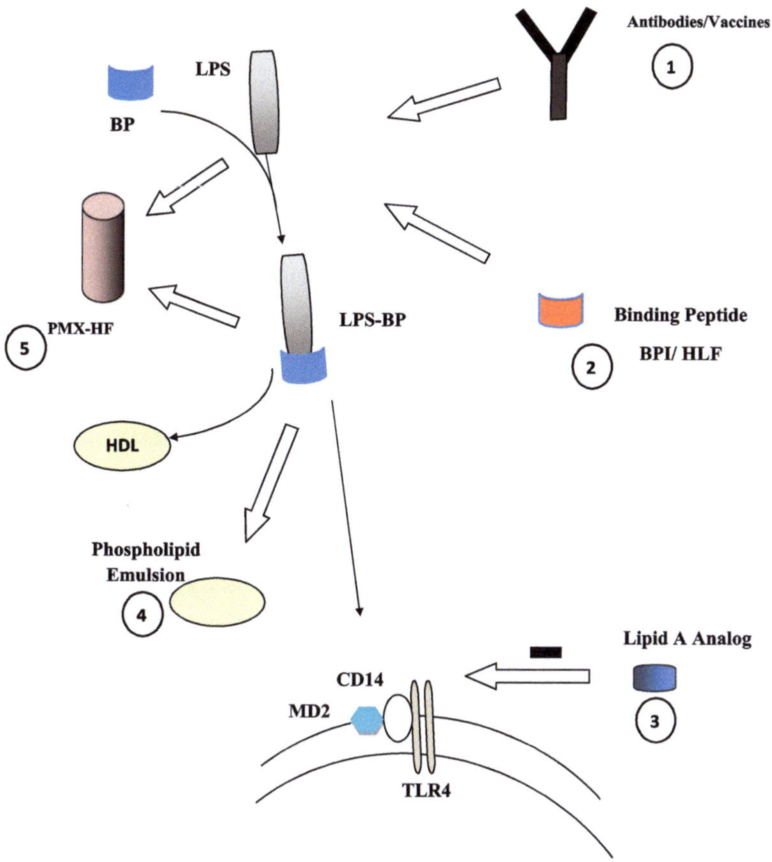

Fig. 15.2 Levels of targeting endotoxin (BP: binding-protein; BPI: bactericidal/permeability increasing protein; HLF: human lactoferrin; PMX-HF: polymyxin B hemoperfusion)

some promise but in the final assessment was judged to have no beneficial effect (The French National Registry of HA-1A in septic shock, 1994; The Intravenous Immunoglobulin Collaborative Study Group, 1992; McCloskey et al., 1994; Ziegler et al., 1991; Ziegler et al., 1982).

Human trials in Gram negative septic shock using human IgG against J5 also failed to show any benefit compared to standard IgG preparations (Calandra et al., 1988). A recurring theme has been demonstrated in other anti-inflammatory approaches to severe sepsis, i.e. early studies showing some promising results but failing to confirm in later larger randomized trials. One explanation for this inconsistency with endotoxin antibody therapy may be the inability to determine the exact quantity of antibody necessary for treatment. Despite the failure of monoclonal antibody targeting endotoxin epitopes, pooled immunoglobulin therapy (particularly that which is IgM enriched) continues to offer promise as potentially beneficial therapy and is in general use in some countries for treatment of severe sepsis (Baumgartner et al., 1985; Greenman et al., 1991).

15.6 Vaccines

There has been a renewed interest in the developpement of vaccines for the treatment of sepsis.

In an animal study, lipopolysaccarides were extracted from detoxified *E. coli* J5, combined with outer membrane protein of *Neisseria meningitides* and injected in rabbits. The serum of the immune animals protected neutropenic rats infected with other Gram-negative organisms (Bhattacharjee et al., 1996; Cross et al., 2001).

This vaccine was tested for safety and immunogenicity in 24 human subjects. The injection elicited local moderate pain with no other notable physiologic abnormalities found. The vaccine increased antibody levels up to 12 months after the initial vaccination. Using functional tests such as activation of white cells ex vivo, cytokine generation, and bacterial clearance, the authors found a higher capacity for neutralizing bacteria and activating white cells (Cross et al., 2003).

Future trials will be needed to elucidate the role of this approach.

15.7 Endotoxin-Binding Peptides

Peptides that can bind endotoxins are another attractive strategic target. These agents must have a high affinity for endotoxin and should not activate the immune response after the binding occurs. Agents with this capability include bactericidal/permeability increasing protein (BPI) and lactoferrin (Appelmelk et al., 1994; Calvano et al., 1994; Elsbach and Weiss, 1993; Elsbach et al., 1994).

15.7.1 Bactericidal/Permeability Increasing Protein

Bactericidal/permeability increasing protein (BPI) is a key element of the azurophilic granules of human granulocytes. This highly cationic protein exerts both a strong attraction/binding to the negatively charged endotoxin and exhibits a high bactericidal activity (Calvano et al., 1994; Elsbach and Weiss, 1993; Elsbach et al., 1994).

In vitro and ex vivo studies of recombinant human variant of the BPI showed strong neutralizing activity against endotoxin and whole gram-negative bacterias (Weiss et al., 1992).

Animal studies in rats with Gram-negative sepsis showed that BPI attenuated organ failure (including renal and cardiovascular systems) and prolonged time to death, while improving 24 hour survival rate (Jiang et al., 1999a, b).

Interestingly, BPI is absent from the blood of newborns (Levy et al., 1999). BPI was therefore logical to be studied in a randomized controlled fashion in a pediatric population with meningococcal sepsis. The results showed a trend towards reducing mortality and improving general status; however, the event rates in this trial were too low and the trial was underpowered to detect a mortality difference (the primary endpoint of the study) making the results difficult to interpret (Giroir et al., 2001; Levin et al., 2000).

In a randomized controlled trial of trauma patients with hemorrhage receiving at least 2 units of blood, the administration of the compound recombinant BPI showed a favorable trend t in reducing respiratory morbidity endpoints (Demetriades et al., 1999). Additional studies are needed to elucidate whether there is a role of this compound in this patient population.

15.7.2 Human Lactoferrin

Human lactoferrin (HL) is a component of innate immunity. Human lactoferrin is an iron-binding protein found in milk, granulocytes and exocrine secretions. It is released during inflammation, has bactericidal effects and reduces cytokine production by binding to the lipid A portion of endotoxin (Appelmelk et al., 1994).

In a rat model of neonatal sepsis, recombinant HL or talactoferrin was shown to improve survival (Venkatesh et al., 2007). In ex vivo studies, HL increased synergy of commonly used antibiotics against coagulase-negative staphylococcus and Candida (Venkatesh and Rong, 2008) and reduced biofilm of infected catheters (Venkatesh et al., 2009).

A recently completed phase 2 trial of talactoferrin versus placebo in patients with severe sepsis showed a statistically significant reduction in mortality with talactoferrin given orally every 6 hours for as long as the subject was in the ICU (personal communication, RPD). This will presumably lead to a phase 3 study.

15.8 Lipid A Analog

Non toxic lipid A analogs were developed from *Rhodobacter capsulatus.* The first generation molecule (E5531) and the second generation molecule (E5564, eritoran) were shown to decrease cytokine production in response to LPS. These drugs antagonize the effects of endotoxin on TLR4. The higher potency and longer duration of action of E5564 make it more suitable for clinical applications. It exerts its action without need for binding protein (Mullarkey et al., 2003).

In a randomized controlled trial 32 healthy individuals challenged with endotoxin were randomized to sequential doses of E5531 vs. placebo. The lipid A analog decreased signs and symptoms of endotoxin administration, decreased TNF alpha and IL-6 levels and diminished the cardiovascular hyperdynamic response (Bunnell et al., 2000). Improvement of cardiac contractility was confirmed in a different study using E5531 (Kumar et al., 2004). Another study using the same placebo-control design in healthy human volunteers showed similar findings for E5564, a second generation compound called eritoran (Lynn et al., 2003).

Three hundred critically-ill septic adults were enrolled in a randomized controlled trial of severe sepsis within 12 hours of diagnosis. Subjects were randomized to either placebo or eritoran (low or high dose E5564). Mortality was not reduced by either dose of the drug. In a pre-specified subgroup analysis, patients with APACHE II in the highest quartile had a trend towards lower mortality (33.3 vs. 56.3% – $p = 0.105$) (Tidswell et al., 2010). A large randomized trial targeting severe sepsis patients with higher APACHE II scores is nearing completion.

15.9 Phospholipid Emulsion

High density lipoprotein (HDL) and other lipoprotein can bind endotoxin, thereby attenuating the immunologic response, a property thought to be related to the phospholipid content (Feingold et al., 1995; Harris et al., 1990; Levine et al., 1993; Wu et al., 2004).

In animal studies on endotoxin-induced sepsis, raising HDL or administering human HDL and/or other lipoproteins reduced mortality (Harris et al., 1990; Levine et al., 1993). Using a hypolipemic animal model an increase in cytokine production and mortality with endotoxin challenge has been demonstrated (Feingold et al., 1995). After lipoprotein administration, the authors showed the mortality return to the same level as the control group.

Since the beneficial effects of lipoprotein were thought to be related to phospholipid content, a lipid emulsion with 10% phospholipids was developed and tested in animal as well as human studies (Gordon et al., 2003).

A porcine sepsis model showed promising results, as the drug decreased mortality, improved cardiovascular function and decreased both endotoxin and TNF levels (Goldfarb et al., 2003). This led to a phase 2 clinical trial, The Lipid Infusion and Patients Outcomes in Sepsis (LIPOS) trial.

The LIPOS trial was a randomized controlled study of 1379 patients to assess the efficacy of a phospholipid emulsion (GR270773) in patients with suspected or confirmed gram-negative sepsis. The emulsion was infused over a course of 3 days. Initially the design consisted of three arms, one being a high dose. The high dose arm was discontinued at interim analysis due to concern for potential toxicity. At the end of the study, no benefit of the lipid emulsion in decreasing mortality or organ failure was observed (Dellinger et al., 2009).

15.10 Hemoperfusion Through a Polymixin B Embedded Cartridge

Polymyxin B (PMXB) has a very strong antimicrobial effect on gram negative bacteria, but its intravenous use is limited owing to toxicity profile. PMX B also binds to endotoxin inactivating it in the process (Morrison and Jacobs, 1976). It has been used as an embedded filter component of hemoperfusion cartridges for removal of endotoxin in Japan since 1994.

15.10.1 Animal Studies

A study using a rat model of sepsis induced by cecal ligation and perforation (CLP) showed that PMXB hemoperfusion performed 24 hours after the CLP procedure reduced plasminogen activator inhibitor -1 expression in the lung, and apoptosis in renal tubular cells (Ito et al., 2009).

The effect of PMX B hemoperfusion on animal survival was examined in a trial of CLP induced sepsis in 3 day old piglets. Although the treatment was associated

Table 15.1 Summary of studies of PMX B hemoperfusion

Author	Setting	Finding
Animal studies		
Ito et al. (2009)	CLP-R	↓PAI-1 in lung, ↓Apoptosis renal tubular cells
Oshima et al. (2008)	Lung I/R-Dogs	↑PaO2/FiO2 ↓Lung water, Better lung tissue preservation
Sato et al. (2009)	Hepatic I/R-Pigs	↑Hepatic circulation, ↑Portal circulation, ↓AST
Sato et al. (2008)	Intestinal I/R-Dogs	↑SMA blood flow, ↑Mucosal Blood flow, ↑IL-10
Hussein et al. (2010)	CLP-P	↑ oxygenation, Better liver function tests, ↑ survival
Cohen et al. (1987)		↓Mortality
Human studies		
Nakamura et al. (2009)	Sepsis-ARDS	Improved oxygenation, ↓Endotoxin, ↓HMGB1
Suyama et al. (2008)	Sepsis ALI/ARDS	Improved oxygenation
Noma et al. (2007), Enomoto et al. (2008)	ILD	Improved oxygenation
Kushi et al. (2006a)	Sepsis	Improved tissue oxygenation
Kushi et al. (2008)	Sepsis	Higher gastric pH
Kushi et al. (2006b, c)	Sepsis	↓IL8, ↓NE
Kushi et al. (2005)	Sepsis	↓IL-8, ↓NE, ↓PAI-1, Improved oxygenation
Kanesaka et al. (2008)	GNR Sepsis	↓IL-6, ↓pyruvate, ↑Vascular resistance, ↑Urine output
Shimizu et al. (2009)	CP sepsis	↓IL-6, ↓IL-8, ↓Endotoxin
Nakamura et al. (2008)	Sepsis	↓BNP, ↓ANP
Cantaluppi et al. (2008)	GNR sepsis	Less proapoptotic activity
Nemoto et al. (2001)	Sepsis	↓Mortality in all patients, ↓Mortality APACHE <30 no apparent effect APACHEII>30
Cruz et al. (2009)	Abdominal Sepsis	↑MAP, ↓vasopressors, ↓SOFA, ↓Mortality

CLP-R CLP induced sepsis-Rats; *CLP-P* CLP-induced sepsis-Piglets; *IVE-R* IV Endotoxin-induced sepsis-Rats; *ILD* interstitial lung disease; *NE* neutrophil elastase

with lower hematocrit values, it improved oxygenation, liver function tests and was associated with a longer survival time (Hussein et al., 2010).

Other animal studies will not be extensively reviewed in this manuscript but are cited in Table 15.1.

15.10.2 Human Studies

PMX B improved oxygenation in patients with acute respiratory distress associated with exacerbation of interstitial lung disease in some but not all patients (Enomoto et al., 2008; Nakamura et al., 2009; Noma et al., 2007; Suyama et al., 2008).

Using a gastric pH measurement and the difference between the gastric and arterial PCO_2 (PCO_2 gap) as a surrogate of tissue oxygen metabolism a group of researchers performed PMX HP therapy in 22 septic patients. Their results showed that the PMX HP therapy performed twice within 24 hours, for a duration of three hours each significantly decreased the arterial-venous PCO_2 gap at 48 and 72 hours, suggesting improved tissue oxygenation (Kushi et al., 2006a). In nearly all of the reported studies, PMXB therapy was dosed for 2 hour treatments. In a study of 16 patients with septic shock, a more prolonged treatment resulted in lower vasopressors requirement, better oxygenation and lower sequential organ failure assessment (SOFA) score (Mitaka et al., 2009). Polymyxin B hemoperfusion therapy has also been shown to decrease the levels of blood endotoxins, as well as those of the proinflammatory cytokine HMGB1 (Nakamura et al., 2009). In a study by Kushi et al., IL-8 was measured sequentially in 15 septic patients that underwent PMX treatment twice, each for three hours within a 24 hour period. The IL-8 levels were significantly lower at 48 and 72 hours (Kushi et al., 2006b). A similar pattern was shown for neutrophil elastase levels by the same group in another study (Kushi et al., 2006c). Both of these mediators in addition to PAI-1 were lowered by PMX B treatment in 36 patients at 48 hours, and correlated with improvement in pulmonary oxygenation which was apparent at 96 hours. In a study of 52 patients with sepsis secondary to colorectal perforation, IL-1, 6 and 8 and endotoxin levels decreased with the PMX B hemoperfusion therapy (Shimizu et al., 2009).

In a randomized controlled trial conducted by Nemoto et al, 98 patients were assigned to either conventional treatment or PMX B (Nemoto et al., 2001). The overall survival was significantly better in the treated group. When looking at subgroups, the authors concluded that this therapeutic approach was more likely to improve survival when applied earlier in sepsis. One criticism of this conclusion is that the subgroups contained a low number of patients to make a definitive statement in this regard.

In a multicenter prospective trial, Cruz et al. randomized 64 patients with abdominal sepsis to either PMX B or conventional therapy (Cruz et al., 2009). The PMX B hemoperfusion procedure was performed twice. At 72 hours, in the treated group, the MAP was higher, vasopressors requirement lower, and SOFA score improved. The 28-day mortality was 32% in the treated group, significantly lower than the conventional group (53%).

In summary, both animal and human sepsis studies suggest a positive impact of PMXB treatment on physiologic parameters and supporting the possibility of improving clinical outcomes. Large randomized studies need to be performed to confirm any beneficial effect. A treatment protocol based on identifying patients with high endotoxin levels (with the goal of removing endotoxin) may offer advantages over previous efforts to neutralize endotoxin based on clinical syndrome only. This therapy is not approved for use in the United States and a randomized multicenter trial is being planned.

15.11 Conclusion

Targeting endotoxin for treatment of Gram-negative sepsis presents several theoretical advantages over other tested therapies. Endotoxin triggers the sepsis cascade by a well characterized mechanism that involves a binding protein, a cell wall receptor and a toll like receptor. Endotoxemia is present not only in Gram negative sepsis but in other infections as well. In addition, elevated levels of endotoxin correlate with higher morbidity and mortality. Finally targeting endotoxin early in the sepsis clinical presentation could help reverse or limit this disease before "the horse is out of the barn".

Conceptually, the action of endotoxin could be blocked at several steps and trials studying different pathways have been (or are currently being) conducted.

Many animal and preclinical studies have showed benefit of a variety of antiendotoxin therapies, but this effect has not been translated into human trials. Since both pre-clinical rationale and studies support the targeting of endotoxin to ameliorate the pro-inflammatory and pro-coagulation response of severe sepsis, we should continue to pursue therapeutic intervention in this area.

References

Angus, D.C., Birmingham, M.C., Balk, R.A., Scannon, P.J., Collins, D., Kruse, J.A., Graham, D.R., Dedhia, H.V., Homann, S., MacIntyre, N. E5 murine monoclonal antiendotoxin antibody in gram-negative sepsis: A randomized controlled trial. E5 Study Investigators. JAMA 283 (2000) 1723–1730.

Appelmelk, B.J., An, Y.Q., Geerts, M., Thijs, B.G., de Boer, H.A., MacLaren, D.M., de Graaff, J., Nuijens, J.H. Lactoferrin is a lipid A-binding protein. Infect Immun 62 (1994) 2628–2632.

Baumgartner, J.D., Glauser, M.P., McCutchan, J.A., Ziegler, E.J., van Melle, G., Klauber, M.R., Vogt, M., Muehlen, E., Luethy, R., Chiolero, R., et al. Prevention of gram-negative shock and death in surgical patients by antibody to endotoxin core glycolipid. Lancet 2 (1985) 59–63.

Bhattacharjee, A.K., Opal, S.M., Taylor, R., Naso, R., Semenuk, M., Zollinger, W.D., Moran, E.E., Young, L., Hammack, C., Sadoff, J.C., Cross, A.S. A noncovalent complex vaccine prepared with detoxified *Escherichia coli* J5 (Rc chemotype) lipopolysaccharide and *Neisseria meningitidis* Group B outer membrane protein produces protective antibodies against gram-negative bacteremia. J Infect Dis 173 (1996) 1157–1163.

Bone, R.C., Balk, R.A., Fein, A.M., Perl, T.M., Wenzel, R.P., Reines, H.D., Quenzer, R.W., Iberti, T.J., Macintyre, N., Schein, R.M. A second large controlled clinical study of E5, a monoclonal antibody to endotoxin: Results of a prospective, multicenter, randomized, controlled trial. The E5 Sepsis Study Group. Crit Care Med 23 (1995) 994–1006.

Bosshart, H., Heinzelmann, M. Targeting bacterial endotoxin: Two sides of a coin. Ann N Y Acad Sci 1096 (2007) 1–17.

Braude, A.I., Jones, J.L., Douglas, H. The behavior of *Escherichia coli* endotoxin (somatic antigen) during infectious arthritis. J Immunol 90 (1963) 297–311.

Bunnell, E., Lynn, M., Habet, K., Neumann, A., Perdomo, C.A., Friedhoff, L.T., Rogers, S.L. and Parrillo, J.E. A lipid A analog, E5531, blocks the endotoxin response in human volunteers with experimental endotoxemia. Crit Care Med 28 (2000) 2713–2720.

Calandra, T., Glauser, M.P., Schellekens, J., Verhoef, J. Treatment of gram-negative septic shock with human IgG antibody to *Escherichia coli* J5: A prospective, double-blind, randomized trial. J Infect Dis 158 (1988) 312–319.

Calvano, S.E., Thompson, W.A., Marra, M.N., Coyle, S.M., de Riesthal, H.F., Trousdale, R.K., Barie, P.S., Scott, R.W., Moldawer, L.L., Lowry, S.F. Changes in polymorphonuclear leukocyte surface and plasma bactericidal/permeability-increasing protein and plasma lipopolysaccharide binding protein during endotoxemia or sepsis. Arch Surg 129 (1994) 220–226.

Cantaluppi, V., Assenzio, B., Pasero, D., Romanazzi, G.M., Pacitti, A., Lanfranco, G., Puntorieri, V., Martin, E.L., Mascia, L., Monti, G., Casella, G., Segoloni, G.P., Camussi, G., Ranieri, V.M. Polymyxin-B hemoperfusion inactivates circulating proapoptotic factors. Intensive Care Med 34 (2008) 1638–1645.

Cinel, I., Dellinger, R.P. Advances in pathogenesis and management of sepsis. Curr Opin Infect Dis 20 (2007) 345–352.

Cohen, J., Aslam, M., Pusey, C.D., Ryan, C.J. Protection from endotoxemia: A rat model of plasmapheresis and specific adsorption with polymyxin B. J Infect Dis 155 (1987) 690–695.

Cross, A.S., Opal, S.M., Palardy, J.E., Drabick, J.J., Warren, H.S., Huber, C., Cook, P., Bhattacharjee, A.K. Phase I study of detoxified *Escherichia coli* J5 lipopolysaccharide (J5dLPS)/group B meningococcal outer membrane protein (OMP) complex vaccine in human subjects. Vaccine 21 (2003) 4576–4587.

Cross, A.S., Opal, S.M., Warren, H.S., Palardy, J.E., Glaser, K., Parejo, N.A., Bhattacharjee, A.K. Active immunization with a detoxified *Escherichia coli* J5 lipopolysaccharide group B meningococcal outer membrane protein complex vaccine protects animals from experimental sepsis. J Infect Dis 183 (2001) 1079–1086.

Cruz, D.N., Antonelli, M., Fumagalli, R., Foltran, F., Brienza, N., Donati, A., Malcangi, V., Petrini, F., Volta, G., Bobbio Pallavicini, F.M., Rottoli, F., Giunta, F., Ronco, C. Early use of polymyxin B hemoperfusion in abdominal septic shock: The EUPHAS randomized controlled trial. JAMA 301 (2009) 2445–2452.

Dellinger, R.P., Tomayko, J.F., Angus, D.C., Opal, S., Cupo, M.A., McDermott, S., Ducher, A., Calandra, T., Cohen, J. Efficacy and safety of a phospholipid emulsion (GR270773) in Gram-negative severe sepsis: Results of a phase II multicenter, randomized, placebo-controlled, dose-finding clinical trial. Crit Care Med 37 (2009) 2929–2938.

Demetriades, D., Smith, J.S., Jacobson, L.E., Moncure, M., Minei, J., Nelson, B.J., Scannon, P.J. Bactericidal/permeability-increasing protein (rBPI21) in patients with hemorrhage due to trauma: Results of a multicenter phase II clinical trial. rBPI21 Acute Hemorrhagic Trauma Study Group. J Trauma 46 (1999) 667–676; discussion 676–677.

Elsbach, P., Weiss, J. The bactericidal/permeability-increasing protein (BPI), a potent element in host-defense against gram-negative bacteria and lipopolysaccharide. Immunobiology 187 (1993) 417–429.

Elsbach, P., Weiss, J., Doerfler, M., Shu, C., Kohn, F., Ammons, W.S., Kung, A.H., Meszaros, K.K., Parent, J.B. The bactericidal/permeability increasing protein of neutrophils is a potent antibacterial and anti-endotoxin agent in vitro and in vivo. Prog Clin Biol Res 388 (1994) 41–51.

Enomoto, N., Suda, T., Uto, T., Kato, M., Kaida, Y., Ozawa, Y., Miyazaki, H., Kuroishi, S., Hashimoto, D., Naito, T., Fujisawa, T., Matsui, T., Inui, N., Nakamura, Y., Sato, J., Mizuguchi,

T., Kato, A., Chida, K. Possible therapeutic effect of direct haemoperfusion with a polymyxin B immobilized fibre column (PMX-DHP) on pulmonary oxygenation in acute exacerbations of interstitial pneumonia. Respirology 13 (2008) 452–460.

Feingold, K.R., Funk, J.L., Moser, A.H., Shigenaga, J.K., Rapp, J.H., Grunfeld, C. Role for circulating lipoproteins in protection from endotoxin toxicity. Infect Immun 63 (1995) 2041–2046.

Fukui, M., Imamura, R., Umemura, M., Kawabe, T., Suda, T. Pathogen-associated molecular patterns sensitize macrophages to Fas ligand-induced apoptosis and IL-1 beta release. J Immunol 171 (2003) 1868–1874.

Giroir, B.P., Scannon, P.J., Levin, M. Bactericidal/permeability-increasing protein--lessons learned from the phase III, randomized, clinical trial of rBPI21 for adjunctive treatment of children with severe meningococcemia. Crit Care Med 29 (2001) S130–S135.

Goldfarb, R.D., Parker, T.S., Levine, D.M., Glock, D., Akhter, I., Alkhudari, A., McCarthy, R.J., David, E.M., Gordon, B.R., Saal, S.D., Rubin, A.L., Trenholme, G.M., Parrillo, J.E. Protein-free phospholipid emulsion treatment improved cardiopulmonary function and survival in porcine sepsis. Am J Physiol Regul Integr Comp Physiol 284 (2003) R550–R557.

Gordon, B.R., Parker, T.S., Levine, D.M., Saal, S.D., Hudgins, L.C., Sloan, B.J., Chu, C., Stenzel, K.H., Rubin, A.L. Safety and pharmacokinetics of an endotoxin-binding phospholipid emulsion. Ann Pharmacother 37 (2003) 943–950.

Greenman, R.L., Schein, R.M., Martin, M.A., Wenzel, R.P., MacIntyre, N.R., Emmanuel, G., Chmel, H., Kohler, R.B., McCarthy, M., Plouffe, J., et al. A controlled clinical trial of E5 murine monoclonal IgM antibody to endotoxin in the treatment of gram-negative sepsis. The XOMA Sepsis Study Group. JAMA 266 (1991) 1097–1102.

Harris, H.W., Grunfeld, C., Feingold, K.R., Rapp, J.H. Human very low density lipoproteins and chylomicrons can protect against endotoxin-induced death in mice. J Clin Invest 86 (1990) 696–702.

Hurley, J.C. Reappraisal with meta-analysis of bacteremia, endotoxemia, and mortality in gram-negative sepsis. J Clin Microbiol 33 (1995) 1278–1282.

Hussein, M.H., Daoud, G.A., Kakita, H., Kato, S., Goto, T., Kamei, M., Goto, K., Ozaki, Y., Ito, T., Fukuda, S., Kato, I., Suzuki, S., Hashimoto, T., Togari, H. Effect of polymyxin B-immobilized fiber hemoperfusion on respiratory impairment, hepatocellular dysfunction, and leucopenia in a neonatal sepsis model. Pediatr Surg Int 26 (2010) 187–193.

Inada, K., Endo, S., Yamashita, H., Suzuki, M., Yoshida, H., Yoshida, M. Inhibitory effect of mouse antiendotoxin monoclonal antibody (E5) on hypotension induced by endotoxin and TNF-alpha. Kansenshogaku Zasshi 67 (1993) 1108–1114.

Ito, M., Kase, H., Shimoyama, O., Takahashi, T. Effects of polymyxin B-immobilized fiber using a rat cecal ligation and perforation model. ASAIO J 55 (2009) 246–250.

Jiang, J., Xie, G., Liu, D., Zhu, P., Wang, Z., He, Y., Zhou, J., Xu, H. Effect of bactericidal /permeability-increasing protein on sepsis induced by intra-abdominal infection in rats. Chin J Traumatol 2 (1999a) 84–86.

Jiang, J., Zhu, P., Wang, Z., Liu, D., He, Y., Xie, G., Xu, H. Protective effects of bactericidal/permeability-increasing protein on the vital organ function after endotoxic shock in rats. Zhonghua Wai Ke Za Zhi 37 (1999b) 120–122.

Kanesaka, S., Sasaki, J., Kuzume, M., Narihara, K., Takahashi, Y. Effect of direct hemoperfusion using polymyxin B immobilized fiber on inflammatory mediators in patients with severe sepsis and septic shock. Int J Artif Organs 31 (2008) 891–897.

Kumar, A., Bunnell, E., Lynn, M., Anel, R., Habet, K., Neumann, A., Parrillo, J.E. Experimental human endotoxemia is associated with depression of load-independent contractility indices: Prevention by the lipid a analogue E5531. Chest 126 (2004) 860–867.

Kushi, H., Miki, T., Nakahara, J., Okamoto, K., Kawahara, Y., Saito, T., Tanjoh, K. Hemoperfusion with an immobilized polymyxin B fiber column improves tissue oxygen metabolism. Ther Apher Dial 10 (2006a) 430–435.

Kushi, H., Miki, T., Nakahara, J., Okamoto, K., Kawahara, Y., Saito, T., Tanjoh, K. Hemoperfusion with an immobilized polymyxin B fiber column reduces circulating interleukin-8 concentrations. Ther Apher Dial 10 (2006b) 425–429.

Kushi, H., Miki, T., Nakahara, J., Okamoto, K., Saito, T., Tanjoh, K. Hemoperfusion with an immobilized polymyxin B column reduces the blood level of neutrophil elastase. Blood Purif 24 (2006c) 212–217.

Kushi, H., Miki, T., Okamaoto, K., Nakahara, J., Saito, T., Tanjoh, K. Early hemoperfusion with an immobilized polymyxin B fiber column eliminates humoral mediators and improves pulmonary oxygenation. Crit Care 9 (2005) R653–R661.

Kushi, H., Miki, T., Sakagami, Y., Sato, J., Saito, T., Tanjoh, K. Early hemoperfusion with a polymyxin b column improves gastric intramucosal pH in sepsis. Ther Apher Dial 12 (2008) 278–284.

Levin, M., Quint, P.A., Goldstein, B., Barton, P., Bradley, J.S., Shemie, S.D., Yeh, T., Kim, S.S., Cafaro, D.P., Scannon, P.J., Giroir, B.P. Recombinant bactericidal/permeability-increasing protein (rBPI21) as adjunctive treatment for children with severe meningococcal sepsis: A randomised trial. rBPI21 Meningococcal Sepsis Study Group. Lancet 356 (2000) 961–967.

Levine, D.M., Parker, T.S., Donnelly, T.M., Walsh, A., Rubin, A.L. In vivo protection against endotoxin by plasma high density lipoprotein. Proc Natl Acad Sci USA 90 (1993) 12040–12044.

Levy, O., Martin, S., Eichenwald, E., Ganz, T., Valore, E., Carroll, S.F., Lee, K., Goldmann, D., Thorne, G.M. Impaired innate immunity in the newborn: Newborn neutrophils are deficient in bactericidal/permeability-increasing protein. Pediatrics 104 (1999) 1327–1333.

Lin, W.J., Yeh, W.C. Implication of Toll-like receptor and tumor necrosis factor alpha signaling in septic shock. Shock 24 (2005) 206–209.

Lynn, M., Rossignol, D.P., Wheeler, J.L., Kao, R.J., Perdomo, C.A., Noveck, R., Vargas, R., D'Angelo, T., Gotzkowsky, S., McMahon, F.G. Blocking of responses to endotoxin by E5564 in healthy volunteers with experimental endotoxemia. J Infect Dis 187 (2003) 631–639.

Marshall, J.C., Foster, D., Vincent, J.L., Cook, D.J., Cohen, J., Dellinger, R.P., Opal, S., Abraham, E., Brett, S.J., Smith, T., Mehta, S., Derzko, A., Romaschin, A. Diagnostic and prognostic implications of endotoxemia in critical illness: Results of the MEDIC study. J Infect Dis 190 (2004) 527–534.

McCloskey, R.V., Straube, R.C., Sanders, C., Smith, S.M., Smith, C.R. Treatment of septic shock with human monoclonal antibody HA-1A. A randomized, double-blind, placebo-controlled trial. CHESS Trial Study Group. Ann Intern Med 121 (1994) 1–5.

Mitaka, C., Tsuchida, N., Kawada, K., Nakajima, Y., Imai, T., Sasaki, S. A longer duration of polymyxin B-immobilized fiber column hemoperfusion improves pulmonary oxygenation in patients with septic shock. Shock 32 (2009) 478–483.

Morrison, D.C., Jacobs, D.M. Binding of polymyxin B to the lipid A portion of bacterial lipopolysaccharides. Immunochemistry 13 (1976) 813–818.

Mullarkey, M., Rose, J.R., Bristol, J., Kawata, T., Kimura, A., Kobayashi, S., Przetak, M., Chow, J., Gusovsky, F., Christ, W.J., Rossignol, D.P. Inhibition of endotoxin response by e5564, a novel Toll-like receptor 4-directed endotoxin antagonist. J Pharmacol Exp Ther 304 (2003) 1093–1102.

Nakamura, T., Fujiwara, N., Sato, E., Kawagoe, Y., Ueda, Y., Yamada, S., Koide, H. Effect of polymyxin B-immobilized fiber hemoperfusion on serum high mobility group box-1 protein levels and oxidative stress in patients with acute respiratory distress syndrome. ASAIO J 55 (2009) 395–399.

Nakamura, T., Suzuki, T., Kawagoe, Y., Koide, H. Polymyxin B-immobilized fiber hemoperfusion attenuates increased plasma atrial natriuretic peptide and brain natriuretic Peptide levels in patients with septic shock. ASAIO J 54 (2008) 210–213.

Nemoto, H., Nakamoto, H., Okada, H., Sugahara, S., Moriwaki, K., Arai, M., Kanno, Y., Suzuki, H. Newly developed immobilized polymyxin B fibers improve the survival of patients with sepsis. Blood Purif 19 (2001) 361–368; discussion 368–369.

Noma, S., Matsuyama, W., Mitsuyama, H., Suetsugu, T., Koreeda, Y., Mizuno, K., Higashimoto, I., Kakihana, Y., Hashiguchi, T., Maruyama, I., Osame, M., Arimura, K. Two cases of acute exacerbation of interstitial pneumonia treated with polymyxin B-immobilized fiber column hemoperfusion treatment. Intern Med 46 (2007) 1447–1454.

Oshima, K., Akao, T., Kobayashi, K., Muraoka, M., Matsumoto, K., Takeyoshi, I. The effect of direct hemoperfusion with a polymyxin B-immobilized fiber column (DHP-PMX therapy) on pulmonary ischemia-reperfusion injury in a canine model. J Invest Surg 21 (2008) 127–132.

Parsons, P.E., Worthen, G.S., Moore, E.E., Tate, R.M., Henson, P.M. The association of circulating endotoxin with the development of the adult respiratory distress syndrome. Am Rev Respir Dis 140 (1989) 294–301.

Pollack, M., Huang, A.I., Prescott, R.K., Young, L.S., Hunter, K.W., Cruess, D.F., Tsai, C.M. Enhanced survival in *Pseudomonas aeruginosa* septicemia associated with high levels of circulating antibody to *Escherichia coli* endotoxin core. J Clin Invest 72 (1983) 1874–1881.

Prophylactic intravenous administration of standard immune globulin as compared with core-lipopolysaccharide immune globulin in patients at high risk of postsurgical infection. The Intravenous Immunoglobulin Collaborative Study Group. N Engl J Med 327 (1992) 234–240.

Raetz, C.R., Reynolds, C.M., Trent, M.S., Bishop, R.E. Lipid A modification systems in gram-negative bacteria. Annu Rev Biochem 76 (2007) 295–329.

Raetz, C.R., Whitfield, C. Lipopolysaccharide endotoxins. Annu Rev Biochem 71 (2002) 635–700.

Romulo, R.L., Palardy, J.E., Opal, S.M. Efficacy of anti-endotoxin monoclonal antibody E5 alone or in combination with ciprofloxacin in neutropenic rats with *Pseudomonas* sepsis. J Infect Dis 167 (1993) 126–130.

Sato, H., Oshima, K., Arakawa, K., Kobayashi, K., Yamazaki, H., Suto, Y., Takeyoshi, I. Direct hemoperfusion with a polymyxin B-immobilized cartridge in intestinal warm ischemia reperfusion. World J Gastroenterol 14 (2008) 5436–5441.

Sato, H., Oshima, K., Kobayashi, K., Yamazaki, H., Suto, Y., Takeyoshi, I. Hemoperfusion with polymyxin B-immobilized fiber column improves liver function after ischemia-reperfusion injury. World J Gastroenterol 15 (2009) 4571–4575.

Shimizu, T., Hanasawa, K., Sato, K., Umeki, M., Koga, N., Naganuma, T., Sato, S., Shimonishi, T., Ikeda, T., Matsuno, N., Ono, S., Saitoh, H., Satoh, K., Otani, Y., Endo, Y., Eguchi, Y., Tani, T. Direct hemoperfusion with polymyxin-B-immobilized fiber columns improves septic hypotension and reduces inflammatory mediators in septic patients with colorectal perforation. Langenbecks Arch Surg 394 (2009) 303–311.

Suyama, H., Kawasaki, Y., Morikawa, S., Kaneko, K., Yamanoue, T. Early induction of PMX-DHP improves oxygenation in severe sepsis patients with acute lung injury. Hiroshima J Med Sci 57 (2008) 79–84.

Taveira da Silva, A.M., Kaulbach, H.C., Chuidian, F.S., Lambert, D.R., Suffredini, A.F., Danner, R.L. Brief report: Shock and multiple-organ dysfunction after self-administration of *Salmonella* endotoxin. N Engl J Med 328 (1993) 1457–1460.

The French National Registry of HA-1A (Centoxin) in septic shock. A cohort study of 600 patients. The National Committee for the Evaluation of Centoxin. Arch Intern Med 154 (1994) 2484–2491.

Tidswell, M., Tillis, W., Larosa, S.P., Lynn, M., Wittek, A.E., Kao, R., Wheeler, J., Gogate, J., Opal, S.M. Phase 2 trial of eritoran tetrasodium (E5564), a Toll-like receptor 4 antagonist, in patients with severe sepsis. Crit Care Med 38 (2010) 72–83.

Townsend, S.R., Schorr, C., Levy, M.M., Dellinger, R.P. Reducing mortality in severe sepsis: The Surviving Sepsis Campaign. Clin Chest Med 29 (2008) 721–733.

Venkatesh, M., Rong, L., Raad, I., Versalovic, J. Novel synergistic antibiofilm combinations for salvage of infected catheters. J Med Microbiol 58 (2009) 936–944.

Venkatesh, M.P., Pham, D., Kong, L., Weisman, L.E. Prophylaxis with lactoferrin, a novel antimicrobial agent, in a neonatal rat model of coinfection. Adv Ther 24 (2007) 941–954.

Venkatesh, M.P., Rong, L. Human recombinant lactoferrin acts synergistically with antimicrobials commonly used in neonatal practice against coagulase-negative staphylococci and Candida albicans causing neonatal sepsis. J Med Microbiol 57 (2008) 1113–1121.

Volk, H.D., Reinke, P., Krausch, D., Zuckermann, H., Asadullah, K., Muller, J.M., Docke, W.D., Kox, W.J. Monocyte deactivation--rationale for a new therapeutic strategy in sepsis. Intensive Care Med 22 Suppl 4 (1996) S474–S481.

Weiss, J., Elsbach, P., Shu, C., Castillo, J., Grinna, L., Horwitz, A., Theofan, G. Human bactericidal/permeability-increasing protein and a recombinant NH_2-terminal fragment cause killing of serum-resistant gram-negative bacteria in whole blood and inhibit tumor necrosis factor release induced by the bacteria. J Clin Invest 90 (1992) 1122–1130.

Wu, A., Hinds, C.J., Thiemermann, C. High-density lipoproteins in sepsis and septic shock: Metabolism, actions, and therapeutic applications. Shock 21 (2004) 210–221.

Zeni, F., Freeman, B., Natanson, C. Anti-inflammatory therapies to treat sepsis and septic shock: A reassessment. Crit Care Med 25 (1997) 1095–1100.

Ziegler, E.J., Fisher, C.J., Jr., Sprung, C.L., Straube, R.C., Sadoff, J.C., Foulke, G.E., Wortel, C.H., Fink, M.P., Dellinger, R.P., Teng, N.N., et al. Treatment of gram-negative bacteremia and septic shock with HA-1A human monoclonal antibody against endotoxin. A randomized, double-blind, placebo-controlled trial. The HA-1A Sepsis Study Group. N Engl J Med 324 (1991) 429–436.

Ziegler, E.J., McCutchan, J.A., Fierer, J., Glauser, M.P., Sadoff, J.C., Douglas, H., Braude, A.I. Treatment of gram-negative bacteremia and shock with human antiserum to a mutant *Escherichia coli.* N Engl J Med 307 (1982) 1225–1230.

Zinner, S.H., McCabe, W.R. Effects of IgM and IgG antibody in patients with bacteremia due to gram-negative bacilli. J Infect Dis 133 (1976) 37–45.

Chapter 16
Lipopolysaccharides in *Rhizobium*-Legume Symbioses

Russell W. Carlson, L. Scott Forsberg, and Elmar L. Kannenberg

Abstract The establishment of nitrogen-fixing symbiosis between a legume plant and its rhizobial symbiont requires that the bacterium adapt to changing conditions that occur with the host plant that both promotes and allows infection of the host root nodule cell, regulates and resists the host defense response, permits the exchange of metabolites, and contributes to the overall health of the host. This adaptive process involves changes to the bacterial cell surface and, therefore, structural modifications to the lipopolysaccharide (LPS). In this chapter, we describe the structures of the LPSs from symbiont members of the *Rhizobiales*, the genetics and mechanism of their biosynthesis, the modifications that occur during symbiosis, and their possible functions.

Keywords *Rhizobium* · Lipopolysaccharides · Structure · Biosynthesis · Symbiosis · Plant defense

Abbreviations

COS	core oligosaccharide
CPS	capsular polysaccharide
EPS	extracellular polysaccharide
ISR	induced systemic resistance
LA	lipid A
LCO	lipochitin oligosaccharide
LPS	lipopolysaccharides
OPS	O-chain polysaccharide
VLCFA	very long chain fatty acid

R.W. Carlson (✉)
Complex Carbohydrate Research Center, University of Georgia, 315 Riverbend Rd., Athens, GA 30602, USA
e-mail: rcarlson@ccrc.uga.edu

X. Wang, P.J. Quinn (eds.), *Endotoxins: Structure, Function and Recognition*, Subcellular Biochemistry 53, DOI 10.1007/978-90-481-9078-2_16,

16.1 Introduction

Bacterial species belonging to the *Rhizobiales* are members of the *Alphaproteobacteria*, which includes the phytopathogen *Agrobacterium*, phylogenetically related animal pathogens such as the intracellular animal pathogens *Bartonella* and *Brucella*, and plant symbionts *Bradyrhizobium*, *Rhizobium*, *Sinorhizobium*, *Azorhizobium*, and *Mesorhizobium* (collectively called rhizobia). These plant symbiont species live and dwell as soil bacteria and as nitrogen-fixing endosymbionts in legume host plants. During symbiosis with legumes, the host plants form a new organ, the root nodule, and within this root nodule the rhizobia develop into specialized symbiotic forms, the bacteroids. Bacteroids are normally found within nodule cells in organelle-like compartments, called symbiosomes. In this environment they fix molecular nitrogen into ammonia and supply it to the host plant as a nitrogen source. Successful survival of rhizobia depends, therefore, on rhizobia being capable of invading and adapting to very different legume plant as well as soil environments. This remarkable adaptability has been fascinating scientists for more than a hundred years ever since nitrogen fixation was discovered by Hellriegel and Wilfarth in 1888 [some recent reviews on various aspects of *Rhizobium* biology are: (Brewin, 2004; Denison and Kiers, 2004; Frugier et al., 2008; Gage, 2004; Gibson et al., 2008; Oldroyd and Downie, 2008; Sanchez-Contreras et al., 2007; Soto et al., 2009; Stacey et al., 2006)].

Besides its agricultural and environmental importance, *Rhizobium*-legume symbiosis provides a powerful model that can teach us a great deal about the molecular cross-talk between a prokaryote and its eukaryotic host. A considerable body of work has established that the culmination of the symbiotic interaction, which is the metabolic exchange of fixed ammonia and reduced carbon compounds between a *Rhizobium* and its host plant, is predicated by a series of complex morphological plant and bacterial developmental steps that begins with the attachment of rhizobia to a newly emerging root hair. The root hair deforms and entraps the rhizobia while the root hair tip growth is redirected placing cell wall material in apposition to the rhizobial attachment point. In a concerted interaction between the host plant and the bacterial microcolony a cell wall invagination is initially formed that develops into an inwardly growing tunnel, the infection thread, filled with a column of rhizobia, and still bordered by a cell wall layer. The growing, eventually branching, infection thread leads the bacteria to a newly formed meristem; on legumes forming indeterminate nodule, like pea, clover and alfalfa, it is formed in the more central parts of the root cortex in layers adjacent to the pericycle; in legumes forming determinate nodules, like those on *Lotus*, soy and *Phaseolus* beans, it is formed in the more outer root cortex layers. Eventually an infection thread makes contact with a meristematic cell and forms what is called an infection droplet. At this stage, the apposition of cell wall material is thinned to the point that colonizing rhizobia make contact with the cytoplasmic membrane of meristematic cells and are endocytosed. Immature symbiosomes are formed that multiply and develop into their mature organelle-like forms that are occupied by nitrogen-fixing bacteroids surrounded by a specialized plant-derived membrane called the peribacteroid membrane. In indeterminate nodules, the symbiosomes are occupied by only an individual bacteroid, while in determinate nodules, symbiosomes are occupied by several bacteroids [for more details, see e.g. (Gibson et al., 2008; Oldroyd and Downie, 2008)].

Nitrogen fixation requires the expression of the highly oxygen-sensitive nitrogenase enzyme complex in aerobically-growing rhizobia. A key function of nodule physiology is to mediate the bacterial metabolic needs for oxygen with the need for protecting the nitrogenase complex from destruction. This is accomplished through a microaerobic nodule environment and a controlled oxygen supply that is simultanously conducive to aerobic metabolism and nitrogen fixation. The bacteroids fix atmospheric nitrogen gas into ammonia, but do not metabolize the fixed ammonium; rather they release it to the host plant. The host plant maintains this rhizobia "service" for weeks, providing necessary nutrients to the rhizobia in form of carboxylic acids and possibly other metabolites.

Successful host colonization by rhizobia requires both reprogramming of plant and bacterial cell development while avoiding infection termination through host defense reactions. The rhizobial cell surface is critical in this process. Initially, following the induction of rhizobial *nod* genes by the host legume-secreted flavonoid signal molecules, the bacteria synthesize and release a Nod-factor, a lipochitin oligosaccharide (LCO), that initiates host cell wall perturbations and causes the above mentioned root hair deformations. This is followed by the formation of infection threads. In the lumen of the infection thread, the rhizobia grow and divide and are morphologically still outside the plant. Modification to the host plant cell wall and the extracellular matrix occurs and hydroxyproline-rich glycoproteins (extensins/AGPs) are targeted into the lumen of the infection threads. The structural peculiarities of these glycoproteins suggest that they could facilitate a fluid-to-solid transition in the extracellular matrix through peroxide-driven cross-linking of proteins. This, and possibly modifications in the cell wall pectic matrix are apparently necessary for infection thread extensibility. In the largely unwalled infection droplet, physical contact of the *Rhizobium* cell surface with the plant membrane glycocalyx seems necessary for endocytosis. The endocytosed rhizobia (or bacteroids) remain within the plant derived membrane (i.e. called the peribacteroid or symbiosome membrane). This membrane also matures and acquires new functions for metabolite exchange between the entrapped nitrogen-fixing bacteroids and the plant cytoplasm. This may also involve actual physical contact between the bacteroid surface and the symbiosome membrane leading to a nitrogen-fixing symbiosome that is stable for weeks. Ultimately, this stable interaction breaks down and the symbiosome-converted lysosomes cause bacteroid senescence. These complex interactions are governed by a range of diffusible signal molecules, in addition to molecular cues in the plant cell wall and on the rhizobial cell surface, and by changes in physiological conditions that occur during symbiosis (Brewin, 2004; Fauvart and Michiels, 2008; Frugier et al., 2008; Gage, 2004; Gibson et al., 2008; Oldroyd and Downie, 2008; Sanchez-Contreras et al., 2007).

As described in the preceeding paragraphs, the interaction between a *Rhizobium* and its host legume, as with other microbe-host interactions, involves bacterial cell surface and, therefore, cell surface polysaccharides. These polysaccharides include extracellular, capsular, and lipopolysaccharides (EPS, CPS, and LPS), cyclic glucans, and LCOs. A recent review (Gibson et al., 2008) describes the roles of EPS, CPS, cyclic glucans, LCOs, and some aspects of LPS in *Rhizobium*-legume symbiosis. Another recent article reviews the function of the LCOs and the molecular basis for the host response to these molecules (Oldroyd and Downie, 2008). The purpose of this chapter is to describe what is known regarding the structures and biological functions of rhizobial LPS. This chapter will describe LPS structures and structural features which are unique to various rhizobial members of the *Rhizobiales* and may be required for symbiosis, structural alterations that occur to the LPS during symbiotic infection, LPS genetics and biosynthesis, and LPS function in symbiosis.

The LPSs from rhizobial members of the *Rhizobiales* consist of the same general structural architecture as LPSs from enteric and animal Gram-negative pathogens. This architecture consists of three structural regions; an O-chain polysaccharide (OPS) that is attached to a core oligosaccharide (COS) which is attached to an acylated saccharide known as the lipid A (LA). Early work showed that an intact LPS is essential for symbiosis in both determinate and indeterminate nodule-forming legume hosts. Mutants that lacked the OPS were defective in determinate nodule-forming hosts in that the infection threads were aborted (Noel et al., 1986), and defective in indeterminate nodule-forming hosts in the endocytotic invasion of the root nodule cells (Brewin et al., 1993; Perotto et al., 1991, 1994). Subsequent work, using LPS monoclonal antibodies (mAbs) showed that structural epitope changes occurred to the OPS region of the LPS during symbiosis (Kannenberg et al., 1994; Tao et al., 1992), and, in some cases, an alternate OPS was produced when the *Rhizobium* was cultured in the presence of *nod* gene-inducing flavonoids (Reuhs et al., 2005). It was also shown that the LA was altered in structure during symbiosis (D'Haeze et al., 2007; Kannenberg and Carlson, 2001). It is apparent from these and other studies that OPS is necessary to form a normal nitrogen-fixing symbiosis, and that structural changes can occur primarily to the OPS and to the LA portion of the LPS during the transition from a vegetative bacterium to the nitrogen-fixing bacteroid. The details of these structural alterations, their genetic basis, and function are discussed in this chapter.

16.2 Rhizobial Lipopolysaccharide Structures

The structures of LPS from rhizobial members of the *Rhizobiales* have recently been reviewed by De Castro et al. (2008). There are only a few cases in which the OPS, COS, and LA structures have all been determined for a single strain and, thereby, allow one to assemble the entire LPS structure for that strain; e.g. *R. etli* CE3, *R. leguminosarum* biovar *viciae* 3841, and *R. leguminosarum* biovar *viciae* 128C53. In other cases either the OPS, COS, or LA, but not all three LPS components, have been structurally determined for a single strain. The LPSs from rhizobia have, for the most part, been extracted using hot phenol/water (Westphal and Jann, 1965) and purified by gel-filtration chromatography in the presence of deoxycholate (DOC) which allows the separation of LPS which contains the OPS (called LPS I or "smooth" LPS) from LPS that lacks the OPS (called LPS II or "rough" LPS) (Carlson et al., 1978). Other protocols include extraction of the cell pellets with a triethylamine-EDTA solution followed by polymyxin-B-Sepharose affinity chromatography (Forsberg and Carlson, 1998; Ridley et al., 2000). In this section, the chemical properties of the LA, COS, and OPS from various *Rhizobiales* LPSs are described.

16.2.1 Lipid A (LA) Structures

Rhizobial members of the *Rhizobiales* have been shown to have various LA structures that differ in significant ways from those of enteric bacteria as well as from each other. Structures have been determined for the LA from strains of *R. leguminosarum* (Bhat et al., 1994; Kannenberg et al., 1998; Que et al., 2000a, b), *R. etli* (Bhat et al., 1994; Kannenberg et al., 1998; Que et al., 2000a, b), *R.* sp. Sin-1 (Jeyaretnam et al., 2002), *R. galegae* (Jeyaretnam University of Georgia PhD thesis, 1998), *Sinorhizobium* species (Ferguson et al., 2005, 2006; Gudlavalleti and Forsberg, 2003), *M. huakuii* (Choma and Sowinski, 2004), and the phytopathogen member of the *Rhizobiales*, *A. tumefaciens* (Silipo et al., 2004). Other LAs have been examined by composition and mass spectrometric analysis.

Structural comparison reveals that the various rhizobial LAs contain both common and variable structural features. In addition, each LA preparation contains a mixture of structures varying in the number and length of their fatty acyl components as well as in their carbohydrate backbone. A common structural feature in all LAs is the very long chain fatty acid (VLCFA), 27-hydroxyoctacosanoic acid (27-OHC28:0), and small amounts of 29-OHC30:0. This was found during the composition analysis of LPSs from every member of the *Rhizobiales* examined, with one exception thus far; the LPS from *A. caulinodans*, which nodulates the stems of *Sesbania*, does not have detectable amounts of VLCFA (Bhat et al., 1991). LA with VLCFA is also found in a number of pathogens that form chronic intracellular infections. These pathogens include *Brucella abortus* (Bhat et al., 1991), *Brucella melitensis* (Basu et al., 2002), *Bartonella hensalae* (Bhat et al., 1991), and *Legionella pneumophila* (Zahringer et al., 1995).

The rhizobial LA can be grouped into four different structural types (Fig. 16.1). *Sinorhizobium meliloti*, *S.* sp. NGR234, and *A. tumefaciens* share the same LA structure, 1, which consists of a bis-phosphorylated glucosamine (GlcN) disaccharide backbone that is pentaacylated with the VLCFA as the single acyloxyacyl residue. There is structural variability in that some molecules in an LA preparation may lack a phosphate group, the β-hydroxybutyryl (3-OHC4:0) moiety, or the fatty acyl residue attached to position 3 of the proximal GlcN residue. In addition, there is variation in the chain length of both the ester and amide-linked fatty acyl residues. The various forms of structure 1 have a molecular weight range from 2010 to 2060. LA structure 2 is shared among *R. leguminosarum* biovars *viciae*, *trifolii*, and *etli*, and *R. etli* strains. This structure is completely devoid of phosphate and the proximal residue can consist of 2-aminogluconate (GlcNonate) rather than GlcN while the

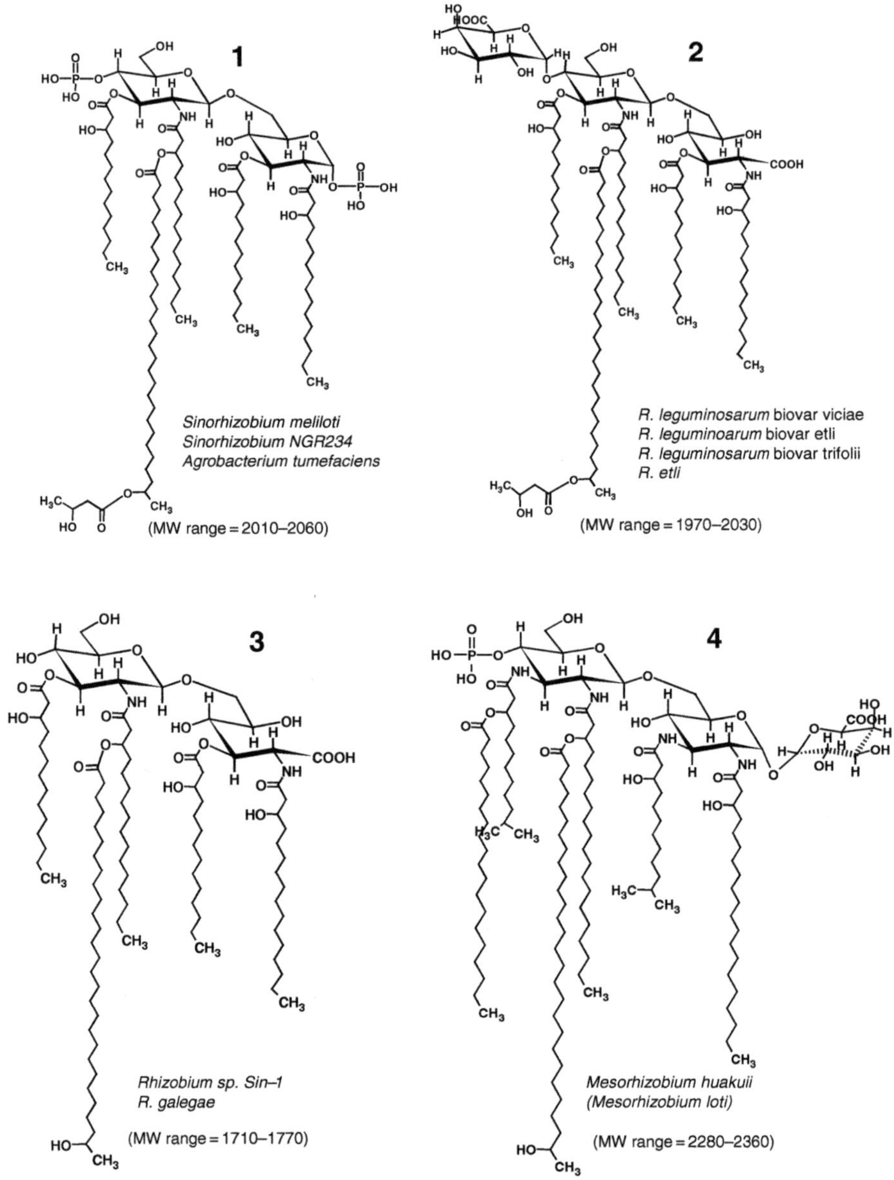

Fig. 16.1 The structures of LA molecules from the indicated members of the *Rhizobiales*. It should be noted that each LA preparation consists of a microhetergeneous mixture of molecules with variation in their *N*-fatty acyl chain lengths as well as each preparation having molecules lacking a β-hydroxymyristyl residue from the structures shown. Other minor structural variations are as described in the text

distal GlcN residue is substituted with a α-D-galacturonosyl (α-D-GalA) residue at the 4′-position. It is also pentaacylated with the VLCFA as the single acyloxyacyl residue and its fatty acyl ation varies as described above for structure 1. In addition, structural variation occurs due to the fact that the proximal glycosyl residue can exist in several forms; as GlcNonate, 2-aminogluconolactone (GlcNono-lactone), or as GlcN. The molecular weight range of thc various forms of structure 2 is 1970–2030. The third LA structure, 3, is shared by *R.* sp. Sin-1 and by *R. galegae*. This structure is essentially the same as structure 2 with the exception that it lacks the GalA substitution of the distal GlcN residue, and the proximal glycosyl residue exists exclusively as GlcNonate (or its lactone). It also lacks the 3-OHC4:0 moiety esterified to the VLCFA 27-OH group. The molecular weight range of structure 3 is from 1710 to 1770. Structure 4 was reported for *M. huakuii* (Choma and Sowinski, 2004). This structure is different from the other structures in that its glycosyl backbone consists of a 2,3-diaminoglucose (GlcN3N) disaccharide which is phosphorylated at the 4′-position, and contains a α-D-GalA glycosidically linked to the 1-position of the proximal GlcN3N residue. As with the other LA structures it contains the VLCFA as an acyloxyacyl residue, but also contains a second acyloxyacyl residue on the 3′-position of the LA backbone, and is, therefore, hexa – rather than pentaacylated. This second acyloxyacyl residue has an extended carbon chain, C20:0. The amide fatty acyl residue at the 2- and 2′-positions of the LA backbone also has an extended carbon chain; it is 3-OHC20:0. In addition, branched 3-OHC13:0 fatty acyl residues are amide linked at positions 3 and 3′ on the LA backbone. Composition and mass spectrometry (MS) analysis shows that the LA of *M. loti* has a very similar fatty acyl composition and MS profile to those of the *M. huakuii* LA (Casabuono et al., 2006), and, therefore, it likely has a vary similar structure. The various forms of structure 4 have a molecular weight range of 2280–2360. While the LA from other members of the *Rhizobiales* have not been fully structurally characterized, partial analysis shows that *Bradyrhizobium* species also have an LA that contain GlcN3N (i.e. *B. japonicum*) or a combination of GlcN3N and GlcN (i.e. *B.* sp. *Lupinus*) (Bhat et al., 1991; Mayer et al., 1989). Recent analysis of the LA from *B. japonicum* USDA I110 showed that the LA contains GlcN3N and mannose (Man) (Muszynski and Carlson, unpublished) likely indicating the presence of an acylated GlcN3N disaccharide that is mannosylated. Mass spectrometric analysis of this LA indicated structural heterogeneity as was observed in other rhizobial LA preparation. The largest molecular weight forms of *B. japonicum* LA were between 3070 and 3140 making these the largest LA molecules observed in any LPS from the *Rhizobiales*. The larger molecular weight of this LA raises the possibility that it is more extensively acylated than the LA structures 1–4. Further structural analysis of this LA is in progress.

16.2.2 Core Oligosaccharide (COS) Structures

The compositions, partial structures, and complete structures for a number of COSs have been determined for the LPSs from members of the *Rhizobiales* (*R. leguminosarum*, *R. etli*, several *Agrobacterium* species, *Sinorhizobium*, *Mesorhizobium*, and *Bradyrhizobium*) (Table 16.1).

Table 16.1 Core oligosaccharides found in the lipopolysacccharides from members of the *Rhizobiales*

Strain	Structure or Composition	Reference
R. leguminosarum biovars viciae trifolii etli *R. etli*	Kdo-(2→6)-α-D-Gal-(1→6)-α-D-Man-(1→5)-Kdo α-D-GalA-(1→4)┘ Kdo-(2→4)┘ α-D-GalA-(1→4)┘ α-D-GalA-(1→5)┘	(Forsberg and Carlson, 1998, Kannenberg, *et al.*, 1998)
A. tumefaciens A1, DSM30150	α-D-Man-(1→6)-α-D-Man-(1→5)-Kdo β-D-Gal-(1→8)-Kdo-(2→4)┘	(De Castro, *et al.*, 2006)
TT111, DSM30204	α-L-Rha-(1→2)-α-L-Rha-(1→2)-α-L-Rha-(1→2)--α-L-Rha-(1→3)-α-L-Rha-(1→3)-α-D-Man-(1→5)-Kdo β-D-GlcN-(1→2)-β-D-Gal-(1→4)-β-D-Gal-(1→8)-Kdo-(2→4)┘	(De Castro, *et al.*, 2006)
A. radiobacter Rv3, DSM30207	α-D-Man-(1→2)-α-D-Man-(1→3)-α-D-Man-(1→3)-α-D-Man-(1→5)-Kdo α-D-Man-(1→6)┘ β-D-Gal-(1→8)-Kdo-(2→4)┘	(Gargiulo, *et al.*, 2008)
A. larrymoorey	β-D-Qui3NAcyl-(1→2)-α-L-Rha-(1→4)-α-D-GalAN-(1→6)-α-D-Gal-(1→5)-Kdo β-D-GlcAN-(1→3)-β-D-Glc-(1→4)┘ Kdo-(2→4)┘	(Molinaro, *et al.*, 2003)
A. rubi^{T}, DSM6772	α-L-Fuc-(1→4)-α-D-GlcN-(1→4)-β-D-GlcA-(1→4)-α-D-GalA-(1→3)-β-D-Glc-(1→6)- 3-OAc-α-D-Glc-(1→5)-Kdo 3/4-OAc-β-D-Gal-(1→3)┘ α-Dha-(2→3)-α-D-GalA-(1→3)-β-D-Glc-(1→4)┘ α-L-Fuc-(1→2)┘ α-D-GalA-(1→7)-Kdo-(2→4)┘	(Gargiulo, *et al.*, 2008)
Sinorhizobium meliloti 102F51	[1]Part. Struct. →4)-Dha-(2→ ; β-D-GlcN-(1→4)-Kdo; α-D-GlcA-(1→4)-Kdo; α-D-GalA-(1→3)-α-D-Glc-(1→5)-Kdo; →6)┘ →4)┘	(Russa, *et al.*, 1996)
Rm1021	[2]Composition: Gal/Glc/Man/GalA/Kdo = 0.50/5.00/0.25/2.50/3.00	(Reuhs, *et al.*, 1998)
Sinorhizobium fredii USDA205	[2]Composition: Gal/Glc/Man/GalA/Kdo = 1.25/5.00/0.75/1.25/2.50	(Reuhs, *et al.*, 1998)
USDA257	[2]Composition: Gal/Glc/Man/GalA/Kdo = 0.25/5.00/0.75/2.50/3.00	(Reuhs, *et al.*, 1998)
Sinorhizobium sp. NGR234	[2]Composition: Gal/Glc/Man/GalA/Kdo = 0.50/5.00/0.25/1.50/3.00	(Reuhs, *et al.*, 1998)

Table 16.1 (continued)

Strain	Structure or Composition		Reference
Mesorhizobium loti			
NZP2213.1	[3]Composition:	Glc/GlcA/Gal/GalA/GlcN/Hep/Kdo = 4.00/0.50/1.10/0.10/1.00/4.20/1.30	(D'Antuono, *et al.*, 2005, Turska-Szewczuk, *et al.*, 2009)
NZP2235	[4]Composition:	Glc/Gal/Hep/Kdo = 1.25/0.32/0.31/1.00	(Townsend, *et al.*, 2006)
MAFF30399	[5]Composition:	Neutral sugars: Glc/Gal = 1.0/0.9; Acidic sugars: GalA/X/Kdo = 1.2/0.6/1.0	(D'Antuono, *et al.*, 2005)
Mesorhizobium huakuii IFO15243^T	[6]Composition:	Rha/Man/Glc/Gal/GlcN/QuiN/Hep/GalA/GlcA/Kdo = 0.14/0.02/0.27/0.33/0.02/0.03/0.11/0.05/0.03/+	(Choma, 2002)
Bradyrhizobium japonicum			
61A101c	Partial Structures:	α-D-4OMeMan-(1→5)-Kdo; α-D-Man-(1→4)-α-D-Glc-(1→4)-Kdo	(Carlson and Krishnaiah, 1992)
61A123	[7]Composition:	4-OMeMan/Man/Glc/Kdo = 0.13/0.46/0.44/+	(Carrion, *et al.*, 1990)
HS123 (a rough mutant of strain I110)	[8]Composition	Man/Glc/GlcN/Kdo = 0.66/0.26/0.08/+	(Puvanesarajah, *et al.*, 1987)

[1]Partial structures included (*inter alia*) 4-linked Dha (deoxyheptulosaric acid); α-D-GlcA-(1→4)-Kdo-(2→ was identified as a terminal (non-reducing) sequence.
[2]Compositions are reported (Reuhs, *et al.*, 1998) as relative mole ratios normalized to Glc. [3]Compositon of R-LPS was reported (Turska-Szewczuk, *et al.*, 2009) as the relative mole ratio. [4]Composition reported (Townsend, *et al.*, 2006) as mole ratio of major glycosyl components in R-LPS chromatographic fraction.
[5]Composition reported (D'Antuono, *et al.*, 2005) as the mole ratio of neutral sugars and relative mole ratio of acidic sugars; and X is present as an unidentified component. [6]This composition is the relative mole fraction of the total estimated core sugars reported (Choma, 2002) for this LPS. [7]This composition is calculated from the values given by Carrion et al (Carrion, *et al.*, 1990), and are the relative mole fraction; a numerical value for Kdo. [8]The composition is the calculated mole fraction from values reported by Puvanesarajah et al. (Puvanesarajah, *et al.*, 1987).

The LPSs from strains of *R. leguminosarum* biovars *viciae*, *trifolii*, and *etli*, and for strains of *R. etli* have the same COS structure. It consists of an octasaccharide of which four glycosyl residues are GalA. Unlike the core region from numerous animal pathogens which have a heptosyl (Hep) residue attached to the 3-deoxy-D-manno-2-octulosonic acid (Kdo) residue that links the COS to the LA, this rhizobial COS structure is devoid of Hep and has a Man residue attached to this Kdo. Another unusual feature is that this COS structure has a third external Kdo residue and, as will be described below, the OPS is attached to this Kdo. The presence of three Kdo residues together with three GalA residues makes this structure quite acidic. The genes and the enzymes involved in the synthesis of this structure have all been identified and are discussed further in the following section of this chapter.

The structures of the COSs have been determined for a number of *Agrobacterium* strains. Unlike the COS structure from *R. leguminosarum* and *R. etli*, these structures vary among different *Agrobacterium* species and strains. However, as with the *R. leguminosarum* and *R. etli* COS structure, the *Agrobacterium* COS structures also are devoid of Hep residues. Three *Agrobacterium* COS structures have the same Man_1Kdo_2 trisaccharide inner structure as found in the *R. leguminosarum* and *R. etli* COS structure; i.e. Man is attached to the Kdo that links the COS to the LA. In *A. larrymoory* COS, this Man is replaced by Gal, and while in *A. rubi* this residue is Glc. Of these *Agrobacterium* COS structures only that of *A. rubi* DSM6772 has acidic glycosyl residues (other than Kdo); two GlcA and one GalA residues. One COS structure, from *A. larrymoorey*, has glycosyluronamide residues. Three structures from *A. tumefaciens* and *A. radiobacter* are enriched in glycosyl residues having a *manno*-configuration; i.e. either rhamnose (Rha) or Man.

In the case of *Sinorhizobium* species, only compositions from several strains and partial COS structures from one strain, *S. meliloti* 102F51, have been reported. As previous mentioned bacterial LPS preparations often consist of two types of LPS molecules, those with the OPS called LPS I and those without OPS called LPS II. For many LPSs from *Sinorhizobium* strains the major form of LPS is LPS II, composed of COS and LA, which also can be the dominant antigen. All of the *Sinorhizobium* LPS IIs contain GalA and Kdo and lack Hep residues. The compositions of the COS from *S. meliloti* Rm1021, *S. fredii* USDA205 and USDA257, and *S.* sp. NGR234 are all very similar containing Gal, Glc, Man, GalA, and Kdo. Even though complete structures from *Sinorhizobium* COSs have not been determined, immunochemical studies have shown that there are several structural groups (Kannenberg et al., 1998; Reuhs et al., 1998). For example, polyclonal antiserum to *S. fredii* USDA 205 reacts with the LPS II from *S. fredii* USDA 191, USDA 205, USDA 208, and *S.* sp. NGR 234 but not with the LPS II from 16 other *S. meliloti* and *S. fredii* strains (Reuhs et al., 1998). However polyclonal antiserum to *S. meliloti* Rm41 reacts with the LPS II from *S. meliloti* AK631, Rm1021, NRG133, NRG247, NRG286, NRG23, NRG53, *S. fredii* USDA192, USDA196, USDA197, USDA257, USDA201, and HH103 (Reuhs et al., 1998). Thus, as indicated by the composition analyses shown in Table 16.1, these immunochemical results suggest that the COSs from *Sinorhizobium* LPSs contain similar, but not identical structures that can be grouped into 3 or 4 related structural types (Kannenberg et al., 1998; Reuhs et al., 1998). Similar results were obtained using a series of monoclonal antibodies prepared against various *S. meliloti* strains (Reuhs et al., 1999). It is also reported that *Sinorhizobium* LPSs are sulfated (Cronan and Keating, 2004; Keating, 2007; Keating et al., 2002; Townsend et al., 2006). Therefore, in addition to the acidic uronosyl residues in their COSs, the presence of sulfate can also add to the negative charge of these LPSs.

As with *Sinorhizobium*, only the compositions of a few COSs of *B. japonicum* LPSs, and partial structures of one COS have been reported (Carlson and Krishnaiah, 1992; Carrion et al., 1990; Puvanesarajah et al., 1987) and are given in Table 16.1. During the initial hot phenol/water extraction of various *B. japonicum* strains, it was found that the LPSs were largely or exclusively extracted into the phenol phase rather that into the water phase as was the case for most other LPSs from the *Rhizobiales* (Carrion et al., 1990). The hydrophobic character of these LPSs may be due to a possible higher degree of fatty acylation in the LA as described above, as well as an OPS that contains predominately deoxyhexosyl residues and appears to be highly O-acetylated (Carrion et al., 1990); the OPSs are discussed further below. The compositions and partial structures of *B. japonicum* COSs were determined from the LPS II of two strains (61A101c and 61A123) which largely, but not completely, lack the OPS, and from the LPS from one rough mutant of strain *B. japonicum* I110 (HS123), which completely lacks the OPS. All of these COSs contain Man, Glc, and Kdo and the isolated LPS II preparations contain small amounts of OPS as well as small amounts of 4-OMeMan. However, the LPS from the rough mutant, HS123, lacks 4-OMeMan. The presence of 4-OMeMan only in structures which contain OPS indicates that this may be the core residue to which the OPS is linked, as suggested by Carlson and Krishnaniah (Carlson and Krishnaiah, 1992). From the partial structures of the COS for strain 61A101c, one can hypothesize that the COS of *B. japonicum* LPSs has the structure:

$$\begin{array}{r}[\text{OPS}\rightarrow]\text{-}\alpha\text{-D-4OMeMan-}(1\rightarrow5)\text{-Kdo}\\ \alpha\text{-D-Man-}(1\rightarrow4)\text{-}\alpha\text{-D-Glc-}(1\rightarrow4)\text{-Kdo-}(2\rightarrow4)\text{-}\lrcorner\end{array}$$

Further structural work is required to determine if this proposed COS structure is correct. However, if correct, then these *B. japonicum* LPSs have a similar Man_1Kdo_2 inner core structural arrangement as described above for the *R. leguminosarum*, *R. etli*, and several *Agrobacterium* LPSs.

16.2.3 O-Chain Polysaccharide (OPS) Structures

Early on in the analysis of LPS from members of the *Rhizobiales*, it was shown that there was (a) significant variability in the OPS even among strains of the same species (Carlson, 1984; Carlson et al., 1978), and (b) that major glycosyl residues commonly found in these OPSs consisted of deoxyhexoses as well as methylated glycosyl residues (Carlson, 1984; Carlson et al., 1978). It was also shown that mutants from numerous members of the *Rhizobiales* that lacked OPS were defective in symbiosis (Carlson et al., 1987; de Maagd et al., 1989; Kannenberg et al., 1992; López-Lara et al., 1995; Noel et al., 1984, 1986; Perotto et al., 1994; Stacey et al., 1991; Turska-Szewczuk et al., 2009; van de Wiel et al., 1990; VandenBosch et al., 1985) and, therefore, it became important to investigate the function of these polysaccharides in symbiosis. Thus, the structures of a number of OPS from rhizobial LPSs have been reported and are shown in Table 16.2. Details of the possible functions of OPSs in symbiosis are described in Section 16.5 of this chapter.

The OPS structures shown in Table 16.2 reveal that the OPSs from *Rhizobium, Mesorhizobium, Sinorhizobium,* and *Agrobacterium* species are enriched in deoxyhexoses as well as methylated deoxyhexoses; either Rha, fucose (Fuc), or 6-deoxytalose (6dTal). Glycosyl composition analysis has also shown that this is the case for *Bradyrhizobium* species. *Bradyrhizobium japonicum* 61A123 LPS has an OPS that contains Fuc, *N*-acetylfucosamine (FucNAc), and Man as the major residues with smaller amounts of 4-OMeMan and Glc (Carrion et al., 1990). Similar components were observed in the LPS of *B. japonicum* I110 with the additional residues of xylose (Xyl) and arabinose (Ara) (Puvanesarajah et al., 1987), while *B. japonicum* 61A101c contained 2,3-di-O-MeMan, Man, FucNAc, and *N*-acetylquinovosamine (QuiNAc) as the major residues (Stacey et al., 1991). These *Bradyrhizobium* OPSs also can be extensively O-acetylated (Carrion et al., 1990). Very few OPSs contain charged glycosyl residues or substituent groups and, therefore, due to the dominance of relatively hydrophobic glycosyl residues the OPSs confer a relatively hydrophobic character to the cell surface. In a few instances charged glycosyl residues are present such as 3-deoxy-2-heptulosaric acid (Dha) in *R. leguminosarum* bv. *trifolii* 24 OPS (Russa et al., 1996), GlcA in the case of *R. etli* CE3 OPS (Forsberg et al., 2000), and Glc3NAmA in *R. leguminosarum* bv. *viciae* 3841 OPS (Forsberg and Carlson, 2008). However, charged glycosyl residues often neutralized; e.g. the GlcA is methyl esterified in *R. etli* CE3 OPS (Forsberg et al., 2000), and the Glc3NAmA residue exists as a neutral zwitterion in *R. leguminosarum* bv. *viciae* 3841 OPS (Forsberg and Carlson, 2008). Thus, the anionic hydrophilic portion of the LPS is located nearest the lipid bilayer surface of the bacterium due to the Kdo, GalA, and/or GlcA in the core region as well as the acidic glycosyl residues and/or phosphate groups on the backbone of the LA, and this hydrophilic area of the LPS is blanketed on either side by the more hydrophobic OPS and the fatty acyl residues of the LA; see Fig. 16.2. Exceptions to this hydrophobic-hydrophilic-hydrophobic arrangement may be the LPS from several of the *Agrobacterium* and *Bradyrhizobium* strains where the COSs do not contain charged glycosyl residues other than Kdo.

In several instances, it was found that a single strain could produce more than one OPS structure. This is the case for *A. tumefaciens* F/1 (De Castro et al., 2004) and *A. radiobacter* DSM30147 (De Castro et al., 2004) and also for two *Sinorhizobium* strains; *S.* sp. NGR234 (Reuhs et al., 2005), and *S. fredii* USDA205 (Reuhs et al., 1994). Many *Sinorhizobium* strains produce LPS that is either devoid of OPS, or produce a low level of OPS-containing LPS. This is true for *S.* sp. NGR234 when

Table 16.2 O-Antigen polysaccharides found in the lipopolysaccharides from members of the *Rhizobiales*

Strain	Structure	Ref
R. leguminosarum biovar trifolii 4S	3)-α-L-Rha-(1→3)-α-L-Rha-(1→4)-β-D-GlcNAc-(1→3)-α-L-Rha-(1→ α-D-ManNAc-(1→2)┘	(Wang and Hollingsworth, 1994)
biovar trifolii 24	3)-α-L-6dTal-(1→2)-α-L-Rha-(1→5)-Dha-(2→	(Russa, *et al.*, 1996)
[1]biovar viciae 128C53	3)-α-L-Rha-(1→3)-α-L-Fuc-(1→3)-α-L-Fuc-(1→ α-D-Man-(1→2)┘	(Kannenberg, *et al.*, 1998)
[2]biovar viciae 3841	R-(1→4)-[β-D-Glc3NAmA-(1→4)-[2-OAc]$_{0.2}$α-L-Fuc-(1→3)-α-L-QuiNAc-(1→4)]$_{n}$-β-D-Glc3NAmA-(1→ 2-OAc-3-OMe-α-D-6dTal-(1→3)┘	(Forsberg and Carlson, 2008)
[3]*R. etli* CE3	2,3,4-tri-OMe-α-L-Fuc-(1-[→4)-α-D-GlcAMe-(1→4)-α-L-Fuc-(1→]$_{n}$-3)-α-L-Fuc-(1→3)-β-D-Man-(1→3)-β-L-QuiNAc-(1→ 2-OAc-3-OMe-α-L-6dTal-(1→3)┘	(Forsberg, *et al.*, 2000)
R. tropici CIAT899	4)-β-D-Glc-(1→3)-2-OAc-α-D-6dTal-(1→3)-α-L-Fuc-(1→	(Gil-Serrano, *et al.*, 1995)
Mesorhiozbium loti NZP2213	3)-2-OAc-α-L-6dTal-(1→	(Russa, *et al.*, 1995)
Mlo-13 (NZP2213 mutant)	2)-α-L-6dTal-(1→3)-4-OAc-α-L-6dTal-(1→2)-[3-OMe]$_{0.25}$-α-L-Rha-(1→	(Turska-Szewczuk, *et al.*, 2008)
NZP2213.1 (NZP2213 mutant)	3)-2-OAc-α-L-6dTal-(1→3)-[2-OMe]$_{0.50}$-α-L-6dTal-(1→3)-α-L-6dTal-(1→3)-α-L-6dTal-(1→	(Turska-Szewczuk, *et al.*, 2008)
Mesorhiozbium huakuii IFO15243T	2)-α-L-6dTal-(1→3)-α-L-6dTal-(1→2)-α-L-Rha-(1→	(Choma, *et al.*, 2000)
Sinorhizobium [4]NGR 234	3)-α-L-Rha-(1→3)-α-L-Rha-(1→2)-[3-OMe]$_{0.20}$-α-L-Rha-(1→	(Reuhs, *et al.*, 2005)
Sinorhizobium fredii SMH12	4)-α-D-GalA-(1→3)-2-OAc-α-L-Rha-(1→3)-2-OAc-[6-OMe]$_{0.50}$-α-D-Man-(1→	(Fernandez de Cordoba, *et al.*, 2008)

Table 16.2 (continued)

Strain	Structure	Ref
Agrobacterium tumefaciens		
[5]C58 (DSM 5178)	3)-4-OAc-α-L-6dTal-(1→	(De Castro, *et al.*, 2003)
B6 (DMS 30205)	3)-α-D-Ara*f*-(1→3)-α-L-Fuc-(1→	(De Castro, *et al.*, 2002)
F/1 (DSM 30206)	3)-α-L-Rha-(1→3)-β-D-GlcNAc-(1→ and 4)-α-L-Rha-(1→3)-β-D-GlcNAc-(1→	(De Castro, *et al.*, 2004)
Agrobacterium radiobacter		
M2/1 (DSM 30199)	3)-α-L,D-Hep-(1→	(De Castro, *et al.*, 2004)
DSM 30147	2)-α-L-Rha-(1→3)-α-L-Rha-(1→3)-α-L-Rha-(1→2)-α-L-Rha-(1→ and 3)-α-D-Fuc-(1→3)-α-D-Fuc-(1→ α-D-Man-(1→2)┘	(De Castro, *et al.*, 2004)

[1]This OPS varies greatly in the number of this repeating unit structure. The isolated intact OPS also has Kdo at the reducing end which is the external Kdo residue shown in the COS structure given in Table 1.
[2]This OPS consists of three to four repeats of this tetrasaccharide (n = 3 to 4). The proximal repeat unit is linked to a Glc3NAmA residue (3-acetimidoylamino-3-deoxy-D-gluco-hexuronic acid) which is linked to O4 of the COS external Kdo residue; the distal repeat unit is capped at O4-by a 4,6-dideoxy-4-formamido hexopyranose residue (R-) of unidentified configuration. The 3-OMe-6dTal residues are non-stoichiometrically methylated at O4 (approx. 20%).
[3]This OPS consists of five repeating trisaccharide units (n = 5) in which the proximal repeat unit is linked via a non-repeating trisaccharide sequence (-Fuc-Man-QuiNAc-) that, in turn, is linked to O4 of the external Kdo of the COS, and the distal repeat unit is capped by a 2,3,4-tri-O-MeFuc residue. In addition, the Fuc residues are non-stoichiometrically methylated at O2.
[4]This rhamnan OPS is induced when the bacterium is cultured in the presence of flavonoids (molecules produced by the host legume that induce expression of the rhizobial nod genes) (Reuhs, et al., 2005), and a similar rhamnan was detected in bacteroids obtained from the host legume root nodule (Fraysse, et al., 2002, Jabbouri, et al., 1996).
[5]Acetylation at O4 is in non-stoichiometric amounts.

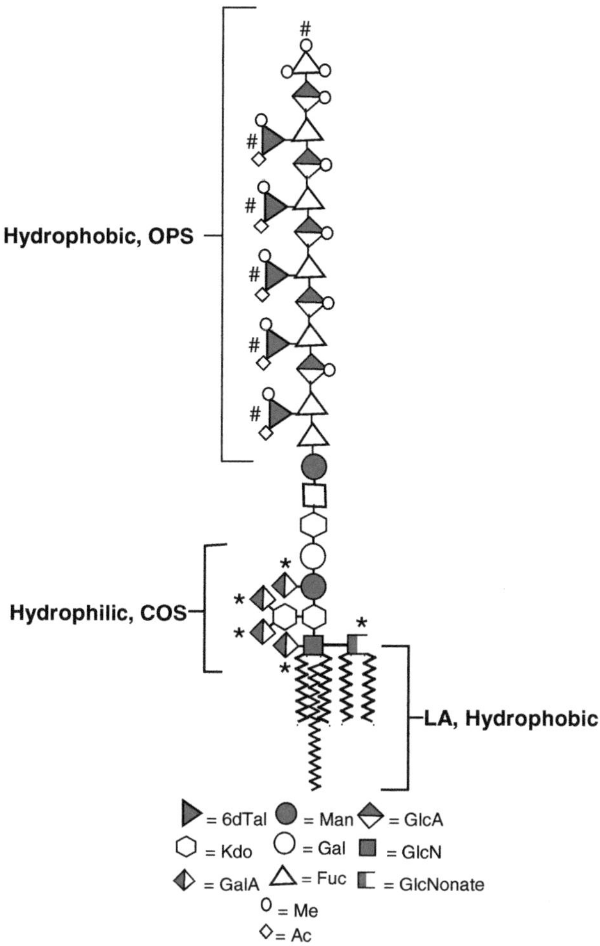

Fig. 16.2 A schematic diagram showing the structure of the LPS from *R. etli* CE3. The various glycosyl residues are as indicated. Those marked with an * are the acidic glycosyl residues, while those marked with # are relatively hydrophobic (e.g. 6-dexoxy hexosyl residue). The hydrophobic and hydrophilic areas of this LPS are indicated

cultured under normal laboratory conditions; however, when the flavone, apigenin, which induces *nod* gene expression is added to the culture a LPS containing a rhamnan OPS (structure is shown in Table 16.2) is produced (Reuhs et al., 2005). In the case of *S. fredii* USDA205, the OPS appears to be a glucan under normal laboratory culture conditions; however, when apigenin or host root exudate is added to the culture there is a significant increase in LPS that contains an OPS with Glc, Xyl, and Man (Reuhs et al., 1994). Thus, at least in these two cases, there exists the ability to produce conditionally-dependent OPS structures. These conditionally-dependent OPSs and OPS modifications are discussed further below.

16.3 Rhizobial Lipopolysaccharide Biosynthesis

Phylogenetic analysis of 16S rDNA placed *Rhizobium* and related bacteria to the *Alphaproteobacteria* and into the order *Rhizobiales* (Sawada et al., 2003; Sprent, 2007). As with other bacteria, the genetic analysis of this group of bacteria has benefited greatly from large scale characterization of numerous genome sequences from strains of the *Alphaproteobacteria.* According to the NCBI data base (as of July 2009) 110 *Alphaproteobacteria* genomes have been sequenced and 54 of those are from members of the *Rhizobiales*. An additional 143 *Alphaproteobacteria* genome sequencing projects are in progress and include another 54 *Rhizobiales* genomes. Of the *Rhizobiales* genomes, there are currently 18 fully characterized symbiont and 17 pathogen genomes. This large amount of information enables a great deal of comparative genomic and functional studies; e.g. transcriptomics, proteomics, glycomics, etc.; including a comparison of genes required for the synthesis LPS.

16.3.1 LA Biosynthesis

Based on the current state of structure analysis, rhizobial LAs seem to group into four basic structural types as (described in the previous section, Fig. 16.1). These structures have unusual features compared to the LA from enteric bacteria. A scheme comparing the biosynthesis of several rhizobial LAs is shown in Fig. 16.3, and the identification of the genes encoding for LA synthesis enzymes are shown in Table 16.3. The first studies on the synthesis of rhizobial LA were done using *R. leguminosarum* and *R. etli* strains since this was the first rhizobial LA structure (structure 2 in Fig. 16.1) that had been determined and it was very unusual compared to that of *E. coli*. Price et al. (Price et al., 1994) observed that *R. leguminosarum* extracts contained the same enzyme activities as those found in *E. coli* that convert UDP-GlcNAc into the LA precursor, Kdo_2lipid-IV_A; i.e. LpxACDHBK and KdtA as shown in Fig. 16.3. Using the genome sequences for these strains, the genes encoding these various enzymes could also be readily identified in the genome of *R. leguminosarum* biovar *viciae* 3841 and *R. etli* CFN42 (strain CE3, which we refer to numerous times in this chapter, is a streptomycin-resistant form of CFN42), and orthologs were found in all members of the *Rhizobiales* for which the genome sequences are available. The only possible exception is *lpxH*. Genes with high homology to *lpxH* are missing. Instead, genes can be found with low to moderate homology to *lpxH* from *E. coli*; but these genes apparently code for phospholipase C enzyme (Zavaleta-Pastor et al., 2010) and do not seem to play a role in LA biosynthesis, raising the question as to what replaces LpxH in rhizobia. Additional phosphoesterases with low similarity to *E. coli lpxH* do exist in all rhizobia, but at this point it is not clear whether one of those functionally replaces *lpxH* in rhizobia. The acyl transferases, LpxA and LpxD, in *E. coli* are quite restricted to the transfer of a C14 chain length fatty acid, β-hydroxymyristic acid. Structural studies have indicated that these acyl

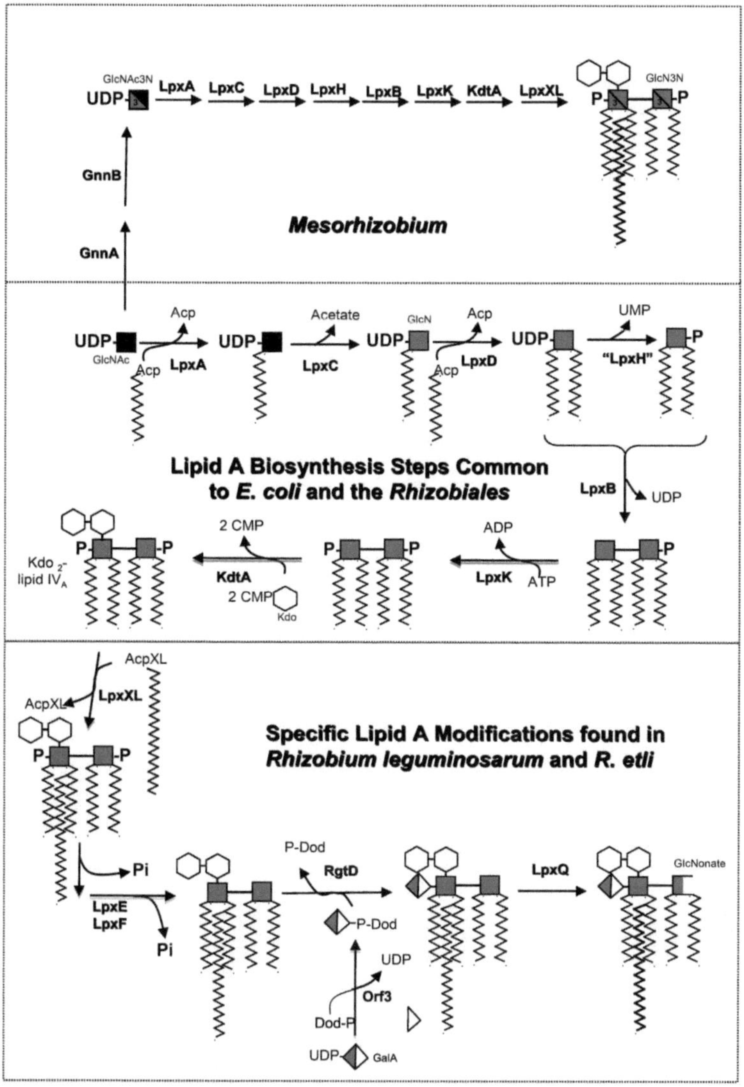

Fig. 16.3 A schematic diagram illustrating the difference in the biosynthesis of various rhizobial LA structures from one another, as well a showing those features shared with one another and with *E. coli*

Table 16.3 A list of genes required for the synthesis of various lipid A from the *Rhizobiales*

	Structure 1 in Fig. 16.1	Structure 1 in Fig. 16.1	Structure 1 in Fig. 16.1	Structure 4 in Fig. 16.1	[8]Structure 5	Structure 2 in Fig. 16.1	Structure 2 in Fig. 16.1	Structure 2 in Fig. 16.1
Gene	Sm1021	At C58	R. sp. NGR234	Ml303099	BjI110	Rlv3841	ReCFN42	Rt WSM2304
lpxA	SMc02091	Atu1384	NGR_c13440	mll0633	bll4849	RL2231	RHE_CH01923	Rleg2_1593
lpxC	SMc01875	Atu2085	NGR_c21000	mll1545	bll6595	RL3297	RHE_CH02839	Rleg2_2585
lpxD	SMc02093	Atu1382	NGR_c13420	mll0635	bll4852	RL2229	RHE_CH01921	Rleg2_1591
[1]*"lpxH"*								
lpxB	SMc02089	Atu1386	NGR_c13460	mll0630	bll4847	RL2233	RHE_CH01925	Rleg2_1595
lpxK	SMc00892	Atu0697	NGR_c04270	mlr8270	bll7514	RL0904	RHE_CH00846	Rleg2_0490
kdtA	SMc00894	Atu0695	NGR_c04250	mlr8268	bll7515	RL0902	RHE_CH00844	Rleg2_0488
acpXL	SMc04278	Atu1600	NGR_c18080	mlr1174	bsl3811	RL2817	RHE_CH02478	Rleg2_2117
lpxXL	SMc04268	Atu1594	NGR_c18030	mlr1179	bsl3807	RL2812	RHE_CH02473	Rleg2_2112
[2]*lpxE*						RL4708	RHE_CH04094	Rleg2_3901
[3]*lpxF*						RL1570	RHE_CH01455	Rleg2_1069
[4]*lpxQ*		Atu0668				RL0868	RHE_CH00812	Rleg2_0454
[5]*rgtD*						RL0684	RHE_CH00635	Rleg2_0294
[6]*Orf3*	SMc02383		NGR_c08000			RL1470	RHE_CH01319	Rleg2_1106
[7]*gnnA (gnnB)*				mll2825	bll5918			

[1]*"lpxH"*: Enzymatically, it has been shown that *Rhizobium* cell extracts contain LpxH activity. *Rhizobium* and related strains contain genes with low to moderate similarity to *lpxH* from *E. coli* (e.g. SMc00171, Atu1649, BLL5904, etc.); however, these genes have recently been characterized as encoding phospholipases C and reclassified as LpxH2 (Zavaleta-Pastor et al.). The genes given in the table are LpxH2 paralogous metalophosphoesterases within the genomes of the different strains. To date, there is no biochemical evidence that these genes display LpxH-like activity.

[2]*lpxE*: Complete orthologs with e^{-90} values or better are found on *R. leguminosarum* and *R. etli* genomes, full orthologs of e^{-75} and e^{-52} were observed in *A. radiobacter* K84 and in *A. vitis* S4, respectively. Other *Rhizobiales* genomes contained partial orthologs with varying degrees of similarity.

[3]*lpxF*: Complete orthologs with e^{-100} or better are found in *R. leguminosarum* and *R. etli* genomes. Partial ortologs of e^{-67} and e^{-40} were present in *A. radiobacter* K84 and in *A. vitis* S4, respectively.

[4]*lpxQ*: Complete ortologs of e^{-110} or better are present in *R. leguminosarum* and *R. etli* genomes. Orthologs are also found in *A. radiobacter* K84 and *A. tumefaciens* C58 at e^{-70} and e^{-72}, respectively; the former has the annotation of a glycolate oxidase.

[5]*rgtD*: Present only in *R. leguminosarum* and *R. etli* genomes.

[6]*orf3*: Other than shown, there are also orthologs in *A. radiobacter* K84 (e^{-148}), *A. vitis* S4 (e^{-109}), as well as other *R. leguminosarum* and *R. etli* strains. Partial orthologs are present in *M. loti* MAFF303099, At C58, and Bj USDA110 but at e-values no better than e^{-45}.

[7]*gnnA* and *gnnB*: The *gnnA* and *gnnB* genes encode for products that oxidize UDP-GlcNAc to produce the 3″-keto product of UDP-GlcNAc and transaminase this position, respectively, forming UDP-GlcNAc3N (Sweet et al., 2004a). This product is required for those LA structures that have GlcN3N in their glycosyl backbone; in the case of the *Rhizobiaceae* this would be the LA from *Bradyrhizobium* and *Mesorhizobium* species. Full length GnnA orthologs, e^{-13}, are found in *Bradyrhizobium* and *Mesorhizobium* only. However orthologs of GnnB, which is the transaminase, occur in numerous bacterial species due to the fact that this type of activity is involved in the synthesis of many aminosugars. Thus, as the Sweet et al. indicate, the presence of GnnA orthologs probably more accurately reflects those species that contain UDP-GlcNAc3N in the LA biosynthetic pathway (Sweet et al., 2004a).

[8]This structure is not shown in Fig. 16.1 as the complete LA structure for *B. japonicum* USDA 110 has not been reported. The *B. japonicum* LA also contains a Man residue indicating other modifications for which the genes and their products have not yet been identified. However, structural information is described in the text and this LA does contain 2,3-diaminoglucose as is found in the *M. loti* structure. The *M. loti* LA also has a GalA residue glycosidically linked to the reducing end of the proximal GlcN3N residue indicating that this strain must have an LpxE ortholog that removes the 1-phosphate and also a GalA transferase.

transferases can act as a chain length "ruler", thereby, restricting fatty acids to certain lengths (Wyckoff et al., 1998). The structures of the rhizobial LAs described in the previous section indicate that there can be variously sized fatty acyl chains on the LA backbone indicating that rhizobial LpxA and LpxD may not have the stringent fatty acyl chain length "ruler" restrictions that are found in these enzymes from *E. coli*. In spite of these slight differences from *E. coli*, genomc sequencing programs in *R. leguminosarum* and related rhizobia (Galibert et al., 2001; Gonzalez et al., 2006; Kaneko et al., 2000a, b; Kaneko et al., 2002a, b; Lee et al., 2008; Schmeisser et al., 2009; Wood et al., 2001; Young et al., 2006) have revealed and corroborated the widespread conservation of these genes required for the biosynthesis of Kdo$_2$lipid-IV$_A$ from UDP-GlcNAc (LpxACDHBK and KdtA) throughout the *Rhizobiales*.

In the case of *Mesorhizobium loti* and *Bradyrhizobium*, the LAs contain 2, 3-diamino-2,3-dideoxyglucose (GlcN3N) as part or all of the glycosyl backbone and, therefore, the synthesis of this LA structure requires some unique enzymes that convert UDP-GlcNAc into UDP-GlcNAc3N. These enzymes are GnnA and GnnB and the UDP-GlcNAc3N product is then N-acylated by the LpxA orthologs in *M. loti* or *B. japonicum* (Sweet et al., 2004a, b). The occurance of the GnnA/GnnB enzymes seems to be restricted to *Mesorhizobium* and *Bradyrhizobium* (Table 16.3), an observation that is consistent with the fact that these are the only rhizobia, thus far, that contain GlcN3N in their LAs.

Once the LA precursor, Kdo$_2$lipid-IV$_A$, is made, *E. coli* adds two secondary fatty acids, myristate and laurate, via LpxM and LpxL, respectively. Initial work on *R. leguminosarum* and *R. etli* LA structures discovered that only one secondary fatty acid was present attached to the LA 2′-hydroxyfatty acyl residue and was unusual in that it was a very long chain fatty acyl (VLCFA) residue, 27-hydroxyoctacosanoic acid. Subsequently it was found that this VLCFA is found in all members of the *Rhizobiales* examined with the possible exception of the *Sesbania* stem nodulator, *A. caulinodans*. It was subsequently found that a unique acyl carrier protein, AcpXL, and a unique acyl tranferase, LpxXL, are required for the synthesis and transfer of the VLCFA (Bhat et al., 1991). The genes which encode these enzymes are part of a five gene cassette that also includes two genes each of which encodes a fatty acyl synthase II, and another that encodes a fatty acyl dehydratase. This five gene cassette is found in all members of the *Rhizobiales* examined (see Table 16.3) as well as in several intracellular pathogens that can survive within modified phageosomes and cause chronic infections such as *Brucella* and *Bartonella* species, whose LAs also contain the VLCFA (Vedam et al., 2003). Interestingly, this cassette of genes is also present in *A. caulinodans* even though the VLCFA has not been detected in its LA. However, the LA from *A. caulinodans* isolated from the plant host, or grown under *in planta*-like conditions has not been examined.

Once the secondary fatty acyl residue(s), VLCFA, has been added to the Kdo$_2$lipid-IV$_A$, a number of the core glycosyl residues are added and then, presumably, the resulting COS-LA molecule is flipped to the periplasmic side of the inner membrane after which the OPS is ligated to the COS-LA which is then translocated to the outer leaflet of the outer membrane. The enzymes which modify the

LA into the unique rhizobial structures are located on either the periplasmic side of the inner membrane or on the outer membrane. *Rhizobium leguminosarum* biovar *viciae* 3841 and *R. etli* CE3, which produce LA structure 2 (see Fig. 16.1), are the most thoroughly examined strains with regard to LA modification. For this LA structure, the modification includes the removal of the 1- and 4′-phosphates, the oxidation of the proximal GlcN to GlcNonate, and the addition of α-D-GalA to the 4′-position. The Raetz laboratory has shown that LpxE and LpxF remove the 1- and 4′-phosphates, respectively (Karbarz et al., 2003, 2009; Ma et al., 2008; Price et al., 1995; Wang et al., 2006, 2007), LpxQ oxidizes the proximal GlcN to GlcNonate (Que-Gewirth et al., 2003b), and RgtD transfers a α-D-GalA from dodecaprenyl-P-GalA to the LA 4′-position (Kanjilal-Kolar et al., 2006; Kanjilal-Kolar and Raetz, 2006). The LpxE and LpxF activities occur on periplasmic side of the inner membrane and the genes which encode these proteins have been identified (Karbarz et al., 2003, 2009; Wang et al., 2006, 2007). Mutations in these genes have been shown to result in the expected LA structures containing either the 1- or 4′-phosphates, or both phosphates in the case of a double mutant (Ingram et al., 2010). *S. meliloti*, which contains a 1,4′-*bis*-phosphorylated LA, does not contain either of these activities nor does its genome contain orthologs to the *lpxE* and *lpxF* genes. It would be expected that *A. tumefaciens*, and *S.* sp. NGR234 would also not contain either of these activities since they also produce *bis*-phosphorylated LA. However, *M. loti*, which produces 4′-monophosphorylated LA with a GalA residue glycosidically linked to the proximal GlcN residue would be expected to have LA 1-phosphatase activity. Interestingly orthologs of LpxE also seem to be present in some intracellular human pathogens, including *Francisella tularensis*, *B. melitensis*, and *L. pneumophila*. In fact, it was shown that *F. tularensis*, which has an LA that lacks the 1- and 4′-phosphates, has a very active 1-phosphatase activity as well as 4′-phosphatase (Wang et al., 2004, 2006). LpxQ, the proximal GlcN oxidase, was shown to be present on the outer membrane of *R. leguminosarum* and *R. etli* (Que-Gewirth et al., 2003b). This location is consistent with the observation that LA preparations from these strains contain LA molecules in which the proximal residue is present as either GlcN or GlcNonate. LpxQ was shown to be a monooxygenase and, therefore, would require O_2 for its activity (Que-Gewirth et al., 2003a, b). As expected orthologs of LpxQ are only observed in the genomes of *R. leguminosarum*, and *R. etli* strains. However, LpxE, LpxF, and LpxQ would presumably be present in *R. galegae* and *R.* sp. Sin-1 since the LA from these species also lack phosphate and contain GlcNonate as the proximal LA residue (see Fig. 16.1). Another structural modification that occurs on rhizobial LA is the removal of a β-hydroxymyristic acid from 3-position of the LA. This is evidenced by mass spectrometric analysis of numerous LA preparations which show two clusters of ions in which the smaller molecular weight cluster differs from the larger by 226 mass units; consistent with the loss of a β-hydroxymyristryl residue (Bhat et al., 1994; Ferguson et al., 2004, 2005; Gudlavalleti and Forsberg, 2003; Que et al., 2000; Vedam et al., 2003). In the case of *R. leguminosarum*, a fatty acylase was discovered which accounts for the removal of this fatty acid and is located in the outer membrane (Basu et al., 1999). This activity is also found in various other bacterial species, such as *Salmonella typhimurium*, and is encoded by the gene, *pagL* (Trent et al., 2001). An ortholog to *S. typhimurium pagL* has not yet been found in *R. leguminosarum* even though this enzyme activity is clearly present.

The transferase that adds the 4′-GalA residue to the LA from *R. leguminosarum* and *R. etli* has been reported by the Raetz laboratory and is present on the periplasmic side of the inner membrane (Kanjilal-Kolar et al., 2006). This enzyme has been designated RgtD and the gene that encodes this enzyme has not yet been identified; however, the most likely candidate is RL0684 in *R. leguminosarum* biovar *viciae* 3841 (Brown et al., unpublished) and the designated candidates for strains *R. etli* CE3 and *R. leguminosarum* biovar trifoli WSM2304 are given in Table 16.3. An interesting aspect of this GalA transferase is that it utilizes dodecaprenyl-P-GalA as the donor and not UDP-GalA (Kanjilal-Kolar and Raetz, 2006). The synthesis of this donor occurs by the transfer of GalA from UDP-GalA to the dodecaprenyl carrier, and this transferase is putatively encoded by an open reading frame, *orf3*, located next to genes coding for the GalA transferases involved in the LPS COS synthesis (Kanjilal-Kolar et al., 2006; Kanjilal-Kolar and Raetz, 2006). Orthologs of *rgtD* are not found in other rhizobial genomes examined, including that of *M. loti* which contains a GalA residue glycosidically linked to the LA proximal GlcN residue. The lack of an *rgtD* ortholog in *M. loti* likely due to the very different location of the GalA residue, and possibly due to a different GalA donor; e.g. UDP-GalA, instead of dodecaprenyl-P-GalA since *M. loti* does not have an ortholog of *orf3*. Interestingly *orf3* orthologs are observed in *S. meliloti* Sm1021 and in *S.* sp. NGR234 indicating that they also may synthesize a prenylated-GalA donor; however, this may be a donor for GalA found in another polysaccharide or, perhaps, in another part of their LPS, such as the COS.

In summary, the biosynthesis of rhizobial LA consists of conserved steps found in numerous Gram-negative bacteria that involve the synthesis of Kdo_2lipid-IV_A from UDP-GlcNAc. This is followed by the addition of a unique VLCFA as a secondary fatty acyl residue which is found in all members of the *Rhizobiales* (except for *A. caulinodans*) as well as in certain intracellular pathogens that form chronic disease, some of which are phylogenetically related to rhizobia. After Kdo_2lipid-IV_A synthesis, there are unique enzymes in different rhizobial species that processes this precursor into the unique rhizobial LA.

16.3.2 Core Biosynthesis

As with LA synthesis, the most thorough examination of rhizobial COS synthesis was done in the laboratory of Raetz and co-workers on *R. leguminosarum* and *R. etli* and is described in a recent review (Raetz et al., 2007). As shown in Table 16.1, these species produce identical COS structures. A schematic diagram showing the synthetic steps for the COS portion of these rhizobial LPSs is shown in Fig. 16.4. The addition of the COS glycosyl residues to the Kdo_2LA of *R. leguminosarum* can be divided into the two “stages”. The first stage involves those glycosyl residues that are added at the cytoplasmic side of the inner membrane

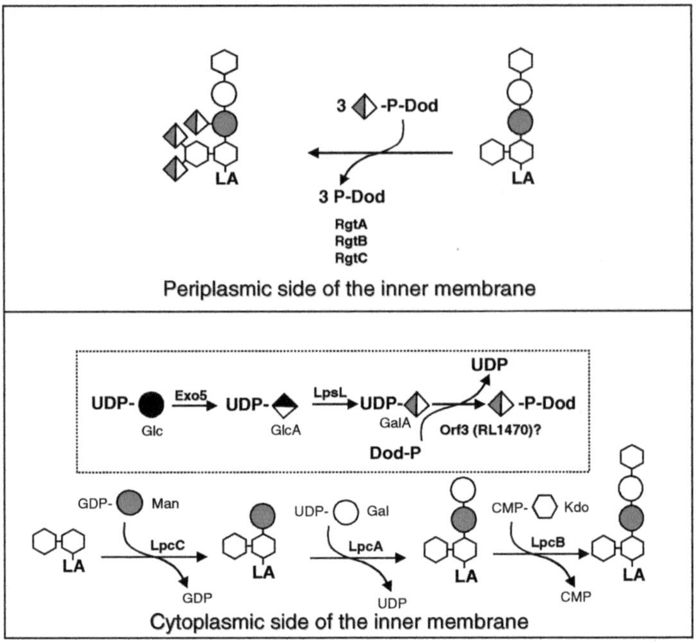

Fig. 16.4 A schematic diagram illustrating the biosynthesis of the COS for *R. leguminosarum* and *R. etli* LPSs

and include, sequentially, Man, Gal, and Kdo (Kadrmas et al., 1998; Raetz et al., 2007) forming Kdo-Gal-Man-Kdo(Kdo)-LA. These glycosyl residues are added by the respective transferases, LpcC (RL3423, RHE_CH02970, Rleg2_2716), LpcA (RL3440, RHE_CH02996, Rleg2_2733), and LpcB (RL3439, RHE_CH02995, Rleg2_2732). An LpcC ortholog is also found in *S. meliloti* 1021 (SMc01219, *lpsB*), *A. tumefaciens* C58 (Atu2183), *S.* sp. NGR234 (NGR_c15710, *lpsB*), and *B. japonicum* USDA 110 (Bll5930). While complete structures of the COS from these strains have not been determined, the presence of LpcC orthologs is consistent with a Man-Kdo structural feature being found in these COSs (e.g. *A. tumefaciens* strains as well as *B. japonicum*) (see Table 16.1). LpcA adds Gal from UDP-Gal to the COS Man residue and orthologs were found only in *R. leguminosarum* and *R. etli* strains. LpcB adds the external Kdo residue from CMP-Kdo to the Gal of the *R. leguminosarum* COS structure. An LpcB ortholog was found on pSymB of *S. meliloti* 1021 (SM_b20803). However, it is quite possible that this ortholog could be involved in the synthesis of the Kdo-containing capsular polysaccharide found in this strain in other Sinorhizobium strains (Kannenberg et al., 1998). At this point, the product, Kdo-Gal-Man-Kdo(Kdo)-LA, is thought to be flipped to the periplasmic side of the inner membrane by MsbA (Raetz et al., 2007), which is followed by the addition of three terminal GalA residues to the COS; two GalA are added to the branching Kdo residue, and one GalA to the Man residue resulting in Kdo-Gal-Man(GalA)-Kdo(Kdo[Gal_2])-LA. The addition of these GalA residues is catalyzed by transferases RgtA (RL1469, RHE_CH01318, Rleg2_2732), RgtB (RL1468, RHE_CH01317, Rleg2_0953), and RgtC (RL1471, RHE_CH01320, Rleg2_0956) (Kanjilal-Kolar et al., 2006). As with RgtD (described above), the donor for these transferases is dodecaprenyl-P-GalA (Kanjilal-Kolar and Raetz, 2006). There is an ortholog of *rgtB* (SMc02383) in *S. meliloti* 1021, but none in the other rhizobial genomes that have been sequenced.

In the case of *R. leguminosarum* and *R. etli*, the OPS is attached to the COS external Kdo residue (described above). Thus, in strains of these species, mutations that affect the synthesis of the Kdo-Gal-Man- region of these COSs result in "rough" LPS that lack the OPS; e.g. a *R. leguminosarum exoB* mutant that lacks Gal lacks LPS I (Laus et al., 2004). However, mutants that lack the GalA residues can still ligate OPS to the COS; recent work on an *R. leguminosarum* mutant of *exo5* which encodes for the synthesis of UDP-Glc dehydrogenase results in LPS that completely lacks GalA, but still contains the OPS (Laus et al., 2004) (Muszyński and Carlson, submitted).

While little is known regarding the biosynthesis of other rhizobial COSs, composition analysis, shown in Table 16.1, reveals that their structures, as with that of *R. leguminosarum* and *R. etli*, are very different from the typical COS found in enteric bacteria. The rhizobial COSs, except for those from two *M. loti* and one *M. huakuii* strains lack Hep and many contain acidic glycosyl residues, GalA, GlcA or both, and a few contain Dha. Of the structures reported, none are reported to contain phosphate-containing groups, and it is possible that the acidic glycosyl residues in these rhizobial LPSs provide some needed ionic character that phosphate-groups provide in enteric bacterial species.

16.3.3 O-Chain Polysaccharide Synthesis

As already stated earlier, the presence of the OPS portion of the LPS is essential for forming an effective nitrogen-fixing symbiosis, and it is known that modifications to the OPS occur during symbiosis. These modifications are described in the next section of this chapter. Because of the importance of the OPS for symbiosis with the host legume, it was important to determine their structures (described above) and understand their biosynthesis. There are many different OPS structures among strains of a single rhizobial species. Thus, different strains of *R. leguminosarum* or *R. etli*, which have the same COS-LA structure vary greatly in their OPS structures. This makes the determination of the OPS biosynthetic pathway a complex process as these pathways will be strain-dependent. However, as with the LA and COS, some detailed work on OPS genetics and biosynthesis has been done on *R. leguminosarum* bv. *viciae* 3841 and *R. etli* CE3, and also the inducible rhamnan OPS of *S.* sp. NGR234. Therefore, the following discussion is focused on these three strains.

Noel and co-workers first identified a region of the *R. etli* CE3 (a streptomycin-resistant derivative of strain CFN42) genome that encodes enzymes responsible for the synthesis of its OPS; the *lps* region (Cava et al., 1990; Duelli et al., 2001). It is a 27.8 kb region consisting of 25 genes. The structure of the CE3 OPS, see

Table 16.1, consists of five repeats of a -GlcAMe-(3OMe6dTal)Fuc- trisaccharide, which is capped by a di- or tri-OMeFuc at the distal end of the OPS, and contains a non-repeating Fuc-Man-QuiNAc sequence at the proximal reducing end. The QuiNAc residue of the OPS is linked to O4 of the COS external Kdo residue. The 25 genes encode for a number of glycosyl transferases, nucleotide sugar synthesis enzymes, polysaccharide transport proteins, and glycosyl modification proteins such as methylases and acetylases. It is also known that some genes required for OPS synthesis are located in regions of the genome other than the *lps* region. The functions of a number of *lps* genes have been published. Near the downstream end of the *lps* region are genes required for addition of the capping tri-O-MeFuc residue to the OPS; *wreB* (RHE_00766), *wreD* (RHE_00767), *wreF* (RHE_00768), *wreA* (RHE_00769), and *wreC* (RHE_00770) (Ojeda et al., 2009). Mutation in this region of *lps*, originally called *lpeA*, resulted in OPS that lacked the capping residue but was otherwise normal in amount and structure (Ojeda et al., 2009). Recently it was shown that mutation in any one of these genes results in the loss of the capping residue and that *wreD*, *wreF*, and *wreA* encode methyltransferases, while *wreB* encodes a glycosyl transferase, and the function of *wreC* is not yet known (Ojeda et al., 2009). Thus, methylation of this capping Fuc residue is apparently required for its transfer to the distal end of the OPS. In addition to methylation of this capping Fuc residue, methylation and acetylation are major features of this OPS. There seems to be two types of methylation that occur to the OPS, variable methylation that is altered during symbiosis (see next section), and constitutive methylation. Variable methylation occurs on the “internal” OPS Fuc residues (i.e. not the capping Fuc residue); one Fuc residue of a possible six is 2-*O*-methylated when the bacteria are cultured under standard laboratory conditions, while two Fuc residues are 2-*O*-methylated in bacteroids isolated from the root nodule or when cultured in the presence of anthocyanins (see further discussion below) (D’Haeze et al., 2007; Noel et al., 2004). Recently, it has been shown that methylase and transferase activities for the 2-OMeFuc residue are present on a single bifunctional protein, WreM (Ojeda et al., 2009). Mutation of the WreM region carrying the methyation activity results in the complete absence of Fuc 2-*O*-methylation and 2-OMeFuc is replaced by Fuc; however, other than this change, the OPS is largely unchanged in amount or structure (Ojeda et al., 2009).

It has been reported that constitutive acetylation and methylation may involve genes present on the upstream end of the *lps* region just prior to ATP-binding cassette (ABC) transporter genes, *wzm* and *wzt*; *orf1*, *orf2*, and *orf3* (Lerouge et al., 2003). *Orf1* and *orf2* putatively encode acetyltransferases, while *orf3* encodes a methyl tranferase (Lerouge et al., 2003). A non-polar mutation in any one of these genes results in the loss of the OPS from the LPS and indicates that the acetylation and methylation by these gene products is required for complete OPS synthesis or transport and ligation to the core region of the LPS. It was proposed that the constitutive acetylation and methylation of the 6dTal residue is required for OPS assembly and, therefore, mutation of these genes disrupts OPS synthesis.

Mutation of *wzm* and *wzt* also produce LPS which is completely deficient in OPS (Lerouge et al., 2001) presumably due to the inability to transport any OPS to the periplasm for ligation to the core region of the LPS. There are two primary pathways by which OPS are synthesized and transported (Raetz and Whitfield, 2002). One involves the synthesis of the OPS on the cytoplasmic side of the inner membrane which is transported to the periplasmic side by an ABC transporter (encoded by *wzm* and *wzt*) and ligated by WaaL to the core region. The second involves the synthesis of the OPS repeat unit on the cytoplasmic side of the inner membrane which is "flipped" to the periplasmic side, polymerized and ligated to the core region; a process encoded by *wzx* and *wzy*. The former pathway usually involves OPS that are homopolysaccharides, while the latter involves heteropolysaccharides. However, since the *R. etli* CE3 *lps* region contains *wzm* and *wzt* and not *wzy* and *wzx*, it likely synthesizes its OPS heteropolysaccharide by the first of these two mechanisms.

The characterization of other genes in the *lps* region has not yet been reported. However, these other genes have similarity to those that encode for a number of glycosyl transferases, as well as for the synthesis of the various nucleotide diphosphate glycosyl residues. For example, upstream from the *wbeABCDF* region, there are genes that have homology to those that encode for GDP-D-mannose 4,6, dehydratase (NoeL) and a NAD-dependent epimerase/dehydratase (NolK), which are required for GDP-L-Fuc synthesis. Other genes in the *lps* region likely encode for proteins required for the synthesis and addition of other glycosyl residues, e.g. 6dTal, and for the methylation of GlcA to form its methyl ester. However, some *lps* genes are not located in the *lps* region. The genes that encode for the synthesis of one of the OPS glycosyl residues, L-QuiNAc, are not all present in the *lps* region. It was shown that the gene encoding for the second step in the conversion of UDP-GlcNAc into UDP-QuiNAc, namely UDP-2-acetamido-2,6-dideoxyhexosyl-4-ulose reductase (LpsQ) is not present on the *lps* region (Forsberg et al., 2003). A mutation in *lpsQ* results in reduced levels of OPS in which a 2-acetamido-2,6-dideoxyhexosyl-4-ulosyl residue has replaced the QuiNAc (Forsberg et al., 2003). Incorporation of multiple copies of *lps* into the *lpsQ* mutant resulted in normal levels of OPS on the LPS with QuiNAc replaced by 2-acetamido-2,6-dideoxyhexosyl-4-ulose (Forsberg et al., 2003). Thus, it is likely that the QuiNAc glycosyl transferase is encoded in the *lps* region and its overproduction allows for the synthesis of "normal" levels of OPS containing the L-QuiNAc precursor. Since L-QuiNAc is also present in *R. leguminosarum* bv *viciae* 3841 OPS, other aspects of its synthesis are described further below.

While the structure of *R. leguminosarum* biovar *viciae* 3841 OPS has been determined, little has been reported on the genetics of its biosynthesis. A likely *lps* gene region required for OPS synthesis, the OPS region, is a 35.3 kb stretch of DNA containing 33 genes (RL0794–RL0826). Mutations in a number of genes in this region result in the total loss of OPS from the LPS (Kannenberg et al., 1992). This *lps* gene region is framed on both ends by putative insertion sequences which may indicate that it was incorporated from some other source via horizontal gene transfer. The functions of many of these genes are not yet known; however, as with the *R. etli* CE3 *lps* region there are a number of genes that encode glycosyl transferases as well as for proteins involved in synthesis of several nucleotide sugars. Among the latter are those that likely encode enzymes required for the synthesis of L-Fuc; GDP-mannose

4,6-dehydratase (RL0825, *gmd*) and GDP-L-fucose synthase (RL0826, *fcl*); which is a component of the OPS. These were also observed in the *R. etli* CE3 *lps* region since this OPS also contains Fuc (described in the previous paragraph).

L-QuiNAc is found in both *R. leguminosarum* biovar *viciae* 3841 and *R. etli* CE3 OPSs. It has been shown for *Vibrio cholerae* that UDP-D-GlcNAc is converted to UDP-L-QuiNAc in three steps. It is converted to UDP-2-acetamido-2,6-dideoxy-L-*lyxo*-4-hexulose by the enzyme WbvB, via dehydration at C-6 and epimerization at C-3, and this intermediate is then converted by the C-4 reductase, WbvR, to UDP-L-RhaNAc, which is then epimerized at C-2 to UDP-L-QuiNAc by WbvD (Kneidinger et al., 2003). A search for orthologs of WbvB, WbvR, and WbvD in the genomes of *R. leguminosarum* biovar *viciae* 3841 and *R. etli* CE3 showed only that strain 3841 contained a gene (RL0810, in the OPS region) that encodes a protein with similarity to WbvB (e^{-33}) and that there were no proteins with significant similarity to WbvR or WbvD. An ortholog to the reductase, LpsQ, is present in *R. leguminosarum* biovar *viciae* 3841. This ortholog is encoded by gene RL3304 (not in the OPS region) which has been annotated as a UDP-Glc 4-epimerase and, as in *R. etli* CE3, is adjacent to *murl*, a putative peptidoglycan hydrolase. As described above, the synthesis of UDP-L-QuiNAc from UDP-D-GlcNAc requires three proteins, WbvB, WbvR, and WbvD; however, the synthesis of UDP-D-QuiNAc is reported to be made by a single protein with both dehydratase and reductase activities, e.g. FlaA1 and WbpM, respectively, in *H. pylori* and *P. aeruginosa* (Creuzenet and Lam, 2001; Creuzenet et al., 2000). Orthologs to WbpM are found in *R. leguminosarum* biovar *viciae* 3841; RL0810 (e^{-171}) in the OPS region and pRL90052 (e^{-172}) in plasmid 9; and in *R. etli* CE3; RHE-PB00001 and RHE-PB00002 in plasmid p42b (e^{-171}). Thus, even though WbpM in *P. aeruginosa* is involved in the synthesis of D- and not L-QuiNAc, perhaps the rhizobial WbpM ortholog and LpsQ carry out the dehydratase and reductase activities required for the synthesis of L-QuiNAc. Further work is required to completely elucidate the L-QuiNAc synthetic pathway in these rhizobial strains.

Another glycosyl residue that is common to both *R. leguminosarum* biovar *viciae* 3841 and *R. etli* CE3 OPS and is also present in a number of other OPSs, is 6dTal. This glycosyl residue is reported as being in the D-configuration in *R. leguminosarum* biovar *viciae* 3841 and in the L-configuration in *R. etli* CE3 OPS. While the assignment of the D-configuration in *R. leguminosarum* biovar *viciae* 3841 OPS was unambiguously determined from nuclear Overhauser effects observed during NMR analysis of the intact OPS (Forsberg and Carlson, 2008), the assignment of the L-configuration in *R. etli* CE3 OPS was technically more complicated and, therefore, less certain. The distinction between D- and L-6dTal is quite important since their syntheses occur by very different mechanisms. GDP-D-6dTal synthesis begins with the conversion of GDP-D-Man to GDP-4-oxo-D-6dMan by Gmd, which is the first enzyme in the GDP-L-Fuc synthetic pathway. This intermediate is the precursor for three possible products; the formation of GDP-L-Fuc catalyzed by Fcl, GDP-D-Rha by Rmd (Rocchetta et al., 1999), or GDP-D-6dTal by Tld (Suzuki et al., 2002). In the case of L-6dTal, synthesis begins with the synthesis of dTDP-D-Glc from D-Glc-1-P catalyzed by RmlA followed by conversion to dTDP-D-*xylo*-6d-4-hexulose by RmlB and then to dTDP-L-*lyxo*-6d-4-hexulose by RmlC. This product is then the precursor to either dTDP-L-Rha formed by RmlD, or dTDP-L-6dTal by Tll (Nakano et al., 2000). Both *R. leguminosarum* bioviar *viciae* 3841 and *R. etli* CE3 contain genes that putatively encode for Gmd in their OPS gene regions and also for RmlABCD in other areas of their genomes. Thus, for each strain the synthesis of D-6dTal would require the presence of a Tld reductase ortholog, and L-6dTal synthesis would require a Tll reductase ortholog. The sequences of Tll and Tld have been reported for *Actinobacillus actinomycetemcomitans* NCTC 9710 (Nakano et al., 2000) and *A. actinomycetemcomitans* SUNYaB 75 (Suzuki et al., 2002), respectively. Orthologs to Tld are present in *R. leguminosarum* biovar *viciae* 3841 (RL4657) and *R. etli* CE3 (RHE_CH03985) at low similarity (e^{-17}) and to Tll at even lower similarity, RL0604 and RHE_CH00569 (e^{-13}). Further work is required to elucidate the synthesis of 6dTal that is present in these OPSs.

The OPS from *R. leguminosarum* biovar *viciae* 3841 contains two unusual and unique glycosyl residues; 3-acetimidoylamino-3-deoxy-D-gluco-hexuronic acid (Glc3NAmA), which is one of the OPS repeat unit glycosyl residues, and a capping glycosyl residue, 3-*O*-methyl-6-deoxy-4-*N*-formyl-hexose (3OMe6dHex4NFo) (see Table 16.1). The genes encoding proteins for the synthesis of Glc3NAmA are not known; however, some candidate genes are present in the OPS gene region. These genes are in the RL0819–RL0824 region. RL0824 has similarity to gene encoding for an oxidoreductase, which may be required for the formation of the 3-oxo-glucose. The RL0821-0822 genes have similarity to that encoding an aminotransferase and, therefore, may form the 3-amino-glucose from 3-oxo-glucose. RL0819 and RL0820 have similarity to genes involved in the synthesis of imidazoyl glycerol phosphate and may be involved in the transfer of acetamidoyl to the 3-amino-glucose. Another required step in the synthesis of this residue would be the oxidation of C6 to a carboxyl group. The formation of UDP-*N*-acetyl-D-glucosaminuronic acid (UDP-GlcNAcA) from UDP-GlcNAc in *P. aeruginosa* takes place via the action of WbpA, a UDP-GlcNAc 6-dehydrogenase (Miller et al., 2004). A WbpA ortholog was found in plasmid 11 of *R. leguminosarum* biovar *viciae* 3841, pRL110056 (e^{-103}) and also in plasmid p42e of *R. etli*, RHE_PE0044 (e^{-109}). Thus, there are certainly likely candidate genes in *R. leguminosarum* biovar *viciae* 3841 for the synthesis of this unique Glc3NAmA residue that need to be investigated.

As described above, some of the putative genes required for the synthesis of these OPSs are located on the plasmids of these organisms. Examination of the plasmids shows that some of them contain gene clusters that could also be involved in the synthesis of OPS. For example pRL90051–p90053 contain genes that encode putative galactosyl transferase, polysaccharide biosynthesis O-antigen-related protein, and O-antigen ligase. Also, pRL90132–pRL90158 contains genes that putatively encode for the synthesis of various glycosyl residues and glycosyl tranferases, some of which are similar to those we described above for the OPS gene region present in the chromosome. Thus, there are likely numerous other plasmid-borne carbohydrate-related genes that could be involved in the synthesis and/or modification of this OPS, or in the synthesis of alternative, perhaps conditionally-dependent, OPS structures.

The genes responsible for the synthesis of the OPS from *S.* sp. NGR234 have been examined in some detail (Broughton et al., 2006; Reuhs et al., 2005). This strain largely produces a "rough" LPS. However, when grown in the presence of flavonoids produced by the host legume or when isolated from host root nodules, it produces an LPS that contains a rhamnan OPS (see Table 16.1 for the structure). More about the induction of this OPS is described in the next section. The synthesis of this OPS involves the genes required for the synthesis of dTDP-L-Rha, *rmlABCD*, as well as *wbgA* which encodes a glycosyl transferase, and *fixF* (Broughton et al., 2006). These latter two genes are thought to be required for the polymerization of dTDP-L-Rha into the rhamnan and its export to the periplasm, respectively. The expression of these genes appears to be under the control of *nod* boxes, indicating symbiosis-related expression of these genes.

In summary, the genes required of OPS synthesis in for *R. leguminosarum* biovar *viciae* 3841 and *R. etli* CE3 are largely clustered in specific regions on the chromosome with some *lps* genes being present at other locations. In addition, there appear to be OPS synthesis gene clusters on some plasmids which may be important for the modification of the OPS, or, possibly the synthesis of alternative OPSs under various conditions. This, in fact, is the case for *S.* sp. NGR234 in which the synthesis of the rhamnan OPS is directed by genes on the symbiotic plasmid whose transcription is under the regulation of *nod* boxes. While little is known regarding the genetics and synthesis of other rhizobial OPSs, it is likely that these general findings will also apply to these strains.

16.4 Structural Modifications to Rhizobial LPSs During Symbiosis

A number of investigations have shown that a *Rhizobium* adapts its cell surface in response to its host during symbiosis. It has been reported that bacteroids isolated from pea root nodules or alfalfa root nodules (pea and alfalfa are indeterminate nodule forming hosts) are significantly increased in their hydrophobicity compared to laboratory cultured rhizobia (Kannenberg and Carlson, 2001). These changes also appear to be reflected in the LPS since the LPSs from *R. leguminosarum* biovar *viciae* 3841 and *S. melilolti* bacteroids also becomes more hydrophobic (Kannenberg and Carlson, 2001; Reuhs et al., 1999). On the other hand, *R. etli* CE3 bacteroids from bean root nodules (bean is a determinate nodule forming host) 28 days after inoculation are more hydrophilic, and more sensitive to cationic peptides (D'Haeze et al., 2007).

Changes to the LPS that occur during symbiosis have also been detected using monoclonal antibodies. LPS-binding MAbs prepared against *R. leguminosarum* biovar *viciae* 3841 bacteroids have been shown to exist in three general classes (Kannenberg and Brewin, 1989; Kannenberg et al., 1994). One class constitutively binds LPS from both bacteroids and laboratory grown cultures and also bacteroids at all stages of development in the pea root nodule. The second class binds only bacteroid LPS and this binding appears to be the function of the low oxygen environment of the root nodule. The third class binds bacteroid LPSs but these mAbs bind differently depending on the developmental stage of the bacteroid. It was also shown that most of these mAbs bind to the form of the LPS that contains the OPS. In the case of *Rhizobium etli* CE3, which forms a symbiosis with bean, it was shown that four mAbs (JIM26, JIM27, JIM28, and JIM29) that bind to the OPS-containing LPS of laboratory grown cultures bind differently to the LPS from bacteroids; JIM26 and JIM27 bind to both laboratory grown bacterial LPS and bacteroid LPS, while bacteroid LPS no longer binds JIM28 and is greatly reduced in its ability to bind JIM29 (Tao et al., 1992; Tao and Noel, 1990). For both *R. leguminosarum* biovar *viciae* 3841 and *R. etli* CE3, the mAb binding properties of bacteroid LPS can be partially mimicked by conditions that are thought to reside inside the root nodule; e.g. low O_2. Thus, in order to fully understand the function of LPS during symbiosis, it was necessary to structurally characterize these changes. Examination of several *Rhizobium*-legume symbioses has revealed that changes can occur to both the OPS and LA during symbiosis. The changes that occur to the OPS portion of the LPS can be large or subtle. Large changes include the replacement of one OPS with another, or the possible addition of a second OPS. Subtle changes usually involve the structural modification of the existing OPS such as changes in methylation or acetylation of certain glycosyl residues. Changes in the LA involve alterations in fatty acylation.

16.4.1 Modifications to the O-Chain Polysaccharide During Symbiosis

It has been observed that some *Sinorhizobium* species produce a secondary OPS during symbiosis or when free-living cells are cultured in the presence of the host plant flavonoid or root exudates. The LPS from *S. fredii* grown in the presence of the flavonoid apigenin or in the presence of root exudate from its plant host (Peking soybean) was analyzed by DOC-PAGE and glycosyl composition. The results showed that apigenin and host root exudate induced the production of a new "smooth" form of LPS (i.e. LPS with OPS) in addition to the "smooth" LPS observed when grown in the absence of these components (Reuhs et al., 1994). Compositional analysis of the isolated LPSs showed that the OPS from normally grown cultures consisted largely of glucose while the induced OPS consisted of xylose and mannose (Reuhs et al., 1994). Other sinorhizobia appear to undergo a pronounced shift in LPS surface chemistry during symbiotic infection, from one in which rough-LPS (lacking OPS) is the dominant form (in free living cells), to one in which smooth-LPS, containing OPS, is the dominant form. In the broad host-range *S.* sp. NGR234, the vegetative form is characterized by an abundant rough-LPS which lacks OPS and consists of a highly branched, anionic core region attached to LA. Growth in the presence of apigenin results in the expression of a smooth-LPS which carries a structurally unique rhamnan O-antigen (Fraysse et al., 2002; Reuhs et al., 2005), with the structure shown in Table 16.2 (Reuhs et al., 2005). Mutants which can not produce this rhamnan OPS are symbiotically defective, Nod^+ Fix^- on *M. atropurpureum*, *C. cearoleum*, and *V. unguiculata*, Nod^- Fix^- on *P. tuberosus*, but are Nod^+ Fix^+ on *L. leucocephala* (Broughton et al., 2006; Reuhs et al., 2005). Thus, for several host plants, the production of this alternative rhamnan OPS is essential for effective symbiosis. Synthesis of this rhamnan involves the products of genes that are regulated by promoter regions called *nod* boxes (Broughton et al., 2006). In the presence of flavonoids, a NodD-flavonoid complex is formed that binds to the *nod* box region and activates gene expression. In the case of this rhamnan, *nod* boxes regulate the expression of genes that encode products required for the synthesis of dTDP-L-Rha (RmlABC) and a putative rhamnosyl transferase (WbgA) (Broughton et al., 2006). The same mechanism also regulates the expression of *y4gm*, which encodes a ABC transporter, and *fixF* whose function is still unknown; however, both genes are required for the production of the rhamnan and are thought to be involved in the polymerization of dTDP-L-Rha and transport of the resulting rhamnan for subsequent attachment to the COS of the LPS (Broughton et al., 2006). Interestingly, the OPS rhamnan produced by *S.* sp. NGR234 has the same primary glycosyl sequence as the A-band LPS produced by the opportunistic pathogen, *Pseudomonas aeruginosa*, when residing within its host (Rocchetta et al., 1999) except that the latter is composed of D-rhamnosyl residues while the former consists of L-rhamnosyl residues.

In addition to *S.* sp. NGR234, *S. meliloti* strains, symbionts of alfalfa, produce LPSs that, in most cases, lack an OPS and largely consist of an oligosaccharide, COS, attached to the LA (Kannenberg et al., 1998). As described above, it is the COS, which is highly anionic, that is the antigenic determinant of these LPSs (Kannenberg et al., 1998; Reuhs et al., 1999). Relatively little structural work has been done on the carbohydrate portions of *S. meliloti* LPSs and there are no reports of a complete structure of either an OPS or COS. However, mAb-binding studies revealed the appearance of novel OPS epitopes in *S. meliloti* smooth LPS isolated from bacteroids (Reuhs et al., 1999) showing that this rhizobial species can also elevate or alter its OPS expression in reponse to the host plant.

The ability of several *Rhizobium* species to modify their LPS structures in response to the host plant has been examined. Noel and co-workers showed that subtle structural modifications occur to the OPS of the LPS from *R. etli* CE3 bacteroids resulting in the loss its ability to bind mAb JIM28 (Tao et al., 1992; Tao and Noel, 1990). Subsequently, it was shown that the loss of binding to JIM28 occurred when the bacteria were cultured in the presence of host root extract or anthocyanins and that this effect was correlated with the addition of a methyl group to position O2 of a fucosyl residue in the OPS (Noel et al., 2004). More recently, the LPS from a large batch of *R. etli* CE3 bacteroids was purified, analyzed and compared to the LPS from laboratory grown cultures. The results showed that the LPS from purified bacteroids completely lost its ability to bind JIM28 and structural analysis revealed that the presence of methyl groups on O2 of fucosyl residues increased from one per OPS to two per OPS (D'Haeze et al., 2007). The structure of the *R. etli* CE3 OPS, shown in Table 16.2, contains three types of fucosyl residues;

a non-reducing terminal capping fucosyl residue that is 2,3-di- or 2,3,4-tri-*O*-methylated, five 3,4-linked fucosyl residues (one in each of the five repeat units), and one 3-linked fucosyl residue which is located near the reducing end of the OPS. Further structural analysis of the bacteroid LPS showed that the methyl group was added to one of the five repeating unit 3,4-linked fucosyl residues (D'Haeze et al., 2007). Since it was shown that the capping 2,3-di- or 2,3,4-tri-*O*-methylfucosyl residue is required JIM28 binding (Duelli et al., 2001), it was hypothesized that it is 2-*O*-methylation of the 3,4-linked fucosyl residue in the distal repeating unit, i.e. the one to which the capping residue is attached, that results in the loss of JIM28 binding (D'Haeze et al., 2007).

The gene encoding WreM, which is responsible for 2-*O*-methylation of these internal OPS fucosyl residues, is in the *lps* region of the *R. etli* chromosome (Noel et al., 2004). This gene was originally identified as *lpeM* and is now known as *wreM* (Ojeda et al., 2009). A mutation which affects both the methylase and glycosyltransferase, and regions, respectively, of *wreM* result in a truncated OPS, while a non-polar mutation in *wreM* results in normal levels of OPS that are completely devoid of 2-*O*-methylfucosyl residues (Ojeda et al., 2009). The lack of 2-*O*-methylation of fucose in the *wreM* mutant is not restored by culturing in the presence of anthocyanin, nor in mutant bacteroids (Noel et al., 2004; Ojeda et al., 2009). The symbiotic phenotype of the *wreM* mutant is a delay in nodulation and a reduction in nodule number at early times after inoculation which is corrected at later times (Noel et al., 2004). Thus, the addition of the 2-*O*-methyl fucosyl residues facilitates but is not required for symbiosis.

The OPS from *R. leguminosarum* biovar *viciae* 3841 LPS is also modified during symbiosis. In this case, both the bacteroids and their LPS become hydrophobic in comparison to the LPS from laboratory cultured cells (Kannenberg and Carlson, 2001). In addition, the LPS isolated from cultures grown at low O_2 or lower pH, conditions thought to mimic those within the host root nodule, is also hydrophobic (Kannenberg and Carlson, 2001). Composition analysis of the LPS extracted into the water or phenol layers from cultures grown under low O_2 or low pH showed changes in acetylation of the OPS (Kannenberg and Carlson, 2001). As mentioned above a number of mAbs generated against this strain isolated from pea nodules specifically bound to the OPS portion of bacteroid LPS (Kannenberg and Brewin, 1994; Kannenberg et al., 1994; Sindhu et al., 1988, 1990; VandenBosch et al., 1989; Wood et al., 1989). In addition, LPS preparations from laboratory cultures grown under normal conditions show small amounts of a separate polysaccharide containing xylose and mannose; however, in low pH grown cells the water layer of the hot/phenol water extraction contained elevated amounts of this xylomannan (Forsberg and Carlson, 2008; Kannenberg and Carlson, 2001). Determining the exact modifications that occur to the OPS during symbiosis requires first knowing the structure of the OPS from laboratory cultured bacteria. This has been determined for strain 3841 (Forsberg and Carlson, 2008) and is shown in Table 16.2. As with the OPS from *R. etli* CE3 (also shown in Table 16.2), this OPS is of uniform size ($\approx$ 3000 Da) and contains 3-*O*-methyl-6-deoxytalosyl, fucosyl, and

N-acetyl-quinovosaminosyl residues, and is heavily O-acetylated. However, it is unusual in that the repeating unit also contains a 3-amino-glucouronosyl residue that is *N*-acetimidoylated. The sugar, 3-acetimidoylamino-3-deoxy-*gluco*-hexuronic acid, was not previously described and is at this point unique to *R. leguminosarum*. The entire 3841 OPS is capped with a 3-*O*-methyl-4,6-dideoxy-4-*N*-formyl-hexosyl residue. Thus, the OPS from both *R. etli* CE3 and *R. leguminosarum* 3841 terminate with a non-repeated "capping" residue at the nonreducing end. For the 3841 strain, the number of non-carbohydrate substituent groups suggests that this OPS is even more subject to variation in response to environmental changes to the bacterium than is the *R. etli* CE3 OPS. For example, the 3-acetimidoylamino-3-deoxyglucuronosyl residue is a zwitterion in which the proportion of ionic forms of the carboxyl and *N*-acetimidoyl groups (negatively and positively charged, respectively) can change readily with pH, and, thereby, change the net charge or hydrophilic character of this LPS when residing within the symbiosome compartment. As noted above, it was also shown that a xylomannan, attached to lipid carrier or anchor, is a secondary polysaccharide that is present in *R. leguminosarum* 3841 bacteroids (Forsberg and Carlson, 2008). Thus, this strain subtly modifies its OPS during symbiosis as observed for *R. etli* CE3 and can also produce a second polysaccharide as was reported for *Sinorhizobium* species. The genetic basis for the modifications to *R. leguminosarum* 3841 LPS is currently under investigation.

As with other rhizobial strains *Mesorhizobium* mutants that are defective in the synthesis of OPS are also symbiotically defective; the development of the infection thread is disturbed and there is premature senescence of symbiosomes (Turska-Szewczuk et al., 2009). The OPS of *M. loti* NZP2213 and *M. huakuii* consist of homoglycan of 6-deoxy-L-talosyl (6dTal) residues (Russa et al., 1995) and a heteroglycan of 6dTal and L-rhamnosyl (Rha) residues (Choma et al., 2000), respectively; see structures in Table 16.2. Little work has been done to determine if any changes to these LPSs are occurring during symbiosis. However, a mutant of *M. loti* NZP2213, strain Mlo-13 is enhanced in symbiosis and produces a 6dTal/Rha heteroglycan OPS (see structure in Table 16.2) rather than the 6dTal homoglycan OPS produced by NZP2213 (Turska-Szewczuk et al., 2008). This result indicates that strain NZP2213 has the capability of making large changes to its OPS as was the case for *Sinorhizobium* species.

In summary, rhizobial strains can dramatically and/or subtly alter their OPS during symbiotic infection, i.e., upon differentiation from the vegetative to bacteroid state. Dramatic alterations include alterations in LPS surface chemistry, from the production of rough- to smooth-LPS, and the expression of secondary (new) OPS; more subtle alterations involve the modification of the existing OPS, e.g., endogenous *O*-methylation of a glycosyl residue. The OPS of the various rhizobial LPSs contain numeous deoxyglycosyl residues that are often modified by methyl and *O*-acetyl groups. When uronosyl residues are present, the carboxyl groups can be methyl esterified, or balanced by a positively charged group forming a neutral zwitterion. Thus, the OPSs are relatively hydrophobic for polysaccharides and, in the case of rhizobial LPSs, changes in hydrophobicity are a likely result of changes in the degree of acetylation, methylation, and pH that occur during symbiosis.

16.4.2 Modifications to the Core Oligosaccharide During Symbiosis

Changes in the core region of the LPS during symbiosis, COS, have only been examined in the case of *R. leguminosarum* and *R. etli* strains. As described above, these species contain the same COS structure; see structures in Table 16.1. The COS of these LPS can be analyzed by HPAEC analysis of the carbohydrates released by mild acid hydrolysis of the LPS (Kannenberg and Carlson, 2001). When this was done on LPS from bacteria and bacteroids or cultures grown at low O_2 or low pH of these two *Rhizobium* species, there was no change in the COS structure indicating that this portion of the LPS is not altered during symbiosis (D'Haeze et al., 2007; Kannenberg and Carlson, 2001).

16.4.3 Modifications to the LA During Symbiosis

As mentioned above the LPS undergoes changes in hydrophobicity during symbiosis. In the case of *R. leguminosarum* and *S. meliloti* bacteroids, the LPSs from bacteroids isolated from the root nodule become hydrophobic in that they are largely extracted into the phenol rather than aqueous phase during hot phenol/water extraction (Kannenberg and Carlson, 2001; Reuhs et al., 1999). This corresponds to an increase in hydrophobicity of the entire bacteroid cell during symbiosis. In the case of *R. etli* CE3, the LPS from bacteroids remains preferentially extracted into the water phase, while the bacteroids (purified from 28 day old nodules) are more hydrophilic than laboratory cultured bacteria (D'Haeze et al., 2007). Thus, in addition to the effect of OPS on LPS hydrophobicity and cell surface hydrophbicity, it would be expected that modifications to the LA may occur that contribute to the changes in LPS hydrophobicity.

The most detailed analysis of the LA from bacteroids or from bacteria grown under low O_2 or low pH conditions has been from *R. leguminosarum* biovar *viciae* 3841 – pea and from *R. etli* CE3 – bean symbioses. The structures of the LA from these two *Rhizobium* species are identical; see Fig. 16.1. The fatty acyl residues in this LA are 3-OHC14:0, 3-OHC15:0, 3-OHC16:0, 3-OHC18:0, and 27-OHC28:0. In the case of *R. leguminosarum* 3841, the LPS from normally cultured bacteria show a shift in the "short" (C14-C16 length fatty acids) to "long" (C18-C28 length fatty acids) chain fatty acid ratio in water phase compared to phenol phase LPS of 1.7 to 0.65 (Kannenberg and Carlson, 2001), respectively. When bacteria are grown at low pH or low O_2, the LPS is almost exclusively extracted into the phenol phase and the short to long fatty acid ratios are 1.0 and 0.83, respectively. Thus, this shift to longer fatty acyl residues is consistent with a more hydrophobic LPS. The most dramatic change in these LPS preparations was an almost two-fold increase in the relative level 27-OHC28:0 in the LPS found in the phenol phase from each growth condition (i.e. low O_2 or low pH) compared to that in the water phase from cells cultured under normal laboratory conditions. However, when the LA from *R. leguminosarum* bacteroids isolated from pea nodules was purified and examined by MALDI-TOF MS analysis and compared to the LA from laboratory cultured cells, no differences in the spectra were observed (Vedam et al., 2006). Thus, these data are in conflict with the data showing an increase in 27-OHC28:0, however, the latter LPSs were isolated using triethylamine/EDTA extraction of the cells rather than hot phenol/water extraction and, therefore, this apparent discrepancy may be due to the different extraction protocols. This discrepancy is under further investigation.

Rhizobium etli CE3 bacteroids isolated from bean nodules contained LPS that was altered in its OPS as described above, and was also altered in its LA. In this case, the LPS from purified bacteroids contained exclusively tetraacylated LA (D'Haeze et al., 2007), while the LA preparation from laboratory cultured bacteria contains a mixture of tetraacylated and pentaacylated LA (D'Haeze et al., 2007; Que et al., 2000). The tetraacylated LA is due to removal of a 3-OHC14:0 residue. A portion of this tetraacylated LA is likely due to acid catalyzed β-elimination of this residue from O3 of the proximal 2-aminogluconolactone moiety during mild acid hydrolysis. However, the complete absence of 3-OHC14:0 in *R. etli* CE3 bacteroid LPS is likely due to its removal by an acylase that has been reported in *R. leguminosarum* (Basu et al., 1999). Our results imply that *R. etli* CE3 contains a *pagL* gene, or PagL protein, that is more active in bacteroids.

16.5 Rhizobial Lipopolysaccharide Function

As soil bacteria and legume symbionts, rhizobia encounter a large number of different physiological conditions. In addition to the changing environment that occurs during infection of the host legume, rhizobia also encounter varying conditions in the soil such as moisture content, temperature, etc. The adaptation of rhizobia to these different conditions most certainly require changes in their cell surface chemistry and, therefore, in their LPS. Some of the LPS structural changes that occur during infection have been described in the previous section. What is less known and, therefore, more speculative, are the functions of the unique structural features of rhizobial LPS and also of the modifications to this structure that occur during symbiosis. These functional aspects can be divided into two general areas. First are those functions that have a direct role in the symbiotic process; e.g. they are involved in the endocytotic infection process, and/or in bacteroid formation. Second are functions that have an indirect role such as protecting the rhizobia from the host defense mechanism, or regulating this mechanism in some manner that permits and enhances symbiosis. These "direct" and "indirect" functions are not mutually exclusive. The effect of various LPS mutants on symbiosis has shed some light on the possible functions of this molecule.

16.5.1 The Symbiotic Function of the Unique Structural Features and Modifications of Rhizobial Lipopolysaccharides

As described above, the general structure of LPS from *R. leguminosarum* and *R. etli* strains, show that the OPS is relatively hydrophobic, the COS is anionic and hydrophilic, and the LA is hydrophobic (Fig. 16.2). During symbiosis significant changes in hydrophobicity occur to the entire rhizobial surface as well as to the LPS and, in the case or *R. leguminosarum* biovar *viciae* 3841, this is accompanied by an increase in the amount of LPS I, i.e. LPS that contains OPS, as well as by changes in methylation and acetylation of the OPS (Kannenberg and Carlson, 2001). This increase in surface hydrophobicity, which OPS appears to play a role, may be important for the infection of the nodule cell which occurs via endocytosis as well as for bacteroid formation within the symbiosome since both of these processes likely involve the interaction of the host cell membrane with the rhizobial cell surface. A role for OPS in this process is supported by the observation that OPS-defective mutants are defective in nodule cell invasion, and the few bacteria that do make it into the host nodule cells form very aberrantly shaped bacteroids with multiple bacteroids within a single symbiosome indicating that there has been a disruption in the synchronous division process between the rhizobial cell and host symbiosome membrane (Perotto et al., 1994). That there is a physical interaction between the bacterial cell surface and host membrane is supported by the finding that a host nodule lectin, PsNLEC-1, as well as host membrane glycolipids were bound to isolated *R. leguminosarum* biovar *viciae* 3841 bacteroids (Bolanos et al., 2004). In that report it was also observed that mutant cells of strain *R. leguminosarum* B659, derivative of strain 3841, which has an altered COS and lacks OPS was unable to bind the host components further supporting that OPS may be involved in the physical interaction between the bacterial cell surface and the host membrane.

The OPS-defective mutant also induces a host defense reponse (Perotto et al., 1994) indicating that its presence may play a role in protecting the rhizobial symbiont from, or inhibiting, the defense response. In the OPS-defective mutant, exposure of the anionic LPS COS may result in an increased host defense response as well as a bacterium that is more sensitive to that response; e.g. exposure of the anionic COS may cause increased sensitivity to host antimicrobial cationic peptides. In the case of *R. leguminosarum* biovar *viciae* 3841, its OPS may also have a direct role in the resistance to cationic peptides. As described above, it contains a unique glycosyl residue called rhizoaminuronic in which the carboxyl group and the *N*-acetimidoyl group would form a zwitterion at neutral pH (Forsberg and Carlson, 2008). However, at acidic pH within the nodule cell that approaches the pKa of the carboxyl group this structure would have a net positive charge which would likely result in increased resistance to the cationic antimicrobial peptides.

In *S.* sp. NGR234, the rhamnan that is induced during symbiosis is also important for symbiosis in some of its host plants, as described above. On the host, *Vigna unguiculata* (Broughton et al., 2006; Reuhs et al., 2005) such a mutant is defective in the release of the bacteria from infection threads (Broughton et al., 2006). While there were some differences in phenotype depending on the type of legume host, microscopic examination revealed that on *V. unguiculata*, only on a few host nodule cortical cells were infected, and the infected cells lacked the peribacteroid membrane, were enlarged and degradation was evident (Broughton et al., 2006). These effects may indicate induction of a defense response and, therefore, imply that the rhamnan may play a role in preventing or regulating the host defense response and/or protecting the bacterium from that response.

A more subtle effect on OPS during symbiosis was the addition of a single methyl group to O2 of a Fuc residue in the distal repeat unit. A mutant in the portion of WreM that encodes the methylation activity but not the glycosyl transferase activity results in the production of normal amounts of OPS that is devoid of these methyl groups under all conditions, including those within its legume host (Noel et al., 2004; Ojeda et al., 2009). Such mutants show a delay in nodulation and less nitrogenase activity, but still form nitrogen-fixing nodules (Noel et al., 2004). This effect was more severe if inoculation with the mutant bacteria was done prior to germination of the seeds (Noel et al., 2004). The function of this fucosyl 2-*O*-methylation is not known, however, the fact that fucosyl 2-*O*-methylation is induced by seed exudates containing anthocyanin suggest that it could act as type of early signal that somehow enhances the symbiotic process.

The VLCFA is a LA substituent that, with the possible exception of *A.caulinodans*, seems to be universally present on rhizobial LPS. As described above, the synthesis of the VLCFA requires a unique acyl carrier protein, AcpXL, and its transfer to the LA requires a unique acyltransferae, LpxXL. Mutation of *acpXL* and/or *lpsXL* in either *R. leguminosarum* biovar *viciae* 3841 of *S. meliloti* 1021 resulted in the production of an LPS that lacked the VLCFA (Sharypova et al., 2003; Vedam et al., 2003; Ferguson et al., 2005). These mutants were still able form nitrogen-fixing nodules on their respective host legume; albeit nodulation and nitrogen-fixation were delayed. This apparent lack of a severe symbiotic phenotype initially indicated that the VLCFA was not a major factor required for symbiosis. However, detailed microscopic examination of pea root nodules showed that the *R. leguminosarum* biovar *viciae* 3841 *acpXL* mutant were dramatically affected in bacteroid formation; bacteroids were greatly enlarged, had very distorted shapes, and the synchronous bacterial cell/symbiosome membrane division was disrupted (Vedam et al., 2004). Further, it was discovered that *acpXL* mutant bacteroids isolated from the host, pea, root nodules were partially restored in their LA VLCFA indicating that the loss of *acpXL* was somehow partially compensated within the host (Vedam et al., 2006). An interesting observation is that the aberrantly shaped bacteroids formed and the disruption of the synchronous cell/symbiosome membrane division by the *acpXL* mutants were similar effects to those observed for the OPS-mutants (described above). It may be that the VLCFA component of the LA is also important in the synchronous bacterial cell/host symbiosome membrane division process. It is possible that a relatively hydrophobic OPS is required for forming the contact between these membranes and promoting endocytosis and synchronous division, while the VLCFA is required to maintain the stability of the bacterial cell during this process. A decrease in membrane stability of a VLCFA mutant is supported by the observation that the *acpXL*, as well as *lpxXL*, mutants are increased in their sensitivity to detergents such as deoxycholate indicating a decrease in the integrity of their membranes (Ferguson et al., 2005). It would be interesting to examine the bacteroids of *acpXL* mutants for the presence of the host membrane glycolipids and PsNLEC-1. Unlike the OPS-mutants, it might be that the *acpXL* mutant bacteroids, which still contain OPS, would contain these host membrane components.

Another symbiotic mutant that is affected in its VLCFA LA component is the *bacA* mutant (Ferguson et al., 2000, 2004, 2005; Tan et al., 2009). These mutants are symbiotically defective in that while they form nodules, bacteria are unable to invade the nodule cells and nitrogen-fixation does not occur. The proportion of LPS from these mutants that contain VLCFA is reduced by about 50%. These results indicate that loss of BacA causes a more severe phenotype than the loss of VLCFA and that expression of BacA may be linked in some manner to the incorporation of VLCFA into the LA. Recent work reports that BacA may be important for the uptake of nodule-specific peptides important for maintaining bacteroids within the pea host (Marlow et al., 2009). Interestingly, *bacA* mutants of *B. abortis*, which forms a chronic infection by surviving within modified phageosomes, are also avirulent and are about 50% reduced in the proportion of LPS molecules that contain the VLCFA (Ferguson et al., 2004).

In addition to a role in symbiosis, the presence of the VLCFA may also be important for the life of the symbiont in the soil. Recently, it was shown that a mutant in genes encoding for the fatty acyl transferases that are in the five gene VLCFA cassette of *R. leguminosarum* biovar *viciae* 3841 also fail to incorporate VLCFA into their LA and they are also less resistance to desiccation, impaired in biofilm formation, and in motility (Vanderlinde et al., 2009). Thus, in addition to promoting rhizobial life inside the plant host, the presence of VLCFA in LPS may also be important for life of the symbiont in the soil.

16.5.2 Rhizobial Lipopolysaccharide, Symbiosis, and the Plant Defense Response

During the past decade, a number of reviews have been published describing and comparing the plant and animal innate immune response (Ausubel, 2005; Dow et al., 2000; Lerouge and Vanderleyden, 2002; Newman et al., 2007; Nürnberger et al., 2004; Nürnberger and Kemmerling, 2006; Shah, 2005). The response of plant hosts to LPSs is also the subject of Chapter 18 of this book (by Mari-Anne Newman) and, therefore, this aspect will not be extensively described here. Rather, here, we briefly describe the response of plants including legume hosts, to rhizobial LPSs.

Plants, as do animals, have an innate defense response to pathogens in which pathogen-associated molecular patterns (PAMPs) are recognized by the plant. This results in a defense response that prevents infection by the pathogen. Thus a successful infection requires that a pathogen have resistance mechanisms by which it can defend itself against the plant defense response and/or modify this response in some manner to reduce its effect. A rhizobial infection of a legume host is more complex in that the rhizobial symbiont must be able to successfully infect its host forming what is, in essence, a chronic infection while at the same time having a vested interest in promoting the health of its host plant. Thus, it might be expected that the rhizobial symbiont would be able to regulate and/or resist the plant's defense response at the site of infection while systemically inducing the defense response (e.g. causing an induced systemic resistance, ISR) in other parts of the plant host and, thereby, promote resistance to attack by pathogens.

We have already described that a number of rhizobial LPS mutants are symbiotically defective because they likely induce an increased defense response by the host and/or are more sensitive to the host defense response. One structural feature of rhizobial LPS that appears to be important is the presence of OPS since its absence appears to result in a more robust plant defense response. We have also suggested (above) that the lack of OPS exposes the anionic COS on the bacterial surface which may make the rhizobial cell more sensitive to antimicrobial cationic peptides. Recent work in our laboratory (Brown, unpublished) has shown that a mutation of *R. leguminosarum* biovar *viciae* 3841 which specifically results in the loss of GalA residues from the core increases resistance to cationic peptides. It has also recently been shown that *R. etli* CE3 mutants in *lpxE* and *lpxF*, which are unable to remove the 1 and 4′-phosphates and, therefore have LA with increased anionic character, show increased sensitivity to cationic peptides (Ingram et al., 2010).

The fatty acylation pattern of rhizobial LA is also likely to be important in suppressing the host plant's defense response. Treatment of suspension-cultured *Medicago truncatula* cells with *S. meliloti* LPSs is reported to suppress the production or reactive oxygen species (ROS) (Heiko Scheidle et al., 2005; Tellstrom et al., 2007) and reduce the expression of genes correlated to a plant defense response, e.g. the gene encoding phenyl alanine ammonia lyase (Tellstrom et al., 2007). While, the structural basis for such responses to rhizobial LPS is not known, the suppression of ROS is reported to be due to the LA portion of the LPS (Heiko Scheidle et al., 2005). It seems likely that structural features such as the fatty acylation pattern may be important in ROS suppression. In the case of the plant pathogen, *Xanothomonas campestris*, it is reported that the innate immune response of *Arabidopsis* is dependent on the fatty acylation and phosphorylation pattern of the *X. campestris* LA (Silipo et al., 2008); a mutant that produces largely penta- rather than hexacylated LA was unable to induce PR1, a defense-related gene product and also did not prevent the hypersensitive response. Interestingly, many rhizobial LA preparations contain a mixture of penta- and tetraacylated molecules (see discussion above). In fact, as we described above, *R. leguminosarum* biovar *viciae* 3841 contains an acylase (Basu et al., 1999) (a PagL homolog) that removes a β-hydroxymyristyl residue from its LA, and in *R. etli* bacteroids, the tetraacylated LA is the major species that is present (D'Haeze et al., 2007). Thus, the fatty acylation pattern of rhizobial LA structures and/or its modification during symbiosis may be important for promoting a localized suppression of the host defense response.

Some early work suggested that symbiotic infection may result in ISR. In that work, a split root system was used and it was discovered that inoculation on one side of the root followed by later inoculation on the second side of the roots resulted in nodules on the first side and inhibition of nodulation of the second side (Kosslak and Bohlool, 1984; Sargent et al., 1987). It is not known if this effect is a result of an LPS caused ISR. However, it has been reported that rice plants inoculated with *R. leguminosarum* biovar *trifolii* have an increased production of phenolics resulting in ISR in these plants under pathogenic stress (Mishra et al., 2006). In another study, it is reported that LPSs from *R. etli* result in an ISR that protects potato roots against nematode infection. Thus, it is possible that *Rhizobium*-legume symbiosis, in addition to providing reduced nitrogen to the host legume, promotes resistance to infection by other potential pathogens via an ISR.

The response of the plant host to rhizobial LPS is an area that requires more investigation. The availability of numerous rhizobial symbiont genome sequences, defined LPS mutants and isolated structures from these mutants, as well as a number of legume host sequences (i.e. *M. truncatula*, *Glcyine max*, and *Lotus japonicus*) offer the tools required to define the structure-function aspects that rhizobial LPS play during symbiosis.

References

Ausubel, F.M. Are innate immune signaling pathways in plants and animals conserved? Nat Immunol 6 (2005) 973–979.

Basu, S.S., Karbarz, M.J., Raetz, C.R.H. Expression cloning and characterization of the C28 acyltransferase of lipid A biosynthesis in *Rhizobium leguminosarum*. J Biol Chem 277 (2002) 28959–28971.

Basu, S.S., White, K.A., Que, N.L.S., Raetz, C.R.H. A deacylase in *Rhizobium leguminosarum* membranes that cleaves the 3- O -linked β-hydroxymyristoyl moiety of lipid A precursors. J Biol Chem 274 (1999) 11150–11158.

Bhat, U.R., Carlson, R.W., Busch, M., Mayer, H. Distribution and phylogenetic significance of 27-hydroxy- octacosanoic acid in lipopolysaccharides from bacteria belonging to the alpha-2 subgroup of Proteobacteria. Int J Sys Bacteriol 41 (1991) 213–217.

Bhat, U.R., Forsberg, L.S., Carlson, R.W. The structure of the lipid A component of *Rhizobium leguminosarum* bv. phaseoli lipopolysaccharide. A unique non-phosphorylated lipid A containing 2-amino-2-deoxy-gluconate, galacturonate, and glucosamine. J Biol Chem 269 (1994) 14402–14410.

Bhat, U.R., Mayer, H., Yokota, A., Hollingsworth, R.I., Carlson, R.W. Occurrence of lipid A variants with 27-hydroxyoctacosanoic acid in lipopolysaccharides from the *Rhizobiaceae* group. J Bacteriol 173 (1991) 2155–2159.

Bolanos, L., Redondo-Nieto, M., Rivilla, R., Brewin, N.J., Bonilla, I. Cell surface interactions of *Rhizobium* bacteroids and other bacterial strains with symbiosomal and peribacteroid membrane components from pea nodules. Mol Plant-Microbe Inter 17 (2004) 216–223.

Brewin, N. Plant cell wall remodelling in the *Rhizobium*-legume symbiosis. Crit Rev Plant Sci 23 (2004) 293–316.

Brewin, N.J., Perotto, S., Kannenberg, E.L., Rae, A.L., Rathbun, E.A., Lucas, M.M., Kardailsky, I., Gunder, A., Bolanos, L., Donovan, N., Drobak, B.K., Nester, E.W., Verma, D.P.S. Mechanisms of cell and tissue invasion by *Rhizobium leguminosarum*: The role of cell surface interactions. Current Plant Science and Biotechnology in Agriculture Advances in Molecular Genetics of Plant-Microbe Interactions. Kluwer Academic Publishers, Dordrecht/Boston/London (1993), pp. 369–380.

Broughton, W.J., Hanin, M., Relic, B., Kopcinska, J., Golinowski, W., Simsek, S., Ojanen-Reuhs, T., Reuhs, B., Marie, C., Kobayashi, H., Bordogna, B., Le Quere, A., Jabbouri, S., Fellay, R., Perret, X., Deakin, W.J. Flavonoid-inducible modifications to rhamnan O antigens are necessary for *Rhizobium* sp. strain NGR234-legume symbioses. J Bacteriol 188 (2006) 3654–3663.

Carlson, R.W. The heterogeneity of *Rhizobium* lipopolysaccharides. J Bacteriol 158 (1984) 1012–1017.
Carlson, R.W., Kalembasa, S., Turowski, D., Pachori, P., Noel, K.D. Characterization of the lipopolysaccharide from a *Rhizobium phaseoli* mutant that is defective in infection thread development. J Bacteriol 169 (1987) 4923–4928.
Carlson, R.W., Krishnaiah, B.S. Structures of the oligosaccharides obtained from the core regions of the lipopolysaccharides of *Bradyrhizobium japonicum* 61A101c and its symbiotically defective lipopolysaccharide mutant, JS314. Carbohydr Res 231 (1992) 205–219.
Carlson, R.W., Sanders, R.E., Napoli, C., Albersheim, P. Host-symbiont interactions III. Purification and characterization of *Rhizobium* lipopolysaccharides. Plant Physiol 62 (1978) 912–917.
Carrion, M., Bhat, U.R., Reuhs, B., Carlson, R.W. Isolation and characterization of the lipopolysaccharides from *Bradyrhizobium japonicum*. J Bacteriol 172 (1990) 1725–1731.
Casabuono, A.C., D'Antuono, A.L., Sato, Y., Nonami, H., Uglade, R., Lepek, V.C., Erra-Balsells, R., Couto, A.S. A matrix-assisted laser desorption/ionization mass spectrometry approach to the lipid A from *Mesorhizobium loti*. Rapid Comm Mass Spectrom 20 (2006) 2175–2182.
Cava, J.R., Tao, H., Noel, K.D. Mapping of complementation groups within a *Rhizobium leguminosarum* CFN42 chromosomal region required for lipopolysaccharide synthesis. Mol Gen Genetics 221 (1990) 125–128.
Choma, A. Lipopolysaccharides from *Mesorhizobium huakuii* and *Mesorhizobium ciceri*: Chemical and immunological comparitive data. Acta Biochim Pol 49 (2002) 1043–1052.
Choma, A., Sowinski, P. Characterization of *Mesorhizobium huakuii* lipid A containing both D-galacturonic acid and phosphate residues. Eur J Biochem/FEBS 271 (2004) 1310–1322.
Choma, A., Sowinski, P., Mayer, H. Structure of the O-specific polysaccharide of *Mesorhizobium huakuii* IFO15243T. Carbohydr Res 329 (2000) 459–464.
Creuzenet, C., Lam, J.S. Topological and functional characterization of WbpM, an inner membrane UDP-GlcNAc C6-dehydratase essential for lipopolysaccharide biosynthesis in *Pseudomonas aeruginosa*. Mol Microbiol 41 (2001) 1295–1310.
Creuzenet, C., Schur, M.J., Li, J., Wakarchuk, W.W., Lam, J.S. FlaA1, a new bifunctional UDP-GlcNAc C6-dehydratase/ C4 reductase from *Helicobacter pylori*. J Biol Chem 275 (2000) 34873–34880.
Cronan, G.E., Keating, D.H. *Sinorhizobium meliloti* sulfotransferase rhat modifies lipopolysaccharide. J Bacteriol 186 (2004) 4168–4176.
D'Antuono, A.L., Casabuono, A., Couto, A., Ugalde, R.A., Lepek, V.C. Nodule development induced by *Mesorhizobium loti* mutant strains affected in polysaccharide synthesis. Mol Plant-Microbe Interact 18 (2005) 446–457.
D'Haeze, W., Leoff, C., Freshour, G., Noel, K.D., Carlson, R.W. *Rhizobium etli* CE3 Bacteroid lipopolysaccharides are structurally similar but not identical to those produced by cultured CE3 bacteria. J Biol Chem 282 (2007) 17101–17113.
De Castro, C., Bedini, E., Garozzo, D., Sturiale, L., Parrilli, M. Structural determination of the O-chain moieties of the lipopolysaccharide fraction from *Agrobacterium radiobacter* DSM 30147. Eur J Org Chem 2004 (2004) 3842–3849.
De Castro, C., Bedini, E., Nunziata, R., Rinaldi, R., Mangoni, L., Parrilli, M. Elucidation of the O-chain structure from the lipopolysaccharide of *Agrobacterium tumefaciens* strain C58. Carbohydr Res 338 (2003) 1891–1894.
De Castro, C., Carannante, A., Lanzetta, R., Liparoti, V., Molinaro, A., Parrilli, M. Core oligosaccharide structure from the highly phytopathogenic *Agrobacterium tumefaciens* TT111 and conformational analysis of the putative rhamnan epitope. Glycobiology 16 (2006) 1272–1280.
De Castro, C., Carannante, A., Lanzetta, R., Nunziata, R., Piscopo, V., Parrilli, M. Elucidation of two O-chain structures from the lipopolysaccharide fraction of *Agrobacterium tumefaciens* F/1. Carbohydr Res 339 (2004) 2451–2455.
De Castro, C., De Castro, O., Molinaro, A., Parrilli, M., Structural determination of the O-chain polysaccharide from *Agrobacterium tumefaciens*, strain DSM 30205. Eur J Biochem/FEBS 269 (2002) 2885–2888.

De Castro, C., Molinaro, A., Lanzetta, R., Silipo, A., Parrilli, M. Lipopolysaccharide structures from *Agrobacterium* and *Rhizobiaceae* species. Carbohydr Res 343 (2008) 1924–1933.

De Castro, C., Sturiale, L., Parrilli, M. Characterisation of the alpha-(1->3) homopolymer of L-*glycero*-D-*manno*-heptose units isolated from the O-chain polysaccharide of *Agrobacterium radiobacter*. Eur J Org Chem 2004 (2004) 2436–2440.

de Maagd, R.A., Rao, A.S., Mulders, I.H.M., Goosen-de Roo, L., van Loosdrecht, M.C.M., Wijffelman, C.A., Lugtenberg, B.J.J. Isolation and characterization of mutants of *Rhizobium leguminosarum* bv. *viciae* 248 with altered lipopolysaccharides: Possible role of surface charge or hydrophobicity in bacterial release from the infection thread. J Bacteriol 171 (1989) 1143–1150.

Denison, R.F., Kiers, E.T. Lifestyle alternatives for rhizobia: Mutualism, parasitism, and forgoing symbiosis. FEMS Microbiol Lett 237 (2004) 187–193.

Dow, M., Newman, M.A., von Roepenack, E. The induction and modulation of plant defense responses by bacterial lipopolysaccharides. Annu Rev Phytopathol 38 (2000) 241–261.

Duelli, D.M., Tobin, A., Box, J.M., Kolli, V.S.K., Carlson, R.W., Noel, K.D. Genetic locus required for antigenic maturation of *Rhizobium etli* CE3 lipopolysaccharide. J Bacteriol 183 (2001) 6054–6064.

Fauvart, M., Michiels, J. Rhizobial secreted proteins as determinants of host specificity in the *Rhizobium*-legume symbiosis. FEMS Microbiol Lett 285 (2008) 1–9.

Ferguson, A.D., Welte, W., Hofmann, E., Lindner, B., Holst, O., Coulton, J.W., Diederichs, K. A conserved structural motif for lipopolysaccharide recognition by procaryotic and eucaryotic proteins. Struct Fold & Design 8 (2000) 585–592.

Ferguson, G.P., Datta, A., Baumgartner, J., Roop, R.M., II, Carlson, R.W., Walker, G.C. Similarity to peroxisomal-membrane protein family reveals that *Sinorhizobium* and *Brucella* BacA affect lipid-A fatty acids. Proc Natl Acad Sci USA 101 (2004) 5012–5017.

Ferguson, G.P., Datta, A., Carlson, R.W., Walker, G.C. Importance of unusually modified lipid A in *Sinorhizobium* stress resistance and legume symbiosis. Mol Microbiol 56 (2005) 68–80.

Ferguson, G.P., Jansen, A., Marlow, V.L., Walker, G.C. BacA-mediated bleomycin sensitivity in *Sinorhizobium meliloti* is independent of the unusual lipid A modification. J Bacteriol 188 (2006) 3143–3148.

Fernandez de Cordoba, F.J., Rodreguez-Carvajal, M.A., Tejero-Mateo, P., Corzo, J., Gil-Serrano, A.M. Structure of the O-antigen of the main lipopolysaccharide isolated from *Sinorhizobium fredii* SMH12. Biomacromolecules 9 (2008) 678–685.

Forsberg, L.S., Bhat, U.R., Carlson, R.W. Structural characterization of the O-antigenic polysaccharide of the lipopolysaccharide from *Rhizobium etli* strain CE3. A unique O-acetylated glycan of discrete size, containing 3-O-methyl-6-deoxy-L-talose and 2,3,4-tri-O-methyl-L-fucose. J Biol Chem 275 (2000) 18851–18863.

Forsberg, L.S., Carlson, R.W. The structures of the lipopolysaccharides from *Rhizobium etli* strains CE358 and CE359 – The complete structure of the core region of *R. etli* lipopolysaccharides. J Biol Chem 273 (1998) 2747–2757.

Forsberg, L.S., Carlson, R.W. Structural characterization of the primary O-antigenic polysaccharide of the *Rhizobium leguminosarum* 3841 lipopolysaccharide and Identification of a new 3-acetimidoylamino-3-deoxyhexuronic acid glycosyl component: A unique O-methylated glycan of uniform size containing 6-deoxy-3-O-methyl-D-talose, N-acetylquinovosamine, and rhizoaminuronic acid (3-acetimidoylamino-3-deoxy-D-*gluco*-hexuronic acid). J Biol Chem 283 (2008) 16037–16050.

Forsberg, L.S., Noel, K.D., Box, J., Carlson, R.W. Genetic locus and structural characterization of the biochemical defect in the O-antigenic polysaccharide of the symbiotically deficient *Rhizobium etli* mutant, CE166: Replacement of N-acetylquinovosamine with its hexosyl-4-ulose precursor. J Biol Chem 278 (2003) 51347–51359.

Fraysse, N., Jabbouri, S., Treilhou, M., Couderc, F., Poinsot, V. Symbiotic conditions induce structural modifications of *Sinorhizobium* sp. NGR234 surface polysaccharides. Glycobiology 12 (2002) 741–748.

Frugier, F., Kosuta, S., Murray, J.D., Crespi, M., Szczyglowski, K. Cytokinin: Secret agent of symbiosis. Trends Plant Sci 13 (2008) 115–120.

Gage, D.J. Infection and invasion of roots by symbiotic, nitrogen-fixing rhizobia during nodulation of temperate legumes. Microbiol Mol Biol Rev 68 (2004) 280–300.

Galibert, F., Finan, T.M., Long, S.R., Puhler, A., Abola, P., Ampe, F., Barloy-Hubler, F., Barnett, M.J., Becker, A., Boistard, P., Bothe, G., Boutry, M., Bowser, L., Buhrmester, J., Cadieu, E., Capela, D., Chain, P., Cowie, A., Davis, R.W., Dreano, S., Federspiel, N.A., Fisher, R.F., Gloux, S., Godrie, T., Goffeau, A., Golding, B., Gouzy, J., Gurjal, M., Hernandez-Lucas, I., Hong, A., Huizar, L., Hyman, R.W., Jones, T., Kahn, D., Kahn, M.L., Kalman, S., Keating, D.H., Kiss, E., Komp, C., Lelaure, V., Masuy, D., Palm, C., Peck, M.C., Pohl, T.M., Portetelle, D., Purnelle, B., Ramsperger, U., Surzycki, R., Thebault, P., Vandenbol, M., Vorholter, F.-J., Weidner, S., Wells, D.H., Wong, K., Yeh, K.-C., Batut, J. The composite genome of the legume symbiont *Sinorhizobium meliloti*. Science 293 (2001) 668–672.

Gargiulo, V., Garozzo, D., Lanzetta, R., Molinaro, A., Sturiale, L., Castro, C.D., Parrilli, M. *Rhizobium rubi*: A gram-negative phytopathogenic bacterium expressing the Lewis B epitope on the outer core of its lipooligosaccharide fraction. ChemBioChem 9 (2008) 1830–1835.

Gibson, K.E., Kobayashi, H., Walker, G.C. Molecular determinants of a symbiotic chronic infection. Ann Rev Gen 42 (2008) 413.

Gil-Serrano, A.M., González-Jiménez, I., Mateo, P.T., Bermabé, M., Jiménez-Barbero, J., Megí as, M., Romero-Vzquez, M.J. Structural analysis of the O-antigen of the lipopolysaccharide of *Rhizobium tropici* CIAT899. Carbohydr Res 275 (1995) 285–294.

Gonzalez, V., Santamaria, R.I., Bustos, P., Hernandez-Gonzalez, I., Medrano-Soto, A., Moreno-Hagelsieb, G., Janga, S.C., Ramirez, M.A., Jimenez-Jacinto, V.N., Collado-Vides, J., Dávila G. The partitioned *Rhizobium etli* genome: Genetic and metabolic redundancy in seven interacting replicons. Proc Natl Acad Sci USA 103 (2006) 3834–3839.

Gudlavalleti, S.K., Forsberg, L.S. Structural Characterization of the lipid A component of *Sinorhizobium* sp. NGR234 rough and smooth form lipopolysaccharide. Demonstration that the distal amide-linked acyloxyacylresidue containing the long chain fatty acid is conserved in *Rhizobium* and *Sinorhizobium* sp.. J Biol Chem 278 (2003) 3957–3968.

Heiko Scheidle, Andrea Groß, Karsten Niehaus. The Lipid A substructure of the *Sinorhizobium meliloti* lipopolysaccharides is sufficient to suppress the oxidative burst in host plants. New Phytol 165 (2005) 559–566.

Ingram, B., Sohlenkamp, C., Geiger, O., Raetz, C.R. Altered lipid A structures and polymyxin hypersensitivity of *Rhizobium etli* mutants lacking the LpxE and LpxF phosphatases. Biochim Biophys Acta. Biochem. Biophys. Acta 1801 (2010) 593–604.

Jabbouri, S., Hanin, M., Fellay, R., Quesada-Vincens, D., Reuhs, B., Carlson, R.W., Perret, X., Freiberg, C., Rosenthal, A., Leclerc, D., Broughton, W.J., Relic, B., Stacey, G., Mullin, B., Gresshoff, P.M. *Rhizobium* species NGR234 host-specificity of nodulation locus III contains nod⁻ and fix⁻ genes. Biology of Plant-Microbe Interactions. ISMPMI, St. Paul (1996), pp. 319–324.

Jeyaretnam, B., Glushka, J., Kolli, V.S.K., Carlson, R.W. Characterization of a novel lipid-A from *Rhizobium* species Sin-1. A unique lipid-A structure that is devoid of phosphate and has a glycosyl backbone consisting of glucosamine and 2-aminogluconic acid. J Biol Chem 277 (2002) 41802–41810.

Kadrmas, J.L., Allaway, D., Studholme, R.E., Sullivan, J.T., Ronson, C.W., Poole, P.S., Raetz, C.R.H. Cloning and overexpression of glycosyltransferases that generate the lipopolysaccharide core of *Rhizobium leguminosarum*. J Biol Chem 273 (1998) 26432–26440.

Kaneko, T., Nakamura, Y., Sato, S., Asamizu, E., Kato, T., Sasamoto, S., Watanabe, A., Idesawa, K., Ishikawa, A., Kawashima, K., Kimura, T., Kishida, Y., Kiyokawa, C., Kohara, M., Matsumoto, M., Matsuno, A., Mochizuki, Y., Nakayama, S., Nakazaki, N., Shimpo, S., Sugimoto, M., Takeuchi, C., Yamada, M., Tabata, S. Complete genome structure of the nitrogen-fixing symbiotic bacterium *Mesorhizobium* loti. DNA Res 7 (2000a) 331–338.

Kaneko, T., Nakamura, Y., Sato, S., Asamizu, E., Kato, T., Sasamoto, S., Watanabe, A., Idesawa, K., Ishikawa, A., Kawashima, K., Kimura, T., Kishida, Y., Kiyokawa, C., Kohara, M., Matsumoto, M., Matsuno, A., Mochizuki, Y., Nakayama, S., Nakazaki, N., Shimpo, S., Sugimoto, M., Takeuchi, C., Yamada, M., Tabata, S. Complete genome structure of the nitrogen-fixing symbiotic bacterium *Mesorhizobium loti* (supplement). DNA Res 7 (2000b) 381–406.

Kaneko, T., Nakamura, Y., Sato, S., Minamisawa, K., Uchiumi, T., Sasamoto, S., Watanabe, A., Idesawa, K., Iriguchi, M., Kawashima, K., Kohara, M., Matsumoto, M., Shimpo, S., Tsuruoka, H., Wada, T., Yamada, M., Tabata, S. Complete genomic sequence of nitrogen-fixing symbiotic bacterium *Bradyrhizobium japonicum* USDA110. DNA Res 9 (2002a) 189–197.

Kaneko, T., Nakamura, Y., Sato, S., Minamisawa, K., Uchiumi, T., Sasamoto, S., Watanabe, A., Idesawa, K., Iriguchi, M., Kawashima, K., Kohara, M., Matsumoto, M., Shimpo, S., Tsuruoka, H., Wada, T., Yamada, M., Tabata, S. Complete genomic sequence of nitrogen-fixing symbiotic bacterium *Bradyrhizobium japonicum* USDA110 (supplement). DNA Res 9 (2002b) 225–256.

Kanjilal-Kolar, S., Basu, S.S., Kanipes, M.I., Guan, Z., Garrett, T.A., Raetz, C.R.H. Expression Cloning of three *Rhizobium leguminosarum* lipopolysaccharide core galacturonosyltransferases. J Biol Chem 281 (2006) 12865–12878.

Kanjilal-Kolar, S., Raetz, C.R.H. Dodecaprenyl phosphate-galacturonic acid as a donor substrate for lipopolysaccharide core glycosylation in *Rhizobium leguminosarum*. J Biol Chem 281 (2006) 12879–12887.

Kannenberg, E.L., Brewin, N.J. Expression of a cell surface antigen from *Rhizobium* leguminosarum 3841 is regulated by oxygen and pH. J Bacteriol 171 (1989) 4543–4548.

Kannenberg, E.L., Brewin, N.J. Host-plant invasion by *Rhizobium*: The role of cell-surface components. Trends Microbiol 2 (1994) 277–283.

Kannenberg, E.L., Carlson, R.W. Lipid A, O-chain modifications cause *Rhizobium* lipopolysaccharides to become hydrophobic during bacteroid development. Mol Microbiol 39 (2001) 379–392.

Kannenberg, E.L., Perotto, S., Bianciotto, V. Rathburn, E.A., Brewin, N.J., Lipopolysaccharide epitope expression of *Rhizobium* bacteroids as revealed by *in situ* immunolabelling of pea root nodule sections. J Bacteriol 176 (1994) 2021–2032.

Kannenberg, E.L., Rathbun, E.A., Brewin, N.J. Molecular dissection of structure and function in the lipopolysaccharide of *Rhizobium leguminosarum* strain 3841 using monoclonal antibodies and genetic analysis. Mol Microbiol 6(17) (1992) 2477–2487.

Kannenberg, E.L., Reuhs, B.L., Forsberg, L.S., Carlson, R.W. Lipopolysaccharides, K-antigens: Their structures, biosynthesis, and function. In: Spaink, H.P., Kondorosi, A., Hooykaas, P.J.J (eds.), The *Rhizobiaceae*: Molecular Biology of Model Plant-Associated Bacteria. Kluwer Academic Publishers, Dordrecht/Boston/London (1998), pp. 119–154.

Karbarz, M.J., Kalb, S.R., Cotter, R.J., Raetz, C.R.H. Expression cloning and biochemical characterization of a *Rhizobium leguminosarum* lipid A 1-phosphatase. J Biol Chem 278 (2003) 39269–39279.

Karbarz, M.J., Six, D.A., Raetz, C.R.H. Purification and characterization of the lipid A 1-phosphatase LpxE of *Rhizobium leguminosarum*. J Biol Chem 284 (2009) 414–425.

Keating, D.H. *Sinorhizobium meliloti* SyrA mediates the transcriptional regulation of genes involved in lipopolysaccharide sulfation and exopolysaccharide biosynthesis. J Bacteriol 189 (2007) 2510–2520.

Keating, D.H., Willits, M.G., Long, S.R. A *Sinorhizobium meliloti* lipopolysaccharide mutant Altered in cell surface sulfation. J Bacteriol 184 (2002) 6681–6689.

Kneidinger, B., Larocque, S., Brisson, J.-R., Cadotte, N., Lam, J.S. Biosynthesis of 2-acetamido-2,6-dideoxy-L-hexoses in bacteria follows a pattern distinct from those of the pathways of 6-deoxy-L-hexoses. Biochem J 371 (2003) 989–995.

Kosslak, R.M., Bohlool, B.B. Suppression of nodule development on one side of a split root system of soybean caused by a prior inoculation of the other side. Plant Physiol 75 (1984) 125–130.

López-Lara, I.M., Orgambide, G., Dazzo, F.B., Olivares, J., Toro, N. Surface polysaccharide mutants of *Rhizobium* sp. (Acacia) strain GRH2: Major requirement of lipopolysaccharide for successful invasion of *Acacia* nodules and host range determination. Microbiology 141 (1995) 573–581.

Laus, M.C., Logman, T.J., van Brussel, A.A.N., Carlson, R.W., Azadi, P., Gao, M.Y., Kijne, J.W. Involvement of *exo5* in production of surface polysaccharides in *Rhizobium leguminosarum* and its role in nodulation of *Vicia sativa* subsp. nigra. J Bacteriol 186 (2004) 6617–6625.

Lee, K.-B., De Backer, P., Aono, T., Liu, C.-T., Suzuki, S., Suzuki, T., Kaneko, T., Yamada, M., Tabata, S., Kupfer, D., Najar, F., Wiley, G., Roe, B., Binnewies, T., Ussery, D., D'Haeze, W., Den Herder, J., Gevers, D., Vereecke, D., Holsters, M., Oyaizu, H. The genome of the versatile nitrogen fixer *Azorhizobium caulinodans* ORS571. BMC Genomics 9 (2008) 271.

Lerouge, I., Laeremans, T., Verreth, C., Vanderleyden, J., van Soom, C., Tobin, A., Carlson, R.W. Identification of an ABC transporter for export of the O-antigen across the inner membrane in *Rhizobium etli* based on the genetic, functional and structural analysis of an lps mutant deficient in O-antigen. J Biol Chem 276 (2001) 17190–17198.

Lerouge, I., Vanderleyden, J. O-antigen structural variation: Mechanisms and possible roles in animal/plant-microbe interactions. FEMS Microbiol Rev 26 (2002) 17–47.

Lerouge, I., Verreth, C., Michiels, J., Carlson, R.W., Datta, A., Gao, M., Vanderleyden, J. Three genes encoding for putative methyl- and acetyltransferases map adjacent to the wzm and wzt genes and are essential for O-antigen biosynthesis in *Rhizobium etli* CE3. Mol Plant-Microbe Interact 16 (2003) 1085–1093.

Ma, B., Reynolds, C.M., Raetz, C.R.H. Periplasmic orientation of nascent lipid A in the inner membrane of an *Escherichia coli* LptA mutant. Proc Natl Acad Sci USA 105 (2008) 13823–13828.

Marlow, V.L., Haag, A.F., Kobayashi, H., Fletcher, V., Scocchi, M., Walker, G.C., Ferguson, G.P. Essential role for the BacA protein in the uptake of a truncated eukaryotic peptide in *Sinorhizobium meliloti*. J Bacteriol 191 (2009) 1519–1527.

Mayer, H., Krauss, J.H., Urbanik-Sypniewska, T., Puvanesarajah, V., Stacey, G., Auling, G. Lipid A with 2,3-diamino-2,3-dideoxyglucose in lipopolysaccharides from slow-growing members of Rhizobiaceae and from *Pseudomonas carboxydovorans*. Arch Microbiol 151 (1989) 111–116.

Miller, W.L., Wenzel, C.Q., Daniels, C., Larocque, S., Brisson, J.-R., Lam, J.S. Biochemical characterization of WbpA, a UDP-N-acetyl-D-glucosamine 6-dehydrogenase involved in O-antigen biosynthesis in *Pseudomonas aeruginosa* PAO1. J Biol Chem 279 (2004) 37551–37558.

Mishra, R., Singh, R., Jaiswal, H., Kumar, V., Maurya, S. *Rhizobium*-mediated induction of phenolics and plant growth promotion in rice (*Oryza sativa* L.). Curr Microbiol 52 (2006) 383–389.

Molinaro, A., De Castro, C., Lanzetta, R., Parrilli, M., Raio, A., Zoina, A. Structural elucidation of a novel core oligosaccharide backbone of the lipopolysaccharide from the new bacterial species *Agrobacterium larrymoorei*. Carbohydr Res 338 (2003) 2721–2730.

Nakano, Y., Suzuki, N., Yoshida, Y., Nezu, T., Yamashita, Y., Koga, T. Thymidine diphosphate-6-deoxy-l-lyxo-4-hexulose reductase synthesizing dTDP-6-deoxy-l-talose from *Actinobacillus actinomycetemcomitans*. J Biol Chem 275 (2000) 6806–6812.

Newman, J., Dow, J.M., Molinaro, A., Parrilli, M. Priming, induction and modulation of plant defence responses by bacterial lipopolysaccharides. J Endotoxin Res 13 (2007) 69 - 84.

Noel, K.D., Box, J.M., Bonne, V.J 2-O-Methylation of fucosyl residues of a rhizobial lipopolysaccharideis increased in response to host exudate and is eliminated in a symbiotically defective mutant. Appl Environ Microbiol 70 (2004) 1537–1544.

Noel, K.D., Sanchez, A., Fernandez, L., Leemans, J., Cevallos, M.A. *Rhizobium phaseoli* symbiotic mutants with transposon Tn5 insertions. J Bacteriol 158 (1984) 148–155.

Noel, K.D., VandenBosch, K.A., Kulpaca, B. Mutations in *Rhizobium phaseoli* that lead to arrested development of infection threads. J Bacteriol 168 (1986) 1392–1401.

Nürnberger, T., Brunner, F., Kemmerling, B., Piater, L. Innate immunity in plants and animals: Striking similarities and obvious differences. Immunol Rev 198 (2004) 249–266.

Nürnberger, T., Kemmerling, B. Receptor protein kinases-pattern recognition receptors in plant immunity. Trends Plant Sci 11 (2006) 519–522.

Ojeda, K.J., Box, J.M., Noel, K.D. Genetic basis for *Rhizobium etli* CE3 O-antigen O-methylated residues that vary according to growth conditions. J Bacteriol (2009) JB.01154–01109.

Oldroyd, G.E.D., Downie, J.A. Coordinating nodule morphogenesis with rhizobial infection in legumes. Ann Rev Plant Biol 59 (2008) 519–546.

Perotto, S., Brewin, N.J., Kannenberg, E.L. Cytological evidence for a host defense response that reduces cell and tissue invasion in pea nodules by lipopolysaccharide- defective mutants of *Rhizobium leguminosarum* strain 3841. Mol Plant-Microbe Interact 7 (1994) 99–112.

Perotto, S., VandenBosch, K.A., Butcher, G.W., Brewin, N.J. Molecular composition and development of the plant glycocalyx associated with the peribacteroid membrane of pea root nodules. Development 112 (1991) 763–773.

Price, N.P.J., Jeyaretnam, B., Carlson, R.W., Kadrmas, J.L., Raetz, C.R.H., Brozek, K.A. Lipid A biosynthesis in *Rhizobium leguminosarum*: Role of a novel Kdo activated 4′-phosphatase. Proc Natl Acad Sci USA 92 (1995) 7352–7356.

Price, N.P.J., Kelly, T.M., Raetz, C.R.H., Carlson, R.W. Biosynthesis of a structurally novel lipid A in *Rhizobium leguminosarum*: Identification and characterization of six metabolic steps leading from UDP-GlcNAc to Kdo_2-lipid IV_A. J Bacteriol 176 (1994) 4646–4655.

Puvanesarajah, V., Schell, F.M., Gerhold, D., Stacey, G. Cell surface polysaccharides from *Bradyrhizobium japonicum* and a nonnodulating mutant. J Bacteriol 169 (1987) 137–141.

Que-Gewirth, N.L.S., Karbarz, M.J., Kalb, S.R., Cotter, R.J., Raetz, C.R.H. Origin of the 2-amino-2-deoxy-gluconate unit in *Rhizobium leguminosarum* Lipid A. Expression cloning of the outer membrane oxidase LpxQ. J Biol Chem 278 (2003a) 12120.

Que-Gewirth, N.L.S., Lin, S., Cotter, R.J., Raetz, C.R.H. An outer membrane enzyme that generates the 2-amino-2- deoxy-gluconate moiety of *Rhizobium leguminosarum* lipid A. J Biol Chem 278 (2003b) 12109.

Que, N.L.S., Lin, S.H., Cotter, R.J., Raetz, C.R.H. Purification and mass spectrometry of six lipid A species from the bacterial endosymbiont *Rhizobium etli* - Demonstration of a conserved distal unit and a variable proximal portion. J Biol Chem 275 (2000a) 28006–28016.

Que, N.L.S., Ribeiro, A.A., Raetz, C.R.H. Two-dimensional NMR spectroscopy and structures of six lipid A species from *Rhizobium etli* CE3-Detection of an acyloxyacyl residue in each component and origin of the aminogluconate moiety. J Biol Chem 275 (2000b) 28017–28027.

Raetz, C.R.H., Reynolds, C.M., Trent, M.S., Bishop, R.E. Lipid Amodification systems in gram-negative bacteria. Ann Rev Biochem 76 (2007) 295–329.

Raetz, C.R.H., Whitfield, C. Lipopolysaccharide endotoxins. Ann Rev Biochem 71 (2002) 635–700.

Reuhs, B.L., Geller, D.P., Kim, J.S., Fox, J.E., Kolli, V.S.K., Pueppke, S.G. *Sinorhizobium fredii* and *Sinorhizobium meliloti* produce structurally conserved lipopolysaccharides and strain-specific K antigens. Appl Environ Microbiol 64 (1998) 4930–4938.

Reuhs, B.L., Kim, J.S., Badgett, A., Carlson, R.W. Production of cell-associated polysaccharides of *Rhizobium fredii* USDA205 is modulated by apigenin and host root extract. Mol Plant Microbe Interact 7 (1994) 240–247.

Reuhs, B.L., Relic, B., Forsberg, L.S., Marie, C., Ojanen-Reuhs, T., Stephens, S.B., Wong, C.-H., Jabbouri, S., Broughton, W.J. Structural characterization of a flavonoid-inducible *Pseudomonas aeruginosa* A-band-Like O antigen of *Rhizobium* sp. strain NGR234, required for the formation of nitrogen-fixing nodules. J Bacteriol 187 (2005) 6479–6487.

Reuhs, B.L., Stephens, S.B., Geller, D.P., Kim, J.S., Glenn, J., Przytycki, J., Ojanen-Reuhs, T. Epitope identification for a panel of anti-*Sinorhizobium meliloti* monoclonal antibodies and application to the analysis of K antigens and lipopolysaccharides from bacteroids. Appl Environ Microbiol 65 (1999) 5186–5191.

Ridley, B.L., Jeyaretnam, B.S., Carlson, R.W. The type and yield of lipopolysaccharide from symbiotically deficient *Rhizobium* lipopolysaccharide mutants vary depending on the extraction method. Glycobiology 10 (2000) 1013–1023.

Rocchetta, H.L., Burrows, L.L., Lam, J.S. Genetics of O-antigen biosynthesis in *Pseudomonas aeruginosa*. Microbiol Mol Biol Rev 63 (1999) 523–553.

Russa, R., Bruneteau, M., Shashkov, A.S., Urbanik-Sypniewska, T., Mayer, H. Characterization of the lipopolysaccharides from *Rhizobium meliloti* strain 102F51 and its nonnodulating mutant WL113. Arch Microbiol 165 (1996) 26–33.

Russa, R., Urbanik-Sypniewska, T., Shashkov, A.S., Banaszek, A., Zamojski, A., Mayer, H. Partial structure of lipopolysaccharide isolated from *Rhizobium leguminosarum* bv. *trifolii* 24 and its GalA-negative Exo⁻ mutant. System Appl Microbiol 19 (1996) 1–8.
Russa, R., Urbanik-Sypniewska, T., Shashkov, A.S., Kochanowski, H., Mayer, H. The structure of the homopolymeric O-specific chain from the phenol soluble LPS of the *Rhizobium loti* type strain NZP2213. Carbohydrate Polymers 27 (1995) 299–303.
Sanchez-Contreras, M., Bauer, W.D., Gao, M., Robinson, J.B., Allan Downie, J. Quorum-sensing regulation in rhizobia and its role in symbiotic interactions with legumes. Phil Trans Royal Soc Biol Sci 362 (2007) 1149–1163.
Sargent, L., Huang, S.Z., Rolfe, B.G., Djordjevic, M. Split-root assays using *Trifolium subterraneum* show that *Rhizobium* infection induces a systemic response that can inhibit nodulation of another invasive *Rhizobium* strain. Appl Environ Microbiol 53 (1987) 1611–1619.
Sawada, H., Kuykendall, L.D., Young, J.M. Changing concepts in the systematics of bacterial nitrogen-fixing legume symbionts. J Gen Appl Microbiol 49 (2003) 155–179.
Schmeisser, C., Liesegang, H., Krysciak, D., Bakkou, N., Le Quere, A., Wollherr, A., Heinemeyer, I., Morgenstern, B., Pommerening-Roser, A., Flores, M., Palacios, R., Brenner, S., Gottschalk, G., Schmitz, R.A., Broughton, W.J., Perret, X., Strittmatter, A.W., Streit, W.R. *Rhizobium* sp. strain NGR234 possesses a remarkable number of secretion systems. Appl Environ Microbiol 75 (2009) 4035–4045.
Shah, J. Lipids, lipases, and lipid-modifying enzymes in plant disease resistance. Ann Rev Phytopathol 43 (2005) 229–260.
Sharypova, L.A., Niehaus, K., Scheidle, H., Holst, O., Becker, A. *Sinorhizobium meliloti acpXL* mutant lacks the C28 hydroxylated fatty acid moiety of lipid A and does not express a slow migrating form of lipopolysaccharide. J Biol Chem 278 (2003) 12946–12954.
Silipo, A., De Castro, C., Lanzetta, R., Molinaro, A., Parrilli, M. Full structural characterization of the lipid A components from the *Agrobacterium tumefaciens* strain C58 lipopolysaccharide fraction. Glycobiology 14 (2004) 805–815.
Silipo, A., Sturiale, L., Garozzo, D., Erbs, G., Tandrup, T., Lanzetta, J.R., Dow, J.M., Parrilli, M., Newman, M.-A., Molinaro, A. The acylation and phosphorylation pattern of lipid A from *Xanthomonas campestris* strongly influence its ability to trigger theinnate immune response in *Arabidopsis*. ChemBioChem 9 (2008) 896–904.
Sindhu, S.S., Brewin, N.J., Kannenberg, E.L. Immunochemical analysis of lipopolysaccharides from free-living and endosymbiotic forms of *Rhizobium leguminosarum*. J Bacteriol 172 (1990) 1804–1813.
Sindhu, S.S., Kannenberg, E.L., Brewin, N.J. Lipopolysaccharide maturation in pea and bean bacteroids. In: Bothe, de, Newton, B. (eds), Nitrogen Fixation: Hundred Years After. Gustav Fischer, Stuttgartand New York (1988), p. 480.
Soto, M.J., Domínguez-Ferreras, A., Pérez-Mendoza, D., Sanjuán, J., Olivares, J. Mutualism versus pathogenesis: The give-and-take in plant-bacteria interactions. Cell Microbiol 11 (2009) 381–388.
Sprent, J.I. Evolving ideas of legume evolution and diversity: A taxonomic perspective on the occurrence of nodulation. New Phytol 174 (2007) 11–25.
Stacey, G., Libault, M., Brechenmacher, L., Wan, J., May, G.D. Genetics and functional genomics of legume nodulation. Curr Op Plant Biol 9 (2006) 110–121.
Stacey, G., So, J.S., Roth, L.E., Bhagya Lakshmi, S.K., Carlson, R.W. A lipopolysaccharide mutant from *Bradyrhizobium japonicum* that uncouples plant from bacterial differentiation. Mol Plant-Microbe Interact 4 (1991) 332–340.
Suzuki, N., Nakano, Y., Yoshida, Y., Nezu, T., Terada, Y., Yamashita, Y., Koga, T. Guanosine diphosphate-4-keto-6-deoxy-D-mannose reductase in the pathway for the synthesis of GDP-6-deoxy-D-talose in *Actinobacillus actinomycetemcomitans*. Eur J Biochem 269 (2002) 5963–5971.
Sweet, C.R., Ribeiro, A.A., Raetz, C.R.H. Oxidation and transamination of the 3″-position of UDP-N-acetylglucosamine by enzymes from *Acidithiobacillus ferrooxidans*: Role in the

formation of lipid A molecules with four amide-linked acyl chains. J Biol Chem 279 (2004a) 25400–25410.

Sweet, C.R., Williams, A.H., Karbarz, M.J., Werts, C., Kalb, S.R., Cotter, R.J., Raetz, C.R.H. Enzymatic synthesis of lipid A molecules with four amide-linked acyl chains: LpxA acyltransferases selective for an analog of UDP-N-acetylglucosamine in which an amine replaces the 3″-hydroxyl group. J Biol Chem 279 (2004b) 25411–25419.

Tan, X.-J., Cheng, Y., Li, Y.-X., Li, Y.-G., Zhou, J.-C. BacA is indispensable for successful *Mesorhizobium–Astragalus* symbiosis. Appl Microbiol Biotech 84 (2009) 519–526.

Tao, H., Brewin, N.J., Noel, K.D. *Rhizobium leguminosarum* CFN42 lipopolysaccharide antigenic changes induced by environmental conditions. J Bacteriol 174 (1992) 2222–2229.

Tao, H., Noel, D. Lipopolysaccharide epitopes missing from bean bacteroids. In: Gresshoff, P.M., Roth, L.E., Stacey, G., Newton, W.E. (eds), Nitrogen Fixation: Achievements and Objectives Proceedings of the 8th International Congress on Nitrogen Fixation. Knoxville, Tennessee, USA, May 20–26, 1990. Chapman and Hall, New York and London (1990), p. 276.

Tellstrom, V., Usadel, B., Thimm, O., Stitt, M., Kuster, H., Niehaus, K. The lipopolysaccharide of *Sinorhizobium meliloti* suppresses defense-associated gene expression in cell cultures of the host plant *Medicago truncatula*. Plant Physiol 143 (2007) 825–837.

Townsend, G.E., II, Forsberg, L.S., Keating, D.H. *Mesorhizobium loti* produces *nodPQ*-dependent sulfated cell surface polysaccharides. J Bacteriol 188 (2006) 8560–8572.

Trent, M.S., Pabich, W., Raetz, C.R.H., Miller, S.I. A PhoP/PhoQ-induced lipase (PagL) that catalyzes 3-O-deacylation of lipid A precursors in membranes of *Salmonella typhimurium*. J Biol Chem 276 (2001) 9083–9092.

Turska-Szewczuk, A., Lotocka, B., Kutkowska, J., Król, J., Urbanik-Sypniewska, T., Russa, R. The incomplete substitution of lipopolysaccharide with O-chain prevents the establishment of effective symbiosis between *Mesorhizobium loti* NZP2213.1 and *Lotus corniculatus*. Microbiol Res 164 (2009) 163–173.

Turska-Szewczuk, A., Palusinska-Szysz, M., Russa, R. Structural studies of the O-polysaccharide chain from the lipopolysaccharide of symbiotically enhanced mutant Mlo-13 of *Mesorhizobium loti* NZP2213. Carbohydr Res 343 (2008) 477–482.

Turska-Szewczuk, A., Pietras, H., Borucki, W., Russa, R. Alteration of O-specific polysaccharide structure of symbiotically defective *Mesorhizobium* 2213.1 derived from strain NZP2213. Acta Biochim Pol 55 (2008) 191–199.

van de Wiel, C., Norris, J.H., Bochenek, B., Dickstein, R., Bisseling, T., Hirsch, A.M. Nodulin gene expression and ENOD2 localization in effective, nitrogen-fixing and ineffective, bacteria-free nodules of alfalfa. Plant Cell 2 (1990) 1009–1017.

VandenBosch, K.A., Brewin, N.J., Kannenberg, E.L. Developmental regulation of a rhizobium cell-surface antigen during growth of pea root-nodules. J Bacteriol 171 (1989) 4537–4542.

VandenBosch, K.A., Noel, K.D., Kaneko, Y., Newcomb, E.H. Nodule initiation elicited by noninfective mutants of *Rhizobium phaseoli*. J Bacteriol 162 (1985) 950–959.

Vanderlinde, E.M., Muszynski, A., Harrison, J.J., Koval, S.F., Foreman, D.L., Ceri, H., Kannenberg, E.L., Carlson, R.W., Yost, C.K. *Rhizobium leguminosarum* biovar *viciae* 3841, deficient in 27-hydroxyoctacosanoate-modified lipopolysaccharide, is impaired in desiccation tolerance, biofilm formation and motility. Microbiology 155 (2009) 3055–3069.

Vedam, V., Haynes, J.G., Kannenberg, E.L., Carlson, R.W., Sherrier, D.J. A *Rhizobium leguminosarum* lipopolysaccharide lipid A mutant induces nitrogen-fixing nodules with delayed and defective bacteroid formation. Mol Plant-Microbe Interact 17 (2004) 283–291.

Vedam, V., Kannenberg, E., Datta, A., Brown, D., Haynes-Gann, J.G., Sherrier, D.J., Carlson, R.W. The pea nodule environment restores the ability of a *Rhizobium leguminosarum* lipopolysaccharide *acpXL* mutant to add 27-hydroxyoctacosanoic acid to its lipid A. J Bacteriol 188 (2006) 2126–2133.

Vedam, V., Kannenberg, E.L., Haynes, J.G., Sherrier, D.J., Datta, A., Carlson, R.W. A *Rhizobium leguminosarum acpXL* mutant produces lipopolysaccharide lacking 27-hydroxyoctacosanoic acid. J Bacteriol 185 (2003) 1841–1850.

Wang, X., Karbarz, M.J., McGrath, S.C., Cotter, R.J., Raetz, C.R.H. MsbA transporter-dependent lipid A 1-dephosphorylation on theperiplasmic surface of the inner membrane: Topography of *Francisella novicida* LpxE expressed in *Escherichia coli*. J Biol Chem 279 (2004) 49470–49478.

Wang, X., McGrath, S.C., Cotter, R.J., Raetz, C.R.H. Expression cloning and periplasmic orientation of the *Francisella novicida* lipid A 4′-phosphatase LpxF. J Biol Chem 281 (2006) 9321–9330.

Wang, X., Ribeiro, A.A., Guan, Z., Abraham, S.N., Raetz, C.R.H. Attenuated virulence of a *Francisella* mutant lacking the lipid A 4′-phosphatase. Proc Natl Acad Sci USA 104 (2007) 4136–4141.

Wang, Y., Hollingsworth, R.I. The structure of the O-antigenic chain of the lipopolysaccharide of *Rhizobium trifolii* 4 s. Carbohydr Res 260 (1994) 305–317.

Westphal, O., Jann, K., Bacterial lipopolysaccharides, Meth Carbohydr Chem, 5 (1965) 83–91.

Wood, D.W., Setubal, J.C., Kaul, R., Monks, D.E., Kitajima, J.P., Okura, V.K., Zhou, Y., Chen, L., Wood, G.E., Almeida, N.F., Jr., Woo, L., Chen, Y., Paulsen, I.T., Eisen, J.A., Karp, P.D., Bovee, D., Sr., Chapman, P., Clendenning, J., Deatherage, G., Gillet, W., Grant, C., Kutyavin, T., Levy, R., Li, M.-J., McClelland, E., Palmieri, A., Raymond, C., Rouse, G., Saenphimmachak, C., Wu, Z., Romero, P., Gordon, D., Zhang, S., Yoo, H., Tao, Y., Biddle, P., Jung, M., Krespan, W., Perry, M., Gordon-Kamm, B., Liao, L., Kim, S., Hendrick, C., Zhao, Z.-Y., Dolan, M., Chumley, F., Tingey, S.V., Tomb, J.-F., Gordon, M.P., Olson, M.V., Nester, E.W. The genome of the natural genetic engineer *Agrobacterium tumefaciens* C58. Science 294 (2001) 2317–2323.

Wood, E.A., Butcher, G.W., Brewin, N.J., Kannenberg, E.L. Genetic derepression of a developmentally regulated lipopolysaccharide antigen from *Rhizobium leguminosarum* 3841. J Bacteriol 171 (1989) 4549–4555.

Wyckoff, T.J.O., Lin, S., Cotter, R.J., Dotson, G.D., Raetz, C.R.H. Hydrocarbon rulers in UDP- N -acetylglucosamine acyltransferases. *J Biol Chem* 273 (1998) 32369–32372.

Young, J.P., Crossman, L., Johnston, A., Thomson, N., Ghazoui, Z., Hull, K., Wexler, M., Curson, A., Todd, J., Poole, P., Mauchline, T., East, A., Quail, M., Churcher, C., Arrowsmith, C., Cherevach, I., Chillingworth, T., Clarke, K., Cronin, A., Davis, P., Fraser, A., Hance, Z., Hauser, H., Jagels, K., Moule, S., Mungall, K., Norbertczak, H., Rabbinowitsch, E., Sanders, M., Simmonds, M., Whitehead, S., Parkhill, J. The genome of *Rhizobium leguminosarum* has recognizable core and accessory components. Genome Biol 7 (2006) R34.

Zahringer, U., Knirel, Y.A., Lindner, B., Helbig, J.H., Sonnesson, A., Marre, R., Rietschel, E.T. The lipopolysaccharide of *Legionella pneumophila* serogroup I (strain Philadelphia 1): Chemical structure and biological significance. Prog Clin Biol Res 392 (1995) 113–139.

Zavaleta-Pastor, M., Sohlenkamp, C., Gao, J.-L., Guan, Z., Zaheer, R., Finan, T.M., Raetz, C.R.H., López-Lara, I.M., Geiger, O. *Sinorhizobium meliloti* phospholipase C required for lipid remodeling during phosphorus limitation. Proc Natl Acad Sci USA 107 (2010) 302–307.

Chapter 17
Lipopolysaccharides and Plant Innate Immunity

G. Erbs, A. Molinaro, J.M. Dow, and M.-A. Newman

Abstract Plants posses an innate immune system that has many parallels with those found in mammals and insects. A range of molecules of microbial origin called Microbe Associated Molecular Patterns (MAMPs) act to trigger basal defense responses in plants. These elicitors include lipopolysaccharides (LPS) from diverse Gram-negative bacteria. Both core oligosaccharide and the lipid A moieties of LPS as well as synthetic O-antigen oligosaccharides have activity in inducing defense responses in the model plant *Arabidopsis thaliana*. Very little is known of the mechanism of LPS perception by plants, although plant receptors for other MAMPs such as flagellin have been described. Recent work has implicated the *Arabidopsis* syntaxin PEN1 as a potential actor in LPS induction of plant defenses, which may suggest a role for vesicle trafficking in the signalling process.

Keywords *Arabidopsis thaliana* · Innate immunity · LPS · MAMPs

Abbreviations

Avr	avirulence
CT	*p*-coumaroyl tyramine
EF-Tu	elongation factor Tu
Flg	flagellin
FT	feruloyl tyramine
HR	hypersensitive response
ISR	induced systemic resistance
LOS	lipo-oligosaccharides
LPS	lipopolysaccharides
LRR	leucine rich repeat
MAMPs	microbe associated molecular patterns
NBS	nucleotide-binding site

M.-A. Newman (✉)
Department of Plant Biology and Biotechnology, Faculty of Life Sciences, University of Copenhagen, Thorvaldsensvej 40, 1871 Frederiksberg, Denmark
e-mail: mari@life.ku.dk

X. Wang, P.J. Quinn (eds.), *Endotoxins: Structure, Function and Recognition*, Subcellular Biochemistry 53, DOI 10.1007/978-90-481-9078-2_17,

NSF	N-ethylmaleimide-sensitive factor
PGN	peptidoglycan
PR	pathogenesis-related
PRRs	pattern recognition receptors
Pst	*Pseudomonas syringae* pv. *tomato*
RLKs	receptor-like kinases
ROS	reactive oxygen species
SA	salicylic acid
SAR	systemic acquired resistance
SNAPs	soluble N-ethylmaleimide-sensitive factor (NSF) adaptor proteins adaptor proteins
SNAREs	soluble NSF adaptor protein (SNAP) receptors
TIR	Toll/IL-1R
TLRs	Toll-like receptors
TTSS	type III secretion system
VAMP	vesicle-associated membrane protein
Xcc	*Xanthomonas campestris* pv. *campestris*

17.1 Introduction

Plants interact with a variety of microorganisms and, like insects and mammals, are able to recognize a range of conserved Microbe- or Pathogen-Associated Molecular Patterns (MAMPs or PAMPs) to trigger defense responses (Nürnberger and Brunner, 2002; Jones and Dangl, 2006). PAMPs have been defined as evolutionarily conserved microbe-derived molecules that distinguish hosts from pathogens (Janeway, 1992; Ausubel, 2005). The term MAMP, which we will use here, was coined to reflect that these elicitor molecules are not restricted to pathogens, but can be found in non-pathogenic and saprophytic organisms. MAMPs described from bacteria include LPS, the major component of the outer membrane of Gram-negative bacteria (Newman et al., 1995; Dow et al., 2000; Bedini et al., 2005; Silipo et al., 2005), peptidoglycan (PGN), which provides rigidity and structure to the cell envelopes of both Gram-negative and Gram-positive bacteria (Erbs et al., 2008a), flagellin, the main component of the bacterial motility organ (Gómez-Gómez and Boller, 2000) and the elongation factor Tu (EF-Tu) which is essential for protein translation and is the most abundant bacterial protein (Zipfel et al., 2006). These examples illustrate common features of MAMPs; they are usually indispensable for microbial fitness and relatively invariant in structure.

Recognition of MAMPs by plant cell surface Pattern Recognition Receptors (PRRs) can lead to induction of a range of plant defense-related responses (Thordal-Christensen, 2003; Jones and Dangl, 2006). Plant PRRs responsible for the recognition of flagellin (Flg) and Ef-Tu and the peptides flg22 and efl18 derived from these proteins have been identified; however PRRs responsible for recognition of LPS and PGN have remained elusive. MAMP-induced defense responses include the production of reactive oxygen species (the oxidative burst), production of reactive nitrogen species such as NO, alterations in the plant cell wall, induction of antimicrobial compounds and the synthesis of pathogenesis-related (PR) proteins. Reactive oxygen species (ROS) and NO can act in signalling and have direct antimicrobial effects. ROS can also drive oxidative cross-linking of polymers in the plant cell wall to strengthen it against degradation, which may restrict pathogen spread. Other alterations in the plant wall include the deposition of the β-(1-3) linked glucan callose. PR proteins comprise a number of families that include enzymes, such as β-(1-3) glucanase and chitinase, which can directly attack pathogen structures, antimicrobial peptides and small proteins, and PR1, which is of unknown function.

In this chapter, we will review the current knowledge of the role of LPS as a MAMP in plant innate immunity. We will give an overview of the range of responses induced by LPS, the sub-structures within LPS that are recognized by plants and variations within the LPS structure that can alter its activity as a MAMP. We will go on to discuss new work that suggests a role for the plasma-membrane resident syntaxin PEN1 in transduction of the LPS signal.

17.2 LPS as a MAMP

17.2.1 LPS as a Direct Inducer of Basal Plant Defenses

As the main surface component of the bacterial cell envelope LPS is thought to contribute to the restrictive Gram-negative membrane permeability, allowing bacterial growth in unfavourable environments such as those that may be encountered within or on plants. The exclusion of antimicrobial substances of plant origin probably contributes to the ability of pathogenic bacteria to parasitize plants. LPS-defective mutants show increased in vitro sensitivity to antibiotics and antimicrobial peptides and the numbers of viable bacteria often decline very rapidly upon introduction into plants. LPS may also promote bacterial adherence to plant surfaces (Newman et al., 2007).

In contrast to this role in promoting plant disease, there are a number of reports detailing the effects of LPS on the induction of basal plant defenses, consistent with its designation as a MAMP (Newman et al., 2007). LPS preparations from a number of bacteria induced NO synthesis in suspension cultures and leaves of *Arabidopsis thaliana* (Zeidler et al., 2004). This common effect of LPS from diverse bacteria suggested the involvement of a shared molecular determinant, the lipid A moiety, and indeed isolated lipid A was also active. LPS can induce the production of ROS, although this is not always observed (Dow et al., 2000; Newman et al., 2007). For example, although LPS from the plant pathogen *Xanthomonas campestris* pv. *campestris* (*Xcc*) induces an oxidative burst in culture cells of tobacco, no effects were seen with LPS from the enteric bacterium *Salmonella typhimurium* (Meyer et al., 2001). Furthermore *Xcc* LPS did not elicit generation of ROS in cultured soybean cells (Dow et al., 2000). Recent studies have revealed that LPS from various pathogenic and nonpathogenic bacteria induce the generation of ROS and defense related gene expression in rice, indicating that the machinery recognizing LPS is evolutionary conserved in monocots and dicots (Desaki et al., 2006).

LPS also has effects on cell wall alterations such as callose deposition (Keshavarzi et al., 2004) and on *PR* gene induction (Zeidler et al., 2004; Silipo et al., 2005). In some cases specific effects of a particular LPS on plant gene induction are observed. LPS from the crucifer pathogen *Xcc* induced expression of a gene encoding a defense-related β-(1-3) glucanase when applied to turnip leaves at 1 μg per ml. In contrast, LPS from *Escherichia coli* and *Salmonella enterica* were ineffective at concentrations up to 50 μg per ml (Newman et al., 1995). Such specific effects may reflect the ability of particular plants to recognize structural features within LPS that are not widely conserved.

17.2.2 LPS as a Primer of Plant Defense Response Induction

LPS has been found not only to induce defense responses directly but also to promote or prime an early triggering of defense responses upon subsequent bacterial inoculation. An example of this is the LPS priming of the synthesis of the antimicrobial compounds feruloyl tyramine (FT) and *p*-coumaroyl tyramine (CT) in pepper plants (Newman et al., 2001, 2002; Prime-A-Plant Group, 2006). LPS treatment of pepper leaves does not lead to synthesis of FT and CT. However, these compounds are synthesized more rapidly upon bacterial inoculation into LPS pre-treated than water pre-treated plants (Newman et al., 2001, 2002; Prime-A-Plant Group, 2006).

17.2.3 LPS can Modulate the Hypersensitive Response, a Programmed Cell Death Associated with Resistance

As part of their virulence strategy, many phytopathogenic bacteria inject a suite of effector proteins directly into the host cell through a type III secretion system (TTSS) (Alfano and Collmer, 2004). These effector molecules contribute to bacterial virulence in susceptible plants by interfering with or subverting host cell processes, including the triggering of innate immunity (Jamir et al., 2004; He et al., 2006; Nomura et al., 2006). However in some plants, effectors can be recognized to trigger the Hypersensitive Response (HR), a programmed cell death associated with plant disease resistance (Jones and Dangl, 2006). This recognition involves the protein products of plant resistance (*R*) genes and the effectors that are recognized have been called avirulence (Avr) proteins, although this term does not reflect their role in promoting virulence in susceptible hosts.

LPS can modulate the triggering of the HR in a number of plants. Pre-treatment of pepper leaves with LPS from different bacteria can prevent the HR caused by *Xcc*, which carries the avirulence gene *avrBs1* (Newman et al., 1997). Similarly, pre-treatment of *Arabidopsis* leaves with LPS from *Xcc* prevents the HR caused by avirulent strains of *Pseudomonas syringae* pv. *tomato* (*Pst*) carrying the *avrRpm1* or the *avrRps4* genes (Newman et al., 2002; Silipo et al., 2005). These effects are localized to the area of LPS inoculation. The effects of LPS on preventing HR (which is associated with plant resistance) while also inducing basal defenses appear to present a conundrum. However measurement of bacterial growth in LPS-treated leaves indicates that prevention of HR does not increase the susceptibility of the plant tissue. This is consistent with the notion that LPS perception allows the plant to express resistance (through enhanced expression of basal defenses) without the catastrophic collapse of the HR. The underlying mechanisms are still unknown. Intriguingly, although LPS has never been shown to elicit HR in dicots (Dow et al., 2000), LPS from various bacteria induces programmed cell death in rice cells PCD (Desaki et al., 2006).

17.2.4 LPS Induces Systemic Effects in Plants

In addition to the effects described above, which are induced in a localized fashion, LPS can elicit systemic resistance responses in plants. Two such systemic resistance responses have been described; these are systemic acquired resistance (SAR) and induced systemic resistance (ISR). SAR involves systemic activation of defense-related responses such as *PR* gene expression upon infection with a locally necrotizing pathogen. SAR is accompanied by a systemic increase in salicylic acid (SA), and SA is required for SAR signalling (Ryals et al., 1996; Schneider et al., 1996). In contrast, ISR does not require necrosis to become established and is associated with jasmonic acid (JA) and ethylene (ET) rather than SA as signals (Pieterse et al., 1998). Plants exhibiting SAR show enhanced expression of defense-related genes in distant leaves in the absence of any pathogen attack on those leaves. This is not seen in plants exhibiting ISR, in which defense responses are only activated after pathogen challenge.

Early studies showed that LPS from the root-colonizing *Pseudomonas fluorescens* induced ISR in carnation and radish, whereas mutant bacteria, lacking the O-antigen side chain could not induce ISR (Leeman et al., 1995; van Loon et al., 1998). Treatment of *Arabidopsis* with *Pseudomonas aeruginosa* LPS, flagellin or bacteria triggering necrosis was shown to be associated with accumulation of SA, expression of the *PR* genes and expression of the SAR marker gene Flavin-dependent monooxygenase 1 in treated as well as in distant leaves (Mishina and Zeier, 2006; 2007). These studies suggest that recognition of the MAMPs, LPS or Flg, rather than the necrotic lesion formation contributed to the bacterial induction of SAR in *Arabidopsis*.

The body of work outlined above demonstrates conclusively that LPS from diverse bacteria can act as a MAMP to either directly induce a range of plant defense responses or prime induction of those responses in both a local and systemic fashion. Differences between the activities of LPS in different plants are however seen. These may relate in part to the plant experimental systems used, the isolation procedures and purity of LPS preparations and to differences in LPS structure. For example global transcriptional profiling of *Arabidopsis* cells treated with *Burkholderia cepacia* LPS revealed, surprisingly, that very few *PR* transcripts were induced, indicating that this cell culture system responds differently than the whole plant (Livaja et al., 2008). A number of the findings suggest that different sub-structures of LPS such as the O-antigen and lipid A moieties are responsible for the MAMP activity, and we will expand on this topic in the following section.

17.3 Different Sub-Structures Within LPS can Act as MAMPs

LPS from plant-associated and plant pathogenic bacteria possess the same tripartite structure comprising lipid A, core oligosaccharide and an O-polysaccharide or O-antigen seen in LPS from other bacteria (Holst and Molinaro, 2009; Raetz and Whitfield, 2002). The lipid A and the core oligosaccharide are linked in the majority of cases by the sugar 3-deoxy-D-manno-2-octulosonate (KDO). LPS molecules that lack an O-antigen are called lipo-oligosaccharides (LOS).

Several laboratories have investigated the contribution of the different moieties within LPS to the MAMP elicitor activity. Silipo and colleagues (2005) determined the complete structure of purified LOS from *Xcc* and the lipid A and core oligosaccharides derived from it by mild acid hydrolysis and in parallel examined the activity of these (structurally-defined) components in defense gene induction in *Arabidopsis*. *Xcc* LOS was found to be a unique molecule with a high negative charge density and a phosphoramide group, which had never been found previously as a component of LPS (Silipo et al., 2005). *Xcc* LOS induced the defense-related *PR1* and *PR2* genes in *Arabidopsis* leaves in two temporal phases; the core oligosaccharide induced only the early phase and the lipid A moiety only the later phase. These findings suggest that although both *Xcc* lipid A and the *Xcc* core oligosaccharide are active in defense gene induction, they may be recognized by different plant receptors (Silipo et al., 2005). This elicitor activity of *Xcc* lipid A correlates with earlier studies by Zeidler et al. (2004) who showed that lipid A preparations from various bacteria induced a rapid burst of NO production that was associated with the induction of defense-related genes in *Arabidopsis*.

Interestingly, the core oligosaccharide from *E. coli* and *Ralstonia solanacearum* does not prevent HR or induce defense-related genes (Newman et al., 1997), indicating that the effect of the *Xcc* core oligosaccharide could be due to the unique phosphoramide group in that particular LPS molecule (Silipo et al., 2005). In contrast, in tobacco cells *Xcc* lipid A could not induce the oxidative burst, but rather it was the inner core part of the LPS molecule that was responsible (Braun et al., 2005). This disparity in outcomes might be a reflection of the use of different plants, the difference in the age of the plants used (plant cell cultures versus seedlings versus fully developed plants) and the different defense responses measured after treatment with LPS and its derivatives.

The role of the O-antigen in eliciting defense responses has been suggested by different ability of LPS derived from wild type *P. fluorescens* and a mutant lacking the O-antigen in induction of ISR (Leeman et al., 1995; van Loon et al., 1998). More recently the role of the O-antigen has been directly examined by studies of the biological activity of synthetic O-antigen polysaccharides. Structural studies of LPS from many phytopathogenic bacteria have revealed that the O-antigen comprises a rhamnan with the trisaccharide repeating unit [α-L-Rha-(1-3)-α-L-Rha-(1-2)-α-L-Rha-(1-3)] (Bedini et al., 2002). This trisaccharide was synthesized and the trimer oligomerised to generate a set of oligosaccharides of increasing chain length. The tri-, hexa- and nona-saccharides synthetic O-antigens were found to suppress the HR and induce *PR1* and *PR2* transcript accumulation in *Arabidopsis*. Interestingly, the efficiency of HR suppression and *PR* gene induction improved with increasing chain length (Bedini et al., 2005). Moreover, this increasing chain length was associated with the development of a coiled structure, suggesting a role for this structure as a MAMP. By extension, these findings suggest a role for the O-antigen from many phytopathogenic bacteria in triggering plant innate immunity (Bedini et al., 2005).

17.4 Structural Variations in LPS Influence its Activity in Plants

LPS is recognized by mammalian cells through the lipid A moiety and this recognition governs the interactions with the innate immune system (Loppnow et al., 1989). *E. coli* lipid A, which is an effective agonistic structure of immune responses in mammalian cells, is composed of a bis-phosphorylated hexa-acylated disaccharide backbone with an asymmetric distribution of the acyl residues. Modifications of lipid A structure influence the biological activity of the molecule in mammals (Raetz et al., 2007). Schromm and colleagues (1998, 2000) showed that the molecular conformation of the lipid A correlated with its biological activity. This molecular shape can be influenced by the net negative charge, usually associated with the degree of phosphorylation (Gutsmann et al., 2007). Molecules with several negative charges adopt a conical shape and have endotoxin activity. Molecules with very few or no negative charges adopt a cylindrical shape, are less potent as endotoxins and can have antagonistic activity. Differences in acylation status of lipid A can also influence biological activity of LPS and agonist or antagonist activity (Munford and Varley, 2006). Recent studies have shown that LPS from *Shigella flexneri* elicits a weaker TLR4-mediated response in mammalian cells than *E. coli* LPS due to differences in the acylation status of their lipid A moieties (Rallabhandi et al., 2008).

Alteration of the lipid A structure also influences the biological activity in plants. De-phosphorylation of *Xcc* LOS leaves only one negative charge on the KDO residue. The resultant molecule is unable to prevent HR in *Arabidopsis* leaves, suggesting that the charged groups present in LOS play a key role in inducing defense responses in plants (Silipo et al., 2005). Other findings suggest that plants are sensitive to the same structures of lipid A that determine biological activity in humans (Erbs et al., 2008b). The lipid A from *Halomonas magadiensis*, a Gram-negative extremophilic and alkaliphilic bacteria is characterized by an unusual and very low degree of acylation (Silipo et al., 2004). This molecule acts as an antagonist of *E. coli* lipid A induced immune responses in human cells (Ialenti et al., 2006). *H. magadiensis* lipid A also antagonises the action of *E. coli* lipid A when inducing *PR1* gene expression in *A. thaliana* (Erbs et al., 2008b).

Structural analysis of LOS from a mutant of *Xcc* defective in core completion revealed that this mutant had modifications in the lipid A moiety, which had reduced acylation and was further derivatised with phosphoryl ethanolamine residues (Dow et al., 1995; Silipo et al., 2008). These changes in lipid A structure abolished the ability to trigger innate immune responses in *Arabidopsis* (Silipo et al., 2008). Importantly these findings indicate that *Xcc* has the capacity to modify the structure of its lipid A to reduce its activity as a MAMP in plants (Silipo et al., 2008). It is not known whether these (or other) modifications to lipid A occur when bacteria are within plants.

17.5 Subversion of LPS-Induced Effects

The effect of MAMPs such as LPS on the induction of basal plant defenses raises the issue of how bacteria can ever cause disease in plants. Successful pathogens have evolved mechanisms to subvert or suppress MAMP-triggered immunity. Many type III secreted effectors act to block induction of basal defenses, thus promoting disease (Jamir et al., 2004; He et al., 2006; Nomura et al., 2006). Other bacterial products such as extracellular cyclic glucans and extracellular polysaccharides have also been shown to suppress defenses (Yun et al., 2006; Rigano et al., 2007; Aslam et al., 2008). Extracellular polysaccharides may exert their suppressive effect through sequestration of Ca^{2+} ions, thus preventing influx from the extracellular apoplastic pool (Yun et al., 2006; Aslam et al., 2008). Ca^{2+} influx occurs as an early local response to pathogen attack and is thought to act as a signal and to activate callose synthetase. The mechanistic basis for suppression of defenses by cyclic glucan is unknown.

17.6 Perception of LPS by Plants

Although plant receptors for the bacterial MAMPs flagellin and Ef-Tu have been described, the mechanisms by which plants perceive LPS is still not understood. The mammalian innate immunity system perceives invading pathogens through Toll-like receptors (TLRs), an interleukin 1 receptor (IL-1R) (Medzhitov et al., 1997), that resembles the Toll receptor found in *Drosophila* (Hashimoto et al., 1988; Lemaitre et al., 1996). The TLRs, one class of PRRs, comprise a family of transmembrane receptors that have an extracellular leucine rich repeat (LRR) domain, by which pathogen components are recognized, and a cytoplasmic Toll/IL-1R (TIR) domain, through which the MAMP signal is transduced. The Toll-like receptor TLR4 is responsible for LPS perception in mammals. Once the TLRs are activated by MAMP recognition, adaptor molecules are recruited to initiate downstream signalling, which involves activation of transcription factors and MAP kinases (Carpenter and O'Neill, 2007). In addition to the surface localised TLR4, a second type of LPS receptor, the intracellular Nod proteins, have been described in animal cells.

Intriguingly proteins with structural similarities to TLRs and Nods are found in plants (Newman et al., 2007). The Toll-like receptor TLR5 is responsible for flagellin perception in mammals. The FLS2 flagellin receptor of *Arabidopsis thaliana* also has extracytoplasmic LRRs, but a cytoplasmic serine/threonine kinase domain replaces the TIR domain found in mammalian TLR5. FLS2 belongs to a large family of plant receptor-like kinases (RLKs) containing extracytoplasmic LRRs, other members of which are responsible for perception of the MAMP Ef-Tu, certain bacterial effector proteins as well as plant signalling molecules and hormones. Some intracellular plant receptors for type III-secreted effectors contain TIR domains and additionally have the nucleotide-binding/apoptotic ATPase (NBS) domain and LRRs, which are also found in the mammalian NOD proteins (Newman et al., 2007). On the basis of these structural similarities between plant and animal receptors for flagellin, it is tempting to speculate that perception of LPS by plants could involve surface-localised LRR-RLKs and/or intracellular TIR-NBS-LRR proteins. Unfortunately this does not substantially narrow the search for an LPS receptor, since many such putative receptor proteins are encoded by plant genomes. In *A. thaliana*, there are at least 135 proteins with a TIR domain, 82 of which have the TIR-NBS-LRR domain organisation. In addition, *A. thaliana* encodes 600 RLKs, many of which like FLS2 have extracellular LRRs.

Despite apparent similarities between innate defense systems in plants, mammals and insects, some differences do occur (Nürnberger and Brunner, 2002). Most plant defense responses thus far described require LPS application at the 5–50 μg per ml level, whereas TLR4-mediated perception of LPS is extremely sensitive and is activated by the ligand at concentrations in the pg to ng per ml range (Miyake, 2004). These considerations have led to suggestions that plants possess only low affinity systems to detect LPS (Zeidler et al., 2004). It is also plausible that high affinity recognition-response systems in plants do not act to directly trigger plant defenses, but prime the plant so that in response to further pathogen-derived signals, such responses are mounted more rapidly or to a greater extent (Newman et al., 2007). This suggestion is open to experimental testing with structurally characterized LPS and pathogen-derived elicitors such as flg22. Available evidence suggests that plants recognize similar structures in lipid A as do mammals but also raise the possibility of different plant receptors for the core oligosaccharide and the lipid A moieties (see above).

17.6.1 SNARE Proteins, Vesicle Trafficking and Plant Defense

In eukaryotes, soluble N-ethylmaleimide-sensitive factor (NSF) adaptor proteins (SNAPs) receptors (SNAREs) are known to be required for docking and fusion of intracellular transport vesicles with acceptor/target membranes. The fusion of vesicles in the secretory pathway involves target-SNAREs (t-SNAREs) on the target membrane and vesicle-SNAREs (v-SNAREs) on vesicle membranes that recognize each other and assemble into trans-SNARE complexes (Söllner et al., 1993).

Specific SNARE proteins have been shown to have roles in defense triggering in plants. Silencing of the *Nicotiana benthamiana* SNARE, NbSYP132, an ortholog of an *Arabidopsis* plasma membrane-resident syntaxin AtSYP132, revealed that NbSYP132 contributes to *R*-gene mediated resistance, basal resistance and SA-associated defense, and is involved in mediating secretion of PR1 into the extracellular space. In contrast to this, PR1 secretion and *R*-gene mediated responses were not affected by silencing NbSYP121, an ortholog of the *Arabidopsis* plasma membrane-resident PEN1 syntaxin (AtSYP121) (Kalde et al., 2007). Nevertheless PEN1 (AtSYP121) is required for non-host penetration resistance in *Arabidopsis* against the barley powdery mildew fungus *Blumeria graminis* f.sp. *hordei* (Collins et al., 2003). The tobacco NbSYP121 has been suggested to be associated with abscisic acid stress, stomatal closing and normal growth and development (Geelen et al., 2002).

Recent studies have revealed that the PEN1 syntaxin, a presumed lipid raft component (Bhat et al., 2005), interacts with a synaptosomal-associated protein of 33 kDa (SNAP33) and vesicle-associated membrane protein (VAMP) 721/722 to form ternary SNARE complexes to mediate vesicle fusion processes (Kwon et al., 2008b). The PEN1-SNAP33-VAMP721/722 complex was found to be required for immune responses, and it was suggested that *Arabidopsis* has appointed the VAMP721/722-dependent secretion machinery for immune response to pathogens, as VAMP721/722 were found to operate in a default secretory pathway (Kwon et al., 2008a, b). AtSYP122, the closest homologue of PEN1 (Sanderfoot et al., 2000) is not engaged in penetration resistance, indicating a functional specialization of these two *Arabidopsis* syntaxins (Assaad et al., 2004). Consistent with these conclusions, overexpression of AtSYP122 in the *pen1*-1 mutant failed to complement the *pen1* immune defect (Pajonk et al., 2008).

A model for SNARE mediated vesicle trafficking in basal immunity has been proposed (Robatzek et al., 2007; Kwon et al., 2008b). PEN1 and AtSNAP33 form a binary t-SNARE complex involved in marking the target membrane for vesicle trafficking. The cognate v-SNAREs are then trafficked to the targeted membrane and together with the t-SNARE complex form a ternary complex resulting in vesicle fusion and antimicrobial compounds secretion (Robatzek et al., 2007; Kwon et al., 2008b).

17.6.2 A Role for the Syntaxin PEN1 in LPS Signalling in Plants?

SNARE proteins play a role in mediating effects of LPS on mammalian cells. Studies by Pagan et al. (2003) showed that a subset of t-SNAREs syntaxin 4/SNAP23/Munc18c, known to control regulated exocytosis in other mammalian cell types than macrophages (Bryant et al., 2002), was up-regulated in response to LPS whereas the level of those involved in endocytosis was decreased or unaffected. The regulation of SNAREs involved in vesicle docking and fusion identified in LPS stimulated macrophages, revealed a SNARE complex necessary for LPS induced exocytosis of TNF-α, indicating again that individual SNAREs are regulated to perform specialized functions in the cell (Murray et al., 2005).

These findings together with the work establishing a role for PEN1 in pathogen resistance in *Arabidopsis* (Collins et al., 2003; Kwon et al., 2008a) prompted us to test the role of this SNARE in induction of defense responses by LPS. The effects of infiltration of LOS from the plant pathogen *Xcc* on *PR1* gene expression, production of ROS and callose deposition in leaves of *A. thaliana* (cv. Columbia) wild type and *pen1* mutants were examined (Erbs and Newman, unpublished). For comparative purposes, parallel experiments with a second MAMP, the flg22 peptide derived from flagellin, were performed. Flg22 had a marked effect on *PR1* gene transcription that was seen in both wild type and *pen1* mutant plants. In contrast, although *Xcc* LOS induced 340-fold accumulation of *PR1* transcripts in wild-type *A. thaliana*, only a very low (2.5 fold) transient accumulation was seen in the *pen1* mutant. Flg22 was active in triggering ROS production, with *pen1* and wild-type showing a similar response. Although wild-type *A. thaliana* responded rapidly to *Xcc* LOS with an oxidative burst, a delayed and substantially reduced response was observed in the *pen1* mutant. Flg22 induced abundant callose deposition in both *pen1* and wild-type *A. thaliana*. In contrast, *Xcc* LOS induced a much lower formation of callose in the *pen1* mutant than the wild type (Erbs and Newman, unpublished). Taken together these results suggest that PEN1 has a role in triggering of immune responses in *Arabidopsis* in response to *Xcc* LOS but not in response to flg22.

A possible function for PEN1 is that it is required for the correct localization at the plant cell plasmamembrane of the putative receptor(s) for LOS, but not for the Flg receptor FLS2. Alternatively, PEN1 may be required for endocytosis of an LOS complex, which may allow signalling to cytoplasmically located proteins to trigger defense responses. Gross et al. (2005) found that, in tobacco cells, *Xcc* LPS was internalized 2 hours after its introduction to the cell suspension, where it co-localized with Ara6, a plant homolog of Rab5 which is known to regulate early endosomal functions in mammals. It was speculated that this endocytosis in tobacco cells was, in correlation with the mammalian system, part of a down regulation of defense responses (Gross et al., 2005). In mammalian macrophage cells, the LPS receptor complex is endocytosed and appears on endosome-like structures. Furthermore, an inhibition of the endosomal pathway increased LPS-induced NF-κB activation (Husebye et al., 2006). Interestingly, it has also been shown that, upon stimulation with flg22, the cell membrane resident flagellin receptor FLS2 is transferred into intracellular mobile vesicles and targeted for degradation (Robatzek et al., 2006).

If *Arabidopsis* PEN1 is a component of the endosomal complex responsible for endocytosis of the LPS/LOS-receptor complex, similar to endocytosis of the LPS-complex in mammals and the observed endocytosis of LPS in tobacco cells discussed above, an increased induction of innate immune responses would, presumably, have been observed in LOS-stimulated *Arabidopsis pen1* mutants compared to the wild-type. Although we favour a model in which PEN1 is involved in exocytosis required for *Xcc* LOS triggered immunity in *Arabidopsis*, we cannot discount an alternate role in endocytosis. Importantly the findings indicate that

PEN1 may have roles in plant disease resistance (i.e. those associated with LPS perception/signalling) that have not been appreciated thus far. The involvement of SNAREs in contributing to fusion specificity is still debated (Kwon et al., 2008a), and our understanding of the regulatory role of PEN1 in fusion of intracellular transport vesicles with target membranes is still limited. Only further experimental work will establish the exact role of PEN1 in secretory pathways acting in LOS triggered immunity.

17.7 Concluding Remarks

A greater understanding of the mechanisms by which LPS elicits defense responses may have considerable impact on the improvement of plant health and disease resistance. Although plant receptors for the bacterial proteinaceous MAMPs flagellin and Ef-Tu elongation factor have been identified, those involved in perception of LPS remain obscure. The cloning and characterization of these genes remains a major goal. The development of a range of molecular genetic tools for model plants such as *Arabidopsis thaliana* affords one approach. Mutagenized plant populations could be screened for variations in a particular response to LPS. Plant mutants showing an altered response may carry mutations in genes involved in LPS perception or signal transduction. The genes responsible could then be isolated by map-based cloning methods. Since the available evidence indicates that *Arabidopsis* can recognize different structures within LPS to trigger the same defense response, any strategy to clone LPS receptors in this fashion has to use a 'minimal' sub-structure, of which lipid A or a synthetic derivative is perhaps the best candidate. Other cloning strategies could employ tagged or immobilized LPS to isolate proteins that physically interact with the ligand. Such binding proteins could then be characterized by peptide mass fingerprinting. Immobilization could involve the construction of 'glycochips' or attachment of LPS or derivatives to nano-gold particles.

Identification of genes encoding the plant receptors for LPS would allow analysis of the contribution that LPS-induced resistance makes to the outcome of plant-microbe interactions. This could be assessed by examining the effects of silencing or mutagenesis of these plant genes on pathogen virulence. Improvement of basal resistance in plants could be achieved either by genetic modification or by breeding programs in which there is a directed selection for specific genes of choice. Knowledge obtained from such experiments may also have a bearing on other aspects of plant-microbe interactions of benefit to the host, such as induction of ISR by beneficial bacteria.

Thus far LPS preparations used for the analysis of plant responses and for structural studies have been derived from bacteria grown in culture. We know almost nothing about the alterations in LPS that occur when bacteria are within plants, although this may be highly relevant to signalling. Changes could occur in both the size distribution of LPS (alteration in the ratio of LOS to LPS) and/or in decoration of LPS with saccharide, fatty acid, phosphate or other constituents. Increases in the sensitivity of mass spectrometric methodologies may allow development of micro-methods to analyze such changes in bacteria isolated from plants. Transcriptome or proteome profiling of bacteria isolated from plants may also give clues as to possible LPS modifications. In conclusion we expect that the next few years will see a substantial increase in our understanding of the processes of LPS perception and signal transduction in plants through the deployment of cross-disciplinary approaches and ever-expanding range of molecular experimental tools.

Acknowledgements G.E. and M.-A.N acknowledge funding by The Danish Council for Independent Research, Technology and Production Sciences (FTP). J.M.D. acknowledges funding by the Science Foundation of Ireland through Principal Investigator Awards 03/IN3/B373 and 07/IN.1/B955.

References

Alfano, J.R., Collmer, A. Type III secretion system effector proteins: Double agents in bacterial disease and plant defense. Annu Rev Phytopathol 42 (2004) 385–414.

Aslam, S.N., Newman, M.A., Erbs, G., Morrissey, K.L., Chinchilla, D., Boller, T., Jensen, T.T., De Castro, C., Ierano, T., Molinaro, A., Jackson, R.W., Knight, M.R., Cooper, R.M. Bacterial polysaccharides suppress induced innate immunity by calcium chelation. Curr Biol 18 (2008) 1078–1083.

Assaad, F.F., Qiu, J.-L., Youngs, H., Ehrhardt, D., Zimmerli, L., Kalde, M., Wanner, G., Peck, S.C., Edwards, H., Ramonell, K., Somerville, C.R., Thordal-Christensen, H. The PEN1 syntaxin defines a novel cellular compartment upon fungal attack and is required for the timely assembly of papillae. Mol Biol Cell 15 (2004) 5118–5129.

Ausubel, F. Are innate immune signalling pathways in plants and animals conserved? Nat Immunol 6 (2005) 973–979.

Bedini, E., De Castro, C., Erbs, G., Mangoni, L., Dow, J.M., Newman, M.-A., Parrilli, M., Unverzagt, C. Structure-dependent modulation of a pathogen response in plants by synthetic O-antigen polysaccharides. J Am Chem Soc 127 (2005) 2414–2416.

Bedini, E., Parrilli, M., Unverzagt, C. Oligomerization of a rhamnanic trisaccharide repeating unit of O-chain polysaccharides from phytopathogenic bacteria. Tetrahedron Lett 43 (2002) 8879–8882.

Bhat, R.A., Miklis, M., Schmelzer, E., Schulze-Lefert, P., Panstruga, R. Recruitment and interaction dynamics of plant penetration resistance components in a plasma membrane microdomain. Plant Biol 102 (2005) 3135–3140.

Braun, S.G., Meyer, A., Holst, O., Pühler, A., Niehaus, K. Characterization of the *Xanthomonas campestris* pv. *campestris* lipopolysaccharide substructures essential for elicitation of an oxidative burst in tobacco cells. Mol Plant-Microbe Interact 18 (2005) 674–681.

Bryant, N.J., Govers, R., David, E., James, D.E. Regulated transport of the glucose transporter GLUT4. Nature 3 (2002) 267–277.

Carpenter, S., O'Neill, L.A.J. How important are Toll-like receptors for antimicrobial responses? Cell Microbiol 9 (2007) 1891–1901.

Collins, N.C., Thordal-Christensen, H., Lipka, V., Bau, S., Kombrink, E., Qiu, J.L., Hückelhoven, R., Stein, M., Freialdenhoven, A., Somerville, S.C., Schulze-Lefert, P. SNARE-protein-mediated disease resistance at the plant cell wall. Nature 425 (2003) 973–977.

Desaki, Y., Miya, A., Venkatesh, B., Tsuyumu, S., Yamane, H., Kaku, H., Minami, E., Shibuya, N. Bacterial lipopolysaccharides induce defense responses associated with programmed cell death in rice cells. Plant Cell Physiol 47 (2006) 1530–1540.

Dow, J.M., Osbourn, A.E., Wilson, T.J.G., Daniels, M.J. A locus determining pathogenicity of *Xanthomonas-campestris* is involved in lipopolysaccharide biosynthesis. Mol Plant-Microbe Interact 8 (1995) 768–777.

Dow, M., Newman, M.A., von Roepenack, E. The induction and modulation of plant defence responses by bacterial lipopolysaccharides. Annu Rev Phytopathol 38 (2000) 241–261.

Erbs, G., Silipo, A., Aslam, S., De Castro, C., Liparoti, V., Flagiello, A., Pucci, P., Lanzetta, R., Parrilli, M., Molinaro, A., Newman, M.-A., Cooper, R.M. Peptidoglycan and muropeptides from pathogens *Agrobacterium* and *Xanthomanas* elicit innate immunity: Structure and activity. Chem Biol 15 (2008a) 438–448.

Erbs, G., Tandrup Jensen, T., Silipo, A., Grant, W., Dow, J.M., Molinaro, A., Parrilli, M., Newman, M.A. An antagonist of lipid A action in mammals has complex effects on lipid A induction of defence responses in the model plant *Arabidopsis thaliana*. Microbes and Infection 10 (2008b) 571–574.

Geelen, D., Leyman, B., Batoko, H., Di Sansabastiano, G.-P., Moore, I., Blatt, M.R. The abscisic acid-related SNARE homolog NtSyr1 contributes to secretion and growth: Evidence from competition with its cytosolic domain. Plant Cell 14 (2002) 387–406.

Gómez-Gómez, L., Boller, T. FLS2: An LRR receptor-like kinase involved in the perception of the bacterial elicitor flagellin in *Arabidopsis*. Mol. Cell 5 (2000) 1003–1011.

Gross, A., Kapp, D., Nielsen, T., Niehaus, K. Endocytosis of *Xathomonas campestris* pathovar *campestris* lipopolysaccharides in non-host plant cells of *N. benthamiana*. New Phytol 165 (2005) 215–226.

Gutsmann, T, Schromm, A.B., Brandenburg, K. The physiochemistry of endotoxins in relation to bioactivity. J Med Microbiol 297 (2007) 341–352.

Hashimoto, C., Hudson, K.L., Anderson, K.V. The Toll gene of *Drosophila*, required for dorsal-ventral embryonic polarity, appears to encode a transmembrane protein. Cell 52 (1988) 269–279.

He, P., Shan, L., Lin, N.C., Martin, G.B., Kemmerling, B., Nürnberger, T., Sheen, J. Specific bacterial suppressors of MAMP signalling upstream of MAPKKK in *Arabidopsis* innate immunity. Cell 125 (2006) 563–575.

Holst, H., Molinaro, A. Core region and lipid A components of lipopolysaccharides. In: Moran, A., Brennan, P., Holst, O., von Itszstein, M. (eds), Microbial Glycobiology: Structures, Relevance and Applications. Oxford, Elsevier (2009), pp. 803–820.

Husebye, H., Halaas, Ø., Stenmark, H., Tunheim, G., Sandanger, Ø., Bogen, B., Brech, A., Latz, E., Espevik, T. Endocytic pathways regulate Toll-like receptor 4 signaling and link innate and adaptive immunity. EMBO J 25 (2006) 683–692.

Ialenti, A., Di Meglio, P., Grassia, G., Maffia, P., Di Rosa, M., Lanzetta, R., Molinaro, A., Silipo, A., Grant, W. and Ianaro, A. A novel lipid A from *Halomonas magadiensis* inhibits enteric LPS-induced human monocyte activation. Eur J Immunol 36 (2006) 354–360.

Jamir, Y., Guo, M., Oh, H.S., Petnicki-Ocwieja, T., Chen, S., Tang, X., Dickman, M.B., Collmer, A., Alfano, J.R. Identification of *Pseudomonas syringae* type III effectors that can suppress programmed cell death in plants and yeast. Plant J 37 (2004) 554–565.

Janeway, C.A. The immune-system evolved to discriminate infectious nonself from non-infectious self. Immunol Today 13 (1992) 11–16.

Jones, J.D.G., Dangl, J.L. The plant immune system. Nature 444 (2006) 323–329.

Kalde, M., Nühse, T.S., Findlay, K., Peck, S.C. The syntaxin SYP132 contributes to plant resistance against bacteria and secretion of pathogenesis-related protein 1. Proc. Natl. Acad. Sci. USA 104 (2007) 11850–11855.

Keshavarzi, M., Soylu, S., Brown, I., Bonas, U., Nicole, M., Rossiter, J., Mansfield, J. Basal defenses induced in pepper by lipopolysaccharides are suppressed by *Xanthomonas campestris* pv. *vesicatoria*. Mol Plant Microbe Interact 17 (2004) 805–815.

Kwon, C., Bednarek, P., Schulze-Lefert, P. Secretory pathways in plant immune responses. Plant Physiol 147 (2008a) 1575–1583.

Kwon, C., Neu, C., Pajonk, S., Yun, H.S., Lipka, U., Humphry, M., Bau, S., Straus, M., Kwaaitaal, M., Rampelt, H., El Kasmi, F., Jürgens, G., Parker, J., Panstruga, R., Lipka, V., Schulze-Lefert, P. Co-option of a default secretory pathway for plant immune responses. Nature 451 (2008b) 835–840.

Lemaitre, B., Nicolas, E., Michaut, L., Reichhart, J.M., Hoffmann, J.A. The dorsoventral regulatory gene cassette spätzle/Toll/cactus controls the potent antifungal response in *Drosophila* adults. Cell 86 (1996) 973–983.

Leeman, M., Vanpelt, J.A., Denouden, F.M., Heinsbroek, M., Pahm, B., Schippers, B. Induction of systemic resistance against fusarium-wilt of radish by lipopolysaccharides of *Pseudomonas fluorescens*. Phytopathology 85 (1995) 1021–1027.

Livaja, M., Zeidler, D., von Rad U., Durner, J. Transcriptional responses of *Arabidopsis thaliana* to the bacteria-derived PAMPs harpin and lipopolysaccharide. Immunobiology 213 (2008) 161–171.

Loppnow, H., Brade, H., Durrbaum, I., Dinarello, C.A., Kusumoto, S., Rietschel, E.T., Flad, H.D. IL-1 induction-capacity of defined lipopolysaccharide partial structures. J Immunol 142 (1989) 3229–3238.

Medzhitov, R., Preston-Hurlburt, P., Janeway, C.A., Jr. A human homologue of the *Drosophila* Toll protein signals activation of adaptive immunity. Nature 388 (1997) 394–397.

Meyer, A., Pühler, A., Niehaus, K. The lipopolysaccharides of the hytopathogen *Xanthomonas campestris* pv. *campestris* induce an oxidative burst reaction in cell cultures of *Nicotiana tabacum*. Planta 213 (2001) 214–222.

Mishina, T.E., Zeier, J. The *Arabidopsis* flavin-dependent monooxygenase FMO1 is an essential component of biologically induced systemic acquired resistance. Plant Physiol 141 (2006) 1666–1675.

Mishina, T.E., Zeier, J. Pathogen-associated molecular pattern recognition rather than development of tissue necrosis contributes to bacterial induction of systemic acquired resistance in *Arabidopsis*. Plant J 50 (2007) 500–513.

Miyake, K. Innate recognition of lipopolysaccharide by Toll-like receptor 4-MD-2. Trends Microbiol 12 (2004) 186–192.

Munford, R.S., Varley, A.W. Shield as signal: Lipopolysaccharide and the evolution of immunity to Gram-negative bacteria. PLos Pathog 2 (2006) 467–471.

Murray, R.Z., Wylie, F.G., Khromykh, T., Hume, D.A., Stow, J.L. Syntaxin 6 and Vti1b form a novel SNARE complex, which is up-regulated in activated macrophages to facilitate exocytosis of tumor necrosis factor-α. J Biol Chem 280 (2005) 10478–10483.

Newman, M.A., Dow, J.M., Molinaro, A., Parrilli, M. Priming, induction and modulation of plant defence responses by bacterial lipopolysaccharides. J Endotoxin Res 13 (2007) 69–84.

Newman, M.-A., von Roepenack-Lahaye, E., Parr, A., Daniels, M.J., Dow, J.M. Prior exposure to lipopolysaccharide potentiates expression of plant defenses in response to bacteria. Plant J 29 (2002) 485–497.

Newman, M.A., von Roepenack-Lahaye, E., Parr, A., Daniels, M.J., Dow, J.M. Induction of hydroxycinnamoyl-tyramine conjugates in pepper by *Xanthomonas campestris*, a plant defense response activated by *hrp* gene-dependent and *hrp* gene-independent mechanisms. Mol Plant Microbe Interact 14 (2001) 785–792.

Newman, M.A., Daniels, M.J., Dow, J.M. The activity of lipid A and core components of bacterial lipopolysaccharides in the prevention of the hypersensitive response in pepper. Mol Plant-Microbe Interact 10 (1997) 926–928.

Newman, M.A., Daniels, M.J., Dow, J.M. Lipopolysaccharide from *Xanthomonas campestris* induces defence-related gene expression in *Brassica campestris*. Mol Plant-Microbe Interact 8 (1995) 778–780.

Nomura, K., DebRoy, S., Lee, Y.H., Pumplin, N., Jones, J., He, S.Y. A bacterial virulence protein suppresses host innate immunity to cause plant disease. Science 313 (2006) 220–223.

Nürnberger, T., Brunner, F. Innate immunity in plants and animals: Emerging parallels between the recognition of general elicitors and pathogen-associated molecular patterns. Curr Opin Plant Biol 5 (2002) 1–7.

Pagan, J.K., Wylie, F.G., Joseph, S., Widberg, C., Bryant, N.J., James, D.E., Stow, J.L. The t-SNARE syntaxin 4 is regulated during macrophage activation to function in membrane traffic and cytokine secretion. Curr Biol 13 (2003) 156–160.

Pajonk, S., Kwon, C., Clemens, N., Panstruga, R., Schulze-Lefert, P. Activity determinants and functional specialization of *Arabidopsis* PEN1 syntaxin in innate immunity. J Biol Chem 283 (2008) 26974–26984.

Pieterse, C.M.J., van Wees, S.C.M., van Pelt, J.A., Knoester, M., Laan, R., Gerrits, N., Weisbeek, P.J., van Loon, L.C. A novel signaling pathway controlling induced systemic resistance in *Arabidopsis*. Plant Cell 10 (1998) 1571–1580.

Prime-A-Plant Group: Conrath, U., Beckers, G.J.M., Flors, V., García-Agustín, P., Jakab, G., Mauch, F., Newman, M.-A., Pieterse, C.M.J., Poinssot, B., Pozo, M.J., Pugin, A., Schaffrath, U., Ton, J., Wendelhenne, D., Zimmerli, L., Mauch-Mani, B. Priming: Getting ready, for battle. Mol Plant Microbe Interact 19 (2006) 1062–1071.

Raetz, C.R.H., Reynolds, C.M., Trent, M.S., Bishop, R.E. Lipid A modification systems in Gram-negative bacteria. Annu Rev Biochem 76 (2007) 295–329.

Raetz, C.R.H., Whitfield, C. Lipopolysaccharide endotoxins. Annu Rev Biochem 71 (2002) 635–700.

Rallabhandi, P., Awomoyi, A., Thomas, K.E., Phalipon, A., Fujimoto, Y., Fukase, K., Kusumoto, S., Qureshi, N., Sztein, M.B., Vogel, S.N. Differential activation of human TLR4 by *Escherichia coli* and *Shigella flexneri* 2a lipopolysaccharide: Combined effects of lipid A acylation state and TLR4 polymorphisms on signaling. J Immunol 180 (2008) 1139–1147.

Rigano, L.A., Payette, C., Brouillard, G., Marano, M. R., Abramowicz, L., Torres, P. S., Yun, M., Castagnaro, A. P., El Oirdi, M., Dufour, V., Malamud, F. Dow, J. M., Bouarab K., Vojnov. A.A. Bacterial Cyclic β-(1, 2) Glucan Acts in Systemic Suppression of Plant Immune Responses. Plant Cell 19 (2007) 2077–2089.

Robatzek, S., Bittel, P., Chinchilla, D., Köchner, P., Felix, G., Shiu, S.H., Boller, T. Molecular identification and characterization of the tomato flagellin LeFLS2, an orthologue of *Arabidopsis* FLS2 exhibiting characteristically different perception specificities. Plant Mol Biol 64 (2007) 539–547.

Robatzek, S., Chinchilla, D., Boller, T. Ligand-induced endocytosis of the pattern recognition receptor FLS2 in *Arabidopsis*. Genes Dev 20 (2006) 537–542.

Ryals, J.A., Neuenschwander, U.H., Willits, M.G., Molina, A., Steiner, H.-Y., Hunt, M.D. Systemic acquired resistance. Plant Cell 8 (1996) 1809–1819.

Sanderfoot, A.A., Assaad, F.F., Raikhel, N.V. The *Arabidopsis* genome. An abundance of soluble N-ethylmaleimide-sensitive factor adaptor protein receptors. Plant Physiol 124 (2000) 1558–1569.

Schneider, M., Schweizer, P., Meuwly, P. and Métraux, J.P. Systemic acquired resistance in plants. Int Rev Cytol 168 (1996) 303–340.

Schromm, A.B., Brandenburg, K., Loppnow, H., Moran, A.P., Koch, M.H.J., Rietschel, E.Th., Seydel, U. Biological activities of lipopolysaccharides are determined by the shape of their lipid A portion. Eur J Biochem 267 (2000) 2008–2013.

Schromm, A.B., Brandenburg, K., Loppnow, H., Zahringer, U., Rietschel, E.T., Carroll, S.F., Koch, M.H.J., Kusumoto, S., Seydel, U. The charge of endotoxin molecules influences their conformation and IL-6-inducing capacity. J Immunol 161 (1998) 5464–5471.

Silipo, A., Sturiale, L., Garozzo, D., Erbs, G., Tandrup Jensen T., Lanzetta, R., Dow, J.M., Parrilli, M., Newman M.A., Molinaro, A. The acylation and phosphorylation pattern of lipid A from *Xanthomonas campestris* strongly influence its ability to trigger the innate immune response in *Arabidopsis*. ChemBioChem 9 (2008) 896–904.

Silipo, A., Sturiale, L., Garozzo D., de Castro, C., Lanzetta, R., Parrilli, M., Grant, W.D., Molinaro, A. Structure elucidation of the highly heterogeneous lipid A from the lipopolysaccharide of the Gram-negative extremophile bacterium *Halomonas magadiensis* strain 21 M1. Eur J Org Chem 2004 (2004) 2263–2271.

Silipo, A., Molinaro, A., Sturiale, L., Dow, J.M., Erbs, G., Lanzetta, R., Newman, M.-A., Parrilli, M. The elicitation of plant innate immunity by lipooligosaccharide of *Xanthomonas campestris*. J Biol Chem 280 (2005) 33660–33668.

Söllner, T., Whiteheart, S.W., Brunner, M., Erdjument-Bromage, H., Geromanos, S., Tempst, P., Rothman, J.E. SNAP receptors implicated in vesicle targeting and fusion. Nature 362 (1993) 318–324.

Thordal-Christensen, H. Fresh insights into processes of nonhost resistance. Curr Opin Plant Biol 6 (2003) 351–357.

Van Loon, L.C., Bakker, P.A., Pieterse, C.M. Systemic resistance induced by rhizosphere bacteria. Annu Rev Phytopathol 36 (1998) 453–483.

Yun, M.H., Torres, P.S., El Oirdi, M., Rigano, L.A., Gonzalez-Lamothe, R., Marano, M.R., Castagnaro, A.P., Dankert, M.A., Bouarab, K., Vojnov, A.A., Xanthan induces plant susceptibility by suppressing callose deposition. Plant Physiol 141 (2006) 178–187.

Zeidler, D., Zahringer, U., Gerber, I., Dubery, I., Hertung, T., Bors, W., Hutzler, P., Durner, J. Innate immunity in *Arabidopsis thaliana*: Lipopolysaccharides activate nitric oxide synthase (NOS) and induce defense genes. Proc Natl Acad Sci USA 101 (2004) 15811–15816.

Zipfel, C., Kunze, G., Chinchilla, D., Caniard, A., Jones, J.D.G., Boller, T., Felix, G. Perception of the bacterial PAMP EF-Tu by the receptor EFR restricts *Agrobacterium*-mediated transformation. Cell 125 (2006) 749–760.

Index

X. Wang, P.J. Quinn (eds.), *Endotoxins: Structure, Function and Recognition*, Subcellular Biochemistry 53, DOI 10.1007/978-90-481-9078-2,

Q

R

MIX
Papier aus verantwortungsvollen Quellen
Paper from responsible sources
FSC® C105338

If you have any concerns about our products,
you can contact us on
ProductSafety@springernature.com

In case Publisher is established outside the EU,
the EU authorized representative is:
Springer Nature Customer Service Center GmbH
Europaplatz 3, 69115 Heidelberg, Germany

Printed by Libri Plureos GmbH
in Hamburg, Germany